Mathematical Statistics *with* Applications

William Mendenhall
Richard L. Scheaffer
University of Florida

DUXBURY PRESS
North Scituate, Massachusetts

DUXBURY PRESS

North Scituate, Massachusetts

A Division of Wadsworth Publishing Company, Inc.

Mathematical Statistics, with Applications was edited and prepared for composition by Service to Publishers, Inc. Interior design was provided by Mr. David Earle, and the cover was designed by Mr. Oliver Kline. The text was set by J. W. Arrowsmith Ltd., and the book was printed and bound by The Maple Press Co., Inc.

ISBN-0-87872-047-2

L. C. CAT. CARD NO. 72-90584

PRINTED IN THE UNITED STATES OF AMERICA

5 6 7 8 9 10—77

Contents

Preface

This text was written for use with a one-year sequence of courses (9 quarter or 6 semester hours) on mathematical statistics for undergraduates. The intent of the text is to present a solid undergraduate foundation in statistical theory and, at the same time, to provide an indication of the relevance and importance of the theory in solving practical problems in the real world. We think a course of this type is suitable for most undergraduate disciplines, including mathematics, where contact with these applications may provide a refreshing and motivating experience. The only mathematical prerequisite is a thorough knowledge of first-year college calculus.

Talking with students taking or having completed a beginning course in mathematical statistics reveals a major flaw that exists in many courses. The student can take the course and leave it without a clear understanding of the nature of statistics. Many see the theory as a collection of topics, weakly or strongly related, but fail to see that statistics is a theory of information with inference as its goal. Further, they may leave the course without an understanding of the important role played by statistics in scientific investigations. Why this is true (assuming that you agree with us) is a matter for conjecture, but several reasons suggest themselves.

First, all mathematical statistics courses require, of necessity, a large amount of time devoted to the theory of probability. Since at least fifty percent of the total course time is devoted to this subject, and it occurs at the beginning of the course, it is not surprising that a student may think of probability and statistics as being synonymous.

A second reason is that the objective of statistics often is not defined clearly at the beginning of the course, and no attempt is made to relate the probabilistic half of the course to the ultimate objective, inference.

Third, the utility of statistics cannot be revealed until late in the course because of the large amount of material that must preface it.

A fourth and final possibility is that the interests of some instructors lead them to present the material as a sequence of topics in applied mathematics rather than as a cohesive course in statistics. In any case, regardless of the reasons why students fail to form a clear picture of their subject, our text is an attempt to cope with the problem.

We think this text differs from others in three ways. First, we have preceded the presentation of probability with a clear statement of the objective of statistics and its role in scientific research, and we hold this objective before the student throughout the text. As the student proceeds through the theory of probability (Chapters 2 through 7), he is reminded frequently of the role that major topics play in the objective of the course, statistical inference. We attempt to strongly emphasize statistical inference as the sole and dominating theme of the course. The second feature of the text is connectivity. We try not only to explain how major topics play a role in statistical inference but also how the topics are related one to another. These connective discussions most frequently appear in chapter introductions and conclusions. Finally, we think the text is unique in its practical emphasis throughout the text, by the exercises, and by the useful statistical methodology presented at the end of the text. We hope to reinforce an elementary but sound theoretical foundation with some very useful methodological topics contained in the last five chapters.

The book can be used in a variety of ways and adjusted to the tastes of the students and instructor. The difficulty of the material can be increased or decreased by controlling the assignment of exercises, by eliminating some topics, and by varying the amount of time to be devoted to each. A stronger applied flavor can be added by the elimination of some topics, for example some sections of Chapter 7, and by devoting more time to the applied chapters at the end.

The authors wish to thank the many colleagues, friends, and students who made helpful suggestions concerning the manuscript for this book. In particular, we are indebted to P. V. Rao, J. Devore, and J. J. Shuster for their technical comments on the writing, and to Catherine Kennedy for her excellent typing. We wish to thank E. S. Pearson, W. H. Beyer, I. Olkin, R. A. Wilcox, C. W. Dunnett, A. Hald, and John Wiley & Sons for their kind permission to use the tables reprinted in Appendix III. Thanks are also due to the reviewers, Mary Lou Harkness, Pennsylvania State University; Stewart Hoover, Northeastern University; Donald Guthrie, Oregon State University; Ralph D'Agostino, Boston University; Paul C. Rogers, University of Maine; and William Harkness, Pennsylvania State University. Finally, we wish to thank our wives for their patience and understanding throughout the time this work was in progress.

William Mendenhall
Richard L. Scheaffer

Note to the Student

As the title *Mathematical Statistics, with Applications* implies, this text is concerned with statistics, both in theory and application, and only deals with mathematics as a necessary tool to give you a firm understanding of statistical techniques. The following suggestions for using the text will increase your learning and save you time.

The connectivity of the textbook is provided by the introductions and summaries in each chapter. These sections explain how each chapter fits into the overall picture of statistical inference and how each chapter relates to the preceding ones.

Within the chapters, important concepts are set off as definitions. These should be read and reread until they are clearly understood, because they form the framework on which everything else is built. The main theoretical results are set off as theorems. Although it is not necessary to understand the proof of each theorem, a clear understanding of the meaning and implications of the theorems is essential.

It is also essential that you work many of the exercises—for at least three reasons. First, you can be certain that you understand what you have read only by putting your knowledge to the test of working problems. Second, many of the exercises are of a practical nature and shed light on the applicability of probability and statistics. Third, some of the exercises present new concepts and thus extend the material covered in the chapter.

W.M.
R.L.S.

1

What Is Statistics?

STATISTICS IS THE ART OR SCIENCE OF KNOWING WHAT TO DO WHEN YOU DON'T KNOW WHAT'S GOING ON.

-Oscar Wesler
1978

1.1 Introduction

Statistical techniques are employed in almost every phase of life. Surveys are designed to collect early returns on election day to forecast the outcome of an election, and consumers are sampled to provide information for predicting product preference. The research physician conducts experiments to determine the effect of various drugs and controlled environmental conditions on humans in order to infer the appropriate method of treatment of a particular disease. The engineer samples a product quality characteristic along with various controllable process variables to assist in locating important variables related to product quality. Newly manufactured fuses are sampled before shipping to decide whether to ship or hold individual lots. The economist observes various indices of economic health over a period of time and uses the information to forecast the condition of the economy next fall. Statistical techniques play an important role in achieving the objective of each of these practical problems, and it is to the theory underlying this methodology that this text is devoted.

A prerequisite to a discussion of theory or methodology is a definition of "statistics" and a statement of its objectives. *Webster's New Collegiate Dictionary* [6] defines statistics as "the science of the collection and classification of facts on the basis of relative number of occurrences as a ground for induction;

systematic compilation of instances for the inference of general truth." Kendall and Stuart [3] state: "Statistics is the branch of scientific method which deals with the data obtained by counting or measuring the properties of populations." Fraser [1], commenting on experimentation and statistical applications, states that "statistics is concerned with methods for drawing conclusions from results of the experiments or processes." Freund [2], among others, views statistics as encompassing "the entire science of decision making in the face of uncertainty," and Mood [5] defines statistics as "the technology of the scientific method" and adds that statistics is concerned with "(1) the design of experiments and investigations, (2) statistical inference." *A superficial examination of these definitions suggests a bewildering lack of agreement, but all possess common elements. Each implies collection of data with inference as the objective. Each requires the selection of a subset of a large collection of data, either existent or conceptual, in order to infer the characteristics of the complete set. Thus statistics is a theory of information with "inference making" as its objective.*

The large body of data that is the target of our interest is called a *population*, and the subset selected from it is a *sample*. The preferences of voters for a gubernatorial candidate, Jones, expressed in quantitative form (1 for "prefer" and 0 for "do not prefer") provide a real, finite, and existing population of great interest to Jones. Indeed he may wish to sample the complete set of eligible voters in order to estimate the true fraction favoring his election. The voltage at a particular point in the guidance system for a spacecraft may be tested in the only three systems that have been built in order to estimate the voltage characteristics for other systems that might be manufactured some time in the future. In this case the population is *conceptual*. We think of the sample of 3 as being representative of a large population of guidance systems that could be built and would possess characteristics similar to the three in the sample. As another example, measurements on patients in a medical experiment employ sampling from a conceptual population consisting of all patients similarly afflicted today as well as those who will be afflicted in the near future. The reader will find it useful to clearly define the populations of interest for each of the statistical problems described earlier in this section and to clarify the inferential objective for each.

It is interesting to note that billions of dollars are spent each year by American industry and government for data from experimentation, sample surveys, or other collection procedures. Consequently, we see that this money is expended solely for information about a phenomenon susceptible to measurement in an area of business, science, or the arts. The implication of this statement provides a key to the nature of the very valuable contribution that statistics makes to research and development in all areas of society. Information useful in inferring some characteristic of a population (either existing or conceptual) can be purchased in a specified quantity and will result in an inference (estimation or decision) with a measured degree of goodness. For example, the greater the number of prospective voters included in a sample, the more accurate will be the estimate of the fraction of all voters favoring candidate Jones. So we can say that *statistics is concerned with the design of experiments or sample surveys to obtain a specified quantity of information at minimum cost and the optimal utilization of this information in making an inference about a population. The objective of statistics is to make an inference about a population based on information contained in a sample.*

1.2 Characterizing a Set of Measurements: Graphical Methods

In the broadest sense, making an inference implies the partial or complete description of a phenomenon or physical object. Little difficulty is encountered when appropriate and meaningful descriptive measures are available, but this is not always the case. For example, it is easy to characterize a person using height, weight, color of hair and eyes, and other descriptive measures of one's physiognomy. Locating a set of descriptive measures to characterize an oil painting would be a comparatively more difficult task; characterizing a population, which consists of a set of measurements, is equally challenging. Consequently, a necessary prelude to a discussion of inference making is the acquisition of a method for characterizing a set of numbers. The characterization must be meaningful so that knowledge of the descriptive measures will enable one to clearly visualize the set of numbers. In addition, we would hope that the characterization possesses practical significance, so that knowledge of the descriptive measures for a population might solve a practical nonstatistical problem. We shall develop our ideas on this subject by examination of a process that generates a population.

Consider a study to determine important variables affecting profit in a business that manufactures custom-made machined devices. Some of these variables might be the dollar size of the contract, the type of industry with whom the contract is negotiated, the degree of competition in acquiring contracts, the salesman who estimates the contract, fixed dollar cost, and the foreman who is assigned the task of organizing and conducting the manufacturing operation. The statistician will wish to measure the profit per contract for a number of jobs (the sample) along with measurements on the variables that might be related to profit. His objective will be to utilize information in the sample to infer the approximate relationship of the independent variables described above with profit, and to measure the strength of this relationship. The manufacturer's objective will be to determine optimum conditions for maximizing the profit in his business.

The population of interest in the manufacturing problem is conceptual and consists of all measurements of profit (per unit of capital and labor invested) that might be made on contracts, now and in the future, for fixed values of the independent variables (size of the contract, measure of competition, and so on). The profit measurements would vary from contract to contract in an apparent random manner as a result of variations in materials, time to complete individual segments of the work, and other uncontrollable variables affecting the job. Consequently, we view the population as a *distribution* of profit measurements, with the form of the distribution changing and depending upon specific values of the independent variables. To say that we seek to determine the relationship between profit and a set of independent variables is therefore translated to mean that we seek to determine the effect of the variables on the conceptual distribution of population measurements.

An individual population (or any set of measurements) can be characterized by a *frequency distribution*, which is also called a *frequency histogram*. A graph is constructed by subdividing the axis of measurement into intervals of equal width. Rectangles are constructed over each interval with the height of the rectangle proportional to the fraction of the total number of measurements falling in each cell. For example, to characterize the ten measurements 2.1, 2.4, 2.2, 2.3, 2.7, 2.5, 2.4, 2.6, 2.6, and 2.10, we could divide the axis of measurement into intervals of equal width (.2 unit) commencing with 2.05. The frequencies (fraction of total number of measurements), calculated for each cell, are shown in Figure 1.1. Note that the figure gives a clear pictorial description of the set of ten measurements.

Observe that we give no precise rule for selecting the width or location of these intervals. None is necessary, because each arbitrary selection will yield

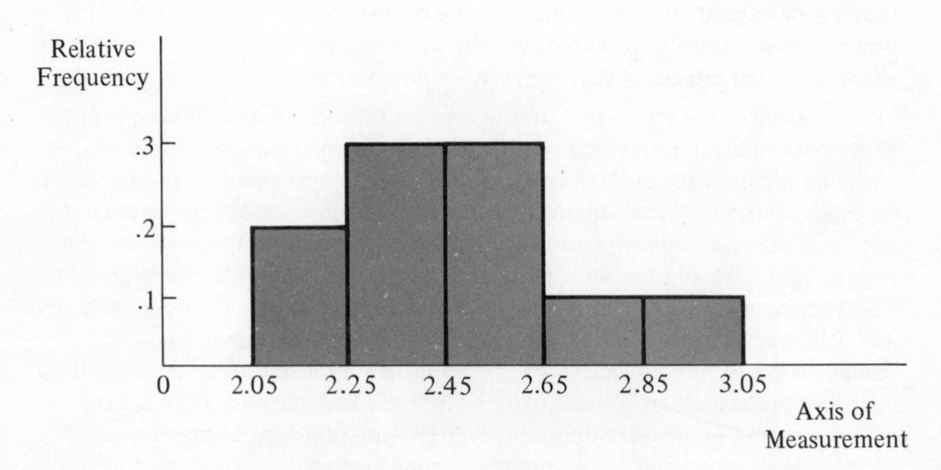

Figure 1.1 Frequency Histogram

a distribution picture that is similar to those for other selections, and all give a reasonably satisfactory pictorial description of the data.

Although arbitrary, adherance to a few guidelines can be very helpful in selecting the intervals. *Points of subdivision of the axis of measurement should be chosen so that it is impossible for a measurement to fall on a point of division.* This eliminates a source of confusion and is easily accomplished, as indicated in Figure 1.1. The second guideline concerns the width and consequently the minimum number of intervals needed to describe the data. Generally speaking, we wish to obtain information on the form of the distribution of the data. Most frequently this will be mound-shaped. A large number of intervals for a small amount of data would result in little summarization and would essentially present a picture very similar to the data in its original form. The larger the amount of data, the greater the number of intervals that can be included and still present a satisfactory picture of the data. *We would suggest spanning the range of the data with from 5 to 20 intervals and using the larger number of intervals for larger quantities of data.*

The objective of description is often misunderstood by many in government and business who like to think that the description of data is primarily an end in itself. No one would debate that man is concerned with history, but the major reason for this interest is to secure a glimpse of the future. We would therefore suggest that the description of a sample, say the national corporate production figures for iron ore, is to be used to forecast ore production next year and farther in the future. Knowledge of today's production may have immediate uses, but the longer-range objective is forecasting (inference).

Finally, we point to a probabilistic interpretation that can be derived from the frequency histogram, Figure 1.1. We have already stated that the area of a rectangle over a given interval is proportional to the fraction of the total number of measurements falling in that interval. We may now wish to extend this idea one step further. If a measurement were selected at random from the

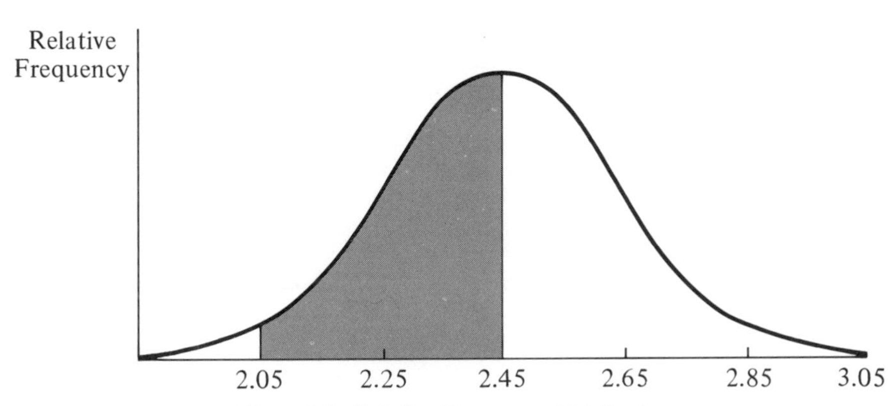

Figure 1.2 Relative Frequency Distribution

original set, the probability that it would fall in a given interval is proportional to the area under the histogram lying over that interval. (At this point, we rely on the layman's concept of probability. This term will be discussed in greater detail in Chapter 2.) For example, the probability of selecting a measurement in the interval 2.05 to 2.45, Figure 1.1, is .5, because half the measurements fall in the specified interval. Correspondingly, the area under the histogram over the interval, 2.05 to 2.45, is .5. It is clear that this interpretation would apply to the distribution for any set of measurements, even populations. If Figure 1.2 gives the frequency distribution for profit (millions) for a conceptual population of profit responses for contracts at a given setting of the independent variables (size of contract, measure of competition, and so on), the probability that the next contract (at the same setting of the independent variables) yields a profit that will lie in the interval 2.05 to 2.45 million is proportional to the shaded area under the distribution curve.

1.3 Characterizing a Set of Measurements: Numerical Methods

The relative frequency histogram presented in Section 1.2 provides useful information regarding a set of measurements, but it is not adequate for purposes of inference, mainly because it is not well defined. That is, many similar histograms could be formed from the same set of measurements. In order to make inferences about a population based on information contained in a sample and measure the goodness of the inferences, we need rigorously defined quantities with which to measure the sample information. We can then mathematically derive certain properties of these sample quantities and make probability statements regarding the goodness of our inferences.

The quantities we define are *numerical descriptive measures* of a set of data. We seek some numbers that describe the frequency distribution for any set of measurements. We will confine our attention to two types of descriptive numbers, *measures of central tendency* and *measures of dispersion or variation.*

The most common measure of central tendency used in statistics is the arithmetic mean. (Since this is the only type of mean discussed in this text, we shall omit the word "arithmetic.")

Definition 1.1: *The* mean *of a set of n measured responses $y_1, y_2, \ldots, y_n$ is given by*

$$\bar{y} = \frac{1}{n}\sum_{i=1}^{n} y_i.$$

The symbol $\bar{y}$ refers to a sample mean. The mean of all responses in a population will be denoted by the symbol μ. Note that we cannot usually measure μ; rather, μ is an unknown constant that we may want to estimate from sample information.

The mean of a set of measurements only locates the center of the distribution of data; by itself it does not provide an adequate description of a set of measurements. Two sets of measurements could have widely different frequency distributions with the same mean, as pictured in Figure 1.3. The difference

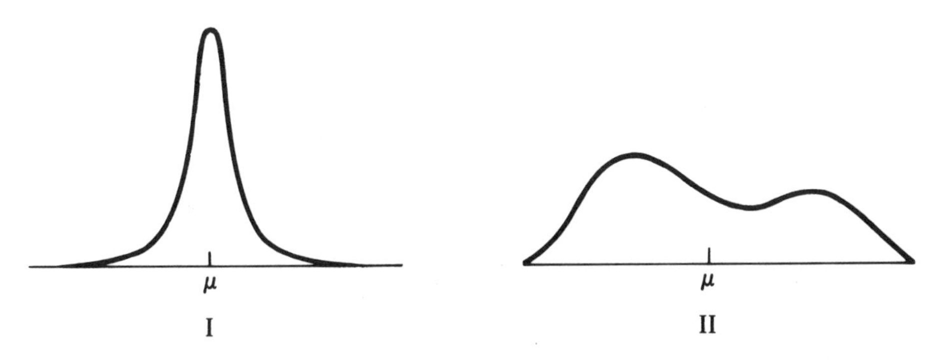

Figure 1.3 Frequency Distributions with Different Amounts of Variation

between distributions I and II in the figure is in the variation or dispersion of measurements to either side of the mean. An adequate description of data requires that we define measures of data variability.

The most common measure of variability used in statistics is the variance, which is a function of the deviations (or distances) of the sample measurements from their mean.

Definition 1.2: *The* variance *of a set of measurements $y_1, \ldots, y_n$ is the average of the square of the deviations of the measurements about their mean. Symbolically, the sample variance is*

$$s'^2 = \frac{1}{n} \sum_{i=1}^{n} (y_i - \bar{y})^2.$$

The corresponding population variance is denoted by the symbol σ^2. The larger the variance of a set of measurements, the greater will be the variation within the set. The variance is of value in comparing the relative variation of two sets of measurements but only gives information concerning the variation in a single set when interpreted in terms of the *standard deviation.*

> ***Definition 1.3:*** *The* standard deviation *of a set of measurements is the square root of the variance; that is,*
>
> $$s' = \sqrt{s'^2}.$$

The corresponding population standard deviation is denoted by σ.

Although closely related to the variance, the standard deviation can be used to give a fairly accurate picture of data variation for a single set of measurements. It can be interpreted using Tchebysheff's theorem, which will be presented in a later chapter, and by the empirical rule, which we will now explain.

Many distributions of data in real life are mound-shaped. That is, they can be approximated by a bell-shaped frequency distribution known as a normal curve. Data possessing this type of distribution will possess very definite characteristics of variation, which are expressed in the following statement:

> ***Empirical Rule:*** *Given a distribution of measurements that is approximately normal (bell-shaped), it then follows that the interval with endpoints*
>
> $\mu \pm \sigma$ *contains approximately* 68 percent *of the measurements,*
> $\mu \pm 2\sigma$ *contains approximately* 95 percent *of the measurements,*
> $\mu \pm 3\sigma$ *contains approximately* 99.7 percent *of the measurements.*

As was mentioned in Section 1.2, once the frequency distribution of a set of measurements is known, probability statements regarding the measurements can be made. These probabilities were shown as areas under a frequency histogram. Thus the probabilities contained in the empirical rule are areas under the normal curve shown in Figure 1.4.

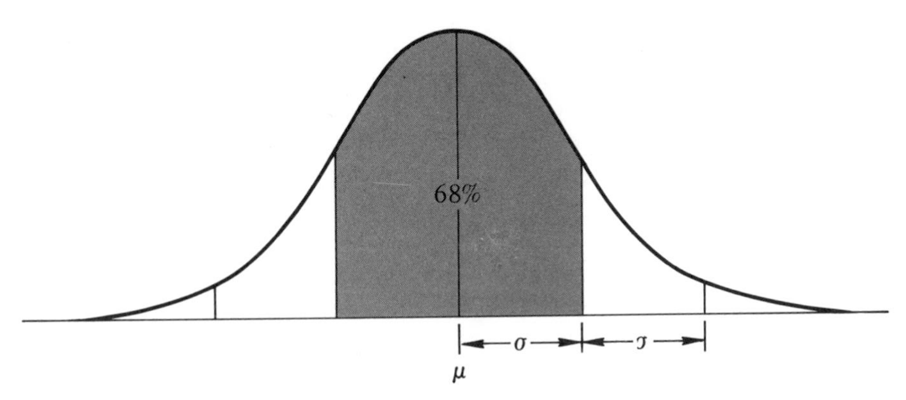

Figure 1.4 Normal Curve

The use of the empirical rule will be illustrated by the following example. Suppose the scores on an achievement test given to all high school seniors in a certain state are known to have, approximately, a normal distribution with a mean, μ, of 64 and a standard deviation, σ, of 10. It can then be deduced that approximately 68 percent of the scores are between 54 and 74, 95 percent of the scores are between 44 and 84, and 99.7 percent of the scores are between 34 and 94. Thus knowledge of the mean and the standard deviation gives us a fairly good picture of the frequency distribution of scores.

Suppose that a single high school student is randomly selected from those who took the test. What is the probability that his score will be between 54 and 74? On looking at the relative frequency of observations in this interval, we find that .68 is a reasonable answer to the probability question.

The utility and value of the empirical rule is great because of the very common occurrence of approximately normal distributions of data in nature—more so because the rule applies to distributions that are not exactly normal but just "mound-shaped." You will find that approximately 95 percent of a set of measurements will be within 2σ of μ for a wide variety of distributions.

1.4 *How Inferences Are Made*

The mechanism instrumental in making inferences can best be seen by analyzing our own intuitive inference-making procedures.

Suppose that two candidates are running for a public office in our community and that we wish to determine whether our candidate, Jones, is favored to win. Thus the population of interest is the set of responses from all eligible voters who will vote on election day, and we wish to determine whether the fraction favoring Jones exceeds .5. For the sake of simplicity, suppose that all eligible voters will go to the polls and that we randomly select a sample of 20 from the courthouse rosters. All 20 are contacted and all favor Jones. What do you conclude with regard to Jones' prospects for winning the election?

There is little doubt that most of us would immediately infer that Jones will win the election and we find that subsequent study will substantiate our good judgment. This was indeed an easy inference to make, but the inference itself is not our immediate goal. Rather, we wish to examine the mental processes that were employed in reaching a conclusion about the prospective behavior of a large voting population on the basis of a sample of only 20 people.

Winning means acquiring more than 50 percent of the votes. Did we conclude that Jones would win because we expect the fraction favoring Jones in the sample to correspond identically to the fraction in the population? We know that this is not true and a simple experiment will verify the fact. Flipping a balanced coin will show that the number of heads in a sample of 20 will vary from sample to sample.

Did we conclude that Jones must win because it would be impossible for 20 out of 20 to favor Jones if in fact less than 50 percent of the electorate intended to vote for him? The answer to this question is certainly "no," but it provides the key to our hidden line of logic. *It is not impossible to draw 20 out of 20 favoring Jones (even though less than 50 percent of the electorate favor him), but it is highly improbable.* Thus our intuitive feel for probability suggested that it was highly improbable that 20 out of 20 in the sample would select Jones if, in fact, he is to be a loser. We therefore concluded that he would win.

The example above illustrates the potent role played by probability in making inferences. A probabilist assumes that he knows the structure of the phenomenon in question, the population, and he uses the theory of probability to make an inference about a sample. Thus he assumes that he knows the structure of a population generated by random drawings of five cards from a standard deck, and he uses probability to make an inference concerning a draw that will yield three aces and two kings. The statistician uses probability to make the trip in reverse, from the sample to the population. Observing a draw of five aces in a sample, he immediately infers that the deck (which generates the population) is loaded and not standard because the probability of drawing five aces from a standard deck is zero. This is an exaggerated case but it makes the point. Probability is the mechanism used in making inferences. Basic to inference is the problem of calculating the probability of the observed sample.

One final comment is in order. The reader who did not think that the sample justified an inference that Jones would win should not feel too chagrined. One can be easily misled when making intuitive evaluations of the probabilities of events. The reader who decided that the probability that 20 out of 20 was very low, assuming Jones to be a loser, was correct. It would not be difficult to concoct an example in which his intuitive assessment of probability would be in error. Thus intuitive assessments of probabilities are unsatisfactory, and one must have a rigorous theory of probability at hand in order to develop the mechanism for inference.

1.5 Theory and Reality

It is essential that the student grasp the difference between theory and reality. Theories are ideas proposed to explain phenomena in the real world and, as such, are approximations or models for reality. These models, or explanations of reality, are presented in verbal forms in some less quantitative fields and as mathematical relationships in others. Whereas a theory of social change might be expressed verbally in sociology, the theory of heat transfer is presented in a precise and deterministic mathematical manner in physics. Neither gives an accurate and unerring explanation for real life. Slight variations from the mathematically expected occur in the observance of heat-transfer phenomena and in other areas of physics. The deviations cannot be blamed solely on the measuring instruments, the explanation that one often hears, but are due in part to a lack of agreement between theory and reality. Anyone who believes that the physical scientist now completely understands the wonders of this world need only look at history to find a contradiction. Theories assumed to be the "final" explanation for nature have been superseded in rapid succession during the past century.

This text is concerned with the theory of statistics and hence a model of reality. We will postulate theoretical frequency distributions for populations and will develop a theory of probability and inference in a precise mathematical manner. The net result will be a theoretical or mathematical model for the acquisition and utilization of information in real life. It will not be an exact representation of nature, but this should not disturb us. Like other theories, its utility will be measured by its ability to assist us in understanding nature and in solving problems in the real world. Such is the role of the theory of heat transfer, the theory of strength of materials, and other models of nature.

1.6 Summary

The objective of statistics is to make an inference about a population based on information contained in a sample. The theory of statistics is a theory of information concerned with its quantification, with the design of experiments or procedures for data collection that will minimize the cost of a specified quantity of information, and with the use of this information in making inferences. Most important, we have viewed the making of an inference about the unknown population as a two-step procedure. First, we seek the best inferential procedure for the given situation and, second, we desire a measure of its goodness. For example, every estimate of a population characteristic based on information contained in the sample might have associated with it a probabilistic bound on the error of estimation.

A necessary prelude to making inferences about a population is the ability to describe a set of numbers. Frequency distributions provide a very graphic and useful method for characterizing a conceptual or real population of numbers. More useful to inference making are numerical descriptive measures.

The mechanism for making inferences is the theory of probability. The probabilist reasons from a known population to the outcome of a single experiment, the sample. In contrast, the statistician utilizes the theory of probability to calculate the probability of an observed sample and to infer from this the characteristics of an unknown population. Thus probability is the foundation of the theory of statistics.

Finally, we have noted the difference between theory and reality. In this text we study the mathematical theory of statistics, which is an idealization of nature. It is rigorous, mathematical, and subject to study in a vacuum completely isolated from the real world. Or, it can be tied very closely to reality and can be useful in making inferences from data in all fields of science. In this text we will be utilitarian. We shall not regard statistics as a branch of mathematics but as an area of science concerned with the development of a practical theory of information. We will consider statistics a separate field analogous to physics, not as a branch of mathematics but as a theory of information that utilizes mathematics heavily.

Subsequent chapters will expand upon the topics that we have just discussed. We shall begin with a study of the mechanism employed in making inferences, the theory of probability. This theory will provide a theoretical model for the generation of experimental data and will provide the basis for our study of statistical inference.

One additional comment might be helpful. Experience would suggest that students rarely read introductions and less frequently return to them later. However, you would be well advised to read this introduction at frequent intervals to refresh your memory concerning the objective of our study. It is all too easy to "miss the forest for the trees." Form a clear picture of statistics as a theory of information and identify its inferential objective. Then relate each topic in the text to the objective. Each topic is relevant. If you do not see the relation to the objective, you are missing an important point. In brief, put the bits and pieces together and see statistics as a meaningful theory of information.

References

1. Fraser, D. A. S., *Statistics, an Introduction*. New York: John Wiley & Sons, Inc., 1958.
2. Freund, J. E., *Mathematical Statistics*. Englewood Cliffs, N.J.: Prentice-Hall, Inc., 1962.
3. Kendall, M. G., and A. Stuart, *The Advanced Theory of Statistics*, Vol. 1. New York: Harper & Row, Inc., 1961.
4. Mendenhall, W., *Introduction to Probability and Statistics*, 3rd ed. North Scituate, Mass.: Duxbury Press, 1971.
5. Mood, A. M., *Introduction to the Theory of Statistics*. New York: McGraw-Hill Book Company, 1950.
6. *Webster's New Collegiate Dictionary*, Springfield, Mass.: G. & C. Merriam Co., Publishers, 1961.

Exercises

1.1. Let c be a constant. Prove

(a) $\sum_{i=1}^{n} c = nc$

(b) $\sum_{i=1}^{n} cy_i = c \sum_{i=1}^{n} y_i$

(c) $\sum_{i=1}^{n} (x_i + y_i) = \sum_{i=1}^{n} x_i + \sum_{i=1}^{n} y_i.$

1.2. Prove that the sum of the deviations of a set of measurements about their mean is equal to zero.

1.3. Show that

$$\sum_{i=1}^{n} (y_i - \bar{y})^2 = \sum_{i=1}^{n} y_i^2 - \frac{\left(\sum_{i=1}^{n} y_i\right)^2}{n}.$$

(Note that this gives a convenient method for computing the sum of squares of deviations needed to calculate the sample variance.)

1.4. Calculate s' for the sample of $n = 6$ measurements 1, 4, 2, 1, 3, and 3. Use the results of Exercise 1.3.

1.5. The following data give the times to failure for $n = 88$ ARC-1 VHF radio transmitter–receivers:

16	224	16	80	96	536	400	80
392	576	128	56	616	224	40	32
408	384	256	246	328	464	448	616
304	16	72	8	80	72	56	608
208	194	136	224	80	16	424	264
256	216	168	184	552	72	184	240
488	120	308	32	272	152	328	480
60	208	440	104	72	168	40	152
360	232	40	112	112	288	168	352
56	72	64	40	184	264	96	224
168	168	114	280	152	208	160	176

The *range* of a set of measurements is defined to be the difference between the largest and smallest members of the set.

(a) If you had to guess the value of s' for a set of measurements, why might you guess $s' = \text{range}/4$? (*Hint*: Consult the empirical rule.)

(b) Use part (a) to guess s' for the $n = 88$ lengths of time to failure shown above.

1.6. Using the data of Exercise 1.5,

(a) Construct a frequency histogram for the data. (Note the skewness in the distribution.)

(b) Use an electric calculator (or computer) to calculate $\bar{y}$ and s'. (Hand calculation is much too tedious for this exercise.)

(c) Calculate the intervals $\bar{y} \pm ks'$, $k = 1, 2, 3$, and count the number of measurements falling in each interval. Compare with the empirical rule. Note that the empirical rule provides a rather good description of this data, even though the distribution is highly skewed.

1.7. Given a sample of $n = 15$ measurements,

8	10	4	3	8
5	6	8	7	6
7	7	5	11	7

(a) Use the range of the measurements to obtain an estimate of the standard deviation.

(b) Construct a frequency histogram for the data. Use the histogram to obtain a visual approximation to $\bar{y}$ and s'.

(c) Calculate $\bar{y}$ and s'. Compare with the calculation checks provided by (a) and (b).

(d) Construct the intervals $\bar{y} \pm ks'$, $k = 1, 2, 3$, and count the number of measurements falling in each interval. Compare the fractions falling in the intervals with the fractions that one would expect according to the empirical rule.

1.8. Compare the ratio of the range to s' for the three sample sizes ($n = 6, 15$, and 88) for Exercises 1.4, 1.5, and 1.7. Note that the ratio tends to increase as the amount of data increases. The greater the amount of data, the greater will be the tendency for it to contain a few extreme values that will inflate the range and have little effect on s'. We ignored this phenomenon and suggest that the student use 4 as the ratio for finding a guessed value of s' for checking calculations.

1.9. A set of 340 examination scores showed a mean of $\bar{y} = 72$ and a standard deviation $s' = 8$. Approximately how many of the scores would you expect to fall in the interval 64 to 80? The interval 56 to 88?

1.10. A machine produces bearings with a mean diameter of 3.00 inches and a standard deviation of .01 inch. Bearings with diameters in excess of 3.02 inches or less than 2.98 inches will fail to meet quality specifications. Approximately what fraction of this machine's production will fail to meet specifications? What assumptions did you make concerning the distribution of bearing diameters in order to answer this question?

1.11. Let $k \geq 1$. Show that for any set of n measurements, the fraction included in the interval $\bar{y} - ks'$ to $\bar{y} + ks'$ is at least $(1 - 1/k^2)$. [*Hint:*

$$s'^2 = \frac{\sum_{i=1}^{n} (y_i - \bar{y})^2}{n}.$$

In this expression, replace all deviations for which $|y_i - \bar{y}| \geq ks'$ with ks'. Simplify.] This result is known as *Tchebysheff's theorem.*

2

Probability

2.1 *Probability*

To most people, "probability" is a loosely defined term employed in everyday conversation to indicate the measure of one's belief in the occurrence of a future event. We accept this as a meaningful and practical interpretation of the term but seek a clearer understanding of the context in which it is used, how it is measured, and how probability assists in making inferences.

The concept of probability is necessary when dealing with physical, biological, or social mechanisms that generate observations which cannot be predicted with certainty. For example, the blood pressure of a human at a given point in time cannot be predicted with certainty, and we never know the exact load that a bridge will endure before collapsing into a river. Such random events cannot be predicted with certainty, but the relative frequency with which they occur in a long series of trials is remarkably stable. Events possessing this property are called *random* or *stochastic* events. This stable relative frequency provides an intuitively meaningful measure of one's belief in the occurrence of a random event for a future observation. It is impossible, for example, to predict with certainty the occurrence of "heads" for the single toss of a coin, but we would be willing to state with a fair measure of confidence that the fraction of heads in a long series of trials would be very near .5. That this relative frequency is used as a measure of belief in the outcome for a single toss is evident when we consider a gambler's objective. He risks his money on the single toss of a coin, not a long series of tosses. The relative frequency of a head in a long series of tosses, which he calls "the probability of a head," gives him a measure of his chances of winning on a single toss. If the coin were unbalanced and gave 90 percent heads in a long series of tosses, the gambler would say that the "probability of a head" is .9, and his belief in the occurrence of a head on a single toss of the coin would be fairly strong.

The preceding example possesses some very realistic and practical analogies. In many respects, all humans are gamblers. The research physician gambles time and money on a research project and he is concerned with his success on the single flip of this symbolic coin. Similarly, the investment of capital in a new manufacturing plant is a gamble that represents a single flip of a coin on which the entrepreneur has high hopes for success. The fraction of similar investments that are successful in a long series of trials is of interest to the entrepreneur only insofar as it provides a measure of belief in the successful outcome of his single individual investment.

The relative frequency concept of probability is intuitively meaningful, but it does not provide a rigorous definition of probability. Many other concepts of probability have been proposed, including that of subjective probability, which allows the probability of an event to vary depending upon the person performing the evaluation.

A clarification of the practical meaning of probability is not essential for the development of a theory, but it is absolutely necessary if it is to be used in achieving the very practical inference-making goal that we have specified as the objective of statistics. We shall devote no more time to this point, because the practical meaning of probability is a matter of philosophy that can be argued endlessly. For our purposes, we accept an interpretation based on relative frequency as a meaningful measure of one's belief in the occurrence of an event, and we shall now examine the link that probability provides between observation and inference.

2.2 *Probability and Inference*

The role that probability plays in making inferences will be discussed in detail after an adequate foundation has been laid in the theory of probability. At this point we will present a preview to motivate an elementary treatment of this theory using an example and an appeal to the reader's intuition.

The example that we have selected is similar to that presented in Section 1.4 but is simpler and less practical. It was chosen because of the ease with which we can visualize the population and sample and because it provides an observation-producing mechanism for which a probabilistic model will be constructed in Section 2.3.

Consider a gambler who wishes to make an inference concerning the balance of a die. The conceptual population of interest is the set of numbers that would be generated if the die were rolled over and over again, ad infinitum. If the die were perfectly balanced, one-sixth of the measurements in the population would be 1s, one-sixth 2s, one-sixth 3s, and so on. The corresponding frequency distribution is shown in Figure 2.1.

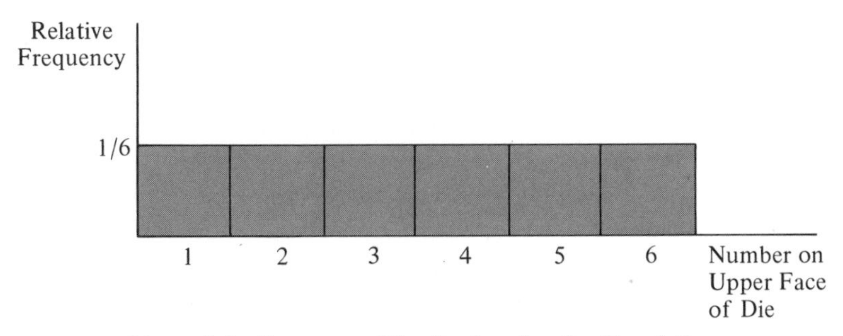

Figure 2.1 Frequency Distribution for the Population Generated by a Balanced Die

Consistent with the scientific method, the gambler proposes a hypothesis *that the die is balanced*, and he seeks observations from nature to contradict the theory, if false. A sample of ten tosses is selected from the population by rolling the die ten times. All ten tosses result in 1s. The gambler looks upon this output of nature with a jaundiced eye and concludes that nature is not in agreement with his hypothesis and hence that the die is unbalanced.

The reasoning employed by the gambler identifies the role that probability plays in making inferences. The gambler rejected his hypothesis (concluding that the die is unbalanced) not because it is *impossible* to throw ten 1s with a balanced die, but because it is highly *improbable*. His evaluation of the probability was most likely subjective. That is, the gambler may not have known how to "calculate" the probability of ten 1s, but he had an intuitive feeling that this event would occur with very low frequency if the ten tosses were repeated over and over again for a long series of trials. But the point to note is that his decision was based on the probability of the observed sample.

The need for a theory of probability that will provide a rigorous method for finding a number (a probability) that will agree with the actual relative frequency of occurrence of an event in a long series of trials is apparent if we imagine a different result for the gambler's sample. Suppose, for example, that instead of ten 1s, he observed five 1s along with a 3, 4, 6, 3, and a 2. Assuming the die to be balanced, is this result so improbable as to cause us to reject our hypothesis and conclude that the die is loaded in favor of 1s? Unlike ten 1s in ten tosses, it is not so easy to decide whether the probability of five 1s in ten tosses is large or small if one must simply rely on experience and intuition to make the evaluation. The probability of throwing four 1s in ten tosses would be even more difficult to guess.

We will not deny that many experimental results are obviously inconsistent with and lead to a rejection of the hypothesis in question, but many samples fall in a gray area that requires a rigorous assessment of probability. Indeed, it is not difficult to show that the intuitive evaluation of probability often leads to an answer that is substantially in error and results in incorrect inferences about the target population.

We need a theory of probability that will permit us to calculate the probability (or a quantity proportional to the probability) of observing specified outcomes, assuming that our hypothesized model is correct. This topic will be developed in detail in subsequent chapters. Our immediate goal is to present an introduction to the theory of probability, which provides the foundation for modern statistical inference. We shall begin by reviewing set notation, which will be used in constructing a probabilistic model for an experiment.

2.3 A Review of Set Notation

In order to proceed with an orderly development of probability theory, we need some basic concepts of set theory. We will use capital letters, $A, B, C, \ldots$, to denote sets of points. If the elements in the set A are a_1, a_2, and a_3, we will write

$$A = \{a_1, a_2, a_3\}.$$

Let S denote an arbitrary set of points. We will say that A is a *subset* of S, or A is contained in S (denoted by $A \subset S$), if every point in A is also in S. The *null* or *empty* set, denoted by $\varnothing$, is the set consisting of no points. Thus $\varnothing$ is a subset of every set.

Consider now two arbitrary sets of points. The *union* of A and B, denoted by $A \cup B$, is the set of all points in A or B or both. That is, the union of A and B contains all points which are in at least one of the sets.

The *intersection* of A and B, denoted by $A \cap B$ or by AB, is the set of all points in both A and B.

Suppose that S contains all points of interest for a particular problem and that A is a subset of S. Then the *complement* of A, denoted by $\bar{A}$, is the set of points that are in S but not in A.

Two sets, A and B, are said to be *disjoint* or *mutually exclusive* if $A \cap B = \varnothing$. That is, mutually exclusive sets have no points in common.

Consider the die-tossing problem of Section 2.2 and let S denote the set of all possible numerical observations for a single toss of a die. That is, $S = \{1, 2, 3, 4, 5, 6\}$. Let $A = \{1, 2\}$, $B = \{1, 3\}$, and $C = \{2, 4, 6\}$. Then $A \cup B = \{1, 2, 3\}$, $A \cap B = \{1\}$, and $\bar{A} = \{3, 4, 5, 6\}$. Also, note that B and C are mutually exclusive, whereas A and C are not.

We shall not attempt a thorough review of set algebra, but we mention two equalities of considerable importance. These are the distributive laws, given by

$$A \cap (B \cup C) = (A \cap B) \cup (A \cap C)$$

and

$$A \cup (B \cap C) = (A \cup B) \cap (A \cup C).$$

We shall now proceed with an elementary discussion of probability theory.

2.4 *A Probabilistic Model for an Experiment: The Discrete Case*

In Section 2.2 we referred to the die-tossing *experiment* when we observed the number appearing on the upper face. We shall use the term "experiment" to include observations obtained from completely uncontrollable situations (such as observations on the daily price of a particular stock) as well as those made under controlled laboratory conditions. We have the following definition:

> ***Definition 2.1:*** *An experiment is the process of making an observation.*

An experiment can result in one, and only one, of a set of distinctly different observable outcomes. In particular, we are interested in experiments that generate outcomes which vary in a random manner and cannot be predicted with certainty. For example, the experiment consisting of tossing a die one time and observing the number on the upper face yields six distinctly different possible outcomes: the integers 1, 2, 3, 4, 5, and 6. Each of these outcomes is different and only one can occur for a single repetition of the experiment.

> ***Definition 2.2:*** *A* sample space, *denoted by S, is a set of points corresponding to all distinctly different possible outcomes of an experiment. Each point corresponds to a particular single outcome.*

Notationally, for the die-tossing experiment we will depict the sample space by $S = \{1, 2, 3, 4, 5, 6\}$.

> ***Definition 2.3:*** *A* sample point *is a single point in a sample space, S.*

Each sample point corresponds to a single outcome of an experiment. There are six sample points in the die-tossing experiment, corresponding to the integers 1 through 6; this forms an example of a *discrete* sample space.

> ***Definition 2.4:*** *A* discrete sample space *is one that contains a finite number or countable infinity of sample points.*

The term "countable infinity of points" refers to a set of points that can be put into a one-to-one correspondence with the positive integers. Examples of such sets will be seen later.

> ***Definition 2.5:*** *For a discrete sample space, S, an* event *is any subset of S.*

We see that *an event can be viewed as a collection of sample points*. For the die-tossing experiment, which has a discrete sample space consisting of six points, one could define the following events:

A: observe an odd number ($A = \{1, 3, 5\}$),
B: observe a number less than 5 ($B = \{1, 2, 3, 4\}$),
C: observe a 2 or a 3 ($C = \{2, 3\}$),
E_1: observe a 1 ($E_1 = \{1\}$),
E_2: observe a 2 ($E_2 = \{2\}$),
E_3: observe a 3 ($E_3 = \{3\}$),
E_4: observe a 4 ($E_4 = \{4\}$),
E_5: observe a 5 ($E_5 = \{5\}$),
E_6: observe a 6 ($E_6 = \{6\}$).

Each of these nine events is a specific collection of sample points.

Note that the events A and B are different from events $E_1, E_2, \ldots, E_6$ in that A and B contain more than one point, whereas the events $E_1, E_2, \ldots, E_6$ each contain a single point. The events $E_1, \ldots, E_6$ are called simple events.

Definition 2.6: *A* simple event *is one that contains a single sample point.*

Events containing two or more sample points can be partitioned into two or more nonempty subsets or, equivalently, two or more new events. *Since you cannot partition a single sample point, we might refer to simple events as events that cannot be decomposed.*

Since any event is a collection of sample points, it follows that any event can be looked upon as a union of simple events. The event A defined above occurs on a single trial of the die-tossing experiment if, and only if, the outcome of that toss is a 1, 3, or 5. That is, A occurs if, and only if, E_1, E_3; or E_5 occurs. Also, it is clear from Definition 2.6 that simple events are mutually exclusive. That is, if a single toss of a die produced a 1, it could not at the same time have produced a 2, 3, 4, 5, or 6.

The sample space and particular events associated with an experiment can be conveniently portrayed by a Venn diagram. For example, the sample space, S, and events A and C are shown in the Venn diagram, Figure 2.2.

The final step in constructing a probabilistic model for an experiment with a discrete sample space is to attach a probability to each simple event. In

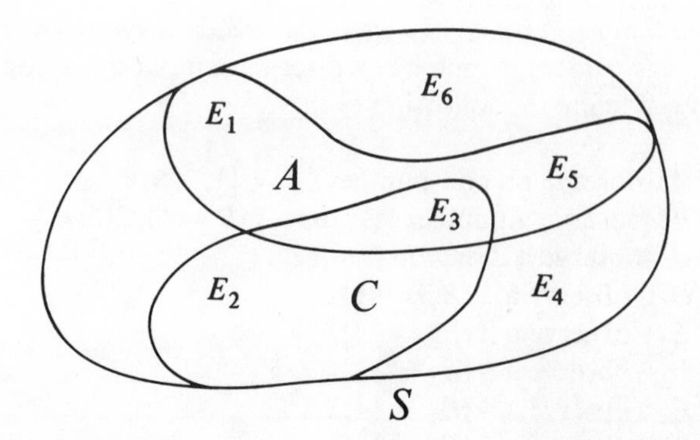

Figure 2.2 Venn Diagram for the Die-Tossing Experiment

doing so, we will select this number, a measure of our belief in its occurrence on a single repetition of the experiment, so that it will be consistent with the relative frequency concept of probability. Although relative frequency does not provide a rigorous definition of probability, any definition that is to be applicable to the real world should agree with our intuitive notion of the behavior of relative frequencies of events.

On analyzing the frequency concept of probability we see that three conditions must hold.

1. *The relative frequency of occurrence of any event must be greater than or equal to zero. A negative frequency does not make sense.*
2. *The relative frequency of the whole sample space, S, must be unity. Since every possible outcome of the experiment is a point in S, it follows that S must occur every time the experiment is run.*
3. *If two events are mutually exclusive, the relative frequency of their union is the sum of their respective relative frequencies.*

As an example of condition 3, if the experiment of tossing a balanced die yields a 1 on 1/6 of the tosses, it should yield a 1 or a 2 on 1/6 + 1/6 or 1/3 of the tosses. These three concepts form the basis of our definition of probability.

Definition 2.7: *Suppose that an experiment has associated with it a sample space S. To every event A in S (A is a subset of S) we assign a number, P(A), called the* probability *of A, so that the following axioms hold:*

Axiom 1: $P(A) \geq 0$.
Axiom 2: $P(S) = 1$.
Axiom 3: If $A_1, A_2, A_3, \ldots$ *form a sequence of pairwise mutually exclusive events in S* ($A_i \cap A_j = \varnothing, i \neq j$), *then*

$$P(A_1 \cup A_2 \cup A_3 \cup \cdots) = \sum_{i=1}^{\infty} P(A_1).$$

One can easily show that Axiom 3, which is stated in terms of an infinite sequence of events, implies a similar property for a finite sequence. Specifically, if $A_1, A_2, \ldots, A_n$ are pairwise mutually exclusive events,

$$P(A_1 \cup A_2 \cup A_3 \cup \cdots \cup A_n) = \sum_{i=1}^{n} P(A_i).$$

Note that the definition only states the properties a probability must satisfy; it does not tell us how to assign specific probabilities to events. For example, suppose that a coin has yielded 8 heads in 10 previous tosses. Consider the experiment of one more toss of the same coin. There are two possible outcomes, head or tail, and hence two simple events. The definition of probability allows us to assign to these simple events any two positive numbers that add to 1. For example, each simple event could have probability 1/2. In light of the past history of this coin, a more reasonable solution would be to give probability of .8 to the outcome involving a head. Specific assignments of probabilities must be done in a manner that is consistent with reality if the probabilistic model is to serve a useful purpose.

For discrete sample spaces it suffices to assign probabilities to each simple event. If a balanced die is used for the die-tossing example, it seems reasonable to assume that all simple events would have the same relative frequency in the long run. We will assign a probability of 1/6 to each simple event; $P(E_i) = 1/6, i = 1, 2, \ldots, 6$. This assignment of probabilities agrees with Axiom 1. To see that Axiom 2 is satisfied, write

$$\begin{aligned} P(S) &= P(E_1 \cup E_2 \cup \cdots \cup E_6) \\ &= P(E_1) + P(E_2) + \cdots + P(E_6) \\ &= 1. \end{aligned}$$

The second equality follows because Axiom 3 must hold. Axiom 3 really tells us that now we can calculate the probability of any event by summing the probabilities of the simple events contained in that event. Event A was defined to be "observe an odd number." Hence

$$P(A) = P(E_1 \cup E_3 \cup E_5) = P(E_1) + P(E_3) + P(E_5)$$
$$= 1/2.$$

To give an example of an experiment with a countable infinity of points, let us define a die-tossing problem as follows: Toss a balanced die until the first appearance of a 1, observing the number of the trial on which this happens. A 1 could appear on the first toss, or on the second, or on the third, but there is no upper limit to the number of trials that may be needed. Thus $S = \{1, 2, 3, \ldots\}$ and the simple events are as follows:

E_1: first 1 appears on first toss,
E_2: first 1 appears on second toss,
$\vdots$
E_n: first 1 appears on nth toss,

where n can be any positive integer. The problem of assigning probabilities to the countable infinity of sample points is slightly more cumbersome than in the first example. Finding the appropriate assignment will be simplified later. We will then show that the probabilities for the sample points satisfy the three axioms of Definition 2.7.

The two preceding examples were associated with sample spaces containing a finite and a countable infinity of sample points, respectively. How can we assign probabilities to sample spaces that are not discrete (for example, intervals)? This problem will be discussed next.

2.5 A Probabilistic Model for an Experiment: The Continuous Case

Many experiments have even more than a countable infinity of sample points, and so we are forced to consider a generalization of the probabilistic models given above. For example, daily rainfall at a specific geographic point may be any number in an interval, say 0 to 20 inches, and there are more than a countable infinity of values in any interval of real numbers. This is not to say that if we observe the daily rainfall long enough, every value between 0 and 20 inches will occur, but we cannot rule out any number between 0 and 20, because there is a *chance* that it may occur. Contrast this situation with the die-tossing problem, in which we know that many values, 5.5 for example, cannot possibly occur.

Since most experiments of interest to us result in numerical outcomes, we make the following definition.

> ***Definition 2.8:*** *A* continuous sample space *is a sample space consisting of a set of real numbers that contains at least one interval.*

In order to assign probabilities to events in a continuous sample space so that the three axioms of Definition 2.7 hold, we must make some restrictions on the types of subsets that are allowable as events. Since we shall usually be interested in intervals, we make the following definition.

> ***Definition 2.9:*** *For a continuous sample space, S, an* event *is any subset of S that can be formed by performing countable set operations on intervals in S.*

Definition 2.9 says that we can start with a collection of intervals in S; perform the operations of union, intersection, and complementation as many times as we want; and regard the new set so formed as an event. Any subset of an interval that is likely to come up in practice can be formed in this way.

As an example of an experiment with a continuous sample space, consider the following. A bus arrives at a certain stop between 8:00 A.M. and 8:10 A.M. A man who rides on this bus has observed that the relative frequency of arrivals in any subinterval is proportional to the length of the subinterval. That is, the relative frequency of arrivals in the interval 8:00 to 8:01 is the same as that for any other 1-minute subinterval. The number of sample points (or simple events) in this experiment is uncountably infinite since the interval 8:00 to 8:10 has an uncountable infinity of points. We cannot, therefore, assign nonzero probabilities to the sample points in a manner consistent with the axioms because, no matter how small the probabilities we assign, we can always find enough sample points so that the sum of the probabilities over these points would be greater than 1. We can, however, assign reasonable probabilities to subintervals. For instance, it would be reasonable to assign a probability of 1/10 to any event which states that the bus will arrive in a specified subinterval 1 minute in length. Suppose that the man arrives at the bus stop on a particular day at 8:04. Let event A be that he misses the bus. That is, event A states that the bus arrives between 8:00 and 8:04. Then $P(A) = .4$, since the interval 8:00 to 8:04 can be looked upon as the union of four mutually exclusive 1-minute intervals.

Sections 2.4 and 2.5 presented general definitions that tell us how to construct probabilistic models for experiments. In the next section we shall review and illustrate these concepts for discrete sample spaces.

2.6 Calculating the Probability of an Event: The Sample-Point Approach

Finding the probability of an event defined on a sample space containing a finite or denumerable (countably infinite) set of sample points can be approached in two ways, which we shall call the *sample-point* and the *event-composition* methods. Both methods utilize the sample-space model and solution via the *sample-point* approach. Despite this theoretical equivalence, the two methods *are* different in the *sequence* of steps necessary to obtain a solution and in the tools that are of assistance in each. Separation of the two procedures may not be palatable to the unity-seeking theorist, but it is extremely useful in finding the probability of an event which, at best, is not an easy task for a beginner. In this section we consider the *sample-point* approach. The event-composition approach requires additional results and will be presented in Section 2.10.

The sample-point approach is outlined in Section 2.4. The steps to find the probability of an event are the following:

1. *Define the experiment.*
2. *List the simple events associated with the experiment and test each to make certain that they cannot be decomposed. This defines the sample space, S.*
3. *Assign reasonable probabilities to the sample points in S, making certain that* $\sum_{S} P(E_i) = 1$.
4. *Define the event of interest, A, as a specific collection of sample points. (A sample point is in A if A occurs when the sample point occurs. Test ALL sample points in S to locate those in A.)*
5. *Find P(A) by summing the probabilities of the sample points in A.*

We will illustrate with two examples.

Example 2.1: *Consider the problem of selecting two applicants for a job out of a group of five and imagine that the applicants vary in competence, 1 being the best; 2, second best; and so on for 3, 4, and 5. These ratings are of course unknown to the employer. Define two events, A and B, as:*

> A: *the employer selects the best and one of the two poorest applicants (applicants 1 and 4 or 1 and 5),*
>
> B: *the employer selects at least one of the two best.*

Find the probabilities of these events.

Solution: *The steps are as follows:*

1. *The experiment involves randomly selecting two applicants out of five.*
2. *The ten simple events with (i, j) denoting the selection of applicants i and j are*

E_1: (1, 2)	E_5: (2, 3)	E_8: (3, 4)	E_{10}: (4, 5)
E_2: (1, 3)	E_6: (2, 4)	E_9: (3, 5)	
E_3: (1, 4)	E_7: (2, 5)		
E_4: (1, 5)			

3. *A random selection of two out of five should give each pair an equal chance for selection. Hence we will assign each sample point a probability equal to 1/10. That is,*

$$P(E_i) = 1/10, \qquad i = 1, 2, \ldots, 10.$$

4. *Checking the sample points, we see that B occurs whenever $E_1, E_2, E_3, E_4, E_5, E_6$, or E_7 occur. Hence these sample points are included in B.*
5. *Finally, $P(B)$ is equal to the sum of the probabilities of the sample points in B, or*

$$P(B) = \sum_{i=1}^{7} P(E_i) = \sum_{i=1}^{7} 1/10 = 7/10.$$

Similarly, we see that event A is the union of E_3 and E_4. Hence P(A) = 2/10. Note that the solution of this and similar problems would be of importance to a company personnel director.

Example 2.2: *Experience has shown that a genetic defect can be transmitted from mother to daughter with probability equal to .5 unless the father possesses a rare, but nondiagnosable, chromosome structure. When this occurs, the probability of transferral from mother to daughter is equal to .85. A mother with the genetic defect has four daughters, and three out of four possess the genetic defect. Calculate the probability that exactly three daughters could acquire the genetic defect if, in fact, the father does not possess the rare chromosome condition. As a matter of intuition, do you think that transferral of the defect in three out of four cases strongly indicates that the probability of transferral for an individual case is greater than .5 (and hence equal to .85)?*

Solution: *The five steps of the sample-point approach are as follows:*

1. *The experiment consists of observing whether each of the four daughters does or does not acquire a genetic defect.*
2. *A simple event for the experiment could be symbolized by a four-letter sequence of N's and D's, representing "nondefective" and "defective," respectively. The first letter in the sequence would represent the condition of the first daughter, the second the second daughter, and so on. Then the 16 sample points in S are*

E_1: *NNNN*	E_5: *NDNN*	E_9: *DNNN*	E_{13}: *DDNN*
E_2: *NNND*	E_6: *NDND*	E_{10}: *DNND*	E_{14}: *DDND*
E_3: *NNDN*	E_7: *NDDN*	E_{11}: *DNDN*	E_{15}: *DDDN*
E_4: *NNDD*	E_8: *NDDD*	E_{12}: *DNDD*	E_{16}: *DDDD*

3. *Since we are assuming that the probability of transferral of the defect is .5 for a single daughter, we would expect any one of the 16 simple events to be equally probable. (We will subsequently have a more direct method for arriving at this conclusion.) Hence we will assign equal probabilities to the sample points, so that*

$$P(E_i) = 1/16, \qquad i = 1, 2, \ldots, 16.$$

4. *The event of interest, A, is that exactly three of the four daughters have the genetic defect transferred to them. Examining the simple events, we see that those implying the occurrence of A are*

$$A: E_8, E_{12}, E_{14}, E_{15}.$$

5. *Then*

$$P(A) = \sum_{E_i \subset A} P(E_i) = P(E_8) + P(E_{12}) + P(E_{14}) + P(E_{15})$$

$$= 1/16 + 1/16 + 1/16 + 1/16$$

or

$$P(A) = 1/4.$$

Deciding whether the probability of transferral exceeds .5 requires an inference about a population of offspring that could conceptually be generated by this mother–father combination. Decision making as a method of inference will be discussed in detail later. For the moment we leave it to the reader to decide whether the observed occurrence is sufficiently improbable to imply the presence of the rare chromosome structure in the father.

Note that the experiment, Example 2.2, is equivalent to tossing four coins and observing a sequence of heads and tails if the probability of transferral to a single daughter is .5. *Also observe that the assignment of equal probabilities to sample points is not always logical or consistent with the observed relative frequency of occurrence of their respective simple events.* For example, if the probability of transferral were only .01 (instead of .5) we would intuitively feel that the occurrence of the simple event $E_{16}: DDDD$ would be much less probable than for $E_1: NNNN$. The simple event, E_{16}, could be regarded as the intersection of four compound events:

D_1: the first daughter possesses the defect,
D_2: the second daughter possesses the defect,
D_3: the third daughter possesses the defect,
D_4: the fourth daughter possesses the defect.

If all four events, D_1, D_2, D_3, and D_4 occur, E_{16} will occur. We will learn in Section 2.9 how to find the probability of an intersection and thereby acquire an easy and logical method for attaching realistic probabilities to the sample points when the probability of transferral differs from .5.

The sample-point method for solving a probability problem is direct, powerful, and is in some respects a bulldozer approach. It can be applied to find the probability of any event defined over a sample space containing a finite or denumerable set of sample points, but it is not resistant to human error. Errors are frequently committed by incorrectly diagnosing the nature of a simple event and by failing to list all the sample points in S. A second complication occurs because most sample spaces contain a very large number of sample points, and a complete itemization of each is both tedious and time consuming. Fortunately, many sample spaces generated by experimental data contain subsets of sample points that are equiprobable. (The sample spaces for Examples 2.1 and 2.2 possess this property.) When this occurs, we need not list the points but only count the number in each subset. If counting methods are unavailable, an orderly method should be concocted for listing the sample points (notice the listing schemes for Examples 2.1 and 2.2). The listing of large numbers of sample points can be accomplished using an electronic computer.

These comments point to the tools which reduce the effort and error associated with the sample-point approach for finding the probability of an event. The tools are orderliness, the electronic computer, and the mathematical theory of counting called combinatorial analysis. Computer programming and applications are a subject in themselves and form a topic for separate study. The mathematical theory of combinatorial analysis is also a broad subject, but some quite useful results can be detailed in a very short space. Hence our next topic concerns some elementary results in combinatorial analysis and their application to the sample-point approach for the solution of probability problems.

2.7 *Tools for Use with the Sample-Point Approach*

The following represent a few elementary but very useful results from the theory of combinatorial analysis. They are applicable to the sample-point approach for calculating the probability of an event if subsets of sample points in S are equiprobable. When this occurs, we need only count the sample points (assuming that the probability assigned to a typical sample point is known). If all points in S are equiprobable and the total number is N, the probability of a sample point is $P(E_i) = 1/N$.

The first theorem is often called the "*mn* rule" and is stated as follows:

> ***Theorem 2.1:*** *With m elements $a_1, a_2, a_3, \ldots, a_m$ and n elements $b_1, b_2, \ldots, b_n$ it is possible to form mn pairs containing one element from each group.*

Proof: *Verification of the theorem can be seen by observing the rectangular table in Figure 2.3. There will be one square in the table for each a_ib_j pair and hence a total of mn squares.*

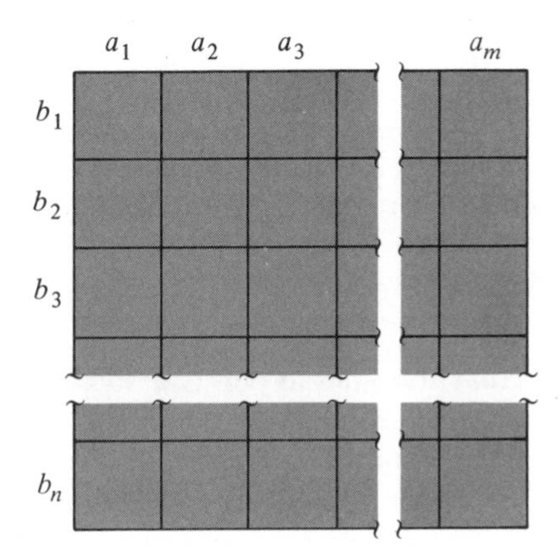

Figure 2.3 Table Indicating the Number of Pairs (a_i, b_j)

The *mn* rule can be extended to any number of sets. Given three sets of elements, $a_1, a_2, \ldots, a_m$; $b_1, b_2 \ldots, b_n$; and $c_1, c_2, \ldots, c_p$, the number of distinct triplets containing one element from each set is equal to *mnp*. The proof of the theorem for three sets involves a reapplication of Theorem 2.1. Thus we think of the first set as an (a_i, b_j) pair. We unite these pairs with elements of the third set, $c_1, c_2, \ldots, c_p$. There are *mn* pairs, (a_i, b_j), and p elements $c_1, c_2, \ldots, c_p$. Then applying Theorem 2.1, the number of ways that they can be paired to form triplets $a_ib_jc_k$ is $(mn)(p) = mnp$.

Example 2.3: *An experiment involves tossing a pair of dice and observing the results. Find the number of sample points in S.*

Solution: *A sample point for this experiment can be represented symbolically as a binary combination representing the outcomes for the first and the second die, respectively. Thus (4, 5) would be the event that the first and second die gave a 4 and a 5, respectively. The totality of points in S would be the total combination of pairs of integers, one from the first die and one from the second.*

The first die can result in one of six numbers. These represent $a_1, a_2, \ldots, a_6$. *Similarly, the second die can fall in one of six ways and these correspond to* $b_1, b_2, \ldots, b_6$. *Then* $m = n = 6$ *and the total number of sample points in S is* $mn = (6)(6) = 36$.

Example 2.4: *Refer to the problem involving the transfer of a genetic defect, Example 2.2. We found for this example that the total number of sample points was 16. Use the mn rule (or its extension) to confirm this result.*

Solution: *Each sample point in S was identified by a sequence of four letters where each position in the sequence could contain one of two letters, an N or a D. The problem therefore involves the formation of quadruplets, selecting an element (an N or a D) from each of four sets. For this example, the four sets are identical and all contain two elements, an N and a D. Then the number of elements in each set is* $m = n = p = q = 2$ *and the total number of quadruplets that can be formed is* $mnpq = (2)^4 = 16$.

We have seen that the sample points associated with an experiment can often be represented symbolically as a sequence of numbers or symbols. In some instances it will be clear that the totality of sample points is the number of distinct ways that the respective symbols can be arranged in sequence. The following two theorems concern arrangements and combinations.

Definition 2.10: *An ordered arrangement of r distinct objects is called a* permutation. *The number of ways of ordering n distinct objects taken r at a time will be designated by the symbol* P^n_r.

Theorem 2.2:

$$P^n_r = n(n-1)(n-2)\cdots(n-r+1)$$

$$= \frac{n!}{(n-r)!}.$$

Proof: *We are concerned with the number of ways of filling r positions with n distinct objects. Applying the extension of the mn rule, the first object can be chosen in one of n ways. After choosing the first, the second can be chosen in $(n - 1)$ ways, the third in $(n - 2)$, and the rth in $(n - r + 1)$ ways. Hence the total number of distinct arrangements is*

$$P_r^n = n(n - 1)(n - 2)\cdots(n - r + 1).$$

Expressed in terms of factorials,

$$P_r^n = n(n - 1)(n - 2)\cdots(n - r + 1)\frac{(n - r)!}{(n - r)!}$$

$$= \frac{n!}{(n - r)!}$$

where $n! = n(n - 1)\cdots(2)(1)$ *and* $0! = 1$.

Example 2.5: *Opening a combination lock requires the selection of the correct set of four different digits in sequence. The digits are set by rotating the tumbler in alternating clockwise and counterclockwise directions. Assume that no digit is used twice. Give the total number of possible combinations.*

Solution: *The total number of lock combinations would equal the number of ways of arranging* $r = 4$ *out of the possible ten digits. Thus*

$$P_4^{10} = \frac{10!}{6!} = (10)(9)(8)(7) = 5040.$$

Example 2.6: *Suppose that an assembly operation in a manufacturing plant involves four steps, which can be performed in any sequence. If the manufacturer wishes to experimentally compare time to assembly for each of the sequences, how many different sequences will be involved in the experiment?*

Solution: *The total number of sequences will be the number of ways of arranging the* $n = 4$ *steps taken* $r = 4$ *at a time. This will equal*

$$P_4^4 = \frac{4!}{(4 - 4)!} = \frac{4!}{0!} = 24.$$

Theorem 2.3: *The number of ways of partitioning n distinct objects into k distinct groups containing $n_1, n_2, \ldots, n_k$ objects, respectively, is*

$$N = \frac{n!}{n_1!n_2!\cdots n_k!} \qquad \text{where } \sum_{i=1}^{k} n_i = n.$$

Proof: *N is the number of distinct arrangements of n objects in a row for a case in which rearrangement of the objects within a group does not count. For example, the letters a to l are arranged in three groups, where $n_1 = 3$, $n_2 = 4$, and $n_3 = 5$:*

abc|defg|hijkl.

The number of distinct arrangements of the n objects, assuming all distinct, *is $P_n^n = n!$ (from Theorem 2.2). Then P_n^n equals the number of ways of partitioning the n objects into k groups (ignoring order within groups) multiplied by the number of ways of ordering the $n_1, n_2, \ldots, n_k$ elements within each group. This application of the extended mn rule gives*

$$P_n^n = (N) \cdot (n_1!n_2!n_3!\cdots n_k!),$$

where $n_i!$ is the number of distinct arrangements of the n_i objects in group i.

Solving for N, we have

$$N = \frac{n!}{n_1!n_2!\cdots n_k!}.$$

We will sometimes use the notation

$$\binom{n}{n_1 n_2 \cdots n_k} = \frac{n!}{n_1!n_2!\cdots n_k!}.$$

Example 2.7: *A labor dispute has arisen concerning the alleged unequal distribution of 20 laborers to four different construction jobs. The first job (considered to be abominable employment) required six laborers; the second, third, and fourth utilized four, five, and five, respectively. The dispute arose over an alleged random distribution of the laborers to the jobs which placed all four members of a particular ethnic group on job 1. In considering whether the assignment represented injustice, a mediation panel desired the probability of the* *observed event. The first step in finding the desired probability is in determining how many sample points will be contained in S. That is, in how many different ways can the 20 laborers be assigned to the four jobs?*

Solution: *The number of ways of assigning the 20 laborers to the four jobs is equal to the number of ways of partitioning the 20 into four groups of* $n_1 = 6$, $n_2 = 4$, $n_3 = n_4 = 5$. *Then*

$$N = \frac{20!}{6!4!5!5!}.$$

By a random assignment *of laborers to the jobs we mean that the N sample points are equiprobable with probability equal to* $1/N$. *If A denotes the event of interest and* n_a *the number of sample points in A, the sum of the probabilities of the sample points in A will be* $P(A) = n_a(1/N) = n_a/N$. *We leave the determination of* n_a *as an exercise for the reader.*

In many situations the sample points are identified by an array of symbols in which the arrangement of symbols is *unimportant*. The sample points for the selection of applicants, Example 2.1, imply a selection of two applicants out of four. Each sample point is identified as a pair of symbols and the order of the symbols identifying the sample points is irrelevant.

Definition 2.11: *The number of combinations of n objects taken r at a time is the number of subsets, each of size r, which can be formed from the n objects. This number will be denoted by* C_r^n *or* $\binom{n}{r}$.

Theorem 2.4:

$$\binom{n}{r} = C_r^n = \frac{P_r^n}{r!} = \frac{n!}{r!(n-r)!}.$$

Proof: *The selection of r objects from a total of n is equivalent to partitioning the n objects into* $k = 2$ *groups, the r selected and the* $(n - r)$ *remaining. It thus is* *a special case of the general partitioning problem of Theorem 2.3, where* $k = 2$, $n_1 = r$, *and* $n_2 = (n - r)$. *Therefore,*

$$\binom{n}{r} = C_r^n = \frac{n!}{r!(n-r)!}.$$

Example 2.8: *Find the number of ways of selecting two applicants out of five and hence the total number of sample points in S for Example 2.1.*

Solution:

$$\binom{5}{2} = \frac{5!}{2!3!} = 10.$$

(Note that this agrees with the number of sample points listed in Example 2.1.)

Example 2.9: *Find the number of ways of selecting exactly one of the two best applicants in a selection of two out of five. Then find the probability of that event.*

Solution: *Let A be the event that exactly one of the two best is selected and let* n_a *denote the number of sample points in A. Then* n_a *equals the number of ways of selecting one of the two best out of a possible two (call this number m) times the number of ways of selecting one of the three low ranking applicants out of a possible three (call this number n). Then* $m = \binom{2}{1}$, $n = \binom{3}{1}$, *and, applying the mn rule,*

$$n_a = \binom{2}{1} \cdot \binom{3}{1} = \frac{2!}{1!1!} \cdot \frac{3!}{1!2!} = 6.$$

(This number can be verified by counting the sample points in A from the listing in Example 2.1.)

In Example 2.8 we found the total number of sample points in S to equal $N = 10$. *Assuming each selection to be equiprobable,* $P(E_i) = 1/10, i = 1, 2, \ldots 10$, *and*

$$P(A) = \sum_{E_i \subset A} P(E_i) = n_a(1/10) = 6/10 = 3/5.$$

Theorems 2.1 through 2.4 provide a few of the many useful counting rules found in the theory of combinatorial analysis. A few additional theorems appear in the exercises at the end of the chapter. The student interested in extending his knowledge of combinatorial analysis is referred to one of the numerous texts on this subject.

We now direct our attention to the concept of conditional probability. Conditional probability will play an important role in the event-composition approach for finding the probability of an event and sometimes will be useful in finding the probabilities of sample points (for sample spaces with unequal probabilities attached to the sample points).

2.8 Conditional Probability and the Independence of Events

The probability of an event will vary depending upon the occurrence or nonoccurrence of one or more related events. For example, Florida sport fishermen are vitally interested in the probability of rain. The probability of rain on a given day, ignoring the daily atmospheric conditions or any other events, is the fraction of days in which rain occurs over a long period of time. This would be called the *unconditional probability* of the event "rain on a given day." Now suppose that we wish to consider the probability of rain tomorrow. It has rained almost continuously for two days in succession and a tropical storm is heading up the coast. What is the probability of rain? This probability is conditional on the occurrence of several events and a Floridian would tell you that it is much larger than the unconditional probability of rain.

The unconditional probability of a 1 in the toss of a single-balance die is 1/6. The conditional probability of a 1, given that an odd number has fallen, is 1/3. That is, 1, 3, and 5 occur with equal frequency. Knowing that an odd number has occurred, the relative frequency of occurrence of a 1 is 1/3. Thus the conditional probability of an event is the probability (relative frequency of occurrence) of the event given the fact that one or more events have already occurred. A careful perusal of this example will indicate agreement of the following definition with the relative frequency concept of probability.

Definition 2.12: *The* conditional probability of an event *A, given that an event B has occurred, is equal to*

$$P(A|B) = \frac{P(A \cap B)}{P(B)}$$

provided $P(B) > 0$. [The symbol $P(A|B)$ is read "probability of A given the occurrence of B."]

Further confirmation of the consistency of Definition 2.12 with the relative frequency concept of probability can be obtained from the following construction. Suppose that an experiment is repeated a large number of times, N, resulting in both A and B, $A \cap B$, n_{11} times; A and not B, $A \cap \bar{B}$, n_{21} times; B and not A, $\bar{A} \cap B$, n_{12} times, and neither A nor B, $\bar{A} \cap \bar{B}$, n_{22} times. We present these results in a two-way table, Table 2.1.

Table 2.1 Two-Way Table for Events A and B

	A	$\bar{A}$
B	n_{11}	n_{12}
$\bar{B}$	n_{21}	n_{22}

Note that $n_{11} + n_{12} + n_{21} + n_{22} = N$. Then it follows that, with $\approx$ read "approximately equal to,"

$$P(A) \approx \frac{n_{11} + n_{21}}{N},$$

$$P(B) \approx \frac{n_{11} + n_{12}}{N},$$

$$P(A|B) \approx \frac{n_{11}}{n_{11} + n_{12}},$$

$$P(B|A) \approx \frac{n_{11}}{n_{11} + n_{21}},$$

and
$$P(A \cap B) \approx \frac{n_{11}}{N}.$$

Using these "probabilities" it is easy to see that

$$P(B|A) \approx \frac{P(A \cap B)}{P(A)}$$

and
$$P(A|B) \approx \frac{P(A \cap B)}{P(B)}.$$

Hence Definition 2.12 is consistent with the relative frequency concept of probability.

Example 2.10: *Use Definition 2.12 to find the probability of a 1, given the occurrence of an odd number, in the toss of a single die.*

Solution: *Define the events*

A: *observe a* 1,
B: *observe an odd number.*

We seek the probability of A given that the event B has occurred. The probability that both A and B occur implies the observance of both a 1 *and an odd number and hence* $P(A \cap B) = 1/6$. *Also,*

$$P(B) = 1/2.$$

Then

$$P(A|B) = \frac{P(A \cap B)}{P(B)} = \frac{1/6}{1/2} = 1/3.$$

Note that this result is in complete agreement with our earlier intuitive evaluation of this probability.

Suppose that the occurrence of an event A is unaffected by the occurrence or nonoccurrence of event B. When this occurs, we would be inclined to say that the event A is independent of B. This event relationship is expressed by the following definition.

Definition 2.13: *Two events, A and B, are said to be* independent *if*

$$P(A \cap B) = P(A)P(B).$$

Otherwise, the events are said to be dependent. *Note that this definition is equivalent to stating that two events, A and B, are independent if* $P(A|B) = P(A)$ *or* $P(B|A) = P(B)$.

Example 2.11: *Consider the following two events in the toss of a single die:*

A: observe an odd number,
B: observe an even number,
C: observe a 1 or 2.

(a) Are A and B independent events?
(b) Are A and C independent events?

Solution:
(a) To decide whether A and B are independent, we must see whether they satisfy the conditions of Definition 2.13. The probability of an odd number given that an even number has occurred is zero. That is,

$$P(A|B) = 0.$$

Also,

$$P(A) = 1/2.$$

Then it is clear that $P(A|B) \neq P(A)$, *and hence A and B* are not *independent events.*
(b) Are A and C independent? Note that

$$P(A|C) = 1/2$$

and

$$P(A) = 1/2.$$

Therefore,

$$P(A|C) = P(A) \text{ and } A \text{ and } C \text{ are independent.}$$

Example 2.12: *Three brands of coffee, X, Y, and Z, are to be ranked according to taste by a judge. Define the following events:*

A: brand X is preferred to Y,
B: brand X is ranked best,
C: brand X is ranked second best,
D: brand X is ranked third best.

If the judge actually has no taste preference and thus randomly assigns ranks to the brands, is event A independent of events B, C, and D?

Solution: *The six equally likely sample points for this experiment are given by*

$$E_1: XYZ \qquad E_3: YXZ \qquad E_5: ZXY$$

$$E_2: XZY \qquad E_4: YZX \qquad E_6: ZYX$$

where XYZ denotes that X is ranked best and Y second best. Then

$$P(A) = P(E_1) + P(E_2) + P(E_5) = 1/2,$$

$$P(A|B) = \frac{P(A \cap B)}{P(B)} = \frac{P(E_1) + P(E_2)}{P(E_1) + P(E_2)} = 1,$$

$$P(A|C) = 1/2,$$

and

$$P(A|D) = 0.$$

Thus A is independent of C but is not independent of B and D.

2.9 *Two Laws of Probability*

The following two laws give the probabilities of unions and intersections. As such, they will play an important role in the event-composition approach to the solution of probability problems.

The Multiplicative Law of Probability: *The probability of the intersection of two events, A and B, is*

$$P(A \cap B) = P(A)P(B|A)$$

$$= P(B)P(A|B).$$

If A and B are independent, then

$$P(A \cap B) = P(A)P(B).$$

Note that the multiplicative law of probability requires no proof because it evolves as a consequence of Definition 2.12, the definition of conditional probability. Hence it is a "law" only insofar as it is consistent with the relative frequency concept of probability. The multiplicative law of probability can be extended to apply to any number of events. Thus by applying the law,

$$\begin{aligned} P(A \cap B \cap C) &= P[(A \cap B) \cap C] \\ &= P(A \cap B)P(C|A \cap B) \\ &= P(A)P(B|A)P(C|A \cap B). \end{aligned}$$

The probability of the intersection of any number, say k events, can be obtained in the same manner:

$$\begin{aligned} &P(A_1 \cap A_2 \cap A_3 \cap \cdots \cap A_k) \\ &\quad = P(A_1)P(A_2|A_1)P(A_3|A_1 \cap A_2)\cdots P(A_k|A_1 \cap A_2 \cap \cdots \cap A_{k-1}). \end{aligned}$$

The additive law of probability gives the probability of the union of two events.

The Additive Law of Probability: *The probability of the union of two events A and B is*

$$P(A \cup B) = P(A) + P(B) - P(A \cap B).$$

If A and B are mutually exclusive events, $P(A \cap B) = 0$ *and*

$$P(A \cup B) = P(A) + P(B).$$

An approach to the proof of the additive law can be seen by viewing the Venn diagram, Figure 2.4.

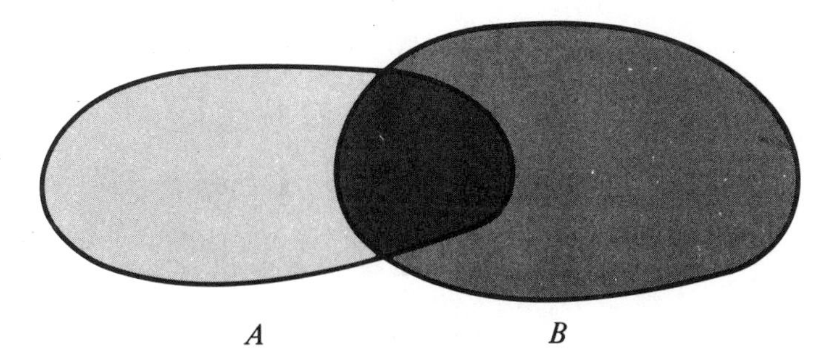

Figure 2.4 Venn Diagram for the Union of A and B

Consider the events $A = (A \cap \bar{B}) \cup (A \cap B)$ and $B = (\bar{A} \cap B) \cup (A \cap B)$, where $[(A \cap \bar{B})$ and $(A \cap B)]$ and $[(\bar{A} \cap B)$ and $(A \cap B)]$ are mutually exclusive pairs of events. Consider also $A \cup B = (A \cap \bar{B}) \cup (A \cap B) \cup (\bar{A} \cap B)$, where all three of these events are mutually exclusive. Then, by Axiom 3,

$$P(A \cup B) = P(A \cap \bar{B}) + P(A \cap B) + P(\bar{A} \cap B),$$

$$P(A) = P(A \cap \bar{B}) + P(A \cap B)$$

$$[\text{or } P(A \cap \bar{B}) = P(A) - P(A \cap B)],$$

and

$$P(B) = P(\bar{A} \cap B) + P(A \cap B)$$

$$[\text{or } P(\bar{A} \cap B) = P(B) - P(A \cap B)].$$

Substituting $P(A \cap \bar{B})$ and $P(\bar{A} \cap B)$ into the expression for $P(A \cup B)$ above we obtain the desired result,

$$P(A \cup B) = P(A) + P(B) - P(A \cap B).$$

The probability of the union of three events can be obtained by making use of the above law. Observe that

$$\begin{aligned} P(A \cup B \cup C) &= P[A \cup (B \cup C)] \\ &= P(A) + P(B \cup C) - P[A \cap (B \cup C)] \end{aligned}$$

$$= P(A) + P(B) + P(C) - P(B \cap C)$$
$$- P[(A \cap B) \cup (A \cap C)]$$
$$= P(A) + P(B) + P(C) - P(B \cap C)$$
$$- P(A \cap B) - P(A \cap C)$$
$$+ P(A \cap B \cap C)$$

since $(A \cap B) \cap (A \cap C) = A \cap B \cap C$.

The formula for the probability of the union of k events, $A_1, A_2, \ldots A_k$, derived in a similar manner, is

$$P(A_1 \cup A_2 \cup \cdots \cup A_{k-1} \cup A_k)$$
$$= \sum_{i=1}^{k} P(A_i) - \sum_{\substack{i \\ i<j}}\sum_{j} P(A_i \cap A_j) + \sum_{\substack{i \\ i<j<m}}\sum_{j}\sum_{m} P(A_i \cap A_j \cap A_m)$$
$$+ \cdots - \cdots + \cdots - \cdots - (-1)^k P(A_1 \cap A_2 \cap \cdots \cap A_k).$$

2.10 Calculating the Probability of an Event: The Event-Composition Method

We learned in Section 2.4 that sets (events) may be combined using the two operations of set algebra to form unions and intersections. The event-composition approach to calculate the probability of an event utilizes this result by expressing the event of interest, say event A, as a composition (unions and/or intersections) of two or more other events. Then the two laws of probability can be applied to the composition to find $P(A)$.

For example, suppose that for some experiment the following composition holds:

$$A = (B \cap C) \cup (D \cap E).$$

This states that the sample points contained in the event A are identically the same as those contained in the union of $B \cap C$ and $D \cap E$. This being true, the probabilities of the left and right sides of the equation must be equal since they would equal the sum of the probabilities of the same set of sample points. Thus

$$P(A) = P[(B \cap C) \cup (D \cap E)],$$

where the probability of the composition can be found by applying the additive and multiplicative laws of probability.

Example 2.13: *Use the additive and multiplicative laws of probability to simplify the expression* $P[(B \cap C) \cup (D \cap E)]$.

Solution: *Applying the additive law of probability,*

$$P[(B \cap C) \cup (D \cap E)] = P(B \cap C) + P(D \cap E) - P(B \cap C \cap D \cap E).$$

Observe that the events on the right are intersections and this calls for the application of the multiplicative law of probability. Then

$$P[(B \cap C) \cup (D \cap E)] = P(B)P(C|B) + P(D)P(E|D) - P(B)P(C|B)P(D|B \cap C)P(E|B \cap C \cap D).$$

The event-composition approach will not be successful unless the probabilities of the events that appear in $P(A)$ (after the additive and multiplicative laws have been applied) are known. If one or more of these probabilities are unknown, the method fails. Also, note that it is frequently desirable to form compositions of mutually exclusive or independent events. Mutually exclusive events simplify the additive law and independence simplifies the multiplicative law of probability.

A summarization of the steps to be followed in the event composition approach is:

1. *Define the experiment.*
2. *Clearly visualize the nature of the sample points. Identify a few to clarify your thinking.*
3. *Write an equation expressing the event of interest, say A, as a composition of two or more events using either or both of the two forms of composition (unions and intersections). Note that this equates point sets. Make certain that the event implied by the composition and event A represent the same set of sample points.*
4. *Apply the additive and multiplicative laws of probability to step 3 and find P(A).*

Step 3 is the most difficult because one can form many compositions that will be equivalent to event A. The trick is to form a composition in which all the probabilities appearing in step 4 are known.

The event-composition approach does not require a listing of the sample points in S, but it does require a clear understanding of the nature of a typical sample point. The major error students tend to make in applying the event-composition approach occurs in writing the composition. That is, the point-set equation that expresses A as union and/or intersection of other events is frequently incorrect. Always test your equality to make certain that the composition implies an event that contains the same set of sample points as those in A.

Some comparison of the sample-point and event-composition methods for calculating the probability of an event can be obtained by applying both methods to the same problem. We will apply the event-composition approach to the problem of selecting applicants that was solved by the sample-point method in Examples 2.8 and 2.9.

Example 2.14: *The experiment involves the selection of two applicants out of five. Find the probability of drawing exactly one of the two best applicants, event A.*

Solution: *Define the following two events:*

B: *draw the best and one of the three poorer applicants,*
C: *draw the second best and one of the three poorer applicants.*

Then B and C are mutually exclusive events and

$$A = B \cup C.$$

Also, let

$$D_1 = B_1 \cap B_2,$$

where B_1 = *draw the best on the first draw*

B_2 = *draw one of the three poorer applicants on the second draw*

and $D_2 = B_3 \cap B_4,$

where B_3 = *draw one of the three poorer applicants on the first draw*

B_4 = *draw the best on the second draw*

Note that $B = D_1 \cup D_2$.

Similarly, we could let $G_1 = C_1 \cap C_2$ *and* $G_2 = C_3 \cap C_4$, *where* C_1, C_2, C_3, *and* C_4 *are defined like* B_1, B_2, B_3, *and* B_4 *with the words "second best" replacing "best." Note that* D_1 *and* D_2 *and* G_1 *and* G_2 *are pairs of mutually exclusive events and*

$$A = B \cup C,$$

$$A = (B_1 \cap B_2) \cup (B_3 \cap B_4) \cup (C_1 \cap C_2) \cup (C_3 \cap C_4).$$

Applying the additive law of probability to these four mutually exclusive events,

$$P(A) = P(B_1 \cap B_2) + P(B_3 \cap B_4) + P(C_1 \cap C_2) + P(C_3 \cap C_4).$$

Applying the multiplicative law,

$$P(B_1 \cap B_2) = P(B_1)P(B_2|B_1).$$

The probability of drawing the best on the first draw is

$$P(B_1) = 1/5.$$

Similarly, the probability of drawing one of the three poorest on the second draw, given that the best was drawn on the first selection, is

$$P(B_2|B_1) = 3/4.$$

Then

$$P(B_1 \cap B_2) = P(B_1)P(B_2|B_1)$$

$$= (1/5)(3/4) = 3/20.$$

The probabilities of all the other intersections in $P(A)$, $P(B_3 \cap B_4)$, $P(C_1 \cap C_2)$, *and* $P(C_3 \cap C_4)$ *are obtained in exactly the same manner and all equal* 3/20. *Then*

$$P(A) = P(B_1 \cap B_2) + P(B_3 \cap B_4) + P(C_1 \cap C_2) + P(C_3 \cap C_4)$$

$$= 3/20 + 3/20 + 3/20 + 3/20$$

$$= 3/5.$$

This answer is identical to that obtained in Example 2.9, where $P(A)$ *was acquired using the sample-point approach.*

Example 2.15: *It is known that a patient will respond to treatment of a particular disease with probability equal to .9. If three patients are treated in an independent manner, find the probability that at least one will respond.*

Solution: *Define the events*

A: *at least one of the three patients will respond,*
B_1: *the first patient will not respond,*
B_2: *the second patient will not respond,*
B_3: *the third patient will not respond.*

Then observe that $\bar{A} = B_1 \cap B_2 \cap B_3$ and that $S = A \cup \bar{A}$, where A and $\bar{A}$ are complementary events and hence mutually exclusive. Then

$$P(S) = P(A) + P(\bar{A})$$

$$1 = P(A) + P(B_1 \cap B_2 \cap B_3)$$

or

$$P(A) = 1 - P(B_1 \cap B_2 \cap B_3).$$

Applying the multiplicative rule,

$$P(B_1 \cap B_2 \cap B_3) = P(B_1)P(B_2|B_1)P(B_3|B_1 \cap B_2)$$

where

$$P(B_2|B_1) = P(B_2) \qquad \text{and} \qquad P(B_3|B_1 \cap B_2) = P(B_3)$$

because the events are independent.

Substituting $P(B_i) = .1$, $i = 1, 2, 3$,

$$P(A) = 1 - (.1)^3$$

$$= .999.$$

Note that we have demonstrated the utility of complementary events. This result is important because it is frequently easier to find the probability of the complement, $P(\bar{A})$, than to find $P(A)$ directly. Then $P(A)$ can be obtained from the simple relationship

$$P(A) = 1 - P(\bar{A}).$$

Example 2.16: *Observation of a waiting line at a medical clinic indicates that the probability that a new arrival will be an emergency case is $p = 1/6$. Find the probability that the rth patient is the first emergency case. (Assume that conditions of arriving patients represent independent events.)*

Solution: *The experiment consists of watching patient arrivals until the first emergency case appears. Then the sample points for the experiment are*

$$E_i\text{: the ith patient is the first emergency case, } i = 1, 2, \ldots.$$

Since only one sample point falls in the event of interest,

$$P(r\text{th patient is the first emergency case}) = P(E_r).$$

Now define A_i to denote the event that the ith arrival is not an emergency case. Then we can represent E_r as the intersection

$$E_r = A_1 \cap A_2 \cap A_3 \cap \cdots \cap A_{r-1} \cap \bar{A}_r.$$

Applying the multiplicative law we have

$$P(E_r) = P(A_1)P(A_2|A_1)P(A_3|A_1 \cap A_2) \cdots P(\bar{A}_r|A_1 \cap \cdots \cap A_{r-1}),$$

and, since the events $A_1, A_2, \ldots, A_{r-1}$, and $\bar{A}_r$ are independent, it follows that

$$\begin{aligned} P(E_r) &= P(A_1)P(A_2) \cdots P(A_{r-1})P(\bar{A}_r) \\ &= (1-p)^{r-1}p \\ &= (5/6)^{r-1}(1/6), \qquad r = 1, 2, 3, \ldots. \end{aligned}$$

Note that

$$\begin{aligned} P(S) &= P(E_1) + P(E_2) + P(E_3) + \cdots + P(E_i) + \cdots \\ &= 1/6 + (5/6)(1/6) + (5/6)^2(1/6) + \cdots \\ &\quad + (5/6)^{i-1}(1/6) + \cdots \\ &= \frac{1/6}{1 - 5/6} = 1. \end{aligned}$$

The above result follows from the formula for the sum of an infinite number of terms of a decreasing geometrical progression, $a/(1-r)$, where a denotes the first term and r the common ratio. This formula is useful in many simple probability problems.

Example 2.17: *A monkey is to be taught to recognize colors by tossing one red, one black, and one white ball into boxes of the same respective colors, one ball to*

a box. If the monkey has not learned the colors, and merely tosses one ball into each box at random, find the probability of
(*a*) *No color matches.*
(*b*) *Exactly one color match.*

Solution: *This problem can be solved by listing sample points since there are only three balls involved, but a more general method of solution will be illustrated. Define the following events:*

A_1: *a color match in the red box,*
A_2: *a color match in the black box,*
A_3: *a color match in the white box.*

There are $3! = 6$ *equally likely ways of randomly tossing the balls into the boxes with one ball in each box. Also, there are only* $2! = 2$ *ways of tossing the balls into the boxes if one particular box is required to have a color match. Hence*

$$P(A_1) = P(A_2) = P(A_3) = 2/6 = 1/3.$$

Similarly, it follows that

$$\begin{aligned} P(A_1 \cap A_2) = P(A_1 \cap A_3) &= P(A_2 \cap A_3) \\ &= P(A_1 \cap A_2 \cap A_3) = 1/6. \end{aligned}$$

We can now answer parts (*a*) *and* (*b*) *by the event-composition method.*
(*a*) *Note that*

$$\begin{aligned} &P(\text{no color matches}) \\ &\quad = 1 - P(\text{at least one color match}) \\ &\quad = 1 - P(A_1 \cup A_2 \cup A_3) \\ &\quad = 1 - [P(A_1) + P(A_2) + P(A_3) - P(A_1 \cap A_2) \\ &\qquad - P(A_1 \cap A_3) - P(A_2 \cap A_3) + P(A_1 \cap A_2 \cap A_3)] \\ &\quad = 1 - [3(1/3) - 3(1/6) + (1/6)] \\ &\quad = 2/6 = 1/3. \end{aligned}$$

(*b*) *We leave it to the reader to show that*

$$\begin{aligned} &P(\text{exactly one match}) \\ &\quad = P(A_1) + P(A_2) + P(A_3) \\ &\qquad - 2[P(A_1 \cap A_2) + P(A_1 \cap A_3) + P(A_2 \cap A_3)] \\ &\qquad + 3[P(A_1 \cap A_2 \cap A_3)] \\ &\quad = (3)(1/3) - (2)(3)(1/6) + (3)(1/6) = 1/2. \end{aligned}$$

The best way to learn how to solve probability problems is to learn by doing. A large number of exercises is provided at the end of the chapter and in the references to assist the reader in developing his ability to diagnose and solve probability problems.

2.11 Bayes' Rule

The event-composition approach to solving probability problems is sometimes facilitated by viewing the sample space, S, as a union of mutually exclusive subsets. That is, assume that $S = B_1 \cup B_2 \cup \cdots \cup B_k$, where $B_i \cap B_j = \varnothing$ for $i \neq j$. Then, any subset, A, of S can be written as

$$\begin{aligned} A &= A \cap S \\ &= A \cap (B_1 \cup B_2 \cup \cdots \cup B_k) \\ &= (A \cap B_1) \cup (A \cap B_2) \cup \cdots \cup (A \cap B_k). \end{aligned}$$

We then observe that

$$\begin{aligned} P(A) &= P(A \cap B_1) + P(A \cap B_2) + \cdots + P(A \cap B_k) \\ &= P(B_1)P(A|B_1) + P(B_2)P(A|B_2) + \cdots + P(B_k)P(A|B_k) \\ &= \sum_{i=1}^{k} P(B_i)P(A|B_i). \end{aligned}$$

A conditional probability of the form $P(B_j|A)$ can then be evaluated as

$$P(B_j|A) = \frac{P(A \cap B_j)}{P(A)}$$

$$= \frac{P(B_j)P(A|B_j)}{\sum_{i=1}^{k} P(B_i)P(A|B_i)}.$$

This formula for conditional probability is called *Bayes' rule*.

Example 2.18: *An electronic fuse is produced by five production lines in a manufacturing operation. The fuses are costly, are quite reliable, and are shipped to suppliers in 100-unit lots. Because testing is destructive, most buyers of the fuses test only a small number of fuses before deciding to accept or reject lots of incoming fuses.*

All five production lines normally produce only 2 percent defective fuses, which are randomly dispersed in the output. Unfortunately, production line 1 suffered mechanical difficulty and produced 5 percent defectives during the month of March. This situation became known to the manufacturer after the fuses had been shipped. A customer received a lot produced in March and tested three fuses. One failed. What is the probability that the lot was produced on line 1? What is the probability that the lot came from one of the four other lines?

Solution: *Let $P(L_i)$ denote the probability that a fuse was drawn from line i and let D denote the event that a fuse was defective. Then*

$$P(L_1) = 1/5,$$

$$P(L_1 \cap D) = P(L_1)P(D|L_1) = (.2)(3)(.05)(.95)^2 = .0270750,$$

$$P(\bar{L}_1 \cap D) = P(\bar{L}_1)P(D|\bar{L}_1) = (.8)(3)(.02)(.98)^2 = .0460992,$$

$$P(D) = P(L_1 \cap D) + P(\bar{L}_1 \cap D)$$

$$= .0270750 + .0460992 = .0731742,$$

$$P(L_1|D) = \frac{P(L_1 \cap D)}{P(D)} = \frac{.0270750}{.0460992} = .37,$$

and $$P(\bar{L}_1|D) = 1 - P(L_1|D) = 1 - .37 = .63.$$

2.12 Numerical Events and Random Variables

Events of major interest to the scientist, engineer, or businessman are those identified by numbers, called *numerical events.* The research physician is interested in the event that ten of ten treated patients survive an illness; the businessman is interested in the event that he have sales next year of $5,000,000. Because the value of a numerical event will vary in repeated samplings, it is called a *random variable.*

To each point in the sample space we shall assign a real number denoting the value of a numerical event. The numbers will vary from one sample point to another, but some points may be assigned the same number. Thus we have defined a variable that is a function of the sample points in S. Letting Y denote this variable, $Y = a$ is the numerical event that contains all sample points assigned the number a. Indeed, the sample space, S, can be partitioned into subsets so that points within a subset are assigned the same value of Y. These subsets are mutually exclusive. The partitioning of S is symbolically indicated in Figure 2.5 for a random variable that can assume values 0, 1, 2, 3, 4.

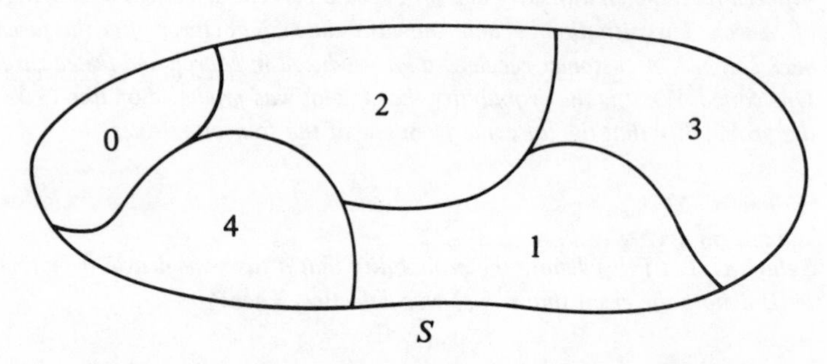

Figure 2.5 Partitioning S into Subsets That Define the Events $Y = 0, 1, 2, 3, 4$

> ***Definition 2.14:*** *A* random variable *is a real-valued function defined over a sample space.*

Example 2.19: *Define an experiment as tossing two coins and observing the results. Let Y equal the number of heads observed. Identify the sample points in S, assign a value of Y to each sample point, and identify the sample points associated with each value of the random variable.*

Solution: *Let H and T represent "head" and "tail," respectively, and let an ordered pair of symbols identify the outcome for the first and second coins, respectively. (Thus HT implies a head on the first coin and a tail on the second.) Then the four sample points in S are* $E_1: HH$, $E_2: HT$, $E_3: TH$, *and* $E_4: TT$. *The values of Y assigned to the sample points depend on the number of heads implied by each point. For* $E_1: HH$, *two heads were observed and* E_1 *is assigned the value* $Y = 2$. *Similarly, we assign the values* $Y = 1$ *to* E_2 *and* E_3 *and* $Y = 0$ *to* E_4. *Summarizing, the random variable Y can take three values,* $Y = 0, 1, 2$, *which are events defined by specific collections of sample points:*

$$Y = 0\colon E_4,$$

$$Y = 1\colon E_2, E_3,$$

$$Y = 2\colon E_1.$$

Let y denote an observed value of the random variable Y. We then set the probability that $Y = y$ equal to the sum of the probabilities of the sample points that are assigned the value y.

Example 2.20: *Compute the probabilities for each value of Y in Example 2.19.*

Solution: $Y = 0$ *results only from sample point* E_4. *If the coins are balanced, the sample points are equally likely and hence*

$$P(Y = 0) = P(E_4) = 1/4.$$

Similarly,

$$P(Y = 1) = P(E_2) + P(E_3) = 1/2$$

and

$$P(Y = 2) = P(E_1) = 1/4.$$

A more detailed examination of random variables will be undertaken in the next two chapters.

2.13 Random Sampling

As our final topic in this chapter, we move from theory to application and examine the type of experiments conducted in statistics. A statistical experiment involves the observation of a sample selected from a larger body of data, existing or conceptual, called a population. The measurements in the sample, viewed as observations on one or more random variables, are then employed to make an inference about the characteristics of the target population.

How are these inferences made? An exact answer to this question is deferred until later but a general observation follows from our discussion in Section 2.2. There we learned that the probability of the observed sample plays a major role in making an inference and evaluating its credibility.

Without belaboring the point, it is clear that the method of sampling will affect the probability of a particular sample outcome. For example, suppose that a fictitious population contains only $N = 5$ elements, from which we plan to sample $n = 2$. You could mix the elements thoroughly and select two in such a way that all pairs of elements possess an equal probability of selection. A second sampling procedure might require selecting a single element, replacing it in the population, and then drawing a single element again. The two methods of sample selection are called sampling without and with replacement, respectively.

If all the $N = 5$ population elements are distinctly different, the probability of drawing a specific pair, sampling without replacement, is 1/10. The probability of drawing the same pair, sampling with replacement, is 2/25. You can easily verify these results.

The point that we make is that the method of sampling, known as the "design of an experiment," affects both the quantity of information in a sample as well as the probability of observing a specific sample result. Hence every sampling procedure must be clearly described if we wish to make valid inferences from sample to population.

The study of the design of experiments, the various types of designs along with their properties, is a course in itself. Hence at this early stage of study, we only introduce the simplest sampling procedure, *simple random sampling*. The notation of simple random sampling will be needed in subsequent discussions of the probabilities associated with random variables, and it will inject some realism into our discussion of statistics. This is because simple random sampling is often employed in practice. Now let us define the term "random sample."

> ***Definition 2.15:*** *Let N and n represent the numbers of elements in the population and sample, respectively. If the sampling is conducted in such a way that each of the $\binom{N}{n}$ samples has an equal probability of being selected, the sampling is said to be random and the result is said to be a* random sample.

Perfect random sampling is difficult to achieve in practice. If the population is not too large, we might write each of the N numbers on a poker chip, mix the total, and select a sample of n chips. The numbers on the poker chips would specify the measurements to appear in the sample. Other techniques are available when the population is large.

In many situations, the population is conceptual, as in an observation made during a laboratory experiment. Here the population is envisioned to be the infinitely large number of measurements obtained when the experiment is repeated over and over again. If we wish a sample of $n = 10$ measurements from this population, we repeat the experiment ten times and hope that the results represent, to a reasonable degree of approximation, a random sample.

Although the primary purpose of this discussion was to clarify the meaning of a random sample, we would like to mention that some sampling techniques are only partly random. For instance, if we wish to determine the voting preference of the nation in a presidential election, we would not likely choose a random sample from the population of voters. By pure chance, all the voters appearing in the sample might be drawn from a single city, say, San Francisco, which might not be at all representative of the population. We would prefer a random selection of voters from smaller political districts, perhaps states, allotting a specified number to each state. The information from the randomly selected subsamples drawn from the respective states would be combined to form a prediction concerning the entire population of voters in the country. In general, one wants to select a sample so as to obtain a specified quantity of information at minimum cost.

2.14 Summary

This chapter has been concerned with providing a model for the repetition of an experiment and, consequently, a model for the population frequency distributions of Chapter 1. The acquisition of a probability distribution is the first step in forming a theory to model reality and to develop the machinery for making inferences.

An experiment was defined as the process of making an observation. The concepts of an event, the simple event, the sample space, and the probability axioms have provided a probabilistic model for calculating the probability of an event. Numerical events and the definition of a random variable were introduced in Section 2.12.

Embedded among our theoretical abstractions of reality are some earthy tools that assist in finding the probability of an event. Inherent in the model is the sample-point approach for calculating the probability of an event (Section 2.6). Counting rules useful in applying the sample-point method were discussed in Section 2.7. The concept of conditional probability, the operations of set algebra, and the two laws of probability set the stage for the event-composition method for calculating the probability of an event (Section 2.10).

Of what value is the theory of probability? It provides the theory and the tools for calculating the probabilities of numerical events and hence the probability distributions for discrete random variables in Chapter 3. The numerical events of interest to us appear in a sample and we will wish to calculate the probability of an observed sample in order to make an inference about a target population. Probability provides both the foundation and the tools for modern statistical inference, and this is the objective of statistics.

References

1. Cramér, H., *The Elements of Probability Theory and Some of Its Applications*. New York: John Wiley & Sons, Inc., 1955; Stockholm: Almqvist and Wiksell, 1954.
2. Feller, W., *An Introduction to Probability Theory and Its Applications*, Vol. 1, 3rd ed. New York: John Wiley & Sons, Inc., 1968.
3. Feller, W., *An Introduction to Probability Theory and Its Applications*, Vol. 2, 2nd ed. New York: John Wiley & Sons, Inc., 1971.
4. Meyer, P. L., *Introductory Probability and Statistical Applications*. Addison-Wesley Publishing Company, Inc., 1965.
5. Parzen, E., *Modern Probability Theory and Its Applications*. New York: John Wiley & Sons, Inc., 1960.
6. Riordan, J., *An Introduction to Combinational Analysis*. New York: John Wiley & Sons, Inc., 1958.

Exercises

2.1. A coin is tossed four times and the outcome is recorded for each toss.
(a) List the sample points for the experiment.
(b) Let A be the event that the experiment yields three heads. List the sample points in A.
(c) Make a reasonable assignment of probabilities to the sample points and find $P(A)$.

2.2. A boxcar contains seven complex electronic systems. Unknown to the purchaser, three are defective. Two of the seven are selected for thorough testing and then classified as defective or nondefective.
(a) List the sample points for this experiment.
(b) Let A be the event that the selection includes no defectives. List the sample points in A.
(c) Assign probabilities to the sample points and find $P(A)$.

2.3. Patients arriving at a hospital outpatient clinic can select one of three stations for service. Suppose that physicians are randomly assigned to the stations and that the patients therefore have no station preference. Three patients arrive at the clinic and their selection of stations is observed.
(a) List the sample points for the experiment.
(b) Let A be the event that each station receives a patient. List the sample points in A.
(c) Make a reasonable assignment of probabilities to the sample points and find $P(A)$.

2.4. A retailer sells two styles of high-priced high-fidelity consoles that experience shows are in equal demand. If he stocks four of each, what is the probability that the first four customers seeking a console all wish to purchase the same style?
(a) Define the experiment.
(b) List the sample points.
(c) Define the event of interest, A, as a specific collection of sample points.
(d) Assign probabilities to the sample points and find $P(A)$.
(Note that this example is illustrative of a very important problem associated with product inventory.)

2.5. An airline has six flights from New York to California and seven flights from California to Hawaii per day. How many different flight arrangements can the airline offer from New York to Hawaii? (Assume that the flights are to be made on separate days.)

2.6. A man is in the process of buying a new car. He has a choice of 3 engine makes, 7 body styles, and 14 colors. How many different cars does he have to choose from?

2.7. An experiment consists of tossing a pair of dice.
(a) Use the combinational theorems to determine the number of sample points in the sample space, S.
(b) Find the probability that the sum of the numbers appearing on the dice is equal to 7.

2.8. Show that $\binom{3}{0} + \binom{3}{1} + \binom{3}{2} + \binom{3}{3} = 2^3$. Note that in general

$$\sum_{i=0}^{n} \binom{n}{i} = 2^n.$$

2.9. Let S contain four sample points, E_1, E_2, E_3, and E_4.
(a) List all possible events in S (include the null event).
(b) Use the results of Exercise 2.8 to give the total number of events in S.
(c) Let A and B be the events $\{E_1, E_2, E_3\}$ and $\{E_2, E_4\}$, respectively. Give the sample points in the following events: $A \cup B, A \cap B, \bar{A} \cap \bar{B}$, and $\bar{A} \cup B$.

2.10. In how many ways can a committee of three be selected from ten people?

2.11. A brand of automobile comes in five different styles, with four types of engines, with two types of transmissions, and in eight colors.
(a) How many autos would a dealer have to stock if he included one for each style–engine–transmission combination?
(b) How many would a distributing center have to carry if all colors of cars were stocked for each combination in part (a)?

2.12. How many different telephone numbers can be formed from a seven-digit number if the first digit cannot be zero?

2.13. An assembly operation in a manufacturing plant requires three steps that can be performed in any sequence. How many different ways can the assembly be performed?

2.14. A personnel director for a corporation has hired ten new engineers. If three (distinctly different) positions are open at a Cleveland plant, in how many ways can he fill the positions?

2.15. An experimenter wishes to investigate the effect of three variables, pressure, temperature, and the type of catalyst, on the yield in a refining process. If the experimenter intends to use three settings each for temperature and pressure and two types of catalysts, how many experimental runs will have to be conducted if he wishes to run all possible combinations of pressure, temperature, and types of catalysts?

2.16. A patient receiving a yearly physical examination must have 18 checks or tests performed. The sequence in which the tests are conducted is important because the time lost between tests will vary and depend on the sequence. If an efficiency expert were to study the sequences to find the one that required the minimum length of time, how many sequences would be included in his study if all possible sequences were admissible?

2.17. A coin is tossed four times and the outcome is recorded for each toss. Use combinational theorems to find the number of sample points in the experiment.

2.18. Refer to Exercise 2.2. Use the combinational theorems to find
(a) The number of sample points in S.
(b) The number of sample points in A. Then find $P(A)$.

2.19. Two cards are drawn from a 52-card deck. What is the probability that the draw will yield an ace and a face card?

2.20. Five cards are drawn from a 52-card deck. What is the probability that all 5 cards will be of the same suit?

2.21. Refer to Example 2.7 and assume a random assignment of laborers to the four jobs. That is, assume that each different assignment of the 20 laborers is equiprobable. Let A be the event that all 4 laborers in the ethnic group are assigned to job 1. How many sample points are in event A? Find $P(A)$.

2.22. An extra-point kicker for a football team is successful approximately 90 percent of the time. If he has three tries, assumed independent, for extra points in a given game,
(a) What is the probability that he will be successful in all three tries?
(b) At least one?
(c) At least two?

2.23. Refer to Exercise 2.3 and find the probability that all three patients select the same station.

2.24. Two men each toss a coin and obtain a "match." That is, both coins are either heads or tails. If the process is repeated three times,
(a) What is the probability of three matches?
(b) What is the probability that all six tosses (three for each man) result in "tails"?
(c) Coin tossing provides a model for many practical experiments. Suppose that the "coin tosses" represented the answers given by two students for three specific true–false questions on an examination. If the two students gave three matches for answers, would the low probability acquired in (a) suggest collusion?

2.25. Refer to Exercise 2.24. What is the probability that the pair of coins are tossed four times before a match occurs (that is, that the match occurs for the first time on the fourth toss)?

2.26. Suppose that the probability of exposure to the flu during an epidemic is .6. Experience has shown that a serum is 80 percent successful in preventing an inoculated person from acquiring the flu, if exposed. A person not inoculated faces a probability of .90 of acquiring the flu if exposed. Two persons, one inoculated and one not, are capable of performing a highly specialized task in a business. Assume that they are not at the same location, are not in contact with the same people, and cannot expose each other. What is the probability that at least one will get the flu?

2.27. A state auto-inspection station utilizes two inspection teams. Team 1 is lenient and passes all automobiles of a recent vintage; team 2 rejects all autos on a first inspection because their "headlights are not properly adjusted." Four unsuspecting drivers take their autos to the station for inspection on four different days and randomly select one of the two stations.

(a) If all four cars are new and in excellent condition, what is the probability that three of the four will be rejected?

(b) What is the probability that all four will pass?

2.28. Two gamblers bet \$1 each on the successive tosses of a coin. Each has a bank of \$6.

(a) What is the probability that they break even after six tosses of the coin?

(b) What is the probability that one player, say Jones, wins all the money on the tenth toss of the coin?

2.29. A machine for producing a new experimental electronic tube generates defectives from time to time in a random manner. The supervising engineer for a particular machine has noticed that defectives seem to be grouping (hence appearing in a nonrandom manner) and thereby suggesting a malfunction in some part of the machine. One test for nonrandomness is based on the number of "runs" of defectives and nondefectives (a run is an unbroken sequence of either defectives or nondefectives). The smaller the number of runs, the greater will be the amount of evidence indicating nonrandomness. Of 12 tubes drawn from the machine, the first 10 were nondefective and the last 2 defective ($NNNNNNNNNNDD$). Assuming randomness,

(a) What is the probability of observing the arrangement shown above (resulting in two runs) given that 10 of the 12 tubes are non-defective?

(b) What is the probability of observing two runs?

2.30. Refer to Exercise 2.29. What is the probability that the number of runs, R, is $R \leq 3$?

2.31. An advertising organization notes that approximately 1 in 50 potential buyers of a product see a given magazine ad and 1 in 5 see a corresponding ad on television. One in 100 see both. One in 3 actually purchase the product if they have seen the ad, 1 in 10 if they have not. Jones is a potential customer. What is the probability that he will purchase the product?

2.32. Suppose that the streets of a city are laid out in a grid with streets running north–south and east–west. Consider the following scheme for patrolling an area of 16 blocks by 16 blocks. A patrolman commences walking at the intersection in the center of the area. At the corner of each block, he randomly elects to go north, south, east, or west.

(a) What is the probability that he will reach the boundary of his patrol area by the time he walks the first eight blocks?

(b) What is the probability that he will return to the starting point after walking exactly four blocks?

2.33. Consider two mutually exclusive events, A and B, such that $P(A) > 0$ and $P(B) > 0$. Are A and B independent? Give a proof for your answer.

2.34. An accident victim will die unless he receives in the next 10 minutes an amount of type A Rh-positive blood which can be supplied by a single donor. It requires 2 minutes to "type" a prospective donor's blood and 2 minutes to complete the transfer of blood. A large number of untyped donors are available and 40 percent of them have type A Rh-positive blood. What is the probability that the accident victim will be saved if there is only one blood-typing kit available?

2.35. An assembler of electric fans uses motors from two sources. Company A supplies 90 percent of the motors and company B supplies the other 10 percent of the motors. Suppose it is known that 5 percent of the motors supplied by company A are defective and 3 percent of the motors supplied by company B are defective. An assembled fan is found to have a defective motor. What is the probability that this motor was supplied by company B?

2.36. Suppose that two defective refrigerators have been included in a shipment of six refrigerators. The buyer begins to test the six refrigerators one at a time.

(a) What is the probability that the last defective refrigerator is found on the fourth test?

(b) What is the probability that no more than four refrigerators need be tested before both of the defective refrigerators are located?

(c) Given that exactly one of the two defective refrigerators has been located in the first two tests, what is the probability that the remaining defective refrigerator is found in the third or fourth test?

2.37. Show that, for three events, A, B, and C,

$$P[(A \cup B)|C] = P(A|C) + P(B|C) - P[(A \cap B)|C].$$

2.38. If A and B are independent events, show that A and $\bar{B}$ are also independent.

2.39. Three events, A, B, and C, are said to be independent if

$$P(AB) = P(A)P(B),$$

$$P(AC) = P(A)P(C),$$

$$P(BC) = P(B)P(C),$$

and

$$P(ABC) = P(A)P(B)P(C).$$

Suppose that a balanced coin is independently tossed two times. Define the following events:
A: head appears on the first toss,
B: head appears on the second toss,
C: both tosses yield the same outcome.
Are A, B, and C independent?

2.40. A line from a to b has midpoint c. A point is chosen at random on the line and marked x (the point x being chosen at random implies that x is equally likely to fall in any subinterval of fixed length l). Find the probability that the line segments ax, bx, and ac can be joined to form a triangle.

2.41. Eight tires of different brands are ranked from one to eight (best to worst) according to mileage performance. If four of these tires are chosen at random by a customer, find the probability that the best tire among those selected by the customer is actually ranked third among the original eight.

2.42. Suppose that n indistinguishable balls are to be arranged in N distinguishable boxes so that each distinguishable arrangement is equally likely. If $n \geq N$, show that the probability that no box will be empty is given by

$$\frac{\binom{n-1}{N-1}}{\binom{N+n-1}{N-1}}.$$

3

Discrete Random Variables and Their Probability Distributions

3.1 *Basic Definition*

As stated in Section 2.12, a random variable is a real-valued function defined over a sample space. The reader will probably recall that a random variable has the effect of changing events in a sample space into numerical events. For example, the event of interest in an opinion poll regarding voter preferences is not the particular people sampled or the order in which preferences were obtained, but the *number* of voters favoring a certain candidate or issue. This gives rise to a random variable, the number of voters in the sample who favor a certain candidate or issue, which can take on only a finite number of values with nonzero probability. That is, the observed value of the random variable of interest must be an integer between zero and the sample size. A random variable of this type is said to be "discrete."

> ***Definition 3.1:*** *A random variable, Y, is said to be* discrete *if there is associated with Y a finite or countably infinite set of points having positive probabilities which sum to unity.*

A less formidable characterization of discrete random variables can be obtained by considering some practical examples. The number of bacteria per unit area in the study of drug control on bacterial growth is a discrete random variable, as is also the number of defective television sets in a shipment of 100. Indeed, discrete random variables most often represent the count of a particular element.

Let us now consider the relation of Chapter 2 to Chapter 3. Why study the theory of probability? The answer is that the probability of an observed sample is needed to make inferences about a population. The sample observations will frequently be numerical counts, that is, values of discrete random variables, and hence it is imperative that we know the probabilities of these numerical events. Since certain types of random variables occur so frequently in practice, it is useful to have at hand the probability for each value of a random variable, and this collection of probabilities is its probability distribution. We will find that many types of experiments exhibit similar identifying characteristics and generate the same random variable. Knowledge of the probability distributions for certain common types of experiments will eliminate the need for solving the same probability problem over and over again.

3.2 The Probability Distribution for a Discrete Random Variable

Notationally, we will use uppercase letters, such as Y, to denote random variables, and lowercase letters, such as y, to denote particular values a random variable may assume. For example, let Y denote any one of the six possible values that could be observed when a die is tossed. After tossing the die, the number actually observed will be denoted by the symbol y. Note that Y is a random variable but the specific observed value, y, is *not* random.

It is now meaningful to talk about the probability that Y takes on the value y, denoted by $P(Y = y)$. As in Section 2.12, this probability is defined as a sum of probabilities of certain sample points.

Definition 3.2: *The probability that Y takes on the value of y, $P(Y = y)$, is defined to be the sum of the probabilities of all sample points in S which are assigned the value y by the function Y. $P(Y = y)$ is sometimes denoted by $p(y)$.*

The expression $(Y = y)$ can be read "the set of all points in S assigned the value y by the random variable Y."

Definition 3.3: *The set of all pairs* $[y, p(y)]$ *for which* $p(y) > 0$ *is called the* probability distribution *for* Y.

Note that $p(y)$ is nothing more than a function which assigns probabilities to each value y, and is sometimes called the *probability function* for Y. We illustrate with the following example.

Example 3.1: *A foreman in a manufacturing plant has three men and three women working for him. He wants to choose two workers for a special job. Not wishing to show any biases in his selection, he decides to select the two workers at random. Let Y denote the number of women in his selection and find the probability distribution for Y.*

Solution: *The foreman can select two workers from six in* $\binom{6}{2} = 15$ *ways. Hence S contains 15 sample points, each equiprobable because random sampling was employed. Thus* $P(E_i) = 1/15$, $i = 1, 2, \ldots, 15$. *The values for Y which have nonzero probability are 0, 1, and 2. The number of ways of selecting* $Y = 0$ *women is* $\binom{3}{0}\binom{3}{2}$ *since the foreman must select zero from the three women and two from the three men. Thus there are* $\binom{3}{0}\binom{3}{2} = 1 \cdot 3 = 3$ *sample points in the event* $Y = 0$ *and*

$$p(0) = P(Y = 0) = \frac{\binom{3}{0}\binom{3}{2}}{15} = \frac{3}{15} = \frac{1}{5}.$$

Similarly,

$$p(1) = P(Y = 1) = \frac{\binom{3}{1}\binom{3}{1}}{15} = \frac{9}{15} = \frac{3}{5}$$

and

$$p(2) = P(Y = 2) = \frac{\binom{3}{2}\binom{3}{0}}{\binom{6}{2}} = \frac{3}{15} = \frac{1}{5}.$$

Note that $Y = 1$ is by far the most likely outcome, which should seem reasonable since the number of women equals the number of men in the original group.

Since probability distributions for discrete random variables are functions with a finite or countably infinite set of values, they can be represented by a table, or its equivalent graph, or by a formula. The table for the probability distribution of Example 3.1 is given as Table 3.1, which can be graphed as

Table 3.1

y	$p(y)$
0	1/5
1	3/5
2	1/5

shown in Figure 3.1. If we regard the width at each bar in Figure 3.1 as one unit, then the area in a bar is equal to the probability that Y takes on the value over which the bar is centered. This concept of areas representing probabilities was introduced in Section 1.2.

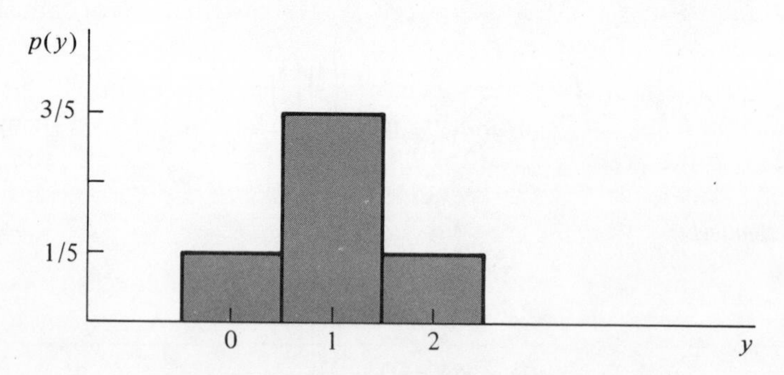

Figure 3.1 Probability Histogram for Table 3.1

The most useful method of representing discrete probability distributions is by means of a formula. For Example 3.1 we see that the formula for

$p(y)$ can be written as

$$p(y) = \frac{\binom{3}{y}\binom{3}{2-y}}{\binom{6}{2}}, \qquad y = 0, 1, 2.$$

Note that the probabilities associated with all values of a discrete random variable must sum to 1. Summarizing, the following properties must hold for all discrete probability distributions:

For any discrete probability distribution it must be true that

1. $0 \leq p(y) \leq 1$

and

2. $\sum_y p(y) = 1,$

where the summation is over all values of y with nonzero probability.

As a final and important parting shot, note that the probability distributions derived in this chapter are *models*, not exact representations, for the frequency distributions of populations of real data that occur (or would be generated) in nature. Thus they are models for real distributions of data similar to those discussed in Chapter 1.

Repetitions of some experiments over and over again will generate measurements on discrete random variables that possess frequency distributions very similar to the probability distributions derived in this chapter and hence reinforce the conviction that our models are quite good characterizations of the physical situations which they are supposed to depict.

3.3 The Binomial Probability Distribution

Some experiments consist of the observation of a sequence of identical and independent trials, each of which can result in one of two outcomes. Each item leaving a manufacturing production line is either defective or nondefective.

Each shot in a sequence of firings at a target can result in a hit or a miss, and each of n persons questioned prior to a local election will either favor candidate Jones or not. In this section we are concerned with experiments, known as binomial experiments, that exhibit the following characteristics:

> ***Definition 3.4:*** *A* binomial experiment *is one that possesses the following properties:*
>
> 1. *The experiment consists of n identical trials.*
> 2. *Each trial results in one of two outcomes. For lack of a better nomenclature, we will call one outcome a success, S, and the other a failure, F.*
> 3. *The probability of success on a single trial is equal to p and remains the same from trial to trial. The probability of a failure is equal to* $(1 - p) = q$.
> 4. *The trials are independent.*
> 5. *The random variable of interest is Y, the number of successes observed during the n trials.*

If you suspect that an experiment is binomial, test to see whether it satisfies the five properties listed above. Suppose that an unknown percentage (suppose 40 percent) of a large group of voters favor candidate Jones, and $n = 10$ people are selected from this population as a sample. The experiment consists of n identical trials that correspond to the questioning of the ten voters. Each question, that is, each trial, can result in one of two outcomes. Either they favor Jones (success) or they do not (failure). The probability of success, $p = .4$, remains approximately the same from trial to trial, and, for all practical purposes, the trials are independent. Finally, we are interested in Y, the number of successes (voters favoring Jones) in the $n = 10$ trials. Note that if the population contained a small number of voters, say 50, and 40 percent favored Jones, the experiment would no longer be binomial. The probability of success on the second trial would depend upon whether a success was observed on the first; similarly, the outcome of succeeding trials would depend on all those preceding. Hence the trials would be dependent and would no longer define a binomial experiment. The reader may wish to practice his ability to identify a binomial experiment by re-examining the exercises at the end of Chapter 2, several of which implied binomial experiments.

The binomial probability distribution, $p(y)$, can be derived by applying the sample-point approach in finding the probability that the experiment yields

y successes. Each sample point in S can be characterized by an n-tuple involving the letters S and F, corresponding to success and failure.

A typical sample point would thus appear as

$$SSFSFFFSFS \cdots FS,$$

where the letter in the ith position (proceeding from left to right) indicates the outcome of the ith trial.

Now let us consider a typical sample point implying y successes and hence contained in the numerical event, $Y = y$. This sample point,

$$\underbrace{SSSSS \cdots SSS}_{y}\underbrace{FFFF \cdots FF}_{n - y},$$

is the intersection of n *independent* events, y successes, and $(n - y)$ failures, and hence its probability is

$$ppppp \cdots pppqqq \cdots qq = p^y q^{n-y}.$$

Every other sample point in the event $Y = y$ will contain the same number of S's and F's and will possess the same probability and will appear as a rearrangement of the symbols in the n-tuple above. Since the number of distinct arrangements of the y S's and $(n - y)$ F's is

$$\frac{n!}{y!(n - y)!} = \binom{n}{y}$$

by Theorem 2.3, we have the following:

The Binomial Probability Distribution

$$p(y) = \binom{n}{y} p^y q^{n-y}, \qquad y = 0, 1, 2, \ldots, n.$$

(Note that the sample points in S are not equiprobable for the binomial experiment.) Figure 3.2 portrays $p(y)$ graphically as a probability histogram for $n = 2$, $p = .5$.

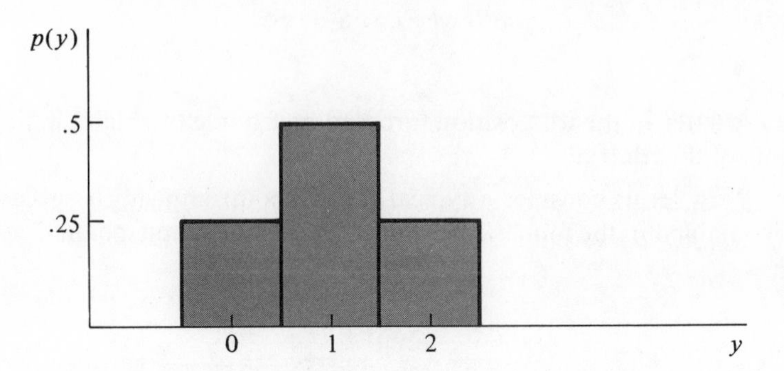

Figure 3.2 The Binomial Probability Distribution: $n = 2$, $p = .5$

The term "binomial experiment" derives from the fact that the probabilities, $p(y)$, $y = 0, 1, 2, \ldots, n$, are terms of the binomial expansion,

$$(q + p)^n = \binom{n}{0}q^n + \binom{n}{1}p^1q^{n-1} + \binom{n}{2}p^2q^{n-2} + \cdots + \binom{n}{n}p^n.$$

The reader will observe that $\binom{n}{0}q^n = p(0)$, $\binom{n}{1}pq^{n-1} = p(1)$, and, in general, $p(y) = \binom{n}{y}p^yq^{n-y}$. It also follows that $p(y)$ satisfies properties 1 and 2, page 71, because $p(y)$ is real, positive, and

$$\sum_y p(y) = \sum_{y=0}^{n} \binom{n}{y}p^yq^{n-y} = (q + p)^n = 1$$

[because $(q + p) = 1$].

The binomial probability distribution has many applications because the binomial experiment occurs in sampling for defectives in industrial quality control, in the sampling of consumer preference or voting populations, in target problems, and in many other physical situations. We shall illustrate with a few examples. Other practical examples will appear in the exercises at the end of the chapter.

Example 3.2: *Experience has shown that* 30 *percent of all persons afflicted by a certain illness recover. A drug company has developed a new vaccine. Ten people with the illness were selected at random and injected with the vaccine; nine recovered shortly thereafter. Suppose that the vaccine was absolutely worthless. What is the probability that at least nine of ten injected by the vaccine will recover?*

Solution: *Let Y denote the number of people who recover. If the vaccine is worthless, the probability that a single ill person will recover is $p = .3$. Then the number of trials is $n = 10$ and the probability of* exactly *nine recoveries is*

$$P(Y = 9) = p(9) = \binom{10}{9}(.3)^9(.7) = .000138.$$

Similarly, the probability of exactly ten recoveries is

$$P(Y = 10) = p(10) = \binom{10}{10}(.3)^{10}(.7)^0 = .000006.$$

Then

$$P(Y \geq 9) = p(9) + p(10) = .000138 + .000006 = .000144.$$

If the vaccine is ineffective, the probability of observing at least nine recoveries is extremely small. Either we have observed a very rare event, or the vaccine is indeed very useful in curing the illness. We would adhere to the latter point of view.

Example 3.3: *Suppose that a lot of* 300 *electrical fuses contains* 5 *percent defectives. If a sample of five fuses is tested, find the probability of observing at least one defective.*

Solution:

$$P(\textit{at least one defective}) = 1 - p(0) = 1 - \binom{5}{0} q^5$$
$$= 1 - (.95)^5 = 1 - .774$$
$$= .226.$$

A frequent source of error in applying the binomial probability distribution to practical problems is the failure to define which of the two possible results of a trial is the success. As a consequence, q may be erroneously used in place of p. Carefully define a success and make certain that p equals the probability of a success for each application.

A tabulation of binomial probabilities in the form $\sum_{y=0}^{a} p(y)$, presented in Table 1, Appendix III, will greatly reduce the computations associated with some of the exercises. The references at the end of the chapter list several more-extensive tabulations of binomial probabilities.

3.4 *The Geometric Probability Distribution*

The random variable having the geometric distribution is defined for an experiment that is very similar to the binomial experiment. It also is concerned with identical and independent trials, each of which can result in one of two outcomes, success or failure, and the probability of success is equal to p and constant from trial to trial. However, instead of the number of successes that occur in n trials, the geometric random variable, Y, is the number of the trial on which the first success occurs. Thus the experiment consists of a series of trials that is concluded when the first success is observed. Consequently, the experiment could end with the first trial if a success is observed or it could go on indefinitely.

The sample space, S, for the experiment contains the denumerable set of sample points,

$E_1: S$ (success on the first trial),
$E_2: FS$ (failure on the first, success on the second),
$E_3: FFS$ (failure on the first and second, success on third),
$E_4: FFFS$,
$\vdots$
$E_k: \underbrace{FFFF \cdots F}_{k-1}S.$

Since the random variable, Y, is the number of trials up to and including the first success, $Y = 1$, $Y = 2$, and $Y = 3$ will contain E_1, E_2 and E_3, respectively, and, in general, the numerical event, $Y = y$, will contain E_y. Therefore,

$$p(y) = P(E_y) = P(\underbrace{FFFF \cdots F}_{y-1}S).$$

The probability of this intersection of y independent events gives the *geometric probability distribution.*

> ***The Geometric Probability Distribution***
>
> $$p(y) = q^{y-1}p, \qquad y = 1, 2, 3, \ldots.$$

A probability histogram for $p(y)$, $p = .5$, is shown in Figure 3.3. Areas over intervals correspond to probabilities as they did for the frequency distributions of data in Chapter 1, except that one must keep in mind that Y can only assume discrete values, $y = 1, 2, \ldots, \infty$.

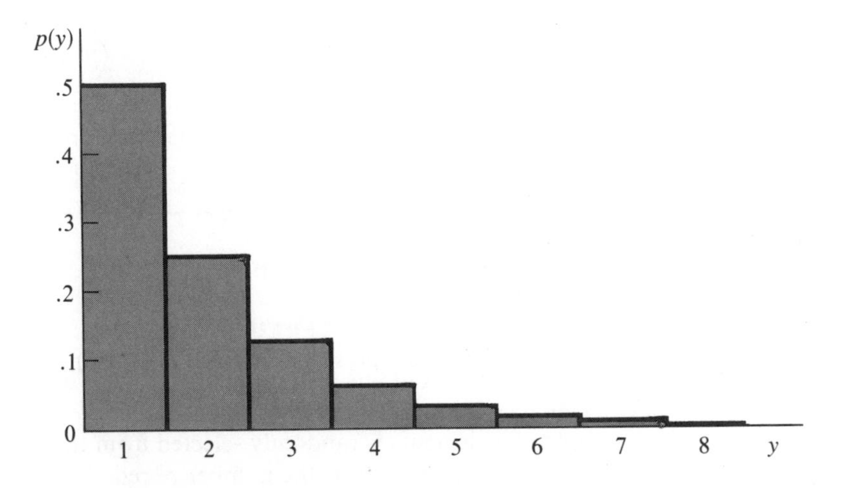

Figure 3.3 The Geometric Probability Distribution, $p = .5$

The geometric probability distribution is frequently used to model distributions of lengths of waiting times. For example, suppose that a commercial aircraft engine is serviced periodically so that its various parts are replaced at different points in time and hence are of varying ages. Then it might be reasonable to assume that the probability of engine malfunction, p, during any 1-hour interval of operation is the same as for any other and the length of time to engine malfunction is the number of 1-hour intervals, Y, until the first malfunction. (For this application, engine malfunction, or failure, in a given 1-hour period is defined to be a "success." Note that either of the two outcomes of a trial can be defined as a success and that this event need not imply success as used in everyday conversation.)

Example 3.4: *Suppose that the probability of engine malfunction during any 1-hour period is $p = .02$. Find the probability that a given engine will survive 2 hours.*

Solution: *Letting Y denote the number of 1-hour intervals until the first malfunction,*

$$P(\text{survive 2 hours}) = P(Y \geq 3) = \sum_{y=3}^{\infty} p(y).$$

Since $\sum_{y=1}^{\infty} p(y) = 1$,

$$\begin{aligned} P(\text{survive 2 hours}) &= 1 - \sum_{y=1}^{2} p(y) \\ &= 1 - p - qp = 1 - .02 - (.98)(.02) \\ &= .9604. \end{aligned}$$

3.5 The Hypergeometric Probability Distribution

Suppose that a population contains a finite number, N, of elements that possess one of two characteristics. Thus r of the elements might be red and $b = N - r$ black. A sample of n elements is randomly selected from the population and the random variable of interest is Y, the number of red elements in the sample. This random variable has what is known as the *hypergeometric probability distribution.* For example, the number of women workers, Y, in Example 3.1 has the hypergeometric distribution.

The hypergeometric probability distribution can be derived using the combinatorial theorems, Section 2.7, and the sample-point approach. A sample point in S will correspond to a unique selection of n elements, some red and the remainder black. Like the binomial experiment, each sample point can be characterized by an n-tuple whose elements correspond to a selection of n elements from the total of N. If each element in the population were numbered from 1 to N, the sample point indicating the selection of items 5, 7, 8, 64, 17, . . . , 87 would appear as the n-tuple

$$(5, 7, 8, 64, 17, \ldots, 87).$$

The total number of sample points in S will therefore equal the number of ways of selecting n elements from a population of N, or $\binom{N}{n}$. Since random selection implies that all sample points are equiprobable, the probability of a sample point in S will equal

$$P(E_i) = \frac{1}{\binom{N}{n}}, \qquad \text{all } E_i \subset S.$$

The total number of sample points in the numerical event $Y = y$ would be the number of sample points in S that contain y red and $(n - y)$ black elements. This number can be obtained by applying the *mn* rule (Section 2.7). The number of ways of selecting y red elements to fill y positions in the n-tuple representing a sample point is the number of ways of selecting y from a total of r, or $\binom{r}{y}$. The total number of ways of selecting $(n - y)$ black elements to fill the $(n - y)$ positions in the n-tuple is the number of ways of selecting $(n - y)$ black elements from a possible $N - r$, or $\binom{N - r}{n - y}$. Then the number of sample points in the numerical event, $Y = y$, is the number of ways of combining a set of y red and $(n - y)$ black elements. By the *mn* rule, this would be the product $\binom{r}{y} \cdot \binom{N - r}{n - y}$. Summing the probabilities of the sample points in the numerical event $Y = y$ (multiplying the number of sample points by the common probability per sample point), we obtain the hypergeometric probability function.

The Hypergeometric Probability Distribution

$$p(y) = \frac{\binom{r}{y}\binom{N-r}{n-y}}{\binom{N}{n}} \qquad \begin{array}{l} y = 0, 1, 2, \ldots, n \text{ if } n < r \\ \text{or } y = 0, 1, 2, \ldots, r \text{ if } n \geq r. \end{array}$$

We will use the convention $\binom{a}{b} = 0$ if $b > a$.

Example 3.5: *An important problem encountered by personnel directors and others faced with the selection of the "best" in a finite set of elements is indicated by the following situation. From a group of 20 Ph.D. engineers, 10 are selected for employment. What is the probability that the 10 selected include all the 5 best engineers in the group of 20?*

Solution: *For this example, $N = 20$, $n = 10$, and $r = 5$. That is, there are only 5 in the set of 5 best engineers, and we seek the probability that $Y = 5$, where Y denotes the number of "best" engineers among the 10 selected. Then*

$$p(5) = \frac{\binom{5}{5}\binom{15}{5}}{\binom{20}{10}} = \frac{15!}{5!10!}\,\frac{10!10!}{20!} = \frac{21}{1292} = .0162.$$

Example 3.6: *A particular industrial product is shipped in lots of 20. Testing to determine whether an item is defective is costly, and hence the manufacturer samples his production rather than using a 100 percent inspection plan. A sampling plan constructed to minimize the number of defectives shipped to customers calls for sampling five items from each lot and rejecting the lot if more than one defective is observed. (If rejected, each item in the lot is tested.) If a lot contains four defectives, what is the probability that it will be rejected?*

Solution: *Let Y equal the number of defectives in the sample. Then $N = 20$, $r = 4$, and $n = 5$. The lot will be rejected if $Y = 2, 3,$ or 4. Then*

$$\begin{aligned} P(\text{rejecting the lot}) &= P(Y \geq 2) = p(2) + p(3) + p(4) \\ &= 1 - p(0) - p(1) \end{aligned}$$

$$= 1 - \frac{\binom{4}{0}\binom{16}{5}}{\binom{20}{5}} - \frac{\binom{4}{1}\binom{16}{4}}{\binom{20}{5}}$$

$$= 1 - .2817 - .4696$$

$$= .2487.$$

The reader will note that Examples 3.3 and 3.6 both involve sampling lots of industrial products for defectives. If the lot size, N, is small, the probability of a defective on the ith draw is dependent on the outcome of the previous draws, and the number of defectives in the sample will follow a hypergeometric probability distribution. If the lot size, N, is large and contains a fraction defective equal to p, the probability of a defective on the ith draw is approximately equal to p and is unaffected by the outcome of the previous draws. Consequently, it can be shown that

$$\lim_{N \to \infty} \frac{\binom{r}{y}\binom{N - r}{n - y}}{\binom{N}{n}} = \binom{n}{y} p^y (1 - p)^{n - y},$$

where

$$\frac{r}{N} = p.$$

Hence, for a fixed fraction defective, $p = r/N$, the hypergeometric probability function converges to the binomial probability function as N becomes large.

3.6 *The Poisson Probability Distribution*

Suppose that we want to find the probability distribution of the number of automobile accidents at a particular intersection during a time period of 1 week. At first glance this random variable, the number of accidents, may not seem even remotely related to a binomial random variable, but we will see that there is an interesting relationship.

Think of the time period, one week in the above example, as being split up into n subintervals, *each of which is so small that at most one accident could occur in it with probability different from zero.* Denoting the probability of an accident in any subinterval by p, we have, for all practical purposes,

$$P(\text{no accidents in a subinterval}) = 1 - p,$$

$$P(\text{one accident in a subinterval}) = p,$$

and

$$P(\text{more than one accident in a subinterval}) = 0.$$

Then the total number of accidents in the week is just the total number of subintervals that contain one accident. If the occurrence of accidents can be regarded as independent from interval to interval, the total number of accidents has a binomial distribution.

Although there is no unique way to choose the subintervals and we therefore know neither n nor p, it seems reasonable that as n increases, p should decrease. Thus we look at the limiting case of the binomial probability distribution as $n \to \infty$ and $p \to 0$ in such a way that np remains fixed at some value, say λ.

Mathematically, with $\lambda = np$,

$$\begin{aligned}
&\lim_{n\to\infty} \binom{n}{y} p^y (1-p)^{n-y} \\
&= \lim_{n\to\infty} \frac{n(n-1)\cdots(n-y+1)}{y!}\left(\frac{\lambda}{n}\right)^y \left(1-\frac{\lambda}{n}\right)^{n-y} \\
&= \lim_{n\to\infty} \frac{\lambda^y}{y!}\left(1-\frac{\lambda}{n}\right)^n \frac{n(n-1)\cdots(n-y+1)}{n^y}\left(1-\frac{\lambda}{n}\right)^{-y} \\
&= \frac{\lambda^y}{y!}\lim_{n\to\infty}\left(1-\frac{\lambda}{n}\right)^n\left(1-\frac{\lambda}{n}\right)^{-y}\left(1-\frac{1}{n}\right)\left(1-\frac{2}{n}\right)\cdots\left(1-\frac{y-1}{n}\right).
\end{aligned}$$

Noting that

$$\lim_{n\to\infty}\left(1-\frac{\lambda}{n}\right)^n = e^{-\lambda}$$

and all other terms to the right of the limit have a limit of 1, we obtain

$$p(y) = \frac{\lambda^y}{y!}e^{-\lambda}(1).$$

(*Note:* $e = 2.718\ldots.$) Random variables possessing this distribution are said to be Poisson random variables. Hence Y, the number of accidents per week, should possess the Poisson distribution given above.

The constant λ can be shown to be the average value of the random variable observed over many intervals of the same size. Thus, if traffic records show that over the past 3 years there is an average of two traffic accidents per week at the intersection in question, we can obtain $p(1)$ by substituting into

$$p(y) = \frac{\lambda^y}{y!}e^{-\lambda}.$$

Then

$$P\text{ (one accident during a particular week)} = \frac{2}{1!}e^{-2} = 2e^{-2}.$$

The number of random events that occur in a unit of time, space, or any other dimension often follows the *Poisson* distribution. The Poisson distribution applies particularly to "rare" events, that is, those which occur infrequently in time, space, volume, or any other dimension.

The Poisson Probability Distribution

$$p(y) = \frac{\lambda^y}{y!}e^{-\lambda}, \qquad y = 0, 1, 2, \ldots.$$

It should be noted that the Poisson probabilities do sum to unity. That is,

$$\sum_{y=0}^{\infty} p(y) = \sum_{y=0}^{\infty} \frac{\lambda^y}{y!}e^{-\lambda}$$

$$= e^{-\lambda} \sum_{y=0}^{\infty} \frac{\lambda^y}{y!}$$

$$= e^{-\lambda}e^{\lambda} = 1,$$

since the infinite sum $\sum_{y=0}^{\infty} \lambda^y/y!$ is a series expansion of e^{λ}. Sums of special series are given in the Chemical Rubber Company handbook [6].

Example 3.7: *Suppose that a random system of police patrol is devised so that a patrolman may visit a given location of his beat $Y = 0, 1, 2, 3, \ldots$ times per half-hour period and that the system is arranged so that he visits each location on an average of once per time period. Assume that Y possesses, approximately, a Poisson probability distribution. Calculate the probability that the patrolman will miss a given location during a half-hour period. What is the probability that he will visit it once? Twice? At least once?*

Solution: *For this example, the time period is 1/2 hour and the mean visits per half-hour interval is $\lambda = 1$. Then*

$$p(y) = \frac{(1)^y e^{-1}}{y!} = \frac{e^{-1}}{y!}.$$

The event that he misses a given location in one half-hour period corresponds to $y = 0$ and

$$p(0) = \frac{e^{-1}}{0!} = e^{-1} = .368.$$

Similarly,

$$p(1) = \frac{e^{-1}}{1!} = e^{-1} = .368$$

and

$$p(2) = \frac{e^{-1}}{2!} = \frac{e^{-1}}{2} = .184.$$

The probability that he visits the location at least *once is the event that $Y \geq 1$. Then*

$$P(Y \geq 1) = \sum_{y=1}^{\infty} p(y) = 1 - p(0)$$

$$= 1 - e^{-1} = .632.$$

The convergence of the binomial probability function to the Poisson is of practical value because the Poisson probabilities can be used to approximate their binomial counterparts for large n, small p, and $\lambda = np$ less than, roughly, 7. Exercise 3.23 will require the calculation of corresponding binomial and Poisson probabilities and will demonstrate the adequacy of the approximation.

3.7 The Expected Value of a Random Variable or a Function of a Random Variable

We have observed that the probability distribution for a random variable is a theoretical model for the empirical distribution of data associated with a real population. If the model is an accurate representation of nature, the theoretical and empirical distributions are equivalent. Consequently, as in Chapter 1, it is natural that we attempt to find the mean and the variance for a random variable and thereby to acquire descriptive measures for the probability distribution, $p(y)$.

Definition 3.5: *Let Y be a discrete random variable with probability function $p(y)$. Then the* expected value *of Y, $E(Y)$, is defined to be*

$$E(Y) = \sum_{y} yp(y).$$

If $p(y)$ is an accurate characterization of the population frequency distribution, then $E(Y) = \mu$, the population mean.

Definition 3.5 is completely consistent with the definition of the mean of a set of measurements, Definition 1.1. For example, consider a discrete random variable, Y, that can assume values 0, 1, 2 with probability distribution, $p(y)$,

y	$p(y)$
0	1/4
1	1/2
2	1/4

and probability histogram as shown in Figure 3.4. A visual inspection will reveal the mean of the distribution to be located at $y = 1$.

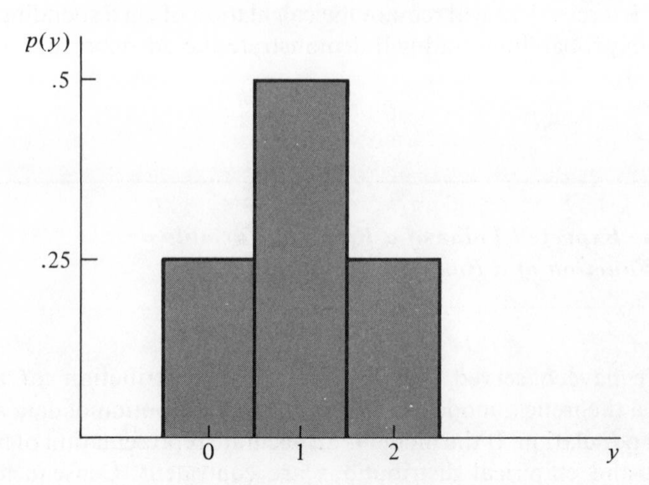

Figure 3.4 Probability Distribution for Y

To show that $E(Y) = \sum_y yp(y)$ is the mean of the probability distribution, $p(y)$, suppose that the experiment were conducted 4,000,000 times and a corresponding number of measurements on Y observed. We wish to find the mean value of Y. Noting $p(y)$, Figure 3.4, we would expect *approximately* one million of the 4,000,000 repetitions to result in the outcome $Y = 0$, two million in $Y = 1$, and one million in $Y = 2$. Averaging the 4,000,000 measurements, we obtain

$$\mu \approx \frac{\sum_{i=1}^{n} y_i}{n}$$

$$= \frac{(1{,}000{,}000)(0) + (2{,}000{,}000)(1) + (1{,}000{,}000)(2)}{4{,}000{,}000}$$

$$= (0)(1/4) + (1)(1/2) + (2)(1/4)$$

$$= \sum_{y=0}^{2} yp(y) = 1.$$

Thus $E(Y)$ is an average and Definition 3.5 is consistent with the definition of a mean.

Similarly, we are frequently interested in the mean or expected value of a function of a random variable Y. For example, molecules in space move at varying velocities where Y, the velocity of a given molecule, is a random variable. The energy imparted upon impact by a moving body is proportional to the square of the velocity. Consequently, to find the mean amount of energy transmitted by a molecule upon impact, it would be necessary to find the mean value of Y^2. More important, we note in Definition 1.2 that the variance of a set of measurements is the mean of the square of the deviation of a set of measurements about their mean, or the mean value of $(Y - \mu)^2$.

Definition 3.6: *Let $g(Y)$ be a function of a discrete random variable, Y, which possesses a probability function $p(y)$. Then the* expected value of *$g(Y)$ is defined to be*

$$E[g(Y)] = \sum_{y} g(y)p(y).$$

The reader can use an empirical argument similar to that employed for $E(Y)$ to show that Definition 3.6 truly implies an averaging of the values of $g(Y)$. Thus $E[g(Y)]$ can be regarded as the mean value of $g(Y)$.

Now let us return to our immediate objective, finding numerical descriptive measures (or *parameters*) to characterize $p(y)$. Since $E(Y)$ provides its mean, we next seek its variance and standard deviation. The reader will recall from Chapter 1 that a variance of a set of measurements is the average of the square of the deviations of a set of measurements about its mean. Thus we wish to find the mean value of the function $g(Y) = (Y - \mu)^2$.

Definition 3.7: *The* variance *of a random variable Y is defined to be the expected value of $(Y - \mu)^2$. That is,*

$$V(Y) = E[(Y - \mu)^2].$$

If $p(y)$ is an accurate characterization of the population frequency distribution (and to simplify notation we will assume this to be true), $E(Y) = \mu$ and $V(Y) = \sigma^2$, the population variance.

Example 3.8: *Given the following probability distribution for a random variable, Y, find its mean and variance:*

y	$p(y)$
0	1/8
1	1/4
2	3/8
3	1/4

Solution: *By Definitions 3.5 and 3.7,*

$$\mu = E(Y) = \sum_{y=0}^{3} yp(y) = (0)(1/8) + (1)(1/4) + (2)(3/8) + (3)(1/4)$$

$$= 1.75,$$

$$\sigma^2 = E[(Y - \mu)^2] = \sum_{y=0}^{3} (y - \mu)^2 p(y)$$

$$= (0 - 1.75)^2(1/8) + (1 - 1.75)^2(1/4)$$

$$+ (2 - 1.75)^2(3/8) + (3 - 1.75)^2(1/4)$$

$$= .9375,$$

and

$$\sigma = \sqrt{\text{variance}} = \sqrt{.9375} = .97.$$

The probability histogram is shown in Figure 3.5. Locate μ on the axis of measurement and observe that it does locate the center of the nonsymmetrical probability distribution, $p(y)$. Also note that the interval $(\mu \pm \sigma)$ contains the discrete points $Y = 1$ and $Y = 2$, which comprise 5/8 of the probability. Thus the empirical rule provides a reasonable approximation to the probability of a measurement falling in this interval. (Keep in mind that the probabilities are concentrated at the points $Y = 0, 1, 2, 3$ because Y cannot take intermediate values.)

It will be helpful to acquire a few additional tools and definitions before attempting to find the expected values and variances of a more complicated discrete random variable such as the binomial or Poisson. Hence we present three useful expectation theorems that follow directly from the theory

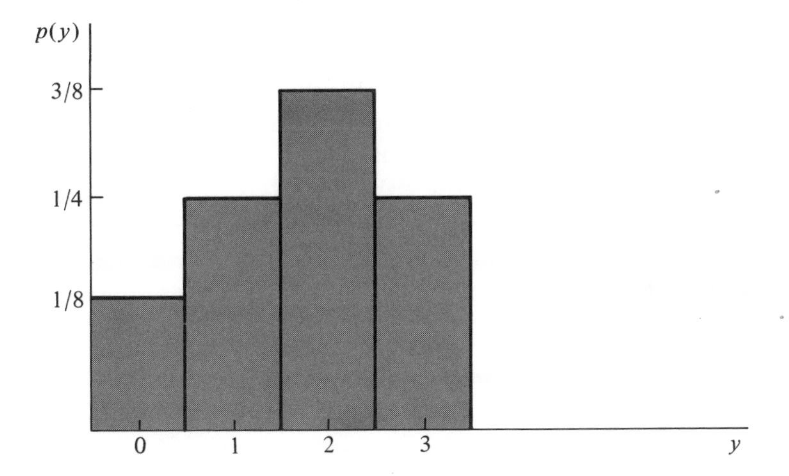

Figure 3.5 A Probability Histogram For Example 3.8

of summation. (The reader unfamiliar with the basic theorems of summation can find a discussion of the subject in Mendenhall [2].) For each theorem we assume that Y is a discrete random variable with probability distribution $p(y)$.

The first theorem states the rather obvious result that the mean or expected value of a nonrandom quantity, c, is equal to c.

> ***Theorem 3.1:*** *Let c be a constant. Then* $E(c) = c$.

Proof: *By Definition 3.6,*

$$E(c) = \sum_y cp(y) = c\sum_y p(y).$$

But $$\sum_y p(y) = 1 \qquad \text{(Section 3.2).}$$

Therefore,

$$E(c) = c(1) = c.$$

The second, Theorem 3.2, states that the expected value of the product of a constant, c, times a function of a random value, is equal to the constant times the expected value of the function.

> ***Theorem 3.2:*** *Let $g(Y)$ be a function of the random variable, Y, and let c be a constant. Then*
>
> $$E[cg(Y)] = cE[g(Y)].$$

Proof: *By Definition 3.6,*

$$E[cg(Y)] = \sum_y cg(y)p(y) = c\sum_y g(y)p(y) = cE[g(Y)].$$

The third theorem, 3.3, states that the mean or expected value of a sum of functions of a random variable, Y, is equal to the sum of their respective expected values.

> ***Theorem 3.3:*** *Let $g_1(Y), g_2(Y), \ldots, g_k(Y)$ be k functions of the random variable, Y. Then*
>
> $$E[g_1(Y) + g_2(Y) + \cdots + g_k(Y)] = E[g_1(Y)] + E[g_2(Y)] + \cdots + E[g_k(Y)].$$

Theorems 3.1, 3.2, and 3.3 can be used immediately to develop a theorem useful in finding the variance of a random variable.

> ***Theorem 3.4:*** $V(Y) = \sigma^2 = E[(Y - \mu)^2] = E(Y^2) - \mu^2.$

Proof: $\sigma^2 = E[(Y - \mu)^2] = E(Y^2 - 2\mu Y + \mu^2)$
$= E(Y^2) - E(2\mu Y) + E(\mu^2)$ *by Theorem 3.3.*
Noting that μ *is a constant and applying Theorems 3.2 and 3.1 to the second and third terms, respectively,*

$$\sigma^2 = E(Y^2) - 2\mu E(Y) + \mu^2.$$

But $\mu = E(Y)$ *and therefore*

$$\sigma^2 = E(Y^2) - 2\mu^2 + \mu^2$$
$$= E(Y^2) - \mu^2.$$

Theorem 3.4 greatly reduces the labor in finding the variance of a discrete random variable. We will demonstrate for the simple probability distribution of Example 3.8.

Example 3.9: *Use Theorem 3.4 to find the variance of* Y, *Example 3.8.*

Solution: *The mean,* $\mu = 1.75$, *was found earlier in Example 3.8. Then*

$$E(Y^2) = \sum_y y^2 p(y) = (0)^2(1/8) + (1)^2(1/4) + (2)^2(3/8) + (3)^2(1/4)$$
$$= 4.$$

Also, by Theorem 3.4,

$$\sigma^2 = E(Y^2) - \mu^2 = 4 - (1.75)^2 = .9375.$$

The purpose of this section was to introduce the concept of an expectation and to develop some useful theorems for finding the expected values and variances of random variables or functions of random variables. Actually finding the expected value of a discrete random variable such as the Poisson or geometric requires skill in the summation of algebraic series and knowledge of a few tricks. Whether it is useful to expend much time on methods for summing series in an introductory statistics theory course is a dubious question. We leave this matter to the instructor and the student.

Table 3.2 Means and Variances For Some Common Random Variables

Distribution	$E(Y)$	$V(Y)$
Binomial	np	npq
Geometric	$\frac{1}{p}$	$\frac{q}{p^2}$
Hypergeometric	$\frac{nr}{N}$	$\frac{nr(N-r)(N-n)}{N^2(N-1)}$
Poisson	λ	λ

The means and variances of the binomial, geometric, hypergeometric, and Poisson random variables are shown in Table 3.2. The reader interested in developing skill in finding expectations can practice with these random variables and compare the results with Table 3.2. A few of the devices useful in summing series are presented in Section 3.8.

Example 3.10: *The number of industrial accidents at a particular manufacturing plant is found to average three per month. During the last month six accidents occurred. Would you regard this number as unusually large (highly improbable if μ were still equal to 3) and indicative of an increase in the mean, μ?*

Solution: *The number of accidents, Y, would likely follow a Poisson probability distribution with $\lambda = 3$. We seek the probability that Y could be as large as 6 given $\mu = \lambda = 3$. Then*

$$P(Y \geq 6) = \sum_{y=6}^{\infty} \frac{3^y e^{-3}}{y!}.$$

The tedious calculations required to find $P(Y \geq 6)$ can be avoided by using the empirical rule. Although $p(y)$ is not symmetrical, it is roughly mound-shaped. Then

$$\mu = \lambda = 3, \qquad \sigma^2 = \lambda = 3, \qquad \sigma = \sqrt{3} = 1.73.$$

Applying the empirical rule, $\mu + 2\sigma = 3 + (2)(1.73) = 6.5$. The observed number of accidents, $Y = 6$, does not lie more than 2σ from μ, but it is close to the boundary. Thus the observed result is not highly improbable, but it may be improbable enough to warrant an investigation. See Exercise 3.46 for the exact probability, $P(|Y - \lambda| \leq 2\sigma)$.

3.8 Some Techniques Useful in Finding Expected Values for Discrete Random Variables

Many useful devices exist for summing series. Examples 3.11, 3.12, and 3.13 illustrate three of the most helpful.

Example 3.11: *Find the expected value of a Poisson random variable.*

Solution: *By definition,*

$$E(Y) = \sum_y yp(y) = \sum_{y=0}^{\infty} y\frac{\lambda^y e^{-\lambda}}{y!}.$$

One of the most helpful devices for finding an expectation, that is, summing the above series, is to use the property (Section 3.2) $\sum_y p(y) = 1$. *To illustrate, note that the first term in this sum will equal zero (when* $y = 0$*) and hence*

$$E(Y) = \sum_{y=1}^{\infty} y\frac{\lambda^y e^{-\lambda}}{y!} = \sum_{y=1}^{\infty} \frac{\lambda^y e^{-\lambda}}{(y-1)!}.$$

As it stands, this quantity is not equal to the sum of a probability function, $p(y)$*, over all values of* y*, but we can change it to the proper form by factoring a* λ *out of the expression and letting* $z = y - 1$*. Then the limits of summation become* $z = 0$ *(when* $y = 1$*) and* $z = \infty$ *(when* $y = \infty$*), and*

$$E(Y) = \lambda \sum_{y=1}^{\infty} \frac{\lambda^{y-1} e^{-\lambda}}{(y-1)!} = \lambda \sum_{z=0}^{\infty} \frac{\lambda^z e^{-\lambda}}{z!}.$$

Note that $p(z) = \lambda^z e^{-\lambda}/z!$ *is the probability function for a Poisson random variable and* $\sum_{z=0}^{\infty} p(z) = 1$*. Therefore,* $E(Y) = \lambda$*. Thus the mean of a Poisson random variable is the single parameter,* λ*, that appears in the expression for the Poisson probability function.*

The identity $\sum_y p(y) = 1$ can be employed in finding the expected values of many random variables.

Example 3.12: *Find the variance of the binomial random variable.*

Solution: *The expected value of the binomial random variable can be obtained using the procedure of Example 3.11. We will take $\mu = np$ as given and find σ^2 from Theorem 3.4, $\sigma^2 = E(Y^2) - \mu^2$. Thus we need to find $E(Y^2)$ to substitute in this expression.*

Finding $E(Y^2)$ directly is very difficult. Thus

$$E(Y^2) = \sum_{y=0}^{n} y^2 p(y) = \sum_{y=0}^{n} y^2 \binom{n}{y} p^y q^{n-y}$$

$$= \sum_{y=0}^{n} y^2 \frac{n!}{y!(n-y)!} p^y q^{n-y}.$$

The quantity y^2 does not appear as a factor of $y!$, and hence it will not divide out. Where do we go from here?

The easy approach is to find the expected value of $y(y-1)$ which will divide nicely into $y!$ Since

$$E[Y(Y-1)] = E(Y^2 - Y) = E(Y^2) - E(Y),$$

we can find

$$E(Y^2) = E[Y(Y-1)] + E(Y) = E[Y(Y-1)] + \mu.$$

Following through,

$$E[Y(Y-1)] = \sum_{y=0}^{n} y(y-1) \frac{n!}{y!(n-y)!} p^y q^{n-y}.$$

The first and second terms of this sum equal zero (when $y = 0$ and $y = 1$). Then

$$E[Y(Y-1)] = \sum_{y=2}^{n} \frac{n!}{(y-2)!(n-y)!} p^y q^{n-y}.$$

Now manipulate to employ the identity $\sum_y p(y) = 1$. Factor $n(n-1)p^2$ out of the sum and let $z = y - 2$. Then

$$E[Y(Y-1)] = n(n-1)p^2 \sum_{y=2}^{n} \frac{(n-2)!}{(y-2)!(n-y)!} p^{y-2} q^{n-y}$$

$$= n(n-1)p^2 \sum_{z=0}^{n-2} \frac{(n-2)!}{z!(n-2-z)!} p^z q^{n-2-z}$$

$$= n(n-1)p^2 \sum_{z=0}^{n-2} \binom{n-2}{z} p^z q^{n-2-z}.$$

Note that $p(z) = \binom{n-2}{z} p^z q^{n-2-z}$ *is the binomial probability function based on* $(n-2)$ *trials. Then* $\sum_{z=0}^{n-2} p(z) = 1$ *and*

$$E[Y(Y-1)] = n(n-1)p^2.$$

Thus
$$E(Y^2) = E[Y(Y-1)] + \mu = n(n-1)p^2 + np$$

and
$$\sigma^2 = E(Y^2) - \mu^2 = n(n-1)p^2 + np - n^2p^2.$$

Letting $q = 1 - p$ *and employing a bit of algebra, the expression reduces to*

$$\sigma^2 = npq.$$

We leave this reduction as an exercise for the reader.

Finding $E(Y^2)$ *circuitously by finding* $E[Y(Y-1)]$ *is also the easy way to find the variance of the Poisson random variable.*

A third device useful in finding expectations is illustrated by the following example.

Example 3.13: *Find* $E(Y)$ *for a geometric random variable.*

Solution:

$$E(Y) = \sum_{y=1}^{\infty} yq^{y-1}p = p \sum_{y=1}^{\infty} yq^{y-1}.$$

This series is difficult to sum directly but can be easily summed by noting that

$$\frac{d}{dq}\{q^y\} = yq^{y-1}$$

and hence

$$\frac{d}{dq}\left\{\sum_{y=1}^{\infty} q^y\right\} = \sum_{y=1}^{\infty} yq^{y-1}.$$

(The interchanging of derivative and sum here can be justified.) Substituting, we obtain

$$E(Y) = p\sum_{y=1}^{\infty} yq^{y-1} = p\frac{d}{dq}\left\{\sum_{y=1}^{\infty} q^y\right\}$$

This latter sum is the summation of the terms of an infinite geometric progression, $q + q^2 + q^3 + \cdots$, *which is equal to* $1/(1 - q)$ *(given in most mathematical handbooks). Therefore,*

$$E(Y) = p\frac{d}{dq}\left\{\frac{1}{1-q}\right\} = p\frac{1}{(1-q)^2} = \frac{p}{p^2} = \frac{1}{p}.$$

To summarize, the device is to express a series that cannot be summed directly as the derivative (or integral) of a series for which the sum can be readily obtained. Once summed, differentiate (or integrate) to complete the process.

3.9 Moments and Moment-Generating Functions

The parameters μ and σ are meaningful numerical descriptive measures that locate the center and describe the spread of $p(y)$, but they do not provide a unique characterization of the distribution. Many distributions possess the same means and standard deviations. Consequently, we might consider a set of numerical descriptive measures that, under rather general conditions, uniquely determine $p(y)$.

> ***Definition 3.8:*** *The ith moment of a random variable Y taken about the origin is defined to be* $E(Y^i)$ *and is denoted by* μ_i'.

Particularly, note that the first moment about the origin is $E(Y) = \mu_1' = \mu$. Also, $\mu_2' = E(Y^2)$ is employed in Theorem 3.4 for finding σ^2.

A second useful moment of a random variable is one taken about its mean.

Definition 3.9: *The* ith moment of a random variable *Y taken about its* mean, *or the* ith *central moment of Y, is defined to be* $E[(Y - \mu)^i]$ *and is denoted by* μ_i.

In particular, $\sigma^2 = \mu_2$.

Let us concentrate on moments about the origin, μ_i', $i = 1, 2, 3, \ldots$. Suppose that two random variables, Y and Z, possess finite moments of all orders and $\mu_{1y}' = \mu_{1z}'$, $\mu_{2y}' = \mu_{2z}', \ldots, \mu_{ky}' = \mu_{kz}'$, where k can assume any integral value. That is, the two random variables possess identical corresponding moments about the origin. Under some fairly general conditions, it can then be shown that Y and Z have identical probability distributions. Thus one of the major uses of moments is to approximate the probability distribution of a random variable (usually an estimator or a decision maker). Consequently, the moments, μ_i', $i = 1, 2, 3, \ldots$, are primarily of theoretical value for $i > 2$.

A third interesting expectation is the moment-generating function for a random variable, which, figuratively speaking, packages all the moments for a random variable in one simple expression. First we will define the moment-generating function and then explain how it works.

Definition 3.10: *The* moment-generating function $m(t)$ for a random variable *Y is defined to be* $E(e^{tY})$. *We say that a moment-generating function for Y exists if there exists a positive constant b such that* $m(t)$ *is finite for* $|t| \leq b$.

Why is $E(e^{tY})$ called the moment-generating function for Y? From a series expansion for e^{ty} we have

$$e^{ty} = 1 + ty + \frac{(ty)^2}{2!} + \frac{(ty)^3}{3!} + \frac{(ty)^4}{4!} + \cdots.$$

Then assuming that μ'_i is finite, $i = 1, 2, 3, \ldots$:

$$E(e^{tY}) = \sum_y e^{ty} p(y)$$

$$= \sum_y \left[1 + ty + \frac{(ty)^2}{2!} + \frac{(ty)^3}{3!} + \cdots\right] p(y)$$

$$= \sum_y p(y) + t \sum_y y p(y) + \frac{t^2}{2!} \sum_y y^2 p(y) + \frac{t^3}{3!} \sum_y y^3 p(y) + \cdots$$

$$= 1 + t\mu'_1 + \frac{t^2}{2!}\mu'_2 + \frac{t^3}{3!}\mu'_3 + \cdots.$$

[The above argument involves an interchange of summations, which is justifiable if $m(t)$ exists.] Thus $E(e^{tY})$ is a function of all the moments about the origin, μ'_i, $i = 1, 2, 3, \ldots$, where μ'_i is the coefficient of $t^i/i!$.

The moment-generating function possesses two important applications. First, if we can find $E(e^{tY})$, we can find any of the moments for Y.

Theorem 3.5: *If $m(t)$ exists, then for any positive integer, k,*

$$\left.\frac{d^k m(t)}{dt^k}\right]_{t=0} = m^{(k)}(0) = \mu'_k.$$

Proof: *$d^k m(t)/dt^k$, or $m^{(k)}(t)$, is the kth derivative of $m(t)$ with respect to t. Now*

$$m(t) = E(e^{tY}) = 1 + t\mu'_1 + \frac{t^2}{2!}\mu'_2 + \frac{t^3}{3!}\mu'_3 + \cdots.$$

Then

$$m^{(1)}(t) = \mu'_1 + \frac{2t}{2!}\mu'_2 + \frac{3t^2}{3!}\mu'_3 + \cdots.$$

Setting $t = 0$,

$$m^{(1)}(0) = \mu_1'.$$

Similarly,

$$m^{(2)}(t) = \mu_2' + \frac{2t}{2!}\mu_3' + \cdots.$$

Setting $t = 0$,

$$m^{(2)}(0) = \mu_2'.$$

Continuing differentiation of $m(t)$, *you can see that*

$$m^{(k)}(0) = \mu_k', \qquad k = 1, 2, \ldots.$$

[*The above operations involve interchanging a derivative and an infinite sum, which can be justified if* $m(t)$ *exists.*]

Example 3.14: *Find the moment-generating function,* $m(t)$, *for the Poisson random variable.*

Solution:

$$m(t) = E(e^{tY}) = \sum_{y=0}^{\infty} e^{ty}p(y) = \sum_{y=0}^{\infty} e^{ty}\frac{\lambda^y e^{-\lambda}}{y!}$$

$$= \sum_{y=0}^{\infty} \frac{(\lambda e^t)^y e^{-\lambda}}{y!} = e^{-\lambda} \sum_{y=0}^{\infty} \frac{(\lambda e^t)^y}{y!}.$$

To complete the summation, consult a mathematical handbook to find

$$\sum_{y=0}^{\infty} \frac{(\lambda e^t)^y}{y!} = e^{\lambda e^t},$$

or employ the method of Example 3.11. Thus multiply and divide by $e^{\lambda e^t}$. *Then*

$$m(t) = e^{-\lambda}e^{\lambda e^t} \sum_{y=0}^{\infty} \frac{(\lambda e^t)^y e^{-\lambda e^t}}{y!}.$$

The quantity inside the summation is the probability function for a Poisson random variable with mean λe^t. Hence

$$\sum_y p(y) = 1 \qquad \text{and} \qquad m(t) = e^{-\lambda}e^{\lambda e^t}(1) = e^{\lambda(e^t - 1)}.$$

Example 3.15: *Use the moment-generating function, Example 3.14, and Theorem 3.5 to find $\mu'_1 = \mu$ for the Poisson random variable.*

Solution:

$$\mu = m^{(1)}(0) = \frac{d}{dt}\{e^{\lambda(e^t - 1)}\}]_{t=0}$$

$$= e^{\lambda(e^t - 1)} \cdot \lambda e^t]_{t=0} = \lambda.$$

The second and primary application of a moment-generating function is in proving that a random variable possesses a particular probability distribution, $p(y)$. If $m(t)$ exists for a probability distribution $p(y)$, it is unique. That is, it is impossible for random variables with different probability distributions to have the same moment-generating functions. Also, it can be shown that if the moment-generating functions for two random variables, Y and Z, are equal, then $p(y) = p(z)$. It follows that if we can recognize the moment-generating function of a random variable Y to be that of a known distribution, Y must have that distribution.

In summary, a moment-generating function is simply a mathematical device that sometimes (but not always) provides an easy way to find μ'_k and to prove the equivalence of two probability distributions.

3.10 Probability-Generating Functions

An important class of discrete random variables is one in which Y takes integral values, $Y = 0, 1, 2, 3, \ldots$, and consequently represents a count. The binomial, geometric, hypergeometric, and Poisson random variables all fall in this class.

The following examples give practical situations that result in integral-valued random variables. One, involving the theory of queues (waiting lines), is concerned with the number of persons (or objects) awaiting service at a particular point in time. Knowledge of the behavior of this random variable is important in designing manufacturing plants where production consists of a sequence of operations each of which requires a different length of time to complete. An insufficient number of service stations for a particular production operation can result in a "bottleneck," the formation of a queue of products ready to be serviced, and a resulting slowdown in the manufacturing operation. Queuing theory is also important in determining the number of checkout counters needed for a supermarket and in the design of hospitals and clinics.

Integral-valued random variables are also very important in studies of population growth. For example, epidemiologists are interested in the growth of bacterial populations and, also, in the growth of the number of persons afflicted by a particular disease. The number of elements in each of these populations will be an integral-valued random variable.

A mathematical device that is very useful in finding the probability distributions and other properties of integral-valued random variables is the probability-generating function.

Definition 3.11: *Let Y be an integral-valued random variable for which $P[Y = i] = p_i, i = 0, 1, 2, \ldots$. The probability-generating function $P(t)$ for Y is defined to be*

$$P(t) = E(t^Y) = p_0 + p_1 t + p_2 t^2 + \cdots$$

for all values of t such that $P(t)$ is finite.

The reason for calling $P(t)$ a probability-generating function is clear when we compare $P(t)$ with the moment-generating function, $m(t)$. Particularly, the coefficient of t^i in $P(t)$ is the probability p_i. Correspondingly, the coefficient of t^i for $m(t)$ is a constant times the ith moment, μ_i'. If we know $P(t)$ and can expand it into a series, we can determine $p(y)$ as the coefficient of t^y.

Repeated differentiation of $P(t)$ yields *factorial moments* for the random variable Y.

Definition 3.12: *The Kth* factorial moment *for a random variable Y is defined to be*

$$\mu_{[K]} = E[Y(Y - 1)(Y - 2) \cdots (Y - K + 1)],$$

where K is a positive integer.

Note that $\mu_{[1]} = E(Y) = \mu$ and that $\mu_{[2]} = E[Y(Y - 1)]$ was useful in finding the variance for the binomial random variable, Example 3.12.

Theorem 3.6:

$$\left.\frac{d^K P(t)}{dt^K}\right]_{t=1} = P^{(K)}(1) = \mu_{[K]}.$$

Proof:

$$P(t) = p_0 + p_1 t + p_2 t^2 + p_3 t^3 + p_4 t^4 + \cdots$$

and

$$P^{(1)}(t) = p_1 + 2p_2 t + 3p_3 t^2 + 4p_4 t^3 + \cdots.$$

Setting $t = 1$,

$$P^{(1)}(1) = p_1 + 2p_2 + 3p_3 + \cdots = \sum_{y=0}^{\infty} y p(y) = E(Y).$$

Then

$$P^{(2)}(t) = (2)(1)p_2 + (3)(2)p_3 t + (4)(3)p_4 t^2 + \cdots,$$

and, in general,

$$P^{(K)}(t) = \sum_{y=K}^{\infty} y(y - 1)(y - 2)\cdots(y - K + 1)p(y)t^{y-K}.$$

Setting $t = 1$, *we obtain*

$$\begin{aligned} P^{(K)}(1) &= \sum_{y=K}^{\infty} y(y - 1)(y - 2)\cdots(y - K + 1)p(y) \\ &= E[Y(Y - 1)(Y - 2)\cdots(Y - K + 1)] \\ &= \mu_{[K]}. \end{aligned}$$

Example 3.16: *Find the probability-generating function for the geometric random variable.*

Solution: *Note that* $p_0 = 0$ *because Y cannot assume this value. Then*

$$P(t) = E(t^Y) = \sum_{y=1}^{\infty} t^y q^{y-1} p = \sum_{y=1}^{\infty} \frac{p}{q}(qt)^y$$

$$= \frac{p}{q}[qt + (qt)^2 + (qt)^3 + \cdots].$$

The terms of the series are those of an infinite geometric progression. We can let $t \le 1$, *so that* $qt < 1$. *Then*

$$P(t) = \frac{p}{q}\left\{\frac{qt}{1 - qt}\right\} = \frac{pt}{1 - qt}.$$

(For summation of the series, consult a mathematical handbook.)

Example 3.17: *Use P(t), Example 3.16, to find the mean of the geometric random variable.*

Solution: *From Theorem 3.6,* $\mu_{[1]} = \mu = P^{(1)}(1)$. *Then*

$$P^{(1)}(t) = \frac{d}{dt}\left\{\frac{pt}{1 - qt}\right\} = \frac{(1 - qt)p - (pt)(-q)}{(1 - qt)^2}.$$

Setting $t = 1$,

$$P^{(1)}(1) = \frac{p^2 + pq}{p^2} = \frac{p(p + q)}{p^2} = \frac{1}{p}.$$

Since we already have the moment-generating function to assist in finding the moments of a random variable, we might ask of what value is $P(t)$. The answer is that it may be exceedingly difficult to find $m(t)$ but easy to find $P(t)$. Thus $P(t)$ simply provides an additional tool for finding the moments of a random variable. It may or may not be useful in a given situation.

Finding the moments of a random variable is not the major use of the probability-generating function. Its primary application is in deriving the probability function (and hence the probability distribution) for other related integral-valued random variables. For these applications, see Feller [1] and Parzen [5].

3.11 Tchebysheff's Theorem

We have seen in Section 1.3 and Example 3.10 that the empirical rule is of great help in approximating probabilities over certain intervals if the mean and variance of the random variable of interest are known. Another result of this same type is given by Tchebysheff's theorem.

> ***Theorem 3.7 Tchebysheff's Theorem:*** *Let Y be a random variable with finite mean μ and variance σ^2. Then, for any positive constant k,*
>
> $$P(|Y - \mu| \leq k\sigma) \geq 1 - \frac{1}{k^2}$$
>
> $$\text{or} \qquad P(|Y - \mu| > k\sigma) \leq \frac{1}{k^2}.$$

The proof of this theorem will be deferred until Section 4.10. Theorem 3.7 states that the probability that Y assumes a value less than $\mu - k\sigma$ or greater than $\mu + k\sigma$ is less than $1/k^2$. We illustrate the usefulness of this theorem with the following example.

Example 3.18: *The number of customers per day at a certain sales counter, denoted by Y, has been observed for a long period of time and found to have a mean of 20 customers with a standard deviation of 2 customers. The probability distribution of Y is not known. What is a good approximation to the probability that Y will be between 16 and 24 tomorrow?*

Solution: *We want to find $P(16 \leq Y \leq 24)$. From Theorem 3.7 we know that $P(|Y - \mu| \leq k\sigma) \geq 1 - 1/k^2$, or*

$$P[(\mu - k\sigma) \leq Y \leq (\mu + k\sigma)] \geq 1 - \frac{1}{k^2}.$$

Since $\mu = 20$ *and* $\sigma = 2$, *then* $\mu - k\sigma = 16$ *and* $\mu + k\sigma = 24$ *if* $k = 2$. *Thus*

$$P(16 \leq Y \leq 24) \geq 1 - \frac{1}{(2)^2} = 3/4.$$

In other words, tomorrow's customer total will be between 16 *and* 24 *with high probability* (*at least* 3/4).

Note that if σ *were only one customer,* k *would be* 4 *and*

$$P(16 \leq Y \leq 24) \geq 1 - \frac{1}{(4)^2} = 15/16.$$

Thus we see that the value of σ *has considerable affect on probabilities associated with intervals.*

3.12 Summary

This chapter has been concerned with discrete random variables and their probability distributions and expectations. Calculating the probability distribution for a discrete random variable implies evaluating the probability of a numerical event and the use of the probabilistic methods of Chapter 2. Probability-distribution functions were derived for the binomial, geometric, hypergeometric, and Poisson random variables.

The expected values of random variables and functions of random variables provided a method for finding the mean and variance of Y and consequently a measure of centrality and variation for $p(y)$. Much of the remaining material in the chapter was devoted to the techniques for acquiring expectations, the sometimes devious manipulations for summing apparently intractable series.

We then presented a discussion of the moment-generating function for a random variable. Although sometimes useful in finding μ and σ, the moment-generating function is of primary value to the theoretical statistician in deriving the probability distribution for a random variable. Appendix II, gives the means, variances, and moment-generating functions for most of the common random variables.

The probability-generating function is another useful device for deriving moments and probability distributions of integer-valued random variables.

Finally, we gave Tchebysheff's theorem, which is useful for approximating probabilities when only the mean and variance are known.

To conclude this summary, we suggest that the reader relate this chapter to the objective of statistics, making an inference about a population based on information contained in a sample. Drawing the sample from the population is the experiment. The sample is a set of measurements on one or more random variables and is the observed event resulting from a single repetition of the experiment. Finally, making the inference about the population requires knowledge of the probability of the observed sample, which, as a basis, requires knowledge of the probability distributions of the random variables that generated the sample.

References

1. Feller, W., *An Introduction to Probability Theory and Its Applications*, Vol. 1, 3rd ed. New York: John Wiley & Sons, Inc., 1968.
2. Mendenhall, W., *An Introduction to Probability and Statistics*, 3rd ed. North Scituate, Mass.: Duxbury Press, 1971.
3. Mosteller, F., R. E. K. Rourke, and G. B. Thomas, *Probability with Statistical Applications*, 2nd ed. Reading, Mass.: Addison-Wesley Publishing Company, Inc., 1970.
4. Parzen, E., *Modern Probability Theory and Its Applications*. New York: John Wiley & Sons, Inc., 1964.
5. Parzen, E., *Stochastic Processes*. San Francisco: Holden-Day, Inc., 1962.
6. *Standard Mathematical Tables*, 17th ed. Cleveland: Chemical Rubber Company, 1969.

Exercises

3.1. In a county containing a large number of rural homes, 60 percent are thought to be insured against a fire. Four rural home owners are chosen at random from the entire population and Y are found to be insured against a fire. Find the probability distribution for Y. What is the probability that at least three of the four will be insured?

3.2. A fire-detection device utilizes three temperature-sensitive cells acting independently of each other in such a manner that any one or more may actuate the alarm. Each cell possesses a probability of $p = .8$ of actuating the alarm when the temperature reaches 100 degrees centigrade or more. Let Y equal the number of cells actuating the alarm when the temperature reaches 100 degrees. Find the probability distribution for Y. Find the probability that the alarm will function when the temperature reaches 100 degrees.

3.3. Construct probability histograms for the binomial probability distribution for $n = 5, p = .1, .5$, and $.9$. (Table 1, Appendix III, will reduce the amount of calculation.) Note the symmetry for $p = .5$ and the direction of skewness for $p = .1$ and $.9$.

3.4. Use Table 1, Appendix III, to construct a probability histogram for the binomial probability distribution for $n = 20$ and $p = .5$. Note that almost all the probability falls in the interval $5 \leqslant y \leqslant 15$.

3.5. The probability that a single radar set will detect an enemy plane is .9. If we have five radar sets, what is the probability that exactly four sets will detect the plane? (Assume that the sets operate independently of each other.) At least one set?

3.6. Suppose that the four engines of a commercial aircraft were arranged to operate independently and that the probability of in-flight failure of a single engine is .01. What is the probability that, on a given flight,

(a) No failures are observed?
(b) No more than one failure is observed?

3.7. Sampling for defectives from large lots of manufactured product yields a number of defectives, Y, that follows a binomial probability distribution. A sampling plan consists in specifying the number of items to be included in a sample, n, and an acceptance number, a. The lot is accepted if $Y \leq a$ and rejected if $Y > a$. Let p denote the proportion of defectives in the lot. For $n = 5$ and $a = 0$ calculate the probability of lot acceptance if (a) $p = 0$, (b) $p = .1$, (c) $p = .3$, (d) $p = .5$, (e) $p = 1.0$. A graph showing the probability of lot acceptance as a function of lot fraction defective is called the *operating characteristic curve* for the sample plan. Construct the operating characteristic curve for the plan, $n = 5$, $a = 0$. Note that a sampling plan is an example of statistical inference. Accepting or rejecting a lot based on information contained in the sample is equivalent to concluding that the lot is either "good" or "bad," respectively. "Good" implies that a low fraction is defective and that the lot is therefore suitable for shipment.

3.8. Refer to Exercise 3.7. Use Table 1, Appendix III, to construct the operating characteristic curve for a sampling plan with
(a) $n = 10, a = 0$.
(b) $n = 10, a = 1$.
(c) $n = 10, a = 2$.
For each, calculate P(lot acceptance) for $p = 0, .05, .1, .3, .5$, and 1.0. Our intuition suggests that sampling plan (a) would be much less likely to accept bad lots than plans (b) and (c). A visual comparison of the operating characteristic curves will confirm this supposition.

3.9. A quality-control engineer wishes to study the alternative sampling plans $n = 5, a = 1$ and $n = 25, a = 5$. On a sheet of graph paper, construct the operating characteristic curves for both plans, making use of acceptance probabilities at $p = .05, p = .10, p = .20, p = .30$, and $p = .40$ in each case.
(a) If you were a seller producing lots with fraction defective ranging from $p = 0$ to $p = .10$, which of the two sampling plans would you prefer?
(b) If you were a buyer wishing to be protected against accepting lots with fraction defective exceeding $p = .30$, which of the two sampling plans would you prefer?

3.10. A manufacturer of floor wax has developed two new brands, A and B, which he wishes to subject to a housewife evaluation to determine which of the two is superior. Both waxes, A and B, are applied to floor surfaces in each of 15 homes.
(a) If there is actually no difference in the quality of the brands, what is the probability that 10 or more housewives would state a preference for brand A?
(b) For either brand A or brand B?

3.11. The probability of a customer arrival at a grocery service counter in any 1 second is equal to .1. Assume that customers arrive in a random stream and hence that the arrival any 1 second is independent of any other:
(a) Find the probability that the first arrival will occur during the third 1-second interval.
(b) Find the probability that the first arrival will not occur until at least the third 1-second interval.

3.12. Sixty percent of a population of consumers is reputed to prefer a particular brand, A, of toothpaste. If a group of consumers are interviewed, what is the probability that exactly five people have to be interviewed before encountering a consumer who prefers brand A? At least five people?

3.13. Suppose that a radio contains six transistors, two of which are defective. Three transistors are selected at random, removed from the radio, and inspected. Let Y equal the number of defectives observed, where $Y = 0, 1$, or 2. Find the probability distribution for Y. Express your results graphically as a probability histogram.

3.14. Simulate the experiment described in Exercise 3.13 by marking six marbles, or coins, so that two represent defectives and four represent nondefectives. Place the marbles in a hat, mix, draw three, and record Y, the number of "defectives observed." Replace the marbles and repeat the process until a total of $n = 100$ observations on Y has been recorded. Construct a relative frequency histogram for this sample and compare it with the population probability distribution, Exercise 3.13.

3.15. The mean number of automobiles entering a mountain tunnel per 2-minute period is one. An excessive number of cars entering the tunnel during a brief period of time produces a hazardous situation. Find the probability that the number of autos entering the tunnel during a 2-minute period exceeds three.

3.16. Assume that the tunnel is observed during ten 2-minute intervals, thus giving ten independent observations, $Y_1, Y_2, \ldots, Y_{10}$, on the Poisson random variable of Exercise 3.15. Find the probability that $Y > 3$ during at least one of the ten 2-minute intervals.

3.17. A starter motor utilized in a space vehicle possesses a high reliability and is reputed to start on any given occasion with probability .99999. What is the probability of at least one failure in the next 10,000 starts?

3.18. Given a random variable, Y, with $p(y)$ as follows:

y	$p(y)$
0	.1
1	.2
2	.4
3	.2
4	.1

(a) Find the expected value and variance of Y.
(b) Construct the interval $(\mu \pm 2\sigma)$. Check $p(y)$ and find the fraction of the theoretical population in the interval $(\mu \pm 2\sigma)$. Compare with Tchebysheff's theorem.

3.19. Refer to Exercise 3.13. Using the probability histogram, find the fraction of the total population lying within two standard deviations of the mean. Compare with Tchebysheff's theorem.

3.20. A balanced coin is tossed three times. Let Y equal the number of heads observed.
(a) Use the formula for the binomial probability distribution to calculate the probabilities associated with $Y = 0, 1, 2$, and 3.

(b) Construct a probability distribution similar to Figure 3.1.
(c) Find the expected value and standard deviation of Y, using the formulas,

$$E(Y) = np, \qquad V(Y) = npq.$$

(d) Using the probability distribution, (b), find the fraction of the population measurements lying within one standard deviation of the mean. Repeat for two standard deviations. How do your results agree with Tchebysheff's theorem and the empirical rule?

3.21. Suppose that a coin was definitely unbalanced and that the probability of a head was equal to $p = .1$. Follow instructions (a), (b), (c), and (d) as stated in Exercise 3.20. Note that the probability distribution loses its symmetry and becomes skewed when p is not equal to 1/2.

3.22. It is known that 10 percent of a brand of television tubes will burn out before their guarantee has expired. If 1000 tubes are sold, find the expected value and variance of Y, the number of original tubes that must be replaced. Within what limits would Y be expected to fall? (*Hint:* Use Tchebysheff's theorem.)

3.23. Consider a binomial experiment for $n = 20$, $p = .05$. Use Table 1, Appendix III, to calculate the binomial probabilities for $Y = 0, 1, 2, 3, 4$. Calculate the same probabilities using the Poisson approximation with $\lambda = np$. Compare.

3.24. A salesman has found that the probability of a sale on a single contact is approximately .03. If the salesman contacts 100 prospects, what is the probability that he will make at least one sale?

3.25. Refer to Exercise 3.24. Would you expect the salesman to make as many as 12 sales in a group of 100 contacts? To answer the question, calculate the interval $\mu \pm 2\sigma$ and employ the reasoning implied in Tchebysheff's theorem and the empirical rule.

3.26. Let Y be a random variable with probability distribution given by

y	$p(y)$
0	1/8
1	1/4
2	1/2
3	1/8

Find the expected value and variance of Y. Construct a graph of the probability distribution.

3.27. Refer to Exercise 3.13. Find μ, the expected value of Y, for the theoretical population by using the probability distribution obtained in Exercise 3.13. Find the sample mean, $\bar{y}$, for the $n = 100$ measurements generated in Exercise 3.14. Does $\bar{y}$ provide a good estimate of μ?

3.28. Find the population variance, σ^2, for Exercise 3.13 and the sample variance, s'^2, for Exercise 3.14. Compare.

3.29. Toss a balanced die and let Y be the number of dots observed on the upper face. Find the mean and variance of Y. Construct a probability histogram and locate the interval $\mu \pm 2\sigma$. Verify that Tchebysheff's theorem holds.

3.30. Draw a sample of $n = 50$ measurements from the die-throw population of Exercise 3.29 by tossing a die 50 times and recording y after each toss. Calculate $\bar{y}$ and s'^2 for the sample. Compare $\bar{y}$ with the expected value of Y and s'^2 with the variance of Y obtained in Exercise 3.29. Do $\bar{y}$ and s'^2 provide good estimates of μ and σ^2?

3.31. A heavy-equipment salesman can contact either one or two customers per day with probability 1/3 and 2/3, respectively. Each contact will result in either no sale or a $50,000 sale with probability 9/10 and 1/10, respectively. Give the probability distribution for daily sales. Find the mean and standard deviation of the daily sales.

3.32. A potential customer for a $20,000 fire insurance policy possesses a home in an area, which, according to experience, may sustain a total loss in a given year with probability of .001 and a 50 percent loss with probability .01. Ignoring all other partial losses, what premium should the insurance company charge for a yearly policy in order to break even?

3.33. The manufacturer of a low-calorie dairy drink wishes to compare the taste appeal of a new formula (formula B) with that of the standard formula (formula A). Each of four judges is given three glasses in random order, two containing formula A and the other containing formula B. Each judge is asked to state which glass he most enjoyed. Suppose that the two formulas are equally attractive. Let Y be the number of judges stating a preference for the new formula.

(a) Find the probability function for Y.

(b) What is the probability that at least three of the four judges state a preference for the new formula?

(c) Find the expected value of Y.

(d) Find the variance of Y.

3.34. Find the moment-generating function for the random variable Y of Exercise 3.18. By differentiating $m(t)$, find μ'_1 and μ'_2. Then find the variance of Y. Compare your answers with those of Exercise 3.18.

3.35. Find the probability-generating function for a binomial random variable Y, and use it to find the mean and variance of Y.

3.36. One concern of a gambler is that he will go broke before he acquires his first win. Suppose that he plays a game in which the probability of winning is .1 (and is unknown to him). It costs him \$10 to play and he receives \$80 for a win. If he commences with \$30, what is the probability that he wins exactly once before he loses his initial capital?

3.37. Find the moment-generating function for the geometric probability distribution. Differentiate to acquire $\mu_1' = \mu$ and μ_2'. Then find the variance of the geometric random variable.

3.38. Find the expected value of $Y(Y-1)$ for the Poisson probability distribution and then find the variance of Y.

3.39. Show that the moment-generating function of the binomial random variable is $(q + pe^t)^n$.

3.40. Refer to Exercise 3.39. Differentiate $m(t)$ to find $\mu_1' = \mu$ and μ_2'. Then find $V(Y) = npq$.

3.41. Refer to Exercise 3.39. Use the fact that

$$e^z = 1 + z + \frac{z^2}{2!} + \frac{z^3}{3!} + \frac{z^4}{4!} + \cdots$$

to expand

$$m(t) = (q + pe^t)^n$$

into a power series in t. (Acquire only the low-order terms in t.) Identify μ_i' as the coefficient of $t^i/i!$ appearing in the series. Specifically, find, μ_1' and μ_2' and compare with the results of Exercise 3.40.

3.42. Consider a sequence of identical and independent trials of the type implied by the binomial experiment and a random variable similar to the geometric random variable. Let Y represent the number of the trial on which the rth success occurs. The parameter r is a constant. For the special case $r = 1$, Y is a geometric random variable. The random variable Y, described above, follows a *negative binomial distribution.* Use the method of Section 3.3 to show that

$$p(y) = \binom{y-1}{r-1} p^r(1-p)^{y-r}, \qquad y = r, r+1, r+2, \ldots.$$

3.43. A supplier of heavy construction equipment has found that new customers are normally obtained by customer request for a sales call and that the probability of a sale of a particular piece of equipment is .3. If the supplier has three pieces of the equipment available for sale, what is the probability that it will take less than five customer contacts to clear his inventory?

3.44. Show that the hypergeometric probability function approaches the binomial in the limit as $N \to \infty$ and $p = r/N$ remains constant. That is, show that

$$\lim_{N \to \infty} \frac{\binom{r}{y}\binom{N-r}{n-y}}{\binom{N}{n}} = \binom{n}{y} p^y q^{n-y}$$

for $p = r/N$ constant.

3.45. A lot of $N = 100$ industrial products contains 40 defectives. Let Y be the number of defectives in a random sample of size 20. Find $p(10)$ using .
(a) The hypergeometric probability distribution.
(b) The binomial probability distribution.
Is N large enough so that the binomial probability function is a good approximation to the hypergeometric probability function?

3.46. Calculate $P(|Y - \lambda| \leq 2\sigma)$ for the Poisson probability distribution of Example 3.10. Does this agree with the empirical rule?

3.47. For simplicity, let us assume that there are two kinds of drivers. The safe drivers, which are 70 percent of the population, have a probability of .1 of causing an accident in a year. The rest of the population are accident makers, who have a probability of .5 of causing an accident in a year. The insurance premium is \$400 times one's probability of causing an accident in the following year. A new subscriber has an accident during the first year. What should be his insurance premium for the next year?

3.48. A merchant stocks a certain perishable item. He knows that on any given day he will have a demand for either 2, 3, or 4 of these items with probabilities .1, .4, and .5, respectively. He buys the items for \$1.00 each and sells them for \$1.20 each. If any are left at the end of the day, they represent a total loss. How many items should the merchant stock so as to maximize his expected daily profit?

3.49. It is known that 5 percent of a population have disease A, which can be discovered by blood test. Suppose that N (a large number) people are to be tested. This can be done in two ways.

(1) Each person is tested separately.

(2) The blood samples of k people are pooled together and analyzed. (Assume that $N = nk$, with n an integer.) If the test is negative, all of them are healthy (that is, just this one test is needed). If the test is positive, each of the k persons must be tested separately (that is, totally $k + 1$ tests are needed).

(a) For fixed k, what is the expected number of tests needed in (2)?

(b) Find the k that will minimize the expected number of tests in (2).

(c) How many tests does (b) save in comparison with (1)?

4

Continuous Random Variables and Their Probability Distributions

4.1 *Introduction*

A moment of reflection on statistical problems encountered in the real world will convince the reader that not all random variables fit the definition for discrete random variables. For example, observing the number of defectives in a lot of n manufactured items leads to a discrete random variable because the number of defectives must be 0, 1, 2, . . . , or n. On the other hand, the daily rainfall at a certain geographic point can theoretically take on any of the uncountable infinity of values in an interval of real numbers. We could have 5.758 inches of rain, but we could not possibly observe 5.758 defectives. The type of random variable that takes on any value in an interval is called *continuous*, and the purpose of this chapter is to study probability distributions for continuous random variables. The yield of an antibiotic in a fermentation process is a continuous random variable, as is the length of life of a washing machine. The line segment over which these two random variables are defined is the positive half of the real line.

4.2 *The Probability Distribution for a Continuous Random Variable*

We will lead to a formal definition for a continuous random variable by first defining a *distribution function* (or *cumulative distribution function*).

Definition 4.1: *Let Y denote any random variable. The distribution function of Y, denoted by $F(y)$, is given by $F(y) = P(Y \leq y)$, $-\infty < y < \infty$.*

Example 4.1: *Suppose that Y has a binomial distribution with $n = 2$ and $p = 1/2$. Find $F(y)$.*

Solution: *The probability distribution for Y is given by*

$$p(y) = \binom{2}{y}(1/2)^y(1/2)^{2-y}, \qquad y = 0, 1, 2,$$

which yields

$$p(0) = 1/4, \qquad p(1) = 1/2, \quad \text{and} \quad p(2) = 1/4.$$

Then

$$\begin{aligned} F(y) &= P(Y \leq y) \\ &= 0 \qquad \text{for } y < 0 \\ &= 1/4 \qquad \text{for } 0 \leq y < 1 \\ &= 3/4 \qquad \text{for } 1 \leq y < 2 \\ &= 1 \qquad \text{for } y \geq 2. \end{aligned}$$

Note, for example, that

$$\begin{aligned} P(Y \leq 1.5) &= P(Y \leq 1) = P(Y = 0) + P(Y = 1) \\ &= (1/4) + (1/2) = 3/4. \end{aligned}$$

$F(y)$ is depicted graphically in Figure 4.1.

Distribution functions for discrete random variables are step functions since such functions increase only at discrete points. In Example 4.1 the points between 0 and 1 or between 1 and 2 contribute nothing to the cumulative

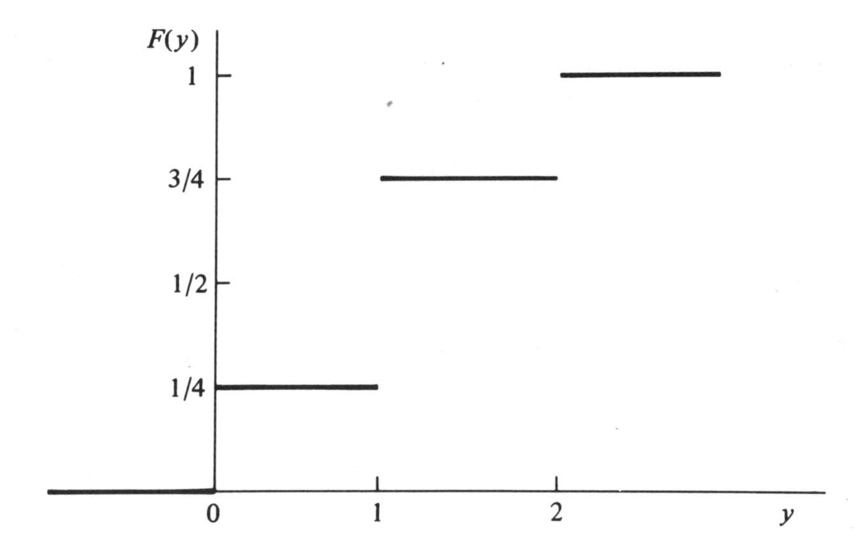

Figure 4.1 Binomial Distribution Function, $n = 2, p = 1/2$

probability. In contrast, distribution functions for continuous random variables, such as the amount of daily rainfall, should be smooth, increasing functions over some interval. That is, if Y, the random variable denoting daily rainfall, has distribution function $F(y)$, then $F(1.1)$ should be greater than $F(1.0)$ because there is some positive probability of observing a daily rainfall between 1.0 and 1.1 inches. Thus we are led to the definition of a continuous random variable.

> ***Definition 4.2:*** *Let Y denote a random variable with distribution function $F(y)$. Y is said to be* continuous *if $F(y)$ is continuous, $-\infty < y < \infty$, and the first derivative of $F(y)$ exists and is continuous except for, at most, a finite number of points in any finite interval.*

To be precise, a distribution function satisfying Definition 4.2 is *absolutely continuous*, but we will call the corresponding random variable merely *continuous*. The statement concerning existence of derivatives will generally be satisfied for distribution functions, $F(y)$, which are continuous. For practical purposes, we can think of continuous random variables as those having continuous distribution functions.

The derivative of $F(y)$ gives rise to another function of prime importance in probability theory and statistics.

Definition 4.3: *Let $F(y)$ be the distribution function of a continuous random variable, Y. Then $f(y)$, given by*

$$f(y) = \frac{dF(y)}{dy} = F'(y),$$

wherever the derivative exists, is called the probability density function *for the random variable Y.*

It follows from Definitions 4.2 and 4.3 that $F(y)$ can be written as

$$F(y) = \int_{-\infty}^{y} f(t)\,dt,$$

where $f(y)$ is the probability density function and t is used as the variable of integration.

Example 4.2: *Suppose that $F(y) = y$, $0 \le y \le 1$, $F(y) = 0$, $y < 0$, $F(y) = 1$, $y > 1$. Find the probability density function for y and graph it.*

Solution: *Since $F(y) = \int_{-\infty}^{y} f(t)\,dt = y, 0 \le y \le 1$,*

$$\frac{dF(y)}{dy} = f(y) = \frac{d(y)}{dy} = \begin{cases} 1 & 0 < y < 1, \\ 0 & \text{for } y < 0 \text{ or } y > 1, \end{cases}$$

and $f(y)$ is undefined at $y = 0$ and $y = 1$. A graph of $F(y)$ is shown in Figure 4.2.

The graph of $f(y)$, Example 4.2, is shown in Figure 4.3. Note that the probability density function is a *theoretical model* for the frequency distribution of a population of measurements. For example, observations of the lengths of life of a particular brand of washing machines will generate measurements that can be characterized by a frequency histogram, as discussed in Chapter 1. Conceptually, the experiment could be repeated ad infinitum, thereby generating a frequency distribution (a smooth curve) that would characterize the population

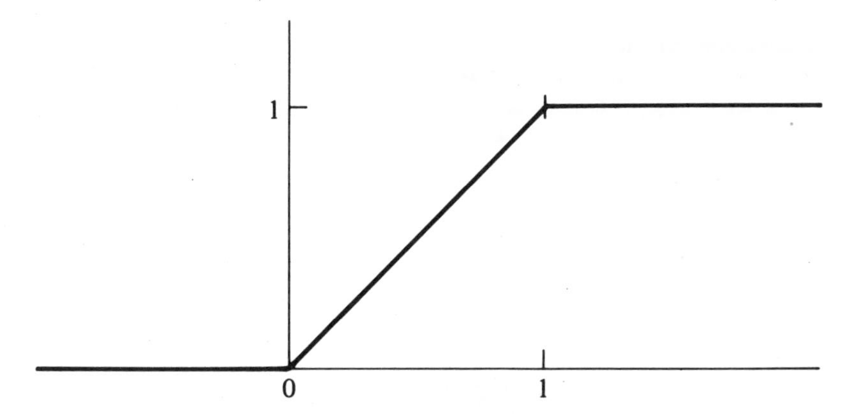

Figure 4.2 Distribution Function, $F(y)$, for Example 4.2

of interest to the manufacturer. This theoretical frequency distribution corresponds to the probability density function for the length of life of a single machine, Y.

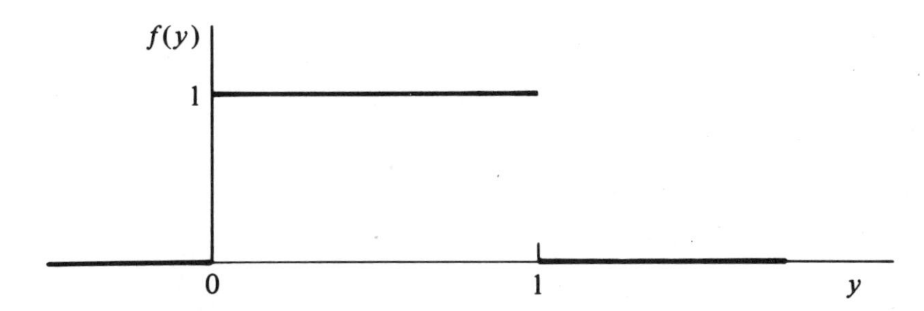

Figure 4.3 Density Function, $f(y)$, for Example 4.2

The real analogue of $P(Y \leq y_0)$ is the relative frequency or the fraction of the total number of measurements in the population less than or equal to y_0. Thus $F(y_0)$ is the area under the frequency distribution over the interval $Y \leq y_0$, and is the shaded area shown in Figure 4.4. From a practical point of view, it is clear that $P(Y \leq -\infty) = F(-\infty)$ must equal zero. As y_0 is increased (or moved to the right in Figure 4.4), the area $F(y_0)$ increases and approaches a maximum, $P(y \leq \infty) = F(\infty) = 1$. Thus $F(y)$ (drop the subscript) is a monotonically increasing function of y. These three characteristics define the properties of a distribution function.

Properties of F(y):

1. $P(Y \le -\infty) = F(-\infty) = 0.$
2. $P(Y \le \infty) = F(\infty) = 1.$
3. $F(y_b) \ge F(y_a)$ *if* $y_b > y_a$.

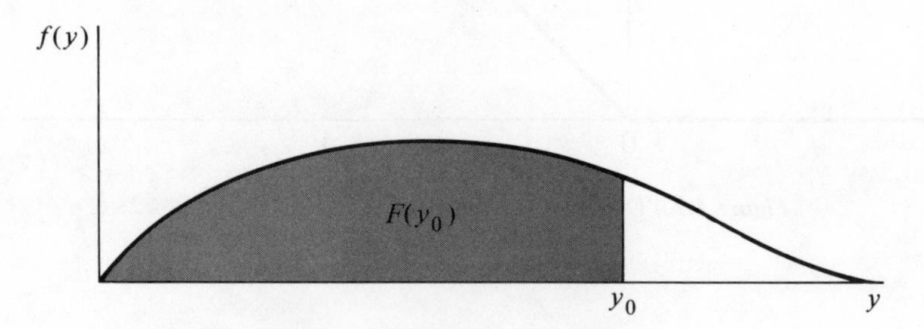

Figure 4.4 The Distribution Function

Example 4.3: *Let Y be a continuous random variable with probability density function given by*

$$f(y) = \begin{cases} 3y^2, & 0 \le y \le 1, \\ 0, & \textit{elsewhere.} \end{cases}$$

Find F(y). Graph both f(y) and F(y).

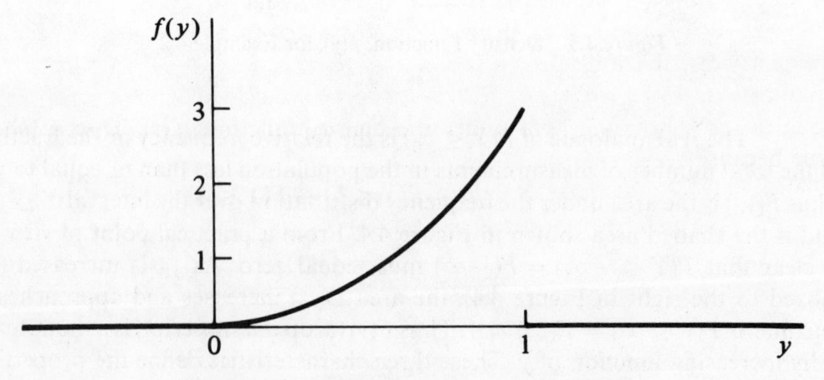

Figure 4.5 Density Function for Example 4.3

Solution: *The graph of $f(y)$ appears in Figure 4.5. Since*

$$F(y) = \int_{-\infty}^{y} f(t)\,dt,$$

we have, for this example,

$$F(y) = \int_{0}^{y} 3t^2\,dt$$

$$= t^3\Big]_0^y = y^3 \qquad \textit{for } 0 < y < 1.$$

Note that

$$F(y) = \begin{cases} 0 & \textit{for } y \le 0, \\ 1 & \textit{for } y \ge 1. \end{cases}$$

The graph of $F(y)$ is given in Figure 4.6.

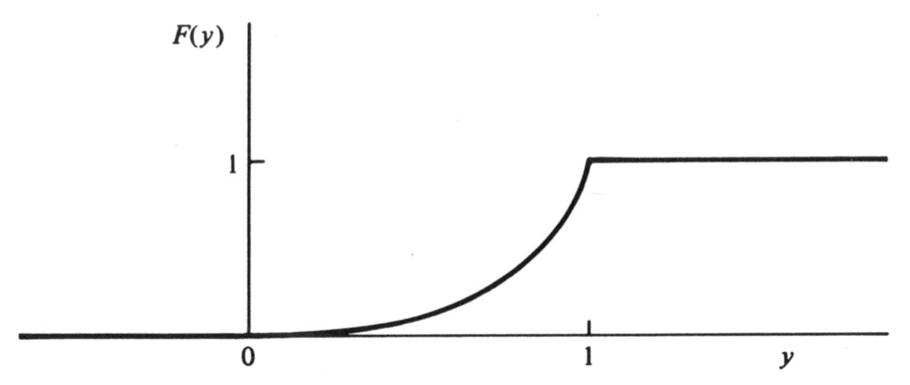

Figure 4.6 Distribution Function for Example 4.3

$F(y_0)$ gives the probability that $Y \le y_0$. The next step is to find the probability that Y falls in a specific interval, that is, $P(a \le Y \le b)$. From Chapter 1 we know that probability corresponds to the area under the frequency distribution over the interval $a \le y \le b$. Since $f(y)$ is the theoretical counterpart of the frequency distribution, we would expect $P(a \le Y \le b)$ to equal a corresponding area under the density function, $f(y)$. This is indeed true because

$$P(a < Y \le b) = P(Y \le b) - P(Y \le a)$$

$$= F(b) - F(a)$$

$$= \int_a^b f(y)\,dy.$$

Since $P(Y = a) = 0$, we have the following result.

The probability that Y falls in the interval [a, b] is

$$P(a \leq Y \leq b) = \int_a^b f(y)\,dy,$$

where $f(y)$ is the probability density function for Y.

This probability is shown graphically in Figure 4.7.

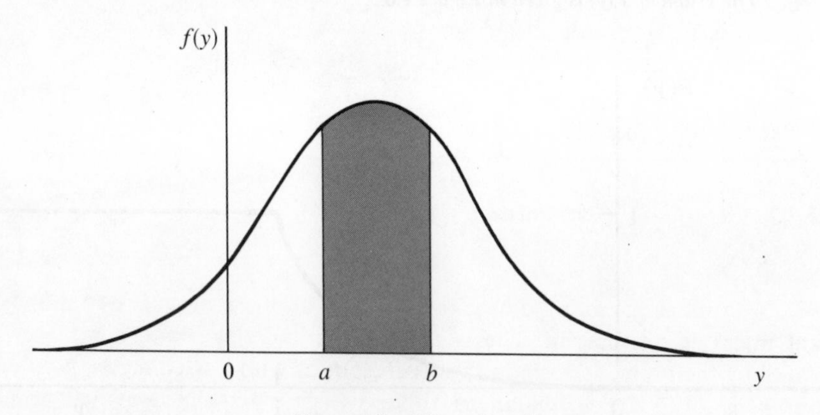

Figure 4.7 $P(a \leq Y \leq b)$

Example 4.4: *Given $f(y) = cy^2, 0 \leq y \leq 2$, and $f(y) = 0$ elsewhere, find the value of c for which $f(y)$ will be a valid density function.*

Solution: *We require a value for c such that*

$$F(\infty) = \int_{-\infty}^{\infty} f(y)\,dy = 1$$

$$= \int_{-\infty}^{\infty} f(y)\,dy = \int_0^2 cy^2\,dy = \frac{cy^3}{3}\bigg]_0^2 = (8/3)c.$$

Setting $F(\infty)$ equal to 1, $F(\infty) = (8/3)c = 1$, we find that $c = 3/8$.

Example 4.5: *Find the probability that $1 \leq Y \leq 2$ for Example 4.4.*

Solution:

$$P(1 \leq Y \leq 2) = \int_1^2 f(y)\,dy = 3/8 \int_1^2 y^2\,dy = (3/8)\frac{y^3}{3}\Big]_1^2 = 7/8.$$

Probability statements with regard to a continuous random variable, Y, will only be meaningful if, first, the integral defining the probability exists and, second, the resulting probabilities agree with the axioms of Chapter 2. These two conditions will always be satisfied if we consider only probabilities associated with a finite or countable collection of intervals. Thus we will almost always make probability statements of the form $P(a \leq Y \leq b)$.

Let us now consider some density functions that provide good models for population frequency distributions encountered in nature.

4.3 The Uniform Distribution

A familiar gambling device is a wheel with numbers marked at equidistant intervals on the rim. The wheel is spun and the winning number is indicated by the point on the rim that corresponds with a fixed arrow pointed at the wheel. If the wheel is balanced, the probability that the arrow will point to any specified interval on the circumference of the wheel is proportional to the length of the interval. Let Y denote the clockwise distance between a fixed point on the circumference of the wheel and the point where the arrow falls. Then the density function for Y, called a uniform probability density function, is as follows:

The Uniform Probability Density Function

$$f(y) = \begin{cases} \dfrac{1}{\theta_2 - \theta_1}, & \theta_1 \leq y \leq \theta_2, \\ 0, & \text{elsewhere.} \end{cases}$$

Note that $\theta_1 = 0$ and θ_2 denotes the circumference of the wheel in the above example. The density function of Figure 4.3 is a uniform distribution with $\theta_1 = 0$ and $\theta_2 = 1$.

***Definition 4.4:** The constants that determine the specific form of a density function are called* parameters *of the density function.*

The quantities θ_1 and θ_2 are parameters of the uniform density function and are clearly numerical descriptive measures of this theoretical density function. Both the range and the probability of Y falling in any given interval are dependent on θ_1 and θ_2.

Equiprobable numerical events defined over an interval on a line segment are frequently called "random numbers." These play an important role in the selection of samples from real populations. Second, some continuous random variables in the physical and biological sciences appear to possess uniform probability distributions. A third and important application of the uniform distribution is of a theoretical nature. We will subsequently show that a particular function of many random variables possesses a uniform probability distribution.

4.4 The Normal Distribution

We have talked about bell-shaped, or normal, distributions of data in connection with the empirical rule, and will subsequently give an argument to justify the common occurrence of bell-shaped distributions of data in nature. Regardless of the reasons for their occurrence, it is a fact that measurements on many random variables appear to have been generated from population frequency distributions that are closely approximated by a normal probability distribution. The normal density function is as follows:

The Normal Probability Density Function

$$f(y) = \frac{e^{-(y-\mu)^2/2\sigma^2}}{\sigma\sqrt{2\pi}} \qquad \sigma > 0,\ -\infty < \mu < \infty,\ -\infty < y < \infty.$$

Note that the normal density function contains two parameters, μ and σ. We will subsequently show that μ and σ are the mean and standard deviation, respectively, of the associated distribution and hence locate the center of the distribution and measure its spread. A graph of the normal density function is shown in Figure 4.8.

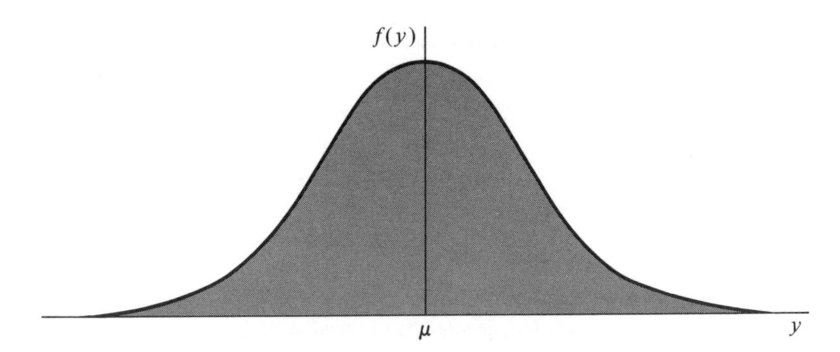

Figure 4.8 The Normal Probability Density Function

Areas under the normal density function corresponding to $P(a \leq Y \leq b)$ require evaluation of the integral,

$$\int_a^b \frac{1}{\sigma\sqrt{2\pi}} e^{-(y-\mu)^2/2\sigma^2}\, dy.$$

Unfortunately, a closed-form expression for the integral does not exist and hence its evaluation can only be obtained by approximate procedures. As a consequence, areas under the normal density function are presented in Table 3, Appendix III.

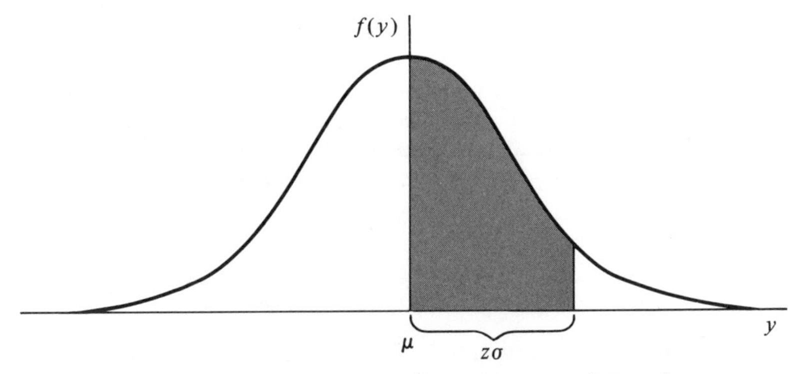

Figure 4.9 Tabulated Area for the Normal Density Function

The normal density function is symmetric with respect to μ, so areas need only be tabulated on one side of the mean. The tabulated value is the area between the mean and a point z, where z is the distance from the mean measured in standard deviations. This area is shaded in Figure 4.9.

Example 4.6: *Let Z denote a normal random variable with mean 0 and variance 1.*

(a) Find $P(0 \leq Z \leq 2)$.
(b) Find $P(-2 \leq Z \leq 2)$.
(c) Find $P(0 \leq Z \leq 1.53)$.

Solution:

(a) Proceed down the first (z) column of Table 3 and read the area opposite $z = 2.0$. *This area, denoted by the symbol* $A(z)$, *is* $A(z = 2.0) = .4772$.

(b) Because the normal density function is symmetric, the area, $P(-2 \leq Z \leq 2)$ *is double the area given in (a) or*

$$P(-2 \leq Z \leq 2) = .9544.$$

Note that this is the area lying within two standard deviations of the mean and is the source of the "95 percent" appearing in the empirical rule.

(c) The area corresponding to $z = 1.53$ *is obtained by proceeding down the z column in Table 3 to the entry "1.5." Then move across the top of the table to the .03 column to read*

$$P(0 \leq Z \leq 1.53) = A(z = 1.53) = .4370.$$

Example 4.7: *The achievement scores for a college entrance examination are normally distributed with mean 75 and standard deviation equal to 10. What fraction of the scores would one expect to lie between 70 and 90? (Assume that the set contains a large number of examination scores.)*

Solution: *Recall that z is a distance measured from the mean of a normal distribution expressed in units of standard deviation. Thus*

$$z = \frac{y - \mu}{\sigma}.$$

Then the desired fraction of the population is the area lying between

$$z_1 = \frac{70 - 75}{10} = -.5$$

and

$$z_2 = \frac{90 - 75}{10} = 1.5.$$

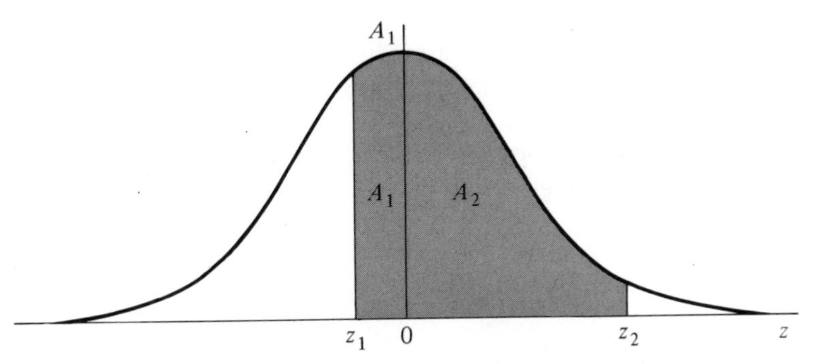

Figure 4.10 Desired Area, Example 4.7

This area, $A = A_1 + A_2$, is the shaded portion of Figure 4.10. From the table,

$$A_1(-.5) = A(.5) = .1915,$$

$$A_2(1.5) = .4332.$$

Then the fraction of scores lying between 70 and 90 is

$$A = A_1 + A_2 = .1915 + .4332 = .6247.$$

One can always transform a normal random variable to Z, and thereby use Table 3, by using the relationship

$$Z = \frac{Y - \mu}{\sigma}.$$

We have noted that Z locates a point measured from the mean of a normal random variable with the distance *expressed in units of standard deviation* of the original normal random variable. Thus the mean value of Z must be 0 and its standard deviation must equal 1. Proof that Z, called a *standardized normal random variable*, is normally distributed with mean equal to 0, standard deviation equal to 1, is given in Chapter 6.

4.5 The Gamma-Type Probability Distribution

Some random variables are always nonnegative and for various reasons yield distributions of data that are skewed (nonsymmetric) to the right. That is, most of the area under the density function is located near the origin and the density function drops gradually as y increases. A skewed probability distribution is shown in Figure 4.11.

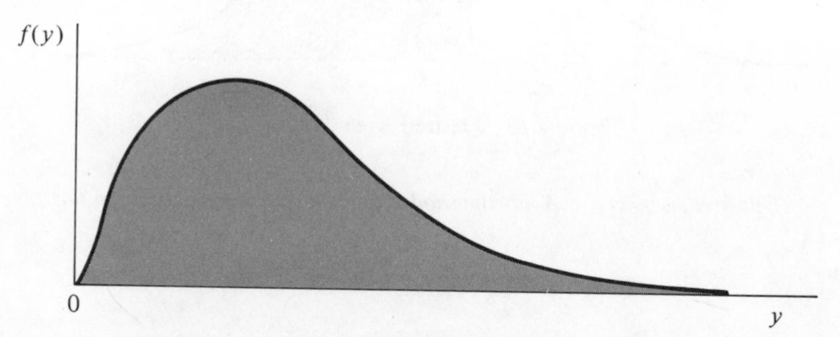

Figure 4.11 A Skewed Probability Distribution

The lengths of time between malfunctions for aircraft engines possess a skewed frequency distribution, as do the lengths of time between arrivals at a supermarket checkout queue (that is, the line at the checkout counter). Similarly, the lengths of time to complete a maintenance checkout for an automobile or aircraft engine form a skewed frequency distribution. The populations associated with these random variables frequently possess distributions that are adequately modeled by the gamma-type density function.

The Probability Density Function for a Gamma-Type Random Variable

$$f(y) = \begin{cases} \dfrac{y^{\alpha-1}e^{-y/\beta}}{\beta^{\alpha}\Gamma(\alpha)}, & \alpha, \beta > 0; \quad 0 \le y \le \infty, \\ 0, & \textit{elsewhere}. \end{cases}$$

For the reader unfamiliar with the gamma function,

$$\Gamma(\alpha) = \int_0^\infty y^{\alpha-1} e^{-y}\, dy.$$

Integrating by parts it can be shown that $\Gamma(n) = (n-1)\Gamma(n-1)$ and that $\Gamma(n) = (n-1)!$, provided n is an integer.

The density function for the special case $\alpha = 1$, called the *exponential distribution*, is

$$f(y) = \frac{e^{-y/\beta}}{\beta}, \qquad \beta > 0;\ y \geq 0.$$

This density function together with those for $\alpha = 2$ and 4 $(\beta = 1)$ are graphically presented in Figure 4.12.

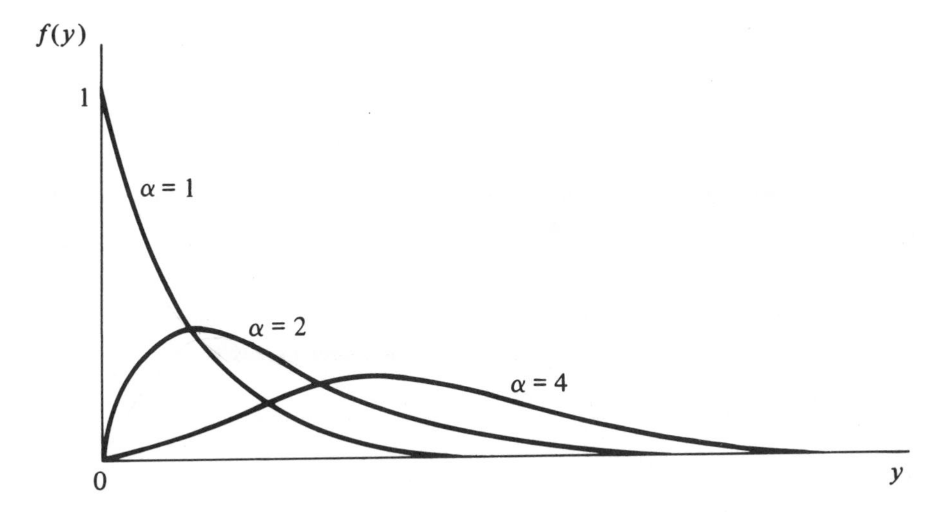

Figure 4.12 The Gamma-Density Function, $\beta = 1$

Like the normal density function, it is impossible to give a closed-form expression for

$$\int_c^d \frac{y^{\alpha-1} e^{-y/\beta}}{\beta^\alpha \Gamma(\alpha)}\, dy, \qquad \text{where } 0 < c < d < \infty,$$

and hence difficult to obtain areas under the gamma-type density function by direct integration. Tabulated values of the integral are given in *Tables of the Incomplete Gamma Function* [7].

One gamma-type density function occurs frequently in statistical theory.

Definition 4.5: *A gamma-type random variable that possesses a density function with parameters $\alpha = \nu/2$ and $\beta = 2$ is called* a chi-square (χ^2) random variable.

The parameter ν is called the "number of degrees of freedom" associated with the chi-square random variable. The reason for this choice of terminology rests on one of the major applications of the distribution and cannot be intuitively explained at this point.

4.6 The Beta Function

The beta density function is a two-parameter density function defined over a closed interval, $0 \le y \le 1$.

The Probability Density Function for the Beta Random Variable

$$f(y) = \begin{cases} \dfrac{y^{\alpha-1}(1-y)^{\beta-1}}{B(\alpha, \beta)}, & \alpha, \beta > 0; \quad 0 \le y \le 1, \\ 0, & \textit{elsewhere}, \end{cases}$$

where

$$B(\alpha, \beta) = \int_0^1 y^{\alpha-1}(1-y)^{\beta-1}dy = \frac{\Gamma(\alpha)\Gamma(\beta)}{\Gamma(\alpha+\beta)}.$$

The graph of the density function will assume widely differing shapes for various values of the two parameters α and β. Some of these are shown in Figure 4.13.

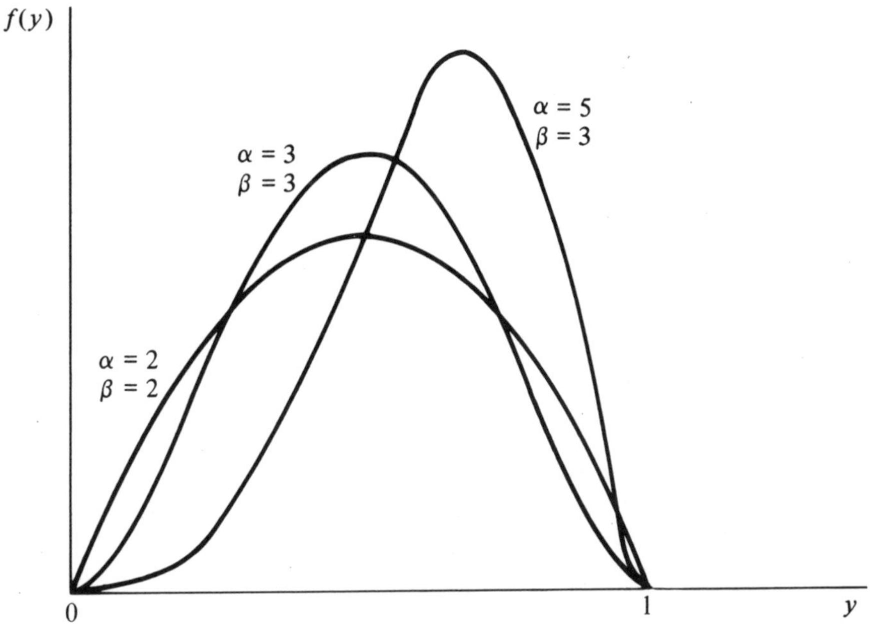

Figure 4.13 The Beta Density Function

Note that defining y over the interval $0 \leq y \leq 1$ does not restrict its use. If $c \leq y \leq d$, $y^* = (y - c)/(d - c)$ will define a new variable defined on the interval $0 \leq y \leq 1$. Thus the beta density function can be applied to a random variable defined on the interval $c \leq y \leq d$ by translation and a change of scale.

The cumulative distribution function for the beta random variable is commonly called the incomplete B-function and is denoted by

$$F(y) = \int_0^y \frac{t^{\alpha-1}(1-t)^{\beta-1}}{B(\alpha, \beta)}\, dt = I_y(\alpha, \beta).$$

A tabulation of $I_y(\alpha, \beta)$ is given in *Tables of the Incomplete Beta Function* [6]. For integral values of α and β, $I_y(\alpha, \beta)$ is related to the binomial probability function. Letting $y = p$, it can be shown that

$$F(p) = \int_0^p \frac{y^{\alpha-1}(1-y)^{\beta-1}}{B(\alpha, \beta)}\, dy = \sum_{y=\alpha}^{n} \binom{n}{y} p^y(1-p)^{n-y}.$$

where $0 < p < 1$ and $n = \alpha + \beta - 1$. The binomial cumulative distribution function is presented in Table 1, Appendix III, for $n = 5, 10, 15, 20, 25$ ($p = .01$,

.05, .10, .20, .30, .40, .50, .60, .70, .80, .90, .95, .99). A more extensive tabulation of the binomial probability function is given in *Tables of the Binomial Probability Distribution* [5].

4.7 Some General Comments

Keep in mind that density functions are theoretical models for populations of real data that occur in nature. How do we know which to use, and to what extent does it matter if we have the "wrong one"?

To answer the last question first, it is unlikely that we will ever select a density function that will provide a perfect representation of nature, but goodness of fit is not the criterion for assessing the adequacy of our model. The purpose of a probabilistic model is to provide the mechanism for making inferences about a population based on information contained in a sample. As noted earlier, the probability of the observed sample (or a quantity proportional to it) will be instrumental in making an inference about the population. It follows that a density function that provides a poor fit to the population frequency distribution could (but does not necessarily) yield incorrect probability statements and lead to erroneous inferences about the population. A good model is one that yields good inferences about the population of interest.

A reasonable selection of a model is sometimes implied by theoretical considerations. For example, a situation for which the discrete Poisson random variable is appropriate will often be indicated by the random behavior of events in time. Knowing this, it is possible to show that the interval between any adjacent pair of events follows an exponential distribution. Similarly, we shall later encounter a theorem (called the "central limit theorem") which will outline some conditions that imply normal distributions of data.

A second way to select a model is to form a frequency histogram (Chapter 1) for some data drawn from the population and to choose a density function that would visually appear to give a similar frequency curve. For example, if a set of $n = 100$ sample measurements yields a bell-shaped frequency distribution, we might conclude that the normal density function would adequately model the population frequency distribution.

Not all model selection is completely subjective. Statistical procedures are available to test a hypothesis that a population frequency distribution is of a particular type. One can also calculate a measure of goodness of fit for several distributions and select the best. Studies of many common inferential methods have been made to determine the magnitude of the errors of inference introduced by incorrect population models. It is comforting to know that many statistical methods of inference are insensitive to assumptions about the form of the underlying population frequency distribution.

The uniform, normal, gamma-type, and the beta distributions give an assortment of density functions that will fit a large number of population frequency distributions. Another, the Weibull distribution, will appear in the exercises at the end of the chapter.

4.8 The Expected Value for a Continuous Random Variable

The next step in the study of continuous random variables is to find their means, variances, and standard deviations and thereby to acquire numerical descriptive measures of their density functions. Particularly, we will learn subsequently that it is sometimes difficult to find the probability distribution for a random variable Y, or a function $g(Y)$, and we have already observed that integration over intervals for many density functions (the normal and gamma type, for example) is very difficult. When this occurs, we can approximately describe the behavior of the random variable using its moments along with the empirical rule and Tchebysheff's theorem (Chapters 1 and 3):

Definition 4.6: *The expected value of a continuous random variable Y is*

$$E(Y) = \int_{-\infty}^{\infty} yf(y)\,dy,$$

provided the integral exists.

If the definition of the expected value for a discrete random variable Y, $E(Y) = \sum_y yp(y)$, is meaningful, then Definition 4.6 should also agree with our intuitive notion of a mean. The quantity $f(y)\,dy$ corresponds to $p(y)$ for the discrete case and integration evolves from and is analogous to summation. Hence $E(Y)$ in Definition 4.6 agrees completely with our notion of an average or mean.

Example 4.8: *Find the expected value of the gamma-type random variable.*

Solution:

$$E(Y) = \int_{-\infty}^{\infty} yf(y)\,dy = \int_{0}^{\infty} y\frac{y^{\alpha-1}e^{-y/\beta}}{\beta^{\alpha}\Gamma(\alpha)}\,dy.$$

By definition, if $f(y)$ is a density function,

$$\int_{-\infty}^{\infty} f(y)\,dy = 1 \quad \text{or} \quad \int_{0}^{\infty} \frac{y^{\alpha-1}e^{-y/\beta}}{\beta^{\alpha}\Gamma(\alpha)}\,dy = 1.$$

Hence

$$\int_{0}^{\infty} y^{\alpha-1}e^{-y/\beta}\,dy = \beta^{\alpha}\Gamma(\alpha) = \beta^{\alpha}(\alpha-1)!.$$

Then

$$E(Y) = \int_{0}^{\infty} \frac{y^{\alpha}e^{-y/\beta}\,dy}{\beta^{\alpha}\Gamma(\alpha)} = \frac{1}{\beta^{\alpha}(\alpha-1)!}\int_{0}^{\infty} y^{\alpha}e^{-y/\beta}\,dy$$

$$= \frac{1}{\beta^{\alpha}(\alpha-1)!}[\beta^{\alpha+1}(\alpha)!],$$

or $E(Y) = \alpha\beta$.

Similarly, we are interested in the expected value of a function of Y.

Definition 4.7: *Let $g(Y)$ be a function of Y. Then the expected value of $g(Y)$ is*

$$E[g(Y)] = \int_{-\infty}^{\infty} g(y)f(y)\,dy,$$

provided the integral exists.

The expected values of three important functions of a continuous random variable, Y, evolve as a consequence of well-known theorems on integration and, as expected, lead to results analogous to those contained in Theorems 3.1, 3.2, and 3.3. As a consequence, proof of Theorem 4.1 will be left as an exercise for the reader.

> ***Theorem 4.1:*** *Let c be a constant and let $g(Y), g_1(Y), g_2(Y), \ldots, g_k(Y)$ be functions of a continuous random variable, Y. Then*
> 1. $E(c) = c$.
> 2. $E[cg(Y)] = cE[g(Y)]$.
> 3. $E[g_1(Y) + g_2(Y) + \cdots + g_k(Y)]$ $= E[g_1(Y)] + E[g_2(Y)] + \cdots + E[g_k(Y)]$.

As in the case of discrete random variables, we seek the expected value of the function $g(Y) = (Y - \mu)^2$, which, intuitively, is the variance of the random variable, Y. That is, $V(Y) = E(Y - \mu)^2$, as in Definition 3.7. Recall from Theorem 3.4 that we also have $V(Y) = E(Y^2) - \mu^2$.

Example 4.9: *Find the variance for the gamma-type random variable.*

Solution: *From Theorem 3.4, $V(Y) = E(Y^2) - [E(Y)]^2$. Then*

$$E(Y^2) = \int_0^\infty y^2 \frac{y^{\alpha-1}e^{-y/\beta}}{\beta^\alpha\Gamma(\alpha)}\,dy = \frac{1}{\beta^\alpha\Gamma(\alpha)}\int_0^\infty y^{\alpha+1}e^{-y/\beta}\,dy$$

$$= \frac{1}{\beta^\alpha\Gamma(\alpha)}[\beta^{\alpha+2}\Gamma(\alpha+2)] = \beta^2\frac{(\alpha+1)!}{(\alpha-1)!} = \alpha(\alpha+1)\beta^2.$$

Then $V(Y) = E(Y^2) - [E(Y)]^2$, where from Example 4.8, $E(Y) = \alpha\beta$. Substituting into this expression for $E(Y^2)$ and $E(Y)$, we have

$$V(Y) = \alpha(\alpha+1)\beta^2 - (\alpha\beta)^2 = \alpha^2\beta^2 + \alpha\beta^2 - \alpha^2\beta^2$$

or

$$V(Y) = \alpha\beta^2.$$

4.9 Other Expected Values

As for discrete random variables, the moments about the origin for continuous random variables are given by

$$\mu'_k = E(Y^k), \qquad k = 1, 2, \ldots.$$

For $k = 1$, $\mu'_1 = \mu$. Also, the moments about the mean, or central moments, are given by

$$\mu_k = E[(Y - \mu)^k], \qquad k = 1, 2, \ldots.$$

Example 4.10: *Find μ'_k for the uniform random variable with $\theta_1 = 0$ and $\theta_2 = \theta$.*

Solution: *By definition,*

$$\mu'_k = E(Y^k) = \int_{-\infty}^{\infty} y^k f(y)\,dy = \int_0^{\theta} y^k \frac{1}{\theta}\,dy = \left.\frac{y^{k+1}}{\theta(k+1)}\right]_0^{\theta}$$

$$= \frac{\theta^k}{k+1}.$$

Thus

$$\mu'_1 = \mu = \frac{\theta}{2}, \mu'_2 = \frac{\theta^2}{3}, \mu'_3 = \frac{\theta^3}{4}, \textit{ and so on.}$$

The moment-generating function $m(t)$ for a continuous random variable Y is defined to be $E(e^{tY})$, as in Definition 3.10. That $m(t)$ generates moments is shown in exactly the same manner as in Section 3.9. If $m(t)$ exists,

$$E(e^{tY}) = \int_{-\infty}^{\infty} e^{ty} f(y)\,dy$$

$$= \int_{-\infty}^{\infty} \left(1 + ty + \frac{t^2y^2}{2!} + \frac{t^3y^3}{3!} + \cdots\right) f(y)\,dy$$

$$= \int_{-\infty}^{\infty} f(y)\,dy + t\int_{-\infty}^{\infty} yf(y)\,dy + \frac{t^2}{2!}\int_{-\infty}^{\infty} y^2 f(y)\,dy + \cdots$$

$$= 1 + t\mu_1' + \frac{t^2}{2!}\mu_2' + \frac{t^3}{3!}\mu_3' + \cdots.$$

The reader will note that the moment-generating function,

$$m(t) = 1 + t\mu_1' + \frac{t^2}{2!}\mu_2' + \cdots,$$

takes the same form for both discrete and continuous random variables. Hence Theorem 3.5 holds for continuous random variables and

$$\left.\frac{d^k m(t)}{dt^k}\right]_{t=0} = \mu_k'.$$

Example 4.11: *Find the moment-generating function for the gamma-type random variable.*

Solution:

$$m(t) = E(e^{tY}) = \int_0^{\infty} \frac{e^{ty}y^{\alpha-1}}{\beta^{\alpha}\Gamma(\alpha)} e^{-y/\beta}\,dy$$

$$= \frac{1}{\beta^{\alpha}\Gamma(\alpha)}\int_0^{\infty} y^{\alpha-1}\exp\left[-y\left(\frac{1}{\beta} - t\right)\right]dy$$

$$= \frac{1}{\beta^{\alpha}\Gamma(\alpha)}\int_0^{\infty} y^{\alpha-1}\exp\{-y/[\beta/(1-\beta t)]\}\,dy.$$

[exp( ) *is simply a more convenient way to write* $e^{(\,)}$ *when the term is long or complex.*]

To complete the integration, note that the integral of the variable factor of any density function must equal the reciprocal of the constant factor. That is, if $f(y) = cg(y)$, *where* c *is a constant,*

$$\int_{-\infty}^{\infty} f(y)\,dy = \int_{-\infty}^{\infty} cg(y)\,dy = 1.$$

Therefore,

$$\int_{-\infty}^{\infty} g(y)\,dy = \frac{1}{c}.$$

Applying this result to the integral in $m(t)$, *and noting that* $g(y) = y^{\alpha-1}\exp\{-y/[\beta/(1-\beta t)]\}$ *is the variable factor of a gamma-type density function with parameters* α *and* $\beta/(1-\beta t)$,

$$m(t) = \frac{1}{\beta^{\alpha}\Gamma(\alpha)}\left[\left(\frac{\beta}{1-\beta t}\right)^{\alpha}\Gamma(\alpha)\right] = \frac{1}{(1-\beta t)^{\alpha}} \qquad \text{for } t < \frac{1}{\beta}.$$

The moments μ'_k *can be extracted from the moment-generating function by differentiating with respect to* t*, Theorem 3.5, or by expanding the function into a power series in* t*. We shall demonstrate the latter approach.*

Example 4.12: *Expand the moment-generating function of Example 4.11 into a power series in* t *and thereby obtain* μ'_k.

Solution: *From Example 4.11,* $m(t) = 1/(1-\beta t)^{\alpha} = (1-\beta t)^{-\alpha}$. *Using the expansion for a binomial,*

$$\begin{aligned} m(t) = (1-\beta t)^{-\alpha} &= 1 + (-\alpha)(1)^{-\alpha-1}(-\beta t) \\ &\quad + \frac{(-\alpha)(-\alpha-1)(1)^{-\alpha-2}(-\beta t)^2}{2!} + \cdots \\ &= 1 + t(\alpha\beta) + \frac{t^2[\alpha(\alpha+1)\beta^2]}{2!} \\ &\quad + \frac{t^3[\alpha(\alpha+1)(\alpha+2)\beta^3]}{3!} + \cdots. \end{aligned}$$

Since μ'_k *will be the coefficient of* $t^k/k!$*, we find, by inspection,*

$$\mu'_1 = \mu = \alpha\beta,$$

$$\mu'_2 = \alpha(\alpha+1)\beta^2,$$

$$\mu'_3 = \alpha(\alpha+1)(\alpha+2)\beta^3,$$

and, in general, $\mu'_k = \alpha(\alpha+1)(\alpha+2)\cdots(\alpha+k-1)\beta^k$. *Note that* μ'_1 *and* μ'_2 *agree with the results of Examples 4.8 and 4.9.*

We have already explained the importance of the expected values of Y^k, $(Y - \mu)^k$ and e^{tY}, all of which are functions of the random variable Y and provide numerical descriptive measures for its density function. However, it is important to note that we are sometimes interested in the expected values of a function of a random variable as an end in itself. (We are also interested in the probability distribution of functions of random variables, but we defer discussion of this topic until Chapter 6.)

Example 4.13: *The force, f, exerted by a mass, m, moving at velocity, v, is*

$$f = \frac{mv^2}{2}.$$

Consider a device that fires a serrated nail into concrete at a mean velocity of 500 feet per second, where V, the random velocity, possesses a density function

$$f(v) = \frac{v^3 e^{-y/b}}{b^4 \Gamma(4)}, \qquad b = 500; \quad v \geq 0.$$

If each nail possesses mass m, find the expected force exerted by a nail.

Solution:

$$E(F) = E\left(\frac{mV^2}{2}\right) = \frac{m}{2}E(V^2)$$

by Theorem 4.1, part 2. The density function for V is a gamma-type function with $\alpha = 4$, $\beta = 500$. Therefore, $E(V^2) = \mu_2'$ for the random variable V. Referring to Example 4.12, $\mu_2' = \alpha(\alpha + 1)\beta^2 = 4(5)(500)^2 = 5{,}000{,}000$. Therefore,

$$E(F) = \frac{m}{2}E(V^2) = \frac{m}{2}(5{,}000{,}000) = 2{,}500{,}000m.$$

Finding the moments of a function of a random variable is frequently facilitated by using moment-generating functions.

Theorem 4.2: *Let $g(Y)$ be a single-valued function of a random variable, Y, with density function, $f(y)$. Then the moment-generating function for $g(Y)$ is*

$$E[e^{tg(Y)}] = \int_{-\infty}^{\infty} e^{tg(y)} f(y)\, dy.$$

This theorem follows directly from Definitions 4.7 and 3.10.

Example 4.14: *Let $g(Y) = Y - \mu$, where Y is a normally distributed random variable with mean μ and variance σ^2. Find the moment-generating function for $(Y - \mu)$.*

Solution:

$$m(t) = E[e^{t(Y-\mu)}] = \int_{-\infty}^{\infty} e^{t(y-\mu)} \frac{\exp[-(y-\mu)^2/2\sigma^2]}{\sigma\sqrt{2\pi}}\, dy.$$

To integrate, let $u = y - \mu$. Then $du = dy$ and

$$m(t) = \frac{1}{\sigma\sqrt{2\pi}} \int_{-\infty}^{\infty} e^{tu} e^{-u^2/2\sigma^2}\, du$$

$$= \frac{1}{\sigma\sqrt{2\pi}} \int_{-\infty}^{\infty} \exp[-(1/2\sigma^2)(u^2 - 2\sigma^2 tu)]\, du.$$

Complete the square in the exponent of e by multiplying and dividing by $e^{t^2\sigma^2/2}$. Then

$$m(t) = e^{t^2\sigma^2/2} \int_{-\infty}^{\infty} \frac{\exp[-(1/2\sigma^2)(u^2 - 2\sigma^2 tu + \sigma^4 t^2)]}{\sigma\sqrt{2\pi}}$$

$$= e^{t^2\sigma^2/2} \int_{-\infty}^{\infty} \frac{\exp[-(u - \sigma^2 t)^2/2\sigma^2]}{\sigma\sqrt{2\pi}}\, du.$$

The function inside the integral is a normal density function with mean and variance equal to $\sigma^2 t$ and σ^2, respectively. (See the equation for the normal density function.) Hence the integral is equal to 1. Then

$$m(t) = e^{(t^2/2)\sigma^2}$$

The moments of $U = Y - \mu$ can be obtained from $m(t)$ by differentiating $m(t)$ in accordance with Theorem 3.5 or by expanding $m(t)$ into a series. We recommend the latter approach.

For the reader's convenience, the probability and density functions, means, variances, and moment-generating functions for some common random variables are presented in Appendix II.

4.10 Tchebysheff's Theorem

The purpose in presenting the preceding discussion of moments is twofold. First, as in the case of a finite set of measurements, they can be used as numerical descriptive measures to describe the data. Second, they can be used in a theoretical sense to prove that a random variable possesses a particular probability distribution. It can be shown that if two random variables, Y and Z, possess identical moment-generating functions, then Y and Z possess identical probability distributions. This latter application of moments was mentioned in discussing moment-generating functions for discrete random variables, Section 3.9; it applies to continuous random variables as well.

The interpretation of μ and σ for continuous random variables utilizes the empirical rule and Tchebysheff's theorem. The justification for their use is that they provide a satisfactory approximation for many distributions.

We would expect Tchebysheff's theorem to hold for probability distributions, both discrete and continuous. We will restate the theorem and give a proof applicable to a continuous random variable.

Theorem 4.3 Tchebysheff's Theorem: *Let Y be a continuous (discrete) random variable with density function $f(y)$ [or probability function $p(y)$]. Then*

$$P(|Y - \mu| \leq k\sigma) \geq 1 - \frac{1}{k^2} \quad \text{or} \quad P(|Y - \mu| > k\sigma) \leq \frac{1}{k^2}.$$

where $E(Y) = \mu$ and $V(Y) = \sigma^2 < \infty$.

Proof: *We will give the proof for a continuous random variable. The proof for the discrete case would proceed in a similar manner.*

$$V(Y) = \sigma^2 = \int_{-\infty}^{\infty} (y - \mu)^2 f(y)\, dy$$

$$= \int_{\mu - k\sigma}^{\mu + k\sigma} (y - \mu)^2 f(y)\, dy$$

$$+ \int_{-\infty}^{\mu - k\sigma} (y - \mu)^2 f(y)\, dy$$

$$+ \int_{\mu + k\sigma}^{\infty} (y - \mu)^2 f(y)\, dy.$$

The first integral is always greater than or equal to zero and $(y - \mu)^2 \geq k^2\sigma^2$ for the second and third integrals [that is, the region of integration is in the tails of the density function and covers only values of y for which $(y - \mu)^2 \geq k^2\sigma^2$]. Let the first integral equal zero and substitute $k^2\sigma^2$ for $(y - \mu)^2$ in the second and third integrals to obtain the inequality

$$V(Y) = \sigma^2 \geq \int_{-\infty}^{\mu - k\sigma} k^2\sigma^2 f(y)\, dy + \int_{\mu + k\sigma}^{\infty} k^2\sigma^2 f(y)\, dy.$$

Then

$$\sigma^2 \geq k^2\sigma^2 \left[\int_{-\infty}^{\mu - k\sigma} f(y)\, dy + \int_{\mu + k\sigma}^{+\infty} f(y)\, dy \right]$$

or

$$\sigma^2 \geq k^2\sigma^2 P(|Y - \mu| \geq k\sigma).$$

Dividing by $k^2\sigma^2$,

$$P(|Y - \mu| \geq k\sigma) \leq \frac{1}{k^2}.$$

Hence it follows that

$$P(|Y - \mu| \leq k\sigma) \geq 1 - \frac{1}{k^2}.$$

Example 4.15: *Suppose that the length of time, Y, to conduct a periodic maintenance check (from previous experience) on a dictating machine follows a gamma-type distribution with $\alpha = 3$ and $\beta = 2$ (minutes). Suppose that a new repairman requires 14 minutes to check a machine. Does it appear that his time to perform a maintenance check disagrees with prior experience?*

Solution: *The mean and variance for the length of maintenance times (prior experience) are (from Examples 4.8 and 4.9)*

$$\mu = \alpha\beta \qquad \text{and} \qquad \sigma^2 = \alpha\beta^2.$$

Then, for our example,

$$\mu = \alpha\beta = (3)(2) = 6,$$

$$\sigma^2 = \alpha\beta^2 = (3)(2)^2 = 12, \qquad \sigma = \sqrt{12} = 3.46$$

and the observed deviation $(Y - \mu)$ *is* $14 - 6 = 8$ *minutes.*

For our example, $y = 14$ *minutes exceeds the mean* $\mu = 6$ *minutes, by* $k = 8/3.46$ *standard deviations. Then, from Tchebysheff's theorem,*

$$P(|Y - \mu| \geq k\sigma) \leq \frac{1}{k^2}$$

or

$$P(|Y - 6| \geq 8) \leq \frac{1}{k^2} = \frac{(3.46)^2}{(8)^2} = 12/64 = 3/16 = .1875.$$

Note that this probability is based on the assumption that the distribution of maintenance times has not changed from prior experience. Then observing that $P(Y \geq 14$ *minutes) is small, we must conclude that either our new maintenance man has generated a lengthy maintenance time that occurs with low probability or he is somewhat slower than his predecessors. Noting the low probability for* $P(Y \geq 14)$, *we would be inclined to favor the latter view.*

The real value of Tchebysheff's theorem is that it enables us to find bounds on probabilities that ordinarily would have to be obtained by tedious integration. For example, the exact probability, $P(Y \geq 14)$, Example 4.15, would require evaluation of the integral,

$$P(Y \geq 14) = \int_{14}^{\infty} \frac{y^2 e^{-y/2}}{8\Gamma(3)}\, dy.$$

(This integral can be shown to equal approximately .02.) Similar integrals are difficult to evaluate for the beta function and for many other density functions. Tchebysheff's theorem often provides a rapid approximate procedure for circumventing laborious integration.

4.11 Expectations of Discontinuous Functions and Mixed Probability Distributions

Problems in probability and statistics frequently involve functions that are partly continuous and partly discrete in one of two ways. First, we may be interested in the properties, perhaps the expectation, of a random variable $g(Y)$ which is a discontinuous function of a discrete or continuous random variable Y. Second, the random variable of interest may itself have a probability distribution made up of isolated points having discrete probabilities and intervals having continuous probability.

We illustrate the first of these two situations with the following example.

Example 4.16: *A certain retailer for a petroleum product sells a random amount, Y, each day. Suppose that Y, measured in hundreds of gallons, has the probability density function*

$$f(y) = \begin{cases} (3/8)y^2, & 0 \le y \le 2, \\ 0, & \text{elsewhere.} \end{cases}$$

The retailer's profit turns out to be \$5 for each 100 gallons sold (5 cents per gallon) if $Y \le 1$ and \$8 per 100 gallons if $Y > 1$. Find the retailer's expected profit for any given day.

Solution: *Let $g(Y)$ denote the retailer's daily profit. Then*

$$g(Y) = \begin{cases} 5Y & \text{if } 0 \le Y \le 1, \\ 8Y & \text{if } 1 < Y \le 2. \end{cases}$$

We want to find expected profit and, by Definition 4.7, the expectation is

$$E[g(Y)] = \int_{-\infty}^{\infty} g(y)f(y)\,dy$$

$$= \int_0^1 5y[(3/8)y^2]\,dy + \int_1^2 8y[(3/8)y^2]\,dy$$

$$= \frac{15}{(8)(4)}[y^4]_0^1 + \frac{24}{(8)(4)}[y^4]_1^2$$

$$= \frac{15}{32}(1) + \frac{24}{32}(15)$$

$$= \frac{(15)(25)}{32} = 11.72.$$

Thus the retailer can expect to profit by \$11.72 *on the daily sale of this particular product.*

A random variable Y, which has some of its probability at discrete points and the remainder spread over intervals, is said to have a *mixed distribution.* Let $F(y)$ denote a distribution function representing a mixed distribution. For all practical purposes, any mixed distribution function $F(y)$ can be written uniquely as

$$F(y) = c_1 F_1(y) + c_2 F_2(y),$$

where $F_1(y)$ is a step distribution function, $F_2(y)$ is a continuous distribution function, c_1 is the accumulated probability of all discrete points, and $c_2 = 1 - c_1$ is the accumulated probability of all continuous portions.

The following example gives an illustration of a mixed distribution.

Example 4.17: *Let Y denote the life length (in hundreds of hours) of a certain type of electronic component. These components frequently fail immediately upon insertion into a system. It has been observed that the probability of immediate failure is 1/4. If a component does not fail immediately, its life-length distribution has the exponential density*

$$f(y) = \begin{cases} e^{-y}, & y > 0, \\ 0, & \text{elsewhere.} \end{cases}$$

Find the distribution function for Y and evaluate $P(Y > 10)$.

Solution: *There is only one discrete point, $Y = 0$, and this point has probability 1/4. Hence $c_1 = 1/4$ and $c_2 = 3/4$. It follows that Y is a mixture of two random variables, X_1 and X_2, where X_1 has probability one at the point zero and X_2 has the given exponential density. That is,*

$$F_1(y) = \begin{cases} 0 & \text{if } y < 0, \\ 1 & \text{if } y \geq 0 \end{cases}$$

and

$$F_2(y) = \int_0^y e^{-x}\,dx = 1 - e^{-y}, \qquad y > 0.$$

Now,

$$F(y) = (1/4)F_1(y) + (3/4)F_2(y)$$

and hence

$$\begin{aligned} P(Y > 10) &= 1 - P(Y \leq 10) \\ &= 1 - F(10) \\ &= 1 - [1/4 + (3/4)(1 - e^{-10})] \\ &= (3/4)[1 - (1 - e^{-10})] = (3/4)e^{-10}. \end{aligned}$$

An easy method for finding expectations of random variables having mixed distributions is given in Definition 4.8.

Definition 4.8: *Let Y have the mixed distribution function*

$$F(y) = c_1 F_1(y) + c_2 F_2(y)$$

and suppose that X_1 is a discrete random variable having distribution function $F_1(y)$ and X_2 is a continuous random variable having distribution function $F_2(y)$. Let $g(Y)$ denote a function of Y. Then

$$E[g(Y)] = c_1 E[g(X_1)] + c_2 E[g(X_2)].$$

Example 4.18: *Find the mean and variance of the random variable defined in Example 4.17.*

Solution: *With all definitions as in Example 4.17, it follows that*

$$E(X_1) = 0$$

and

$$E(X_2) = \int_0^\infty y e^{-y}\, dy = 1.$$

Therefore,

$$\begin{aligned} \mu = E(Y) &= (1/4)E(X_1) + (3/4)E(X_2) \\ &= 3/4. \end{aligned}$$

Also,

$$E(X_1^2) = 0$$

and

$$E(X_2^2) = \int_0^\infty y^2 e^{-y}\, dy = 2.$$

Therefore,

$$E(Y^2) = (1/4)E(X_1^2) + (3/4)E(X_2^2)$$
$$= (1/4)(0) + (3/4)(2) = 3/2.$$

Then

$$V(Y) = E(Y^2) - \mu^2$$
$$= 3/2 - (3/4)^2 = 15/16.$$

4.12 Summary

This chapter has been related to inference because it presents a probabilistic model for continuous random variables. The density function, which provides a model for a population frequency distribution associated with a continuous random variable, will subsequently yield the mechanism for inferring characteristics of the population based on measurements contained in a sample. As a consequence, the density function provides a model for a real distribution of data that exists or could be generated by repeated experimentation. Similar distributions for small sets of data (samples from populations) were discussed in Chapter 1.

Four density functions, the uniform, normal, gamma type, and beta, were presented to provide the reader with a wide assortment of models for population frequency distributions. Many other density functions could be employed to fit real situations, but the four described above adequately suit most situations. A few other density functions are presented in the exercises at the end of the chapter.

The adequacy of a density function to model the frequency distribution for a random variable depends upon the inference-making technique to be employed. If modest disagreement between the model and the real population frequency distribution does not affect the goodness of the inferential procedure, the model is adequate.

The latter part of the chapter concerned expectations, particularly moments and moment-generating functions. It is important that the reader focus his attention on the reason for presenting these quantities and avoid excessive concentration on the mathematical aspects of the material. Moments, particularly the mean and variance, are numerical descriptive measures for random variables. Particularly, we will subsequently learn that it is sometimes difficult to find the probability distribution for a random variable, Y, or a function, $g(Y)$, and we have already observed that integration over intervals for many density functions (the normal and gamma type, for example) is very difficult. When this occurs, we can approximately describe the behavior of the random variable using its moments along with Tchebysheff's theorem and the empirical rule (Chapter 1).

References

1. Hogg, R. V., and A. T. Craig, *Introduction to Mathematical Statistics*, 3rd ed. New York: The Macmillan Company, 1970.
2. Mendenhall, W., *An Introduction to Probability and Statistics*, 3rd ed. North Scituate, Mass.: Duxbury Press, 1971.
3. Parzen, E., *Modern Probability Theory and Its Applications*. New York: John Wiley & Sons, Inc., 1964.
4. *Standard Mathematical Tables*, 17th ed. Cleveland: Chemical Rubber Company, 1969.
5. *Tables of the Binomial Probability Distribution*, Department of Commerce, National Bureau of Standards, Applied Mathematics Series 6, 1950.
6. *Tables of the Incomplete Beta Function*, edited by K. Pearson. New York: Cambridge University Press, 1956.
7. *Tables of the Incomplete Gamma Function*, edited by K. Pearson. New York: Cambridge University Press, 1956.

Exercises

4.1. Let Y possess a density function

$$f(y) = \begin{cases} cy, & 0 \leq y \leq 2, \\ 0, & \text{elsewhere.} \end{cases}$$

(a) Find c.
(b) Find $F(y)$.
(c) Graph $f(y)$ and $F(y)$.
(d) Use $F(y)$ in (b) to find $P(1 \leq Y \leq 2)$.
(e) Use the geometric figure for $f(y)$ to calculate $P(1 \leq Y \leq 2)$.

4.2. Let Y have the density function given by

$$f(y) = \begin{cases} cy^2 + y, & 0 \leq y \leq 1, \\ 0, & \text{elsewhere.} \end{cases}$$

(a) Find c.
(b) Find $F(y)$.
(c) Graph $f(y)$ and $F(y)$.
(d) Use $F(y)$ in (b) to find $F(-1)$, $F(0)$, and $F(1)$.
(e) Find $P(0 \leq Y \leq .5)$.

4.3. Let Y have the density function given by

$$f(y) = \begin{cases} .2 & -1 < y \leq 0, \\ .2 + cy, & 0 < y \leq 1, \\ 0, & \text{elsewhere.} \end{cases}$$

Answer (a) through (e), Exercise 4.2.

4.4. Using Table 3, Appendix III, calculate the area under the normal curve between
(a) $z = 0$ and $z = 1.2$.
(b) $z = 0$ and $z = -.9$.

4.5. Repeat Exercise 4.4 for
(a) $z = 0$ and $z = 1.46$.
(b) $z = 0$ and $z = -.42$.

4.6. Repeat Exercise 4.4 for
(a) $z = .3$ and $z = 1.56$.
(b) $z = .2$ and $z \doteq -.2$.
(c) $z = -1.56$ and $z = -.2$.

4.7. Find the probability that Z is greater than $-.75$.

4.8. Find the probability that Z is less than 1.35.

4.9. Find a z_0 such that $P(Z > z_0) = .5$.

4.10. Find a z_0 such that $P(Z < z_0) = .8643$.

4.11. Find the probability that Z lies between $z = .6$ and $z = 1.67$.

4.12. Find a z_0 such that $P(-z_0 < Z < z_0) = .90$.

4.13. Find a z_0 such that $P(-z_0 < Z < z_0) = .99$.

4.14. Let Y be a normally distributed random variable with mean equal to 7 and standard deviation equal to 1.5. If a value of Y is chosen at random from the population, find the probability that Y falls between $Y = 8$ and $Y = 9$.

4.15. The grade-point averages of a large population of college students are approximately normally distributed with mean equal to 2.4 and a standard deviation equal to .8. What fraction of the students will possess a grade-point average in excess of 3.0?

4.16. Refer to Exercise 4.15. If students possessing a grade-point average equal to or less than 1.9 are dropped from college, what percentage of the students will be dropped?

4.17. Refer to Exercise 4.15. Suppose that three students are randomly selected from the student body. What is the probability that all three will possess a grade-point average in excess of 3.0?

4.18. The length of life of a type of automatic washer is approximately normally distributed with mean and standard deviation equal to 3.1 and 1.2 years, respectively. If this type of washer is guaranteed for 1 year, what fraction of original sales will require replacement?

4.19. The average length of time required for a college achievement test was found to equal 70 minutes, with a standard deviation of 12 minutes. When should the test be terminated if we wish to allow sufficient time for 90 percent of the students to complete the test? (Assume that the time required to complete the test is normally distributed.)

4.20. A soft-drink machine can be regulated so that it discharges an average of μ ounces per cup. If the ounces of fill are normally distributed with standard deviation equal to .3 ounce, give the setting for μ so that 8-ounce cups will overflow only 1 percent of the time.

4.21. A manufacturing plant utilizes 3000 electric light bulbs that have a length of life which is normally distributed with mean and standard deviation equal to 500 and 50 hours, respectively. In order to minimize the number of bulbs that burn out during operating hours, all the bulbs are replaced after a given period of operation. How often should the bulbs be replaced if we wish not more than 1 percent of the bulbs to burn out between replacement periods?

4.22. A machining operation produces bearings with a diameter that is normally distributed with mean and standard deviation equal to 3.0005 and .0010, respectively. Customer specifications require the bearing diameters to lie in the interval $3.000 \pm .0020$. Those outside the interval are considered scrap and must be remachined or used as stock for smaller bearings. With the existing machine setting, what fraction of total production will be scrap?

4.23. Refer to Exercise 4.22. Suppose that five bearings are drawn from production. What is the probability that at least one will be defective?

4.24. Find the mean and variance for the random variable, Exercise 4.1.

4.25. Find the mean and variance for the random variable, Exercise 4.2.

4.26. Find the mean and variance for the random variable, Exercise 4.3.

4.27. Find the moment-generating function for the uniform random variable. Differentiate to find μ'_1 and μ'_2.

4.28. Let Y have density function

$$f(y) = \begin{cases} cye^{-2y}, & 0 \leq y \leq \infty, \\ 0, & \text{elsewhere.} \end{cases}$$

(a) Give the mean and variance for Y.
(b) Give the moment-generating function for Y.
(c) Find the value of c.

4.29. Use the fact that

$$e^z = 1 + z + \frac{z^2}{2!} + \frac{z^3}{3!} + \frac{z^4}{4!} + \cdots$$

to expand the moment-generating function, Example 4.14, into a series to find μ_1, μ_2, μ_3, and μ_4 for the normal random variable.

4.30. Give the mean, variance, and moment-generating function for the exponential density

$$f(y) = \begin{cases} \frac{1}{b} e^{-y/b}, & y > 0, \\ 0, & \text{elsewhere.} \end{cases}$$

4.31. Find $\mu_k' = E(X^k)$ for the beta random variable. Then find the mean and variance for the beta random variable.

4.32. Find $P(|Y - \mu| \leq 2\sigma)$ for Exercise 4.1. Compare with the comparable probabilistic statements given by Tchebysheff's theorem and the empirical rule.

4.33. Find $P(|Y - \mu| \leq 2\sigma)$ for the uniform random variable. Compare with the comparable probabilistic statements given by Tchebysheff's theorem and the empirical rule.

4.34. Find $P(|Y - \mu| \leq 2\sigma)$ for the exponential random variable. Compare with the comparable probabilistic statements given by Tchebysheff's theorem and the empirical rule.

4.35. Let Y have the exponential density shown in Exercise 4.30. For any two positive constants c and d, show that

$$P[Y > (c + d) | Y > c] = P(Y > d).$$

This is a unique property of the exponential density.

4.36. The number of arrivals, n, at a supermarket checkout counter in the time interval 0 to t follows a Poisson probability distribution with mean λt. Let T denote the length of time until the first arrival. Find the density function for T. [*Note:* $P(T > t_0) = P(n = 0; t = t_0)$.]

4.37. Find the moment-generating function for the normally distributed random variable with mean μ and variance σ^2.

4.38. The median value of a continuous random variable is that value y such that $F(y) = .5$. Find the median value of the random variable in Exercise 4.1.

4.39. The random variable Y, where

$$f(y) = \frac{my^{m-1}}{\alpha} e^{-y^m/\alpha}, \qquad 0 \le y < \infty;\alpha, m > 0,$$

is called a *Weibull random variable.* The Weibull density function provides a good model for the distribution of lengths of time to failure for many mechanical devices and biological plants and animals. Find the mean and variance for a Weibull random variable with $m = 2$.

4.40. The life length Y of a certain component being used in a complex electronic system is known to have an exponential density, Exercise 4.30, with $b = 100$ hours. The component is replaced at failure or at age 200 hours, whichever comes first.
(a) Find the distribution function for X, the length of time that the component is in use.
(b) Find $E(X)$.

4.41. Refer to Exercise 4.36. The average number of auto accidents at a given intersection per month is $\lambda = 3$. Suppose that no accidents occur for a whole month. Is this a "rare" event? (Compute the probability of this event.)

4.42. Graph the beta function for $\alpha = 3$, $\beta = 2$. If Y has this beta density, find $P(.1 \le Y \le .2)$ using binomial probabilities to evaluate $F(y)$.

4.43. A retail grocer has a daily demand, Y, for a certain food sold by the pound, where Y, measured in hundreds of pounds, has probability density

$$f(y) = \begin{cases} 3y^2, & 0 \le y \le 1, \\ 0, & \text{elsewhere.} \end{cases}$$

(He cannot stock over 100 pounds.) The grocer wants to order $100k$ pounds of food. He buys the food at 6 cents per pound and sells it at 10 cents per pound. What value of k will maximize his expected daily profit?

4.44. Let X have a moment-generating function given by $m_X(t)$ and let Y be defined as $aX + b$ for finite constants a and b. Show that the moment-generating function for Y is given by

$$m_Y(t) = e^{bt} m_X(at).$$

4.45. Suppose that Y is a normally distributed random variable with mean μ and variance σ^2. Use the results of Exercise 4.44 and Example 4.14 to find the moment-generating function, mean, and variance of
(a) Y.
(b) $Z = \dfrac{Y - \mu}{\sigma}$.

4.46. We can show that the normal density function integrates to unity by showing that

$$\frac{1}{\sqrt{2\pi}} \int_{-\infty}^{\infty} e^{-(1/2)uy^2}\, dy = \frac{1}{\sqrt{u}}.$$

This, in turn, can be shown by considering the product of two such integrals,

$$\frac{1}{2\pi}\left(\int_{-\infty}^{\infty} e^{-(1/2)uy^2}\, dy\right)\left(\int_{-\infty}^{\infty} e^{-(1/2)ux^2}\, dx\right)$$

$$= \frac{1}{2\pi}\int_{-\infty}^{\infty}\int_{-\infty}^{\infty} e^{-(1/2)u(x^2+y^2)}\, dx\, dy.$$

By transforming to polar coordinates, show that the above double integral is equal to $1/u$.

4.47. The function $\Gamma(u)$ is defined by

$$\Gamma(u) = \int_0^{\infty} y^{u-1} e^{-y}\, dy.$$

Integrate by parts to show that

$$\Gamma(u) = (u - 1)\Gamma(u - 1).$$

Hence, if n is a positive integer, it follows that $\Gamma(n) = (n - 1)!$.

4.48. Show that $\Gamma(1/2) = \sqrt{\pi}$ by writing

$$\Gamma(1/2) = \int_0^\infty y^{-1/2} e^{-y}\, dy,$$

making the transformation $y = (1/2)x^2$, and employing the result of Exercise 4.46.

4.49. The function $B(\alpha, \beta)$ is defined by

$$B(\alpha, \beta) = \int_0^1 y^{\alpha-1}(1-y)^{\beta-1}\, dy.$$

(a) Letting $y = \sin^2\theta$, show that

$$B(\alpha, \beta) = 2\int_0^{\pi/2} \sin^{2\alpha-1}\theta \cos^{2\beta-1}\theta\, d\theta.$$

(b) Write $\Gamma(\alpha)\Gamma(\beta)$ as a double integral, transform to polar coordinates, and conclude that

$$B(\alpha, \beta) = \frac{\Gamma(\alpha)\Gamma(\beta)}{\Gamma(\alpha+\beta)}.$$

4.50. The lifetime, X, of a certain electronic component is a random variable with density function

$$f(x) = \begin{cases} \dfrac{1}{100} e^{-x/100}, & x > 0, \\ 0, & \text{elsewhere.} \end{cases}$$

Three of these components operate independently in a piece of equipment. The equipment fails if at least two of the components fail. Find the probability that the equipment operates for at least 200 hours without failure.

4.51. The weekly repair cost, Y, for a certain machine has a probability density function given by

$$f(y) = \begin{cases} 3(1 - y)^2, & 0 < y < 1, \\ 0, & \text{elsewhere,} \end{cases}$$

with measurements in hundreds of dollars. How much money should be budgeted each week for repair costs so that the actual cost will only exceed the budgeted amount 10 percent of the time?

5

Multivariate Probability Distributions

5.1 *Introduction*

The intersection of two or more events is frequently of interest to an experimenter. For example, the gambler playing black-jack is interested in the event that he draws both an "ace" and a "face card" from a 52-card deck. The biologist, observing the number of animals surviving in a litter, is concerned with the intersection of

A: the litter contains n animals,
B: y animals survive.

Similarly, the observation of both height and weight on an individual represent the intersection of a specific pair of height–weight measurements.

Most important to statisticians are the intersections that occur when sampling. Suppose that $Y_1, Y_2, \ldots, Y_n$ denote the outcomes on n successive trials of an experiment. For example, this sequence could represent the weights of n people or the measurements of n physical characteristics of a single person. A specific set of outcomes, or sample measurements, gives rise to the intersection of the n events $(Y_1 = y_1, Y_2 = y_2, \ldots, Y_n = y_n)$, which we will denote as $(y_1, y_2, \ldots, y_n)$. Then in order to make inferences about the population from which the sample was drawn, we will wish to calculate the probability of the intersection $(y_1, y_2, \ldots, y_n)$.

A review of the role probability plays in making inferences, Section 2.2, emphasizes the need for acquiring the probability of the observed sample or, equivalently, the probability of the intersection of a set of numerical events. Knowledge of this probability is fundamental to making an inference about the population from which the sample was drawn. Indeed, this need motivates the discussion of multivariate probability distributions.

5.2 Bivariate and Multivariate Probability Distributions

Many random variables can be defined over the same sample space. For example, consider the experiment of tossing a pair of dice. The sample space contains 36 sample points, corresponding to the $mn = (6)(6) = 36$ ways in which numbers may appear on the faces of the dice. Any one of the following random variables could be defined over the sample space and might be of interest to the experimenter:

Y_1: the number of dots appearing on die 1,
Y_2: the number of dots appearing on die 2,
Y_3: the sum of the number of dots on the dice,
Y_4: the product of the numbers of dots appearing on the dice.

Many other random variables could be defined. Although not particularly pertinent, the reader will note that

$$Y_3 = Y_1 + Y_2 \quad \text{and} \quad Y_4 = Y_1 Y_2.$$

The 36 sample points associated with the experiment are equiprobable and correspond to the 36 numerical events (y_1, y_2). Thus throwing a pair of 1s would be the simple event, (1, 1). Throwing a 2 on die 1 and a 3 on die 2 would be the simple event, (2, 3). Because all pairs (y_1, y_2) occur with the same relative frequency, we would assign a probability of 1/36 to each sample point. For this simple example, the intersection (y_1, y_2) contains only one sample point. Hence the bivariate probability function is

$$p(y_1, y_2) = 1/36 \qquad y_1 = 1, 2, \ldots, 6; \quad y_2 = 1, 2, \ldots, 6.$$

The bivariate probability distribution for the die-tossing experiment, which allocates $p(y_1, y_2) = 1/36$ to all pairs of values of y_1 and y_2, is shown in Figure 5.1.

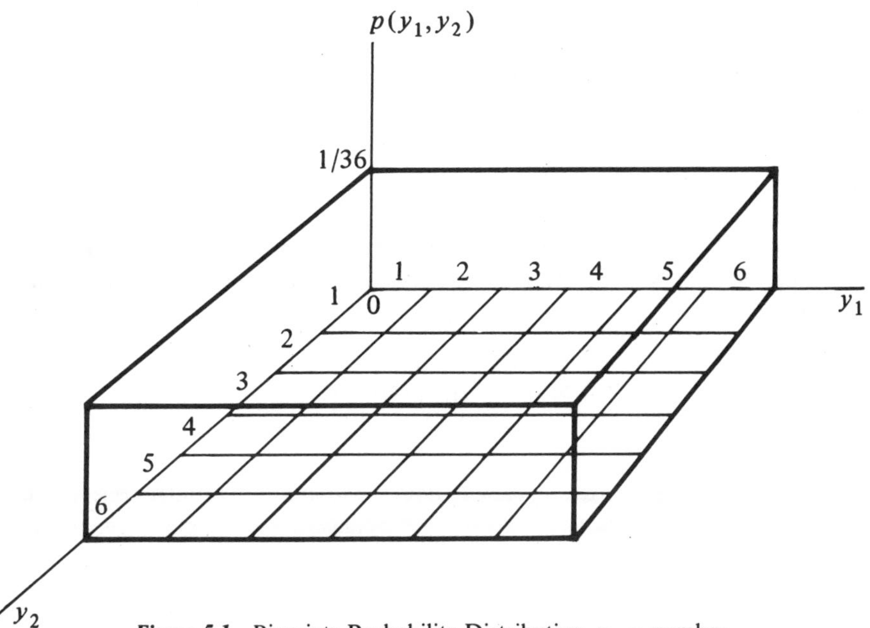

Figure 5.1 Bivariate Probability Distribution, y_1 = number of dots on die 1, y_2 = number of dots on die 2

> ***Definition 5.1:*** *Let Y_1 and Y_2 be discrete random variables. The* joint *(or bivariate)* probability distribution *for Y_1 and Y_2 is given by*
>
> $$p(y_1, y_2) = P(Y_1 = y_1, Y_2 = y_2)$$
>
> *defined for all real numbers y_1 and y_2. The function $p(y_1, y_2)$ will be referred to as the* joint probability function.

Similar to the univariate case, a joint (bivariate) distribution function is defined by

$$F(a, b) = P(Y_1 \leq a, Y_2 \leq b).$$

For two discrete variables, Y_1 and Y_2, $F(a, b)$ has the form

$$F(a, b) = \sum_{y_1=-\infty}^{a} \sum_{y_2=-\infty}^{b} p(y_1, y_2).$$

Two random variables will be said to be jointly continuous if their joint distribution function is continuous in both arguments.

Definition 5.2: *Let Y_1 and Y_2 be continuous random variables with joint distribution function, $F(a, b)$. If there exists a nonnegative function $f(a, b)$ such that*

$$F(a, b) = \int_{-\infty}^{a} \int_{-\infty}^{b} f(y_1, y_2)\, dy_2\, dy_1,$$

for any real numbers a and b, then Y_1 and Y_2 are said to be jointly continuous random variables. *The function $f(y_1, y_2)$ is called the* joint density function.

Bivariate cumulative distribution functions satisfy a set of properties similar to those specified for the univariate cumulative distribution function. That is,

1. $F(-\infty, -\infty) = 0.$
2. $F(\infty, \infty) = 1.$
3. *If $a_2 \geq a_1$ and $b_2 \geq b_1$, then*

$$F(a_2, b_2) - F(a_2, b_1) - F(a_1, b_2) + F(a_1, b_1) \geq 0.$$

Note that the expression in property 3 is simply $P(a_1 < Y_1 \leq a_2, b_1 < Y_2 \leq b_2)$.

For the die-tossing experiment,

$$F(2, 3) = P(Y_1 \leq 2, Y_2 \leq 3) = p(1, 1) + p(1, 2) + p(1, 3)$$
$$+ p(2, 1) + p(2, 2) + p(2, 3).$$

Since $p(y_1, y_2) = 1/36$ for all pairs of values of y_1 and y_2, $F(2, 3) = 6/36 = 1/6$. Similarly, $P(2 \leq Y_1 \leq 3, 1 \leq Y_2 \leq 2)$ is the sum of the probabilities of the intersections of the numerical events, (y_1, y_2), which satisfy the designated inequality. Thus

$$P(2 \leq Y_1 \leq 3, 1 \leq Y_2 \leq 2)$$
$$= p(2, 1) + p(2, 2) + p(3, 1) + p(3, 2)$$
$$= 4/36 = 1/9.$$

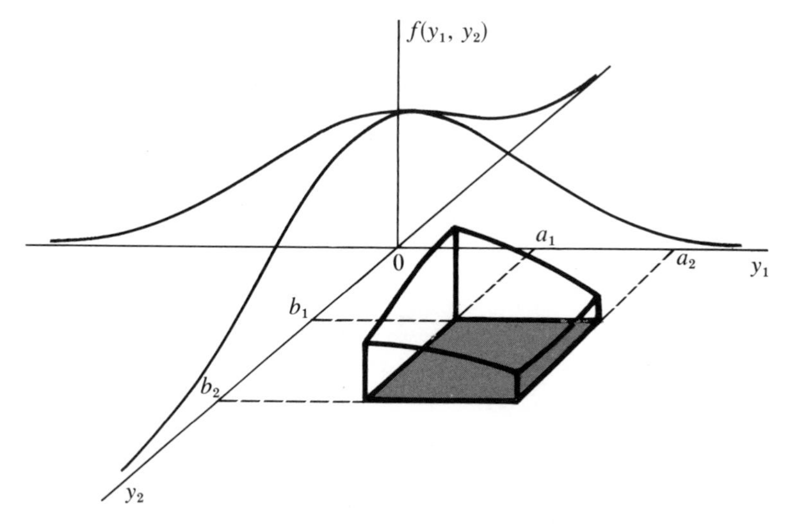

Figure 5.2 A Bivariate Density Function $f(y_1, y_2)$

For the univariate case, areas under the probability density over an interval correspond to probabilities. Similarly, the bivariate probability density function, $f(y_1, y_2)$, traces a probability density surface over the (y_1, y_2) plane (Figure 5.2). Volumes under this surface correspond to probabilities. Thus $P(a_1 \leq Y_1 \leq a_2, b_1 \leq Y_2 \leq b_2)$ is the shaded volume shown in Figure 5.2 and is equal to

$$\int_{b_1}^{b_2} \int_{a_1}^{a_2} f(y_1 y_2)\, dy_1\, dy_2 .$$

Example 5.1: *Consider a model to represent the distribution of particles, say particles of radioactive material, over an area where the distribution is equiprobable from one unit of area to another. That is, let* $f(y_1, y_2) = 1$, *where the area of interest is* $0 \leq y_1 \leq 1, 0 \leq y_2 \leq 1$, *and* $f(y_1, y_2) = 0$ *elsewhere.*
(a) Sketch the probability density surface.
(b) Find $F(.2, .4)$.
(c) Find $P(.1 \leq Y_1 \leq .3, 0 \leq Y_2 \leq .5)$.

Solution:
(a) The sketch is shown in Figure 5.3.

$$\text{(b)}\quad F(.2, .4) = \int_{-\infty}^{.4} \int_{-\infty}^{.2} f(y_1, y_2)\, dy_1\, dy_2$$

$$= \int_0^{.4} \int_0^{.2} (1)\, dy_1\, dy_2$$

$$= \int_0^{.4} y_1 \Big]_0^{.2} dy_2 = .08.$$

The probability, $F(.2, .4)$, *would correspond to the volume under* $f(y_1, y_2) = 1$, *part (a), over the shaded region of the* (y_1, y_2) *plane. From geometrical considerations, the reader will observe that the desired probability (volume) is equal to .08.*

$$\text{(c)}\quad P(.1 \leq Y_1 \leq .3, 0 \leq Y_2 \leq .5)$$

$$= \int_0^{.5} \int_{.1}^{.3} f(y_1, y_2)\, dy_1\, dy_2$$

$$= \int_0^{.5} \int_{.1}^{.3} dy_1\, dy_2 = .10.$$

This probability would correspond to the volume under $f(y_1, y_2) = 1$ *over the region* $.1 \leq y_1 \leq .3, 0 \leq y_2 \leq .5$. *Like (b), the solution for (c) can be obtained using the elementary concepts of geometry. The density or "height" of*

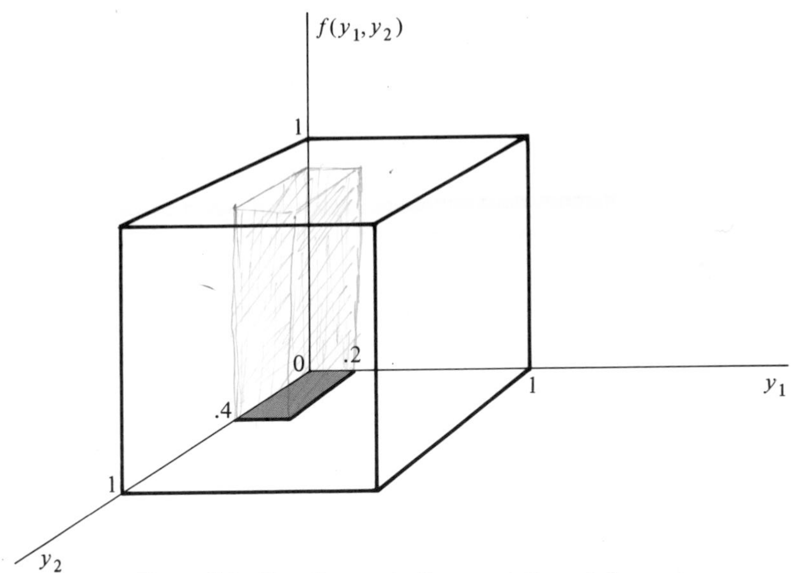

Figure 5.3 The Geometric Representation of $f(y_1, y_2)$, Example 5.1

the probability is equal to 1 *and hence the desired probability* (*volume*) *is*

$$P(.1 \leq Y_1 \leq .3, 0 \leq Y_2 \leq .5) = (.2)(.5)(1) = .1.$$

The intersection of two random variables is a special case (bivariate) of the general *multivariate* case. Thus we can define a probability function (or probability density) for the intersection of n events $(Y_1 = y_1, Y_2 = y_2, \ldots, Y_n = y_n)$. The probability function corresponding to the discrete case is given by

$$p(y_1, y_2, y_3, \ldots, y_n) = P(Y_1 = y_1, Y_2 = y_2, \ldots, Y_n = y_n).$$

The joint density function of $Y_1, \ldots, Y_n$, if it exists, is the function $f(y_1, \ldots, y_n)$, which satisfies

$$P(Y_1 \leq y_1, Y_2 \leq y_2, \ldots, Y_n \leq y_n)$$

$$= F(y_1, \ldots, y_n)$$

$$= \int_{-\infty}^{y_1} \int_{-\infty}^{y_2} \cdots \int_{-\infty}^{y_n} f(t_1, t_2, \ldots, t_n)\, dt_n \cdots dt_1$$

for every set of real numbers $(y_1, \ldots, y_n)$. Multivariate distribution functions, defined by the above equality, will satisfy properties similar to those specified for the bivariate case.

5.3 Marginal and Conditional Probability Distributions

The reader will recall that the values assumed by a discrete random variable represent mutually exclusive events. Consequently, all of the bivariate events $(Y_1 = y_1, Y_2 = y_2)$, represented by (y_1, y_2), are mutually exclusive events. It then follows that the univariate event $(Y_1 = y_1)$ is the union of bivariate events of the type $(Y_1 = y_1, Y_2 = y_2)$, with the union being taken over all possible values for y_2.

For example, reconsider the die-tossing experiment of Section 5.2, where

$$Y_1 = \text{number of dots on the upper face of die 1}$$

and

$$Y_2 = \text{number of dots on the upper face of die 2.}$$

Then

$$\begin{aligned}
P(Y_1 = 1) &= p(1,1) + p(1,2) + p(1,3) + \cdots + p(1,6) \\
&= 1/36 + 1/36 + 1/36 + \cdots + 1/36 = 6/36 = 1/6, \\
P(Y_1 = 2) &= p(2,1) + p(2,2) + p(2,3) + \cdots + p(2,6) = 1/6, \\
P(Y_1 = 6) &= p(6,1) + p(6,2) + p(6,3) + \cdots + p(6,6) = 1/6.
\end{aligned}$$

Expressed in summation notation,

$$p(y_1) = \sum_{y_2=1}^{6} p(y_1, y_2).$$

Similarly,

$$p(y_2) = \sum_{y_1=1}^{6} p(y_1, y_2).$$

Summation in the discrete case corresponds to integration in the continuous case, and so we are led to the following definition.

Definition 5.3:

(*a*) *Let Y_1 and Y_2 be jointly discrete random variables with probability function $p(y_1, y_2)$. Then, the* marginal probability functions *of Y_1 and Y_2, respectively, are given by*

$$p_1(y_1) = \sum_{y_2} p(y_1, y_2)$$

and

$$p_2(y_2) = \sum_{y_1} p(y_1, y_2).$$

(*b*) *Let Y_1 and Y_2 be jointly continuous random variables with joint density function $f(y_1, y_2)$. Then the* marginal density functions *of Y_1 and Y_2, respectively, are given by*

$$f_1(y_1) = \int_{-\infty}^{\infty} f(y_1, y_2)\, dy_2$$

and

$$f_2(y_2) = \int_{-\infty}^{\infty} f(y_1, y_2)\, dy_1.$$

The term "marginal," as applied to the univariate probability distributions of Y_1 and Y_2, has intuitive meaning. Finding $p(y_1)$ implies summing $p(y_1, y_2)$ over all values of y_2 and hence accumulating the probabilities on the y_1 axis (or margin).

Example 5.2: *Let*

$$f(y_1, y_2) = \begin{cases} 2y_1, & 0 \le y_1 \le 1;\quad 0 \le y_2 \le 1, \\ 0, & \text{elsewhere.} \end{cases}$$

Sketch $f(y_1, y_2)$ and find the marginal density functions for Y_1 and Y_2.

Solution: *Viewed geometrically, $f(y_1, y_2)$ traces a surface shaped like a wedge. A sketch is presented in Figure 5.4.*

Before applying Definition 5.3 to find $f_1(y_1)$ and $f_2(y_2)$, we will use Figure 5.4 to visualize the result. If the probability represented by the "wedge" were accumulated on the y_1 axis (summing the probability along lines parallel to the

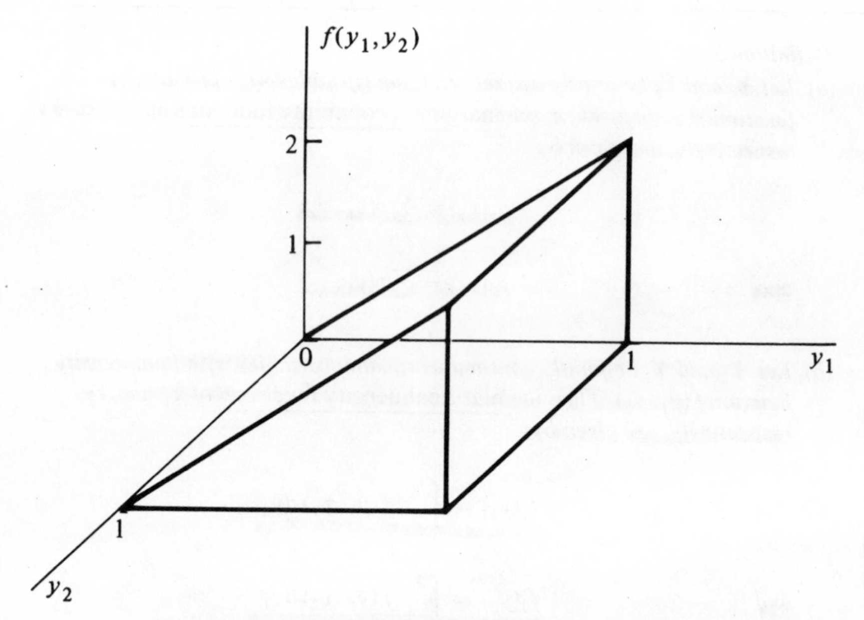

Figure 5.4 The Geometric Representation of $f(y_1, y_2)$, Example 5.2

y_2 axis), the result would be a triangular probability distribution that would look like the side of the wedge in Figure 5.4. If accumulated along the y_2 axis (summing along lines parallel to the y_1 axis), the resulting distribution would be uniform. We will confirm the visual solution by applying Definition 5.3. Then

$$f_1(y_1) = \int_{-\infty}^{\infty} f(y_1, y_2)\, dy_2 = \int_0^1 2y_1\, dy_2 = 2y_1y_2\Big]_0^1$$

$$= 2y_1, \qquad 0 \le y_1 \le 1.$$

Similarly,

$$f_2(y_2) = \int_{-\infty}^{\infty} f(y_1, y_2)\, dy_1 = \int_0^1 2y_1\, dy_1 = y_1^2\Big]_0^1$$

$$= 1, \qquad 0 \le y_2 \le 1.$$

The reader will note that graphs of $f_1(y_1)$ and $f_2(y_2)$ trace triangular and uniform probability distributions, respectively, as expected.

The multiplicative law (Section 2.9) gives the probability of the intersection, $A \cap B$, as

$$P(A \cap B) = P(A)P(B|A),$$

where $P(A)$ is the unconditional probability of A and $P(B|A)$ is the probability of B given A has occurred. Now consider the intersection of the two numerical events, $(Y_1 = y_1) \cap (Y_2 = y_2)$, represented by the bivariate event, (y_1, y_2). Then it follows directly from the multiplicative law of probability that the bivariate probability for the intersection (y_1, y_2) is

$$\begin{aligned} p(y_1, y_2) &= p_1(y_1)p(y_2|y_1) \\ &= p_2(y_2)p(y_1|y_2). \end{aligned}$$

The probabilities $p_1(y_1)$ and $p_2(y_2)$ are those associated with the univariate probability distributions for Y_1 and Y_2 (content of Chapter 3). The meaning of the conditional probabilities $p(y_1|y_2)$ and $p(y_2|y_1)$ is given in Chapter 2. Thus $p(y_1|y_2)$ is the probability that the random variable, Y_1, takes a specific value, given a value of Y_2. That is,

The Conditional Discrete Probability Function for Y_1 Given Y_2

$$\begin{aligned} p(y_1|y_2) &= P(Y_1 = y_1|Y_2 = y_2) \\ &= \frac{P(Y_1 = y_1, Y_2 = y_2)}{P(Y_2 = y_2)} = \frac{p(y_1, y_2)}{p_2(y_2)}, \end{aligned}$$

provided $p_2(y_2) > 0$.

Thus $p(y_1 = 2|y_2 = 3)$ is the conditional probability that $Y_1 = 2$ given that $Y_2 = 3$. A similar interpretation can be attached to the conditional probability $p(y_2|y_1)$.

In the continuous case we can obtain an appropriate analogue of the conditional probability function $p(y_1|y_2)$, but it is not obtained in such a straightforward manner. If Y_1 and Y_2 are continuous, $P(Y_1 = y_1|Y_2 = y_2)$ cannot be defined as in the discrete case since both events involved have zero probability. The following considerations, however, do lead to a useful and consistent definition for a condition density function.

Assuming that Y_1 and Y_2 are jointly continuous with density $f(y_1, y_2)$, we might be interested in a probability of the form

$$P(Y_1 \leq y_1 | Y_2 = y_2) = F(y_1 | y_2),$$

which, as a function of y_1 for fixed y_2, we will call the *conditional distribution function* of Y_1 given $Y_2 = y_2$. If we could take $F(y_1|y_2)$, multiply by $P(Y_2 = y_2)$ for each possible value of Y_2, and sum all the resulting probabilities, we would obtain $F(y_1)$. This is not possible since the number of values for y_2 is uncountable and all probabilities $P(Y_2 = y_2)$ are zero. But we can do something analogous by multiplying by $f_2(y_2)$ and then integrating to obtain

$$F(y_1) = \int_{-\infty}^{\infty} F(y_1|y_2) f_2(y_2)\, dy_2.$$

The quantity $f_2(y_2)\, dy_2$ can be thought of as the approximate probability that Y_2 takes on a value in a small interval about y_2, and the integral is a generalized sum.

Now, from previous considerations we know that

$$\begin{aligned} F(y_1) &= \int_{-\infty}^{y_1} f_1(t_1)\, dt_1 \\ &= \int_{-\infty}^{y_1} \left[\int_{-\infty}^{\infty} f(t_1, y_2)\, dy_2 \right] dt_1 \\ &= \int_{-\infty}^{\infty} \int_{-\infty}^{y_1} f(t_1, y_2)\, dt_1\, dy_2. \end{aligned}$$

From the expressions for $F(y_1)$ we must have

$$F(y_1|y_2) f_2(y_2) = \int_{-\infty}^{y_1} f(t_1, y_2)\, dt_1$$

or

$$F(y_1|y_2) = \int_{-\infty}^{y_1} \frac{f(t_1, y_2)}{f_2(y_2)}\, dt_1.$$

We will call the integrand of the above expression the "conditional density function" of Y_1 given $Y_2 = y_2$ and denote it by $f(y_1|y_2)$.

Definition 5.4: *Let Y_1 and Y_2 be jointly continuous random variables with joint density $f(y_1, y_2)$ and marginal densities $f_1(y_1)$ and $f_2(y_2)$, respectively. Then the conditional density of Y_1 given $Y_2 = y_2$ is given by*

$$f(y_1|y_2) = \begin{cases} \dfrac{f(y_1, y_2)}{f_2(y_2)}, & f(y_2) > 0, \\ 0, & \text{elsewhere}, \end{cases}$$

and the conditional density of Y_2 given $Y_1 = y_1$ is given by

$$f(y_2|y_1) = \begin{cases} \dfrac{f(y_1, y_2)}{f_1(y_1)}, & f(y_1) > 0, \\ 0, & \text{elsewhere}. \end{cases}$$

Example 5.3: *A soft-drink machine has a random amount Y_2 in supply at the beginning of a given day and dispenses a random amount Y_1 during the day (with measurements in gallons). It is not resupplied during the day and hence $Y_1 \leq Y_2$. It has been observed that Y_1 and Y_2 have joint density*

$$f(y_1, y_2) = \begin{cases} 1/2, & 0 \leq y_1 \leq y_2; 0 \leq y_2 \leq 2, \\ 0, & \text{elsewhere}. \end{cases}$$

That is, the points (y_1, y_2) are uniformly distributed over the triangle with the given boundaries. Find the conditional density of Y_1 given $Y_2 = y_2$. Evaluate the probability that less than 1/2 gallon is sold, given that the machine contains 1 gallon at the start of the day.

Solution: *The marginal density of Y_2 is given by*

$$f_2(y_2) = \int_{-\infty}^{\infty} f(y_1, y_2)\, dy_1,$$

$$f_2(y_2) = \begin{cases} \displaystyle\int_0^{y_2} (1/2)\, dy_1 = (1/2)y_2, & y_1 \leq y_2 \leq 2, \\ 0, & \text{elsewhere}. \end{cases}$$

By Definition 5.4,

$$f_1(y_1|y_2) = \frac{f(y_1, y_2)}{f_2(y_2)},$$

$$f(y_1|y_2) = \begin{cases} \dfrac{(1/2)}{(1/2)(y_2)} = \dfrac{1}{y_2}, & 0 < y_1 \le y_2 \le 2, \\ 0, & \text{elsewhere.} \end{cases}$$

The probability of interest is

$$P(Y_1 \le 1/2 | Y_2 = 1) = \int_{-\infty}^{1/2} f(y_1|y_2 = 1)\, dy_1$$

$$= \int_0^{1/2} (1)\, dy_1 = 1/2.$$

Note that if the machine had contained 2 gallons at the start of the day, then

$$P(Y_1 \le 1/2 | Y_2 = 2) = \int_0^{1/2} (1/2)\, dy_1$$

$$= 1/4.$$

Thus the amount sold is highly dependent on the amount in supply.

5.4 Independent Random Variables

In Example 5.3 we saw two random variables that were dependent in the sense that probabilities associated with Y_1 depended on the value of Y_2. We now will present a formal definition of independence and dependence of random variables.

Two events, A and B, are independent if $P(AB) = P(A)P(B)$. When discussing random variables we are often concerned with events of the type $(a \le Y_1 \le b)$ and $(c \le Y_2 \le d)$. To be consistent with the earlier definition of independent events we would like to have

$$P(a \le Y_1 \le b, c \le Y_2 \le d) = P(a \le Y_1 \le b)P(c \le Y_2 \le d)$$

if Y_1 and Y_2 are independent. All such events of this type can be accounted for in the following definition.

Definition 5.5: *Let Y_1 have distribution function $F_1(y_1)$, Y_2 have distribution function $F_2(y_2)$ and Y_1 and Y_2 have joint distribution function $F(y_1, y_2)$. Then Y_1 and Y_2 are said to be* independent *if*

$$F(y_1, y_2) = F_1(y_1)F_2(y_2)$$

for every pair of real numbers (y_1, y_2).

If Y_1 and Y_2 are discrete with joint probability function $p(y_1, y_2)$ and marginals $p_1(y_1)$ and $p_2(y_2)$, respectively, then the above definition implies

$$p(y_1, y_2) = p_1(y_1)p_2(y_2)$$

for all real numbers (y_1, y_2).

If Y_1 and Y_2 are continuous with joint density function $f(y_1, y_2)$ and marginal densities $f_1(y_1)$ and $f_2(y_2)$, respectively, then the above definition implies

$$f(y_1, y_2) = f_1(y_1)f_2(y_2)$$

for all real numbers (y_1, y_2).

If Y_1 and Y_2 are not independent, they are said to be dependent.

We now illustrate the concept of independence with some examples.

Example 5.4: *For the die-tossing problem of Section 5.2, show that Y_1 and Y_2 are independent.*

Solution: *In this problem, each of the 36 sample points was given equal probability of 1/36. Consider, for example, the point (1, 2). We know that $p(1, 2) = 1/36$. Also, $p_1(1) = P(Y_1 = 1) = 1/6$ and $p_2(2) = P(Y_2 = 2) = 1/6$. Hence*

$$p(1, 2) = p_1(1)p_2(2).$$

The same is true for all other points, and it follows that Y_1 and Y_2 are independent.

Example 5.5: *Let*

$$f(y_1 y_2) = \begin{cases} 4y_1 y_2, & 0 \le y_1 \le 1; \quad 0 \le y_2 \le 1, \\ 0, & \textit{elsewhere.} \end{cases}$$

Show that Y_1 and Y_2 are either independent or dependent.

Solution:

$$f_1(y_1) = \int_0^1 f(y_1, y_2)\, dy_2 = \int_0^1 4y_1 y_2\, dy_2$$

$$= 4y_1 \frac{y_2^2}{2}\Big]_0^1 = 2y_1, \qquad 0 \le y_1 \le 1.$$

Similarly,

$$f_2(y_2) = \int_0^1 f(y_1, y_2)\, dy_1 = 2y_2, \qquad 0 \le y_2 \le 1.$$

Hence

$$f(y_1, y_2) = f_1(y_1) f_2(y_2),$$

for any real numbers (y_1, y_2) and Y_1 and Y_2 are independent.

Example 5.6: *Let*

$$f(y_1, y_2) = \begin{cases} 2, & 0 \le y_2 \le y_1; 0 \le y_1 \le 1, \\ 0, & \textit{elsewhere.} \end{cases}$$

Show that Y_1 and Y_2 are dependent or independent.

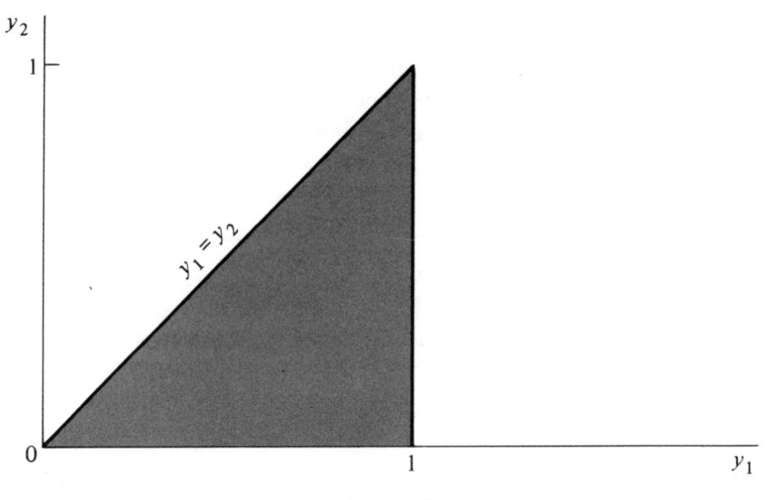

Figure 5.5

Solution: *$f(y_1, y_2) = 2$ over the shaded region shown in Figure 5.5. Therefore,*

$$f_1(y_1) = \int_0^{y_1} f(y_1, y_2)\, dy_2 = \int_0^{y_1} 2\, dy_2 = 2y_2 \Big]_0^{y_1}$$

$$= 2y_1, \qquad 0 \leq y_1 \leq 1.$$

Similarly,

$$f_2(y_2) = \int_{y_2}^{1} f(y_1, y_2)\, dy_1 = \int_{y_2}^{1} 2\, dy_1 = 2y_1 \Big]_{y_2}^{1}$$

$$= 2(1 - y_2), \qquad 0 \leq y_2 \leq 1.$$

Hence

$$f(y_1, y_2) \neq f_1(y_1) f_2(y_2),$$

for some real numbers (y_1, y_2) and Y_1 and Y_2 are dependent.

The reader will note a distinct difference in the limits of integration employed in finding the marginal density functions obtained in Examples 5.5 and 5.6. In finding $f_1(y_1)$, Example 5.6, the limits of integration for y_2 depended on y_1. In contrast, the limits of integration involved in finding the marginal density functions, Example 5.5, were constants. If the limits of integration are constants, the following theorem provides an easy way to show independence of two random variables.

> ***Theorem 5.1:*** *Let Y_1 and Y_2 have a joint density $f(y_1, y_2)$, which is positive, over a region of the form $a < y_1 < b$, $c < y_2 < d$, for constants a, b, c, and d. Then Y_1 and Y_2 are independent random variables if and only if*
>
> $$f(y_1, y_2) = kg(y_1)h(y_2),$$
>
> *where $g(y_1)$ is a nonnegative function of y_1 alone, $h(y_2)$ is a nonnegative function of y_2 alone, and k is a positive constant.*

The proof of this theorem is omitted. (See the references at the end of the chapter.)

Example 5.7: *Let Y_1 and Y_2 have joint density*

$$f(y_1, y_2) = \begin{cases} 2y_1, & 0 \le y_1 \le 1;\ \ 0 \le y_2 \le 1. \\ 0, & \textit{elsewhere.} \end{cases}$$

Are Y_1 and Y_2 independent variables?

Solution: *$f(y_1, y_2)$ is defined over a region with constant boundaries $f(y_1, y_2) = g(y_1)h(y_2)$, where $g(y_1) = 2y_1$ and $h(y_2) = 1$. Therefore, Y_1 and Y_2 are independent random variables.*

Definition 5.5 can easily be generalized to n dimensions. Suppose we have n random variables, $Y_1, \ldots, Y_n$, with Y_i having distribution function $F_i(y)$, $i = 1, \ldots, n$; and $Y_1, \ldots, Y_n$ having joint distribution function $F(y_1, \ldots, y_n)$. Then $Y_1, \ldots, Y_n$ are independent if

$$F(y_1, \ldots, y_n) = F_1(y_1) \cdots F_n(y_n)$$

for all real numbers $(y_1, \ldots, y_n)$, with the obvious equivalent forms for the discrete and continuous cases.

5.5 *The Expected Value of a Function of Random Variables*

The reader need only construct the multivariate analogy to the univariate situation to justify the following definition concerning the expected value of a function of random variables.

Definition 5.6: *Let* $g(Y_1, Y_2, \ldots, Y_k)$ *be a function of the random variables,* $Y_1, Y_2, \ldots, Y_k$, *which possess a probability function* $p(y_1, y_2, \ldots, y_k)$. *Then the expected value of* $g(Y_1, Y_2, \ldots, Y_k)$ *is*

$$E[g(Y_1, Y_2, \ldots, Y_k)] = \sum_{y_k} \cdots \sum_{y_2} \sum_{y_1} g(y_1, y_2, \ldots, y_k) p(y_1, y_2, \ldots, y_k).$$

If $Y_1, Y_2, \ldots, Y_k$ *are continuous random variables with joint density function,* $f(y_1, y_2, \ldots, y_k)$, *then*

$$E[g(Y_1, Y_2, \ldots, Y_k)]$$

$$= \int_{y_k} \cdots \int_{y_2} \int_{y_1} g(y_1, y_2, \ldots, y_k) f(y_1, y_2, \ldots, y_k)\, dy_1\, dy_2 \cdots dy_k.$$

(*Note:* For some functions the expectations as defined above may not exist.)

Example 5.8: *Let* Y_1 *and* Y_2 *have joint density*

$$f(y_1, y_2) = \begin{cases} 2y_1, & 0 \le y_1 \le 1; \quad 0 \le y_2 \le 1. \\ 0, & \textit{elsewhere.} \end{cases}$$

Find $E(Y_1 Y_2)$.

Solution: *From Definition 5.6,*

$$E(Y_1 Y_2) = \int_0^1 \int_0^1 y_1 y_2 f(y_1, y_2)\, dy_1\, dy_2$$

$$= \int_0^1 \int_0^1 y_1 y_2 (2y_1)\, dy_1\, dy_2$$

$$= \int_0^1 y_2 \frac{2y_1^3}{3}\Big]_0^1 dy_2 = \int_0^1 (2/3) y_2\, dy_2$$

$$= \frac{2}{3}\frac{y_2^2}{2}\Big]_0^1 = 1/3.$$

We will show that Definition 5.6 is consistent with Definition 4.6, which defines the expected value of a univariate random variable. Consider two random variables Y_1 and Y_2, with density function $f(y_1, y_2)$. We wish to find the expected value of $g(Y_1, Y_2) = Y_1$.

Then from Definition 5.6,

$$E(Y_1) = \int_{-\infty}^{\infty} \int_{-\infty}^{\infty} y_1 f(y_1, y_2)\, dy_1\, dy_2$$

$$= \int_{-\infty}^{\infty} y_1 \left[\int_{-\infty}^{\infty} f(y_1, y_2)\, dy_2 \right] dy_1 .$$

The quantity within the brackets is by definition the marginal density function for Y_1. Therefore,

$$E(Y_1) = \int_{-\infty}^{\infty} y_1 f_1(y_1)\, dy_1,$$

which agrees with Definition 4.6.

Example 5.9: *Let Y_1 and Y_2 have joint density*

$$f(y_1, y_2) = \begin{cases} 2y_1, & 0 \le y_1 \le 1;\ \ 0 \le y_2 \le 1, \\ 0, & \textit{elsewhere.} \end{cases}$$

Find the expected value of Y_1.

Solution:

$$E(Y_1) = \int_0^1 \int_0^1 y_1(2y_1)\, dy_1\, dy_2$$

$$= \int_0^1 \frac{2y_1^3}{3}\Big]_0^1 dy_2 = \int_0^1 \frac{2}{3} dy_2 = \frac{2}{3} y_2\Big]_0^1 = 2/3.$$

Refer to Figure 5.4 and estimate the expected value of Y_1. The value $E(Y_1) = 2/3$ would appear to be quite reasonable.

Example 5.10: *Examine Figure 5.4 and note that the mean value of Y_2 would appear to equal .5. Let us confirm this visual estimate. Find $E(Y_2)$.*

Solution:

$$E(Y_2) = \int_0^1 \int_0^1 y_2(2y_1)\, dy_1\, dy_2 = \int_0^1 y_2 \frac{2y_1^2}{2}\Big]_0^1 dy_2$$

$$= \int_0^1 y_2\, dy_2 = \frac{y_2^2}{2}\Big]_0^1 = \frac{1}{2}.$$

Example 5.11: *Let Y_1 and Y_2 be random variables with density function*

$$f(y_1, y_2) = \begin{cases} 2y_1, & 0 \le y_1 \le 1;\ \ 0 \le y_2 \le 1, \\ 0, & \text{elsewhere.} \end{cases}$$

Find $V(Y_1)$.

Solution: *The marginal density for Y_1 was obtained in Example 5.2 as $f_1(y_1) = 2y_1, 0 \le y_1 \le 1$.*

Then $V(Y_1) = E(Y_1^2) - [E(Y_1)]^2$

$$E(Y_1^k) = \int_0^1 y_1^k f(y_1)\, dy_1 = \int_0^1 y_1^k(2y_1)\, dy_1,$$

$$= \frac{2y_1^{k+2}}{k+2}\Big]_0^1 = \frac{2}{k+2}.$$

Letting $k = 1$ and $k = 2$, $E(Y_1)$ and $E(Y_1^2)$ are 2/3 and 1/2, respectively. Then $V(Y_1) = E(Y_1^2) - [E(Y_1)]^2 = 1/2 - (2/3)^2 = 1/18$.

5.6 Special Theorems

Theorems concerning the expected value of a constant, the expected value of a constant times a function of random variables, and the expected value of the sum of functions of random variables are similar to those for the univariate case.

Theorem 5.2: *Let c be a constant. Then*

$$E(c) = c.$$

The proof is left to the reader. Follow the steps employed in proving Theorem 4.1.

Theorem 5.3: *Let $g(Y_1, Y_2)$ be a function of the random variables Y_1, Y_2, and let c be a constant. Then*

$$E[cg(Y_1, Y_2)] = cE[g(Y_1, Y_2)].$$

Proof: *The proof follows directly from the fact that the integral (or sum) of a constant times a function of Y_1 and Y_2 is equal to the constant times the integral (or sum). The proof is identical to that for Theorem 4.1.*

Theorem 5.4: *Let Y_1 and Y_2 be random variables with density function $f(y_1, y_2)$, and let $g_1(Y_1, Y_2), g_2(Y_1, Y_2), \ldots, g_k(Y_1, Y_2)$ be functions of Y_1 and Y_2. Then*

$$E[g_1(Y_1, Y_2) + g_2(Y_1, Y_2) + \cdots + g_k(Y_1, Y_2)]$$

$$= E[g_1(Y_1, Y_2)] + E[g_2(Y_1, Y_2)] + \cdots + E[g_k(Y_1, Y_2)].$$

Proof:

$$E[g_1(Y_1, Y_2) + \cdots + g_k(Y_1, Y_2)]$$
$$= \int_{-\infty}^{\infty} \int_{-\infty}^{\infty} [g_1(y_1, y_2) + g_2(y_1, y_2) + \cdots$$
$$+ g_k(y_1, y_2)] f(y_1, y_2)\, dy_1\, dy_2.$$

Because the integral of a sum is equal to the sum of the integrals,

$$E[g_1(Y_1, Y_2) + g_2(Y_1, Y_2) + \cdots + g_k(Y_1, Y_2)]$$
$$= \int_{-\infty}^{\infty} \int_{-\infty}^{\infty} g_1(y_1, y_2) f(y_1, y_2) + \cdots$$
$$+ \int_{-\infty}^{\infty} \int_{-\infty}^{\infty} g_k(y_1, y_2) f(y_1, y_2)\, dy_1\, dy_2$$
$$= E[g_1(Y_1, Y_2)] + E[g_2(Y_1, Y_2)] + \cdots + E[g_k(Y_1, Y_2)].$$

Theorem 5.5: *Let Y_1 and Y_2 be independent random variables with joint density $f(y_1, y_2)$. Let $g(Y_1)$ and $h(Y_2)$ be functions of Y_1 and Y_2, respectively. Then*

$$E[g(Y_1)h(Y_2)] = E[g(Y_1)]E[h(Y_2)]$$

provided the expectations exist.

Proof: *The product $g(Y_1)h(Y_2)$ is a function of Y_1 and Y_2. Hence, by Definition 5.6,*

$$E[g(Y_1)h(Y_2)]$$
$$= \int_{-\infty}^{\infty} \int_{-\infty}^{\infty} g(y_1)h(y_2) f(y_1, y_2)\, dy_2\, dy_1$$
$$= \int_{-\infty}^{\infty} \int_{-\infty}^{\infty} g(y_1)h(y_2) f_1(y_1) f_2(y_2)\, dy_2\, dy_1$$

$$= \int_{-\infty}^{\infty} g(y_1)f_1(y_1)\left[\int_{-\infty}^{\infty} h(y_2)f_2(y_2)\,dy_2\right]dy_1$$

$$= \int_{-\infty}^{\infty} g(y_1)f_1(y_1)E[h(Y_2)]\,dy_1$$

$$= E[g(Y_1)]E[h(Y_2)].$$

5.7 *The Covariance of Two Random Variables*

Intuitively, we think of the dependence of two random variables, Y_1 and Y_2, as implying that one variable, say Y_1, either increases or decreases as Y_2 changes. We shall confine our attention to two measures of dependence, the covariance and the simple coefficient of linear correlation, and will utilize Figure 5.6(a) and (b) to justify their choice as measures of dependence. Figure 5.6(a) and (b) represents plotted points of two random samples of $n = 10$ experimental units drawn from a population. Measurements on Y_1 and Y_2 were made on each experimental unit. If all the points fell along a straight line as indicated in Figure 5.6(a), Y_1 and Y_2 are obviously dependent. In contrast, Figure 5.6(b) would indicate little or no dependence between Y_1 and Y_2.

Suppose that one really knew the values of $E(Y_1) = \mu_1$ and $E(Y_2) = \mu_2$ and located this point on the graph, Figure 5.6. Now locate a plotted point on Figure 5.6(a) and measure the deviations, $(y_1 - \mu_1)$ and $(y_2 - \mu_2)$. Note that

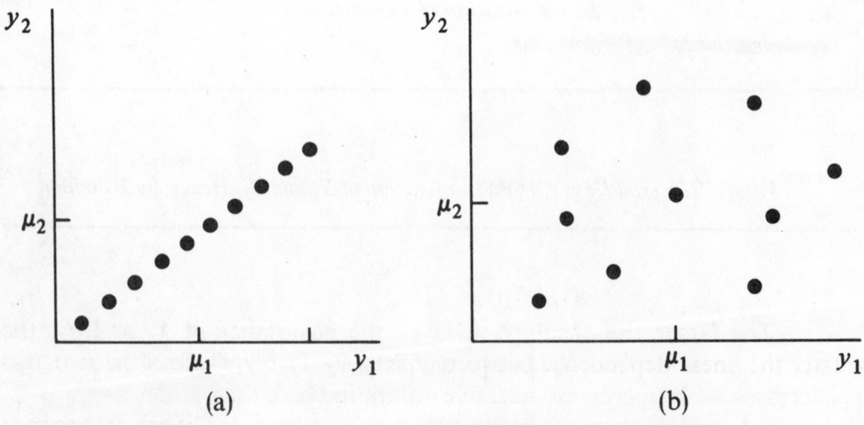

Figure 5.6 Dependent and Independent Observations, (y_1, y_2)

both deviations will assume the same algebraic sign for a particular point and hence that their product, $(y_1 - \mu_1)(y_2 - \mu_2)$, will be positive. This will be true for *all* plotted points on Figure 5.6(a). Points to the right of μ_1 will yield pairs of positive deviations, points to the left will produce pairs of negative deviations, and the average of the product of the deviations, $(y_1 - \mu_1)(y_2 - \mu_2)$, will be "large" and positive. If the linear relation indicated in Figure 5.6(a) had sloped downward to the right, all corresponding pairs of deviations would have been of the opposite sign, and the average value of $(y_1 - \mu_1)(y_2 - \mu_2)$ would have been a large negative number.

The situation described above will not occur for Figure 5.6(b), where little dependence exists between Y_1 and Y_2. Corresponding deviations, $(y_1 - \mu_1)$ and $(y_2 - \mu_2)$, will assume the same algebraic sign for some points and opposite signs for others. Thus the product

$$(y_1 - \mu_1)(y_2 - \mu_2)$$

will be positive for some points, negative for others, and will average to some value near zero.

It is clear from the foregoing discussion that the average value of $(y_1 - \mu_1)(y_2 - \mu_2)$ will provide a measure of the linear dependence of Y_1 and Y_2. This quantity, defined over the bivariate population associated with Y_1 and Y_2, is called the *covariance* of Y_1 and Y_2.

Definition 5.7: *The covariance of Y_1 and Y_2 is defined to be the expected value of $(Y_1 - \mu_1)(Y_2 - \mu_2)$. In the notation of expectation, the covariance will equal*

$$\mathrm{Cov}(Y_1, Y_2) = E[(Y_1 - \mu_1)(Y_2 - \mu_2)],$$

where $\mu_1 = E(Y_1)$ and $\mu_2 = E(Y_2)$.

The larger the absolute value of the covariance of Y_1 and Y_2, the greater the linear dependence between Y_1 and Y_2. Positive values indicate that Y_1 increases as Y_2 increases; negative values indicate that Y_1 decreases as Y_2 increases. A zero value of the covariance would indicate no linear dependence between Y_1 and Y_2.

Unfortunately, it is difficult to employ the covariance as an absolute measure of dependence because its value depends upon the scale of measurement and so it is hard to determine whether a particular covariance is "large" at first glance. This problem can be eliminated by standardizing its value, using the familiar simple coefficient of linear correlation. The population linear coefficient of correlation, ρ, is related to the covariance and is defined as

$$\rho = \frac{\text{Cov}(Y_1, Y_2)}{\sigma_1 \sigma_2},$$

where σ_1 and σ_2 are the standard deviations of Y_1 and Y_2, respectively. Supplemental discussion of correlation may be found in Mendenhall [3].

It can be shown that the coefficient of correlation, ρ, satisfies the inequality $-1 \leq \rho \leq 1$. The proof is outlined in Exercise 5.29. Thus -1 or $+1$ imply perfect correlation, with all points falling on a straight line. A value $\rho = 0$ implies a zero covariance and no correlation. The sign of the correlation coefficient is dependent on the sign of the covariance. Thus a positive coefficient or correlation indicates that Y_2 increases as Y_1 increases. A negative coefficient of correlation implies a decrease in Y_2 as Y_1 increases.

Theorem 5.6: *Let Y_1 and Y_2 be random variables with joint density function $f(y_1, y_2)$. Then*

$$\begin{aligned}\text{Cov}(Y_1, Y_2) &= E[(Y_1 - \mu_1)(Y_2 - \mu_2)] \\ &= E(Y_1 Y_2) - E(Y_1)E(Y_2).\end{aligned}$$

Proof:

$$\begin{aligned}\text{Cov}(Y_1, Y_2) &= E[(Y_1 - \mu_1)(Y_2 - \mu_2)] \\ &= E(Y_1 Y_2 - \mu_1 Y_2 - \mu_2 Y_1 + \mu_1 \mu_2).\end{aligned}$$

From Theorem 5.4, the expected value of a sum is equal to the sum of the expected values, or

$$\text{Cov}(Y_1, Y_2) = E(Y_1 Y_2) - \mu_1 E(Y_2) - \mu_2 E(Y_1) + \mu_1 \mu_2.$$

Because $E(Y_1) = \mu_1$ *and* $E(Y_2) = \mu_2$,

$$\text{Cov}(Y_1, Y_2) = E(Y_1 Y_2) - E(Y_1)E(Y_2)$$

or

$$\text{Cov}(Y_1, Y_2) = E(Y_1 Y_2) - \mu_1\mu_2.$$

Example 5.12: *Let* Y_1 *and* Y_2 *have joint density*

$$f(y_1, y_2) = \begin{cases} 2y_1, & 0 \leq y_1 \leq 1, \quad 0 \leq y_2 \leq 1, \\ 0, & \text{elsewhere.} \end{cases}$$

Find the covariance of Y_1 *and* Y_2.

Solution: *From Example 5.8,* $E(Y_1 Y_2) = 1/3$. *Also from Examples 5.9* and *5.10,* $E(Y_1)$ *and* $E(Y_2)$ *equal* 2/3 *and* 1/2, *respectively. Then*

$$\text{Cov}(Y_1, Y_2) = E(Y_1 Y_2) - \mu_1\mu_2 = 1/3 - (2/3)(1/2) = 0.$$

Example 5.12 furnishes an illustration of a general concept given in Theorem 5.7.

Theorem 5.7: *If* Y_1 *and* Y_2 *are independent random variables, then*

$$\text{Cov}(Y_1, Y_2) = 0.$$

Proof: *We know that*

$$\text{Cov}(Y_1, Y_2) = E(Y_1 Y_2) - \mu_1\mu_2.$$

But, if Y_1 *and* Y_2 *are independent,*

$$E(Y_1 Y_2) = E(Y_1)E(Y_2) = \mu_1\mu_2,$$

from Theorem 5.5. The desired result follows immediately.

Note that the Y_1 and Y_2 of Example 5.12 are independent, and hence their covariance must be zero. The converse of Theorem 5.7 is not true, as will be illustrated in the following example.

Example 5.13: *Let Y_1 and Y_2 be discrete random variables with joint probability distribution as follows.*

	Y_1		
Y_2	-1	0	$+1$
-1	1/16	3/16	1/16
0	3/16	0	3/16
$+1$	1/16	3/16	1/16

Show that Y_1 and Y_2 are dependent and have zero covariance.

Solution: *Calculation of marginal probabilities yields $p_1(-1) = p_2(-1) = 5/16$, $p_1(0) = p_2(0) = 6/16$, and $p_1(1) = p_2(1) = 5/16$. Looking at the upper left-hand cell, we have $p(-1, -1) = 1/16$. Obviously,*

$$p(-1, -1) \neq p_1(-1)p_2(-1),$$

and this is sufficient to show that Y_1 and Y_2 are dependent.

Again looking at the marginal probabilities, we see that $E(Y_1) = E(Y_2) = 0$. Also,

$$\begin{aligned} E(Y_1 Y_2) &= \sum_{y_1} \sum_{y_2} y_1 y_2 p(y_1, y_2) \\ &= (-1)(-1)(1/16) + (0)(-1)(3/16) \\ &\quad + \cdots + (1)(1)(1/16) \\ &= 0. \end{aligned}$$

Thus

$$\begin{aligned} \mathrm{Cov}(Y_1, Y_2) &= E(Y_1 Y_2) - E(Y_1)E(Y_2) \\ &= 0. \end{aligned}$$

This example shows that the converse of Theorem 5.7 is not true. That is, zero covariance does not imply independence.

5.8 The Expected Value and Variance of Linear Functions of Random Variables

We will frequently be concerned with parameter estimators that are linear functions of the sample measurements, $Y_1, Y_2, \ldots, Y_n$. Consequently, we will wish to find the expected value and variance of a linear function,

$$U_1 = a_1 Y_1 + a_2 Y_2 + a_3 Y_3 + \cdots + a_n Y_n = \sum_{i=1}^{n} a_i Y_i,$$

when $a_1, a_2, \ldots, a_n$ are constants and $Y_1, Y_2, \ldots, Y_n$ are random variables.

We also may be interested in the covariance between two such linear functions. The necessary results are summarized in the following theorem.

> ***Theorem 5.8:*** *Let $Y_1, \ldots, Y_n$ and $X_1, \ldots, X_m$ be random variables with $E(Y_i) = \mu_i$ and $E(X_j) = \xi_j$. Define*
>
> $$U_1 = \sum_{i=1}^{n} a_i Y_i, \qquad U_2 = \sum_{j=1}^{m} b_j X_j$$
>
> *for constants $a_1, \ldots, a_n, b_1, \ldots, b_m$. Then*
>
> *(a)* $E(U_1) = \sum_{i=1}^{n} a_i \mu_i,$
>
> *(b)* $V(U_1) = \sum_{i=1}^{n} a_i^2 V(Y_i) + 2 \sum_{i<j}\sum a_i a_j \operatorname{Cov}(Y_i, Y_j),$
>
> *where the double sum is over all pairs (i, j) with $i < j$ and*
>
> *(c)* $\operatorname{Cov}(U_1, U_2) = \sum_{i=1}^{n} \sum_{j=1}^{m} a_i b_j \operatorname{Cov}(Y_i, X_j).$

Proof: *Part (a) follows directly from Theorems 5.3 and 5.4.*

To prove part (b) we appeal to the definition of variance and write

$$\begin{aligned}
V(U_1) &= E[U_1 - E(U_1)]^2 \\
&= E\left[\sum_{i=1}^{n} a_i Y_i - \sum_{i=1}^{n} a_i \mu_i\right]^2 \\
&= E\left[\sum_{i=1}^{n} a_i (Y_i - \mu_i)\right]^2 \\
&= E\left[\sum_{i=1}^{n} a_i^2 (Y_i - \mu_i)^2 + \sum_{i \neq j}\sum a_i a_j (Y_i - \mu_i)(Y_j - \mu_j)\right] \\
&= \sum_{i=1}^{n} a_i^2 E(Y_i - \mu_i)^2 + \sum_{i \neq j}\sum a_i a_j E[(Y_i - \mu_i)(Y_j - \mu_j)].
\end{aligned}$$

By definition of variance and covariance, we then have

$$V(U_1) = \sum_{i=1}^{n} a_i^2 V(Y_i) + \sum_{i \neq j}\sum a_i a_j \operatorname{Cov}(Y_i, Y_j).$$

Note that $\operatorname{Cov}(Y_i, Y_j) = \operatorname{Cov}(Y_j, Y_i)$, *and hence we can write*

$$V(U_1) = \sum_{i=1}^{n} a_i^2 V(Y_i) + 2\sum_{i<j}\sum \operatorname{Cov}(Y_i, Y_j).$$

Part (c) is obtained by similar steps. We have

$$\begin{aligned}
\operatorname{Cov}(U_1, U_2) &= E\{[U_1 - E(U_1)][U_2 - E(U_2)]\} \\
&= E\left[\left(\sum_{i=1}^{n} a_i Y_i - \sum_{i=1}^{n} a_i \mu_i\right)\left(\sum_{j=1}^{m} b_j X_j - \sum_{j=1}^{m} b_j \xi_j\right)\right] \\
&= E\left\{\left[\sum_{i=1}^{m} a_i (Y_i - \mu_i)\right]\left[\sum_{j=1}^{m} b_j (X_j - \xi_j)\right]\right\}
\end{aligned}$$

$$= E\left[\sum_{i=1}^{n}\sum_{j=1}^{m} a_i b_j (Y_i - \mu_i)(X_j - \xi_j)\right]$$

$$= \sum_{i=1}^{n}\sum_{j=1}^{m} a_i b_j E(Y_i - \mu_i)(X_j - \xi_j)$$

$$= \sum_{i=1}^{n}\sum_{j=1}^{m} a_i b_j \operatorname{Cov}(Y_i, X_j).$$

On observing that $\operatorname{Cov}(Y_i, Y_i) = V(Y_i)$, *we can see that part* (b) *is a special case of part* (c).

Example 5.14: *Let* $Y_1, Y_2, \ldots, Y_n$ *be independent random variables with* $E(Y_i) = \mu$ *and* $V(Y_i) = \sigma^2$. *(These variables may denote the outcomes on n independent trials of an experiment.) Defining*

$$\bar{Y} = \frac{1}{n}\sum_{i=1}^{n} Y_i,$$

show that $E(\bar{Y}) = \mu$ *and* $V(\bar{Y}) = \sigma^2/n$.

Solution: *Note that* $\bar{Y}$ *is a linear function with all constants,* a_i, *equal to* $1/n$. *That is,*

$$\bar{Y} = \left(\frac{1}{n}\right)Y_1 + \cdots + \left(\frac{1}{n}\right)Y_n.$$

By Theorem 5.8, part (a),

$$E(\bar{Y}) = \sum_{i=1}^{n} a_i\mu = \mu\sum_{i=1}^{n} a_i$$

$$= \mu\sum_{i=1}^{n}\frac{1}{n} = \frac{n\mu}{n} = \mu.$$

By Theorem 5.8, part (b),

$$V(\bar{Y}) = \sum_{i=1}^{n} a_i^2 V(Y_1) + 2\sum_{i<j}\sum a_i a_j \operatorname{Cov}(Y_i, Y_j),$$

but the covariance terms are all zero since the random variables are independent. Thus

$$V(\bar{Y}) = \sum_{i=1}^{n} \left(\frac{1}{n}\right)^2 \sigma^2 = \frac{n\sigma^2}{n^2} = \frac{\sigma^2}{n}.$$

Example 5.15: *Let Y_1, Y_2, and Y_3 be random variables where $E(Y_1) = 1$, $E(Y_2) = 2$, $E(Y_3) = -1$, $V(Y_1) = 1$, $V(Y_2) = 3$, $V(Y_3) = 5$, $\text{Cov}(Y_1, Y_2) = -4$, $\text{Cov}(Y_1, Y_3) = 1/2$, and $\text{Cov}(Y_2, Y_3) = 2$. Find the expected value and variance of $U = Y_1 - 2Y_2 + Y_3$.*

Solution: *$U = a_1Y_1 + a_2Y_2 + a_3Y_3$, where $a_1 = 1$, $a_2 = -2$, and $a_3 = 1$. Then, by Theorem 5.8,*

$$E(U) = a_1E(Y_1) + a_2E(Y_2) + a_3E(Y_3)$$
$$= (1)(1) + (-2)(2) + (1)(-1) = -4.$$

Similarly,

$$\begin{aligned} V(U) &= a_1^2V(Y_1) + a_2^2V(Y_2) + a_3^2V(Y_3) \\ &\quad + 2a_1a_2\,\text{Cov}(Y_1, Y_2) + 2a_1a_3\,\text{Cov}(Y_1, Y_3) \\ &\quad + 2a_2a_3\,\text{Cov}(Y_2, Y_3) \\ &= (1)^2(1) + (-2)^2(3) + (1)^2(5) + (2)(1)(-2)(-4) \\ &\quad + (2)(1)(1)(1/2) + (2)(-2)(1)(2) \\ &= 27. \end{aligned}$$

Example 5.16: *The number of defectives, Y, in a sample of $n = 10$ items selected from a manufacturing process follows a binomial probability distribution. An estimator of the fraction defective in the lot is the random variable, $\hat{p} = Y/n$. Find the expected value and variance of $\hat{p}$.*

Solution: *$\hat{p}$ is a linear function of a single random variable, Y, where $\hat{p} = a_1Y$ and $a_1 = 1/n$. Then, by Theorem 5.8,*

$$E(\hat{p}) = a_1E(Y) = \frac{1}{n}E(Y).$$

The expected value and variance of a binomial random variable are np and npq, respectively. Substituting for E(Y),

$$E(\hat{p}) = \frac{1}{n}np = p.$$

Thus the expected value of the number of defectives Y, divided by the sample size, is p. Similarly,

$$V(\hat{p}) = a_1^2 V(Y) = \left(\frac{1}{n}\right)^2 npq = \frac{pq}{n}.$$

Example 5.17: *Suppose an urn contains r white balls and (N − r) black balls. A random sample of n balls is drawn without replacement and Y, the number of white balls in the sample, is observed. From Chapter 3 we know that Y has a hypergeometric probability distribution. Find the mean and variance of Y.*

Solution: *We will first observe some characteristics of sampling without replacement. Suppose the sampling is done sequentially and we observe outcomes for $X_1, X_2, \ldots, X_n$, where*

$$X_i = \begin{cases} 1, & \text{if the } i\text{th draw results in a white ball,} \\ 0, & \text{otherwise.} \end{cases}$$

Unquestionably, $P(X_1 = 1) = r/N$. But it is also true that $P(X_2 = 1) = r/N$ since

$$\begin{aligned} P(X_2 = 1) &= P(X_1 = 1, X_2 = 1) + P(X_1 = 0, X_2 = 1) \\ &= P(X_1 = 1)P(X_2 = 1|X_1 = 1) \\ &\quad + P(X_1 = 0)P(X_2 = 1|X_1 = 0) \\ &= \frac{r}{N}\frac{r-1}{N-1} + \frac{N-r}{N}\frac{r}{N-1} \\ &= \frac{r(N-1)}{N(N-1)} = \frac{r}{N}. \end{aligned}$$

The same is true for X_k; *that is,*

$$P(X_k = 1) = \frac{r}{N}, \qquad k = 1, \ldots, n.$$

Thus the probability of drawing a white ball on any draw, given no knowledge of the outcomes on previous draws, is r/N.

In a similar way it can be shown that

$$P(X_j = 1, X_k = 1) = \frac{r(r-1)}{N(N-1)}, \qquad j \neq k.$$

Now, observe that $Y = \sum_{i=1}^{n} X_i$, *and hence*

$$E(Y) = \sum_{i=1}^{n} E(X_i) = n\left(\frac{r}{N}\right).$$

In order to find $V(Y)$ *we need* $V(X_i)$ *and* $\text{Cov}(X_i, X_j)$. *Since* X_i *is* 1 *with probability* r/N *and* 0 *with probability* $1 - r/N$, *it follows that*

$$V(X_i) = \frac{r}{N}\left(1 - \frac{r}{N}\right).$$

Also,

$$\begin{aligned} \text{Cov}(X_i, X_j) &= E(X_iX_j) - E(X_i)E(X_j) \\ &= \frac{r(r-1)}{N(N-1)} - \left(\frac{r}{N}\right)^2 \\ &= -\frac{r}{N}\left(1 - \frac{r}{N}\right)\frac{1}{N-1} \end{aligned}$$

since $X_iX_j = 1$ *if and only if* $X_i = 1$ *and* $X_j = 1$. *From Theorem 5.8 we have that*

$$\begin{aligned} V(Y) &= \sum_{i=1}^{n} V(X_i) + 2\sum\sum_{i<j} \text{Cov}(X_i, X_j) \\ &= n\frac{r}{N}\left(1 - \frac{r}{N}\right) + 2\sum\sum_{i<j}\left[-\frac{r}{N}\left(1 - \frac{r}{N}\right)\frac{1}{N-1}\right] \\ &= n\frac{r}{N}\left(1 - \frac{r}{N}\right) - n(n-1)\frac{r}{N}\left(1 - \frac{r}{N}\right)\frac{1}{N-1} \end{aligned}$$

since there are $n(n-1)/2$ terms in the double summation. A little algebra yields

$$V(Y) = n\frac{r}{N}\left(1 - \frac{r}{N}\right)\frac{N-n}{N-1}.$$

An appreciation for the usefulness of Theorem 5.8 can be gained by trying to find the expected value and variance for the hypergeometric random variable by proceeding directly from the definition of an expectation. The necessary summations are exceedingly difficult to obtain.

5.9 The Multinomial Probability Distribution

A very useful multivariate probability distribution for discrete random variables $Y_1, Y_2, \ldots, Y_k$ is

$$p(y_1, y_2, \ldots, y_k) = \frac{n!}{y_1!y_2!\cdots y_k!}p_1^{y_1}p_2^{y_2}\cdots p_k^{y_k},$$

where $\sum_{i=1}^{k} p_i = 1$ and $\sum_{i=1}^{k} y_i = n$.

The experiment underlying the probability distribution called a multinomial experiment is a generalization of the binomial experiment and is defined by the following five characteristics:

The Multinomial Experiment

1. *The experiment consists of n identical trials.*
2. *The outcome of each trial falls into one of k classes or cells.*
3. *The probability that the outcome of a single trial will fall in a particular cell, say cell i, is p_i $(i = 1, 2, \ldots, k)$, and remains the same from trial to trial. Note that*

$$p_1 + p_2 + p_3 + \cdots + p_k = 1.$$

4. *The trials are independent.*
5. *The random variables of interest are $Y_1, Y_2, \ldots, Y_k$, where Y_i $(i = 1, 2, \ldots, k)$ is equal to the number of trials in which the outcome falls in cell i. Note that $Y_1 + Y_2 + Y_3 + \cdots + Y_k = n$.*

Finding the probability that the n trials result in $Y_1, Y_2, \ldots, Y_k$, where Y_i is the number of trials falling in cell i, is an excellent application of the probabilistic methods of Chapter 2. We leave this problem as an exercise for the reader.

Example 5.18: *Find $E(Y_i)$ and $V(Y_i)$ for the multinomial probability distribution.*

Solution: *We are concerned with the marginal distribution of Y_i, the number of trials falling in cell i. Imagine all the cells, excluding cell i, combined into a single large cell. Hence every trial will result in cell i or not with probabilities p_i and $1 - p_i$, respectively, and Y_i possesses a binomial marginal probability distribution. Consequently,*

$$E(Y_i) = np_i,$$

$$V(Y_i) = np_iq_i, \quad \text{where } q_i = 1 - p_i.$$

Note: *The same results can be obtained by setting up the expectations and evaluating. For example,*

$$E(Y_1) = \sum_{y_1}\sum_{y_2}\cdots\sum_{y_k} y_1 \frac{n!}{y_1!y_2!\cdots y_k!} p_1^{y_1}p_2^{y_2}\cdots p_k^{y_k}.$$

Since we have already derived the expected value and variance of Y_i, we leave the tedious summation of this expectation to the interested reader.

Example 5.19: *If $Y_1, \ldots, Y_k$ have the multinomial distribution given above, find* $\text{Cov}(Y_s, Y_t)$, $s \neq t$.

Solution: *Thinking of the multinomial experiment as a sequence of n independent trials we define*

$$U_i = \begin{cases} 1, & \text{if trial } i \text{ results in class } s, \\ 0, & \text{otherwise} \end{cases}$$

and

$$W_i = \begin{cases} 1, & \text{if trial } i \text{ results in class } t, \\ 0, & \text{otherwise.} \end{cases}$$

Then

$$Y_s = \sum_{i=1}^{n} U_i \qquad \text{and} \qquad Y_t = \sum_{j=1}^{n} W_j.$$

To evaluate $\text{Cov}(Y_s, Y_t)$ *we need the following results:*

$$E(U_i) = p_s,$$

$$E(W_j) = p_t,$$

$$\text{Cov}(U_i, W_j) = 0 \quad \text{if } i \neq j \text{ since the trials are independent,}$$

and

$$\text{Cov}(U_i, W_i) = E(U_i W_i) - E(U_i)E(W_i)$$
$$= 0 - p_s p_t$$

since $U_i W_i$ *always equals zero. From Theorem 5.8 we then have*

$$\text{Cov}(Y_s, Y_t) = \sum_{i=1}^{n} \sum_{j=1}^{n} \text{Cov}(U_i, W_j)$$
$$= \sum_{i=1}^{n} \text{Cov}(U_i, W_i) + \sum_{i \neq j}\sum \text{Cov}(U_i, W_j)$$
$$= \sum_{i=1}^{n} (-p_s p_t) + 0$$
$$= -np_s p_t.$$

Note that the covariance is negative, which is to be expected since a large number of outcomes in cell s would force the number in cell t to be small.

Many experiments involving classification yield multinomial experiments. For example, the classification of people into five income brackets would result in an enumeration or count corresponding to each of five income classes. Or, we might be interested in studying the reaction of a mouse to a particular stimulus in a psychological experiment. If a mouse will react in one of three ways when the stimulus is applied and if a large number of mice were subjected to the stimulus, the experiment would yield three counts indicating the number of mice falling in each of the reaction classes. Similarly, a traffic study might require a count and classification of the type of motor vehicles using a section of highway. An industrial process manufactures items that fall into one of three quality classes: acceptable, seconds, and rejects. A student of the arts might classify paintings in one of k categories according to style and period in order to study trends in style over time. We might wish to classify ideas in a philosophical study or style in the field of literature. The result of an advertising campaign would yield count data indicating a classification of consumer reaction. Indeed, many observations in the physical sciences are not amenable to measurement on a continuous scale and hence result in enumerative or classificatory data.

Inferential problems associated with the multinomial experiment will be discussed later.

5.10 The Multivariate Normal Distribution

No discussion of multivariate probability distributions would be complete without reference to the multivariate normal distribution, which is a keystone to much modern statistical theory. In general the multivariate normal density function would be defined for k continuous random variables, $Y_1, Y_2, \ldots, Y_k$. Because of its complexity we will present only the bivariate density function ($k = 2$).

$$f(y_1, y_2) = \frac{e^{-Q/2}}{2\pi\sigma_1\sigma_2\sqrt{1-\rho^2}}, \qquad \begin{array}{l} -\infty < y_1 < \infty, \\ -\infty < y_2 < \infty, \end{array}$$

where

$$Q = \frac{1}{1-\rho^2}\left[\frac{(y_1-\mu_1)^2}{\sigma_1^2} - 2\rho\frac{(y_1-\mu_1)(y_2-\mu_2)}{\sigma_1\sigma_2} + \frac{(y_1-\mu_2)^2}{\sigma_2^2}\right].$$

The bivariate normal density is a function of five parameters, μ_1, μ_2, σ_1^2, σ_2^2, and ρ. The choice of symbols employed for these parameters is not coincidental. With a bit of tedious integration the reader can show that

$E(Y_i) = \mu_i$, $i = 1, 2$, $V(Y_i) = \sigma_i^2$, $i = 1, 2$, and the covariance of Y_1 and Y_2 is $\rho\sigma_1\sigma_2$.

Note that if $\text{Cov}(Y_1, Y_2) = 0$, or equivalently $\rho = 0$, then

$$f(y_1, y_2) = g(y_1)h(y_2),$$

and hence Y_1 and Y_2 are independent by Theorem 5.1. Recall that zero covariance for two random variables does not generally imply independence. Normal random variables provide an exception to this rule.

The expression for the joint density function, $k > 2$, is most easily expressed using the matrix algebra. A discussion of the general case can be found in the references.

5.11 Conditional Expectations

Section 5.3 contains a discussion of conditional probability functions and conditional density functions which we will now relate to conditional expectations. Conditional expectations are defined in the same manner as univariate expectations except that the conditional density is used in place of the marginal density function.

Definition 5.8: *If Y_1 and Y_2 are any two random variables, the* conditional expectation *of Y_1 given that $Y_2 = y_2$ is defined to be*

$$E(Y_1|Y_2 = y_2) = \int_{-\infty}^{\infty} y_1 f(y_1|y_2)\,dy_1$$

if Y_1 and Y_2 are jointly continuous and

$$E(Y_1|Y_2 = y_2) = \sum_{y_1} y_1 p(y_1|y_2)$$

if Y_1 and Y_2 are jointly discrete.

Example 5.20: *Refer to the Y_1 and Y_2 of Example 5.3, where*

$$f(y_1, y_2) = \begin{cases} 1/2, & 0 \le y_1 \le y_2; \quad 0 \le y_2 \le 2, \\ 0, & \textit{elsewhere.} \end{cases}$$

Find the conditional expectation of amount of sales, Y_1, given that $Y_2 = 1$.

Solution: *In Example 5.3 we found that*

$$f(y_1|y_2) = \begin{cases} \dfrac{1}{y_2}, & 0 < y_1 \le y_2 \le 2, \\ 0, & \textit{elsewhere.} \end{cases}$$

Thus, from Definition 5.8,

$$\begin{aligned} E(Y_1|Y_2 = 1) &= \int_{-\infty}^{\infty} y_1 f(y_1|y_2)\, dy_1 \\ &= \int_0^1 y_1(1)\, dy_1 \\ &= \frac{y_1^2}{2}\bigg]_0^1 = 1/2. \end{aligned}$$

That is, if the soft-drink machine contains 1 *gallon at the start of the day, the expected sales for that day is* 1/2 *gallon.*

The conditional expectation of Y_1 given $Y_2 = y_2$ is a function of y_2. If we now let Y_2 range over all its possible values, we can think of the conditional expectation as a function of the random variable Y_2 and hence we can find the expected value of the conditional expectation. The result of this type of iterated expectation is given in Theorem 5.9.

Theorem 5.9: *Let Y_1 and Y_2 denote random variables. Then*

$$E(Y_1) = E[E(Y_1|Y_2 = y_2)],$$

where, on the right-hand side, the inside expectation is with respect to the conditional distribution of Y_1 given Y_2, and the outside expectation is with respect to the distribution of Y_2.

Proof: *Let Y_1 and Y_2 have joint density function $f(y_1, y_2)$ and marginal densities $f_1(y_1)$ and $f_2(y_2)$, respectively. Then*

$$
\begin{aligned}
E(Y_1) &= \int_{-\infty}^{\infty} y_1 f_1(y_1)\, dy_1 \\
&= \int_{-\infty}^{\infty} \int_{-\infty}^{\infty} y_1 f_1(y_1, y_2)\, dy_1\, dy_2 \\
&= \int_{-\infty}^{\infty} \int_{-\infty}^{\infty} y_1 f(y_1 | y_2) f_2(y_2)\, dy_1\, dy_2 \\
&= \int_{-\infty}^{\infty} \left[\int_{-\infty}^{\infty} y_1 f(y_1 | y_2)\, dy_1 \right] f_2(y_2)\, dy_2 \\
&= \int_{-\infty}^{\infty} E(Y_1 | Y_2 = y_2) f_2(y_2)\, dy_2 \\
&= E[E(Y_1 | Y_2 = y_2)].
\end{aligned}
$$

The proof is similar for the discrete case.

Example 5.21: *A quality-control plan for an assembly line involves sampling $n = 10$ finished items per day and counting Y, the number of defectives. If p denotes the probability of observing a defective, then Y has a binomial distribution, assuming that the number of items produced by the line is large. But p varies from day to day and is assumed to have a uniform distribution on the interval 0 to 1/4. Find the expected value of Y for any given day.*

Solution: *From Theorem 5.9 we know that*

$$E(Y) = E[E(Y|p)].$$

For a given p, Y has a binomial distribution and hence

$$E(Y|p) = np.$$

Thus

$$
\begin{aligned}
E(Y) &= E(np) = nE(p) \\
&= n \int_{0}^{1/4} 4p\, dp \\
&= n(1/8)
\end{aligned}
$$

$E(p) = \int_0^{\frac{1}{4}} p\, f(p)\, dp$
where: $f(p) = 4$

and, for $n = 10$,

$$E(Y) = 10/8 = 5/4.$$

This inspection policy should average 5/4 defectives per day, in the long run. The calculations above could be checked by actually finding the unconditional distribution of Y and computing E(Y) directly.

5.12 Summary

The multinomial experiment, Section 5.9, and its associated multinomial probability distribution convey the theme of this chapter. Most experiments yield sample measurements, $y_1, y_2, \ldots, y_k$, which may be regarded as observations on k random variables. Inferences about the underlying structure that generates the observations, the probabilities of falling in cells, $1, 2, \ldots, k$, are based on knowledge of the probabilities associated with various samples $(y_1, y_2, \ldots, y_k)$. Joint, marginal, and conditional distributions are essential concepts in finding the probabilities of various sample outcomes.

Generally we draw n sample observations from a population which are specific values of $Y_1, Y_2, \ldots, Y_k$. Most often the first, second, and nth random variables are independent and have the same probability distribution. As a consequence, the concept of independence is useful in finding the probability of observing the given sample.

The reader should note that the objective of this chapter has been to convey the ideas contained in the two preceding paragraphs. The details contained in the chapter are numerous and are essential in providing a solid background for a study of inference. At the same time, the reader should be careful to avoid overemphasis on details; he should be sure to keep the broader inferential objectives in mind.

References

1. Hoel, P. G., *Introduction to Mathematical Statistics*, 4th ed. New York: John Wiley & Sons, Inc., 1971.

2. Hogg, R. V., and A. T. Craig, *Introduction to Mathematical Statistics*, 3rd ed. New York: The Macmillan Company, 1970.
3. Mendenhall, W., *Introduction to Probability and Statistics*, 3rd ed. North Scituate, Mass.: Duxbury Press, 1971.
4. Mood, A. M., and F. A. Graybill, *Introduction to the Theory of Statistics*, 2nd ed. New York: McGraw-Hill Book Company, 1963.
5. Parzen, E., *Modern Probability Theory and Its Applications*. New York: John Wiley & Sons, Inc., 1964.

Exercises

5.1. Let Y_1 and Y_2 have the joint probability density function given by

$$f(y_1, y_2) = \begin{cases} Ky_1y_2, & 0 \le y_1 \le 1; \quad 0 \le y_2 \le 1, \\ 0, & \text{elsewhere.} \end{cases}$$

(a) Find the value of K which makes this a probability density function.
(b) Find the marginal densities of Y_1 and Y_2.
(c) Find the joint distribution function for Y_1 and Y_2.
(d) Find the probability $P(Y_1 < 1/2, Y_2 < 3/4)$.
(e) Find the probability $P(Y_1 \le 1/2 | Y_2 > 3/4)$.

5.2. Let Y_1 and Y_2 have the joint density function given by

$$f(y_1, y_2) = \begin{cases} 3y_1, & 0 \le y_2 \le y_1 \le 1, \\ 0, & \text{elsewhere.} \end{cases}$$

(a) Find the marginal density functions of Y_1 and Y_2.
(b) Find $P(Y_1 \le 3/4, Y_2 \le 1/2)$.
(c) Find $P(Y_1 \le 1/2 | Y_2 \ge 3/4)$.

5.3. From a group consisting of 4 Republicans, 3 Democrats, and 2 Independents, a committee of 3 persons is to be randomly selected. Let Y_1 denote the number of Republicans and Y_2 the number of Democrats on the committee.

(a) Find the joint probability distribution of Y_1 and Y_2.
(b) Find the marginal densities of Y_1 and Y_2.
(c) Find the probability $P(Y_1 = 1 | Y_2 \geq 1)$.

5.4. For Exercise 5.1, find the conditional density of Y_1 given $Y_2 = y_2$. Are Y_1 and Y_2 independent?

5.5. For Exercise 5.2,
(a) Find the conditional density of Y_1 given $Y_2 = y_2$.
(b) Find the conditional density of Y_2 given $Y_1 = y_1$.
(c) Show that Y_1 and Y_2 are dependent.
(d) Find the probability $P(Y_1 \leq 3/4 | Y_2 = 1/2)$.

5.6. Let Y_1 denote the amount of a certain bulk item stocked by a supplier at the beginning of a week and suppose that Y_1 has a uniform distribution over the interval $0 \leq y_1 \leq 1$. Let Y_2 denote the amount of this item sold by the supplier during the week and suppose that Y_2 has a uniform distribution over the interval $0 \leq y_2 \leq y_1$, where y_1 is a specific value of Y_1.
(a) Find the joint density function for y_1 and y_2.
(b) If the supplier stocks an amount of 1/2, what is the probability that he sells an amount greater than 1/4?
(c) If it is known that the supplier sold an amount equal to 1/4, what is the probability that he had stocked an amount greater than 1/2?

5.7. Let (Y_1, Y_2) denote the coordinates of a point dropped at random inside a unit circle with center at the origin. That is, Y_1 and Y_2 have joint density function given by

$$f(y_1, y_2) = \begin{cases} \dfrac{1}{\pi}, & y_1^2 + y_2^2 \leq 1, \\ 0, & \text{elsewhere.} \end{cases}$$

(a) Find the marginal density function of Y_1.
(b) Find $P(Y_1 \leq Y_2)$.

5.8. Let Y_1 and Y_2 have the joint density function given by

$$f(y_1, y_2) = \begin{cases} y_1 + y_2, & 0 \leq y_1 \leq 1; \quad 0 \leq y_2 \leq 1, \\ 0, & \text{elsewhere.} \end{cases}$$

(a) Find the marginal density functions of Y_1 and Y_2.
(b) Are Y_1 and Y_2 independent?
(c) Find the conditional density of Y_1 given $Y_2 = y_2$.

5.9. Let Y_1 and Y_2 have the joint density function given by

$$f(y_1, y_2) = \begin{cases} K, & 0 \le y_1 \le 2;\quad 0 \le y_2 \le 1; \\ & 2y_2 \le y_1, \\ 0, & \text{elsewhere.} \end{cases}$$

(a) Find the value of K which makes the above function a probability density.
(b) Find the marginal densities of Y_1 and Y_2.
(c) Find the conditional density of Y_1 given $Y_2 = y_2$.
(d) Find the conditional density of Y_2 given $Y_1 = y_1$.
(e) Find $P(Y_1 \le 1.5, Y_2 \le .5)$.
(f) Find $P(Y_2 \le .5 | Y_1 \le 1.5)$.

5.10. Let Y_1 and Y_2 have a joint distribution that is uniform over the region shaded in the diagram.

(a) Find the marginal density for Y_2.
(b) Find the marginal density for Y_1.
(c) Find $P[(Y_1 - Y_2) \ge 0]$.

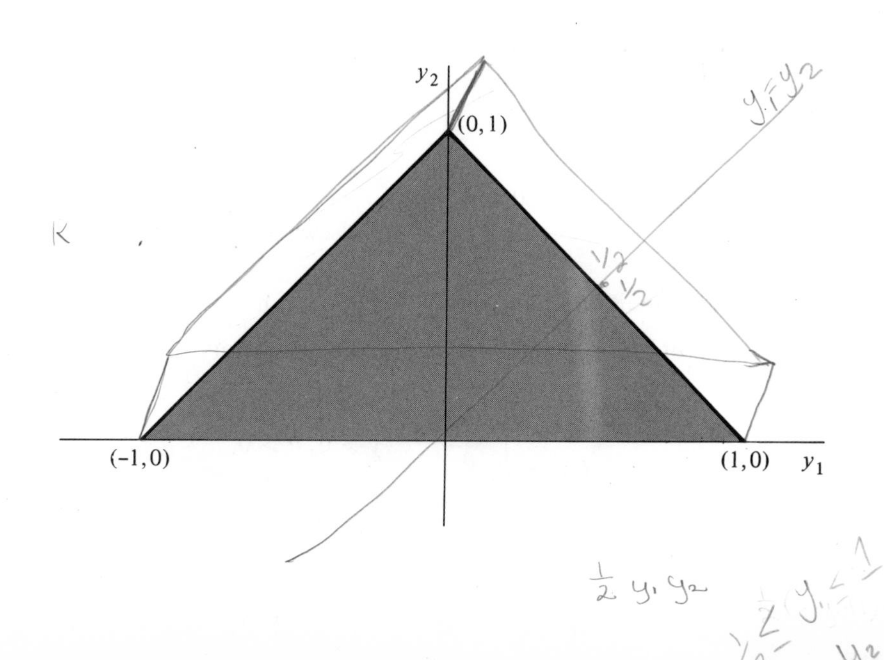

5.11. Refer to Exercise 5.1.
(a) Find $E(Y_1)$.
(b) Find $V(Y_1)$.
(c) Find $\text{Cov}(Y_1, Y_2)$.

5.12. Refer to Exercise 5.2. Find $\text{Cov}(Y_1, Y_2)$.

5.13. Refer to Exercise 5.3.
(a) Find $\text{Cov}(Y_1, Y_2)$.
(b) Find $E(Y_1 + Y_2)$ and $V(Y_1 + Y_2)$ by first finding the probability distribution of $Y_1 + Y_2$.
(c) Find $E(Y_1 + Y_2)$ and $V(Y_1 + Y_2)$ by using Theorem 5.8.

5.14. Refer to Exercise 5.8.
(a) Find $\text{Cov}(Y_1, Y_2)$.
(b) Find $E(3Y_1 - 2Y_2)$.
(c) Find $V(3Y_1 - 2Y_2)$.

5.15. Refer to Exercise 5.9.
(a) Find $E(Y_1 + 2Y_2)$.
(b) Find $V(Y_1 + 2Y_2)$.

5.16. A quality-control plan calls for randomly selecting three items from the daily production (assumed large) of a certain machine and observing the number of defectives. However, the proportion, p, of defectives produced by the machine varies from day to day and is assumed to have a uniform distribution on the interval (0, 1). For a randomly chosen day, find the unconditional probability that exactly two defectives are observed in the sample.

5.17. The number of defects per yard, denoted by Y, for a certain fabric is known to have a Poisson distribution with parameter λ. However, λ is not known and is assumed to be random with its probability density function given by

$$f(\lambda) = \begin{cases} e^{-\lambda}, & \lambda \geq 0, \\ 0, & \text{elsewhere.} \end{cases}$$

Find the unconditional probability function for Y.

5.18. The length of life, Y, of a fuse has a probability density,

$$f(y) = \begin{cases} \dfrac{e^{-y/\theta}}{\theta}, & y > 0; \theta > 0, \\ 0, & \text{elsewhere.} \end{cases}$$

Three such fuses operate independently. Find the joint density of their lengths of life, Y_1, Y_2, and Y_3.

5.19. A retail grocery merchant figures that his daily gain from sales, X, is a normally distributed random variable with $\mu = 50$ and $\sigma^2 = 10$ (measurements in dollars). X could be negative if he is forced to dispose of perishable goods. Also, he figures daily overhead costs, Y, to have a gamma distribution with $\alpha = 4$ and $\beta = 2$. If X and Y are independent, find the expected value and variance of his net daily *gain*. Would you expect his net gain for tomorrow to go above \$70?

5.20. A coin has probability p of coming up "heads" when tossed. In n independent tosses of the coin, let $X_i = 1$ if the ith toss results in "heads" and $X_i = 0$ if the ith toss results in "tails." Then Y, the number of "heads" in the n tosses, has a binomial distribution and can be represented as

$$Y = \sum_{i=1}^{n} X_i.$$

Find $E(Y)$ and $V(Y)$ using Theorem 5.8.

5.21. Refer to Exercises 5.2 and 5.5.
(a) Find $E(Y_2 | Y_1 = y_1)$.
(b) Use Theorem 5.9 to find $E(Y_2)$.
(c) Find $E(Y_2)$ directly from the marginal density of Y_2.

5.22. Refer to Exercise 5.17.
(a) Find $E(Y)$ by first finding the conditional expectation of Y for given λ and then using Theorem 5.9.
(b) Find $E(Y)$ directly from the probability distribution of Y.

5.23. Refer to Exercise 5.6. If the supplier stocks an amount equal to 3/4, what is the expected amount sold during the week?

5.24. Let Y be a continuous random variable with distribution function $F(y)$ and density function $f(y)$. We can then write, for $y_1 \leq y_2$,

$$P(Y \leq y_2 | Y \geq y_1) = \frac{F(y_2) - F(y_1)}{1 - F(y_1)}.$$

As a function of y_2 for fixed y_1, the right-hand side of the above expression is called the conditional distribution function of Y given that $Y \geq y_1$.

On taking the derivative with respect to y_2, we see that the corresponding conditional density function is given by

$$\frac{f(y_2)}{1 - F(y_1)}, \quad y_2 \geq y_1.$$

Suppose that a certain type of electronic component has life length, Y, with the density function (life length measured in hours)

$$f(y) = \begin{cases} \frac{1}{200}e^{-y/200}, & y \geq 0, \\ 0, & \text{elsewhere.} \end{cases}$$

Find the expected length of life for a component of this type which has already been in use for 100 hours.

5.25. Let X_1, X_2, and X_3 be random variables, either continuous or discrete. The joint moment-generating function of X_1, X_2, and X_3 is defined by

$$m(t_1, t_2, t_3) = E(e^{t_1X_1 + t_2X_2 + t_3X_3}).$$

(a) Show that $m(t, t, t)$ gives the moment-generating function of $X_1 + X_2 + X_3$.
(b) Show that $m(t, t, 0)$ gives the moment-generating function of $X_1 + X_2$.
(c) Show that

$$\left.\frac{\partial^{k_1+k_2+k_3} m(t_1, t_2, t_3)}{\partial t_1^{k_1} \partial t_2^{k_2} \partial t_3^{k_3}}\right]_{t_1=t_2=t_3=0} = E(X_1^{k_1} X_2^{k_2} X_3^{k_3}).$$

5.26. Let X_1, X_2, and X_3 have a multinomial distribution with probability function

$$p(x_1, x_2, x_3) = \frac{n!}{x_1!x_2!x_3!} p_1^{x_1} p_2^{x_2} p_3^{x_3}, \quad \sum_{i=1}^{n} x_i = n.$$

Employ the results of Exercise 5.25 to answer the following.
(a) Find the joint moment-generating function of X_1, X_2, and X_3.
(b) Use the answer to part (a) to show that the marginal distribution of X_1 is binomial with parameter p_1.
(c) Use the joint moment-generating function to find $\text{Cov}(X_1, X_2)$.

5.27. The negative binomial random variable, Y, was defined in Exercise 3.42 as the number of the trial on which the rth success occurs, in a sequence of independent trials with constant probability, p, of success on each trial. Let X_i denote a geometric random variable, defined as the number of the trial on which the first success occurs. Then we can write

$$Y = \sum_{i=1}^{r} X_i$$

for independent random variables $X_1, \ldots, X_r$. Use Theorem 5.8 to show that $E(Y) = r/p$ and $V(Y) = r(1 - p)/p^2$.

5.28. A box contains four balls, numbered 1 through 4. One ball is selected at random from this box. Let

$$X_1 = 1 \text{ if ball number 1 or ball number 2 is drawn,}$$
$$X_2 = 1 \text{ if ball number 1 or ball number 3 is drawn,}$$
$$X_3 = 1 \text{ if ball number 1 or ball number 4 is drawn,}$$

and the X_i's are zero otherwise. Show that any two of the random variables X_1, X_2, and X_3 are independent but the three together are not.

5.29. Let Y_1 and Y_2 be jointly distributed random variables with finite variances.
(a) Show that $[E(Y_1 Y_2)]^2 \le E(Y_1^2)E(Y_2^2)$. *Hint*: Observe that, for any real number t, $E[(tY_1 - Y_2)^2] \ge 0$ or, equivalently,

$$t^2 E(Y_1^2) - 2tE(Y_1 Y_2) + E(Y_2^2) = 0.$$

This is a quadratic expression of the form $At^2 + Bt + C$ and, since it is nonnegative, we must have $B^2 - 4AC \le 0$. The above inequality follows directly.
(b) Let ρ denote the correlation coefficient of Y_1 and Y_2. Using the inequality of part (a), show that $\rho^2 \le 1$.

5.30. A box contains N_1 white balls, N_2 black balls, and N_3 red balls ($N_1 + N_2 + N_3 = N$). A random sample of n balls is selected from the box (without replacement). Let Y_1, Y_2, and Y_3 denote the number of white, black, and red balls, respectively, observed in the sample. Find the correlation coefficient for Y_1 and Y_2. (Let $p_i = N_i/N$, $i = 1, 2, 3$.)

5.31. Let $Y_1, Y_2, \ldots, Y_n$ be independent random variables with $E(Y_i) = \mu$ and $V(Y_i) = \sigma^2$, $i = 1, \ldots, n$. Let

$$U_1 = \sum_{i=1}^{n} a_i Y_i \quad \text{and} \quad U_2 = \sum_{i=1}^{n} b_i Y_i,$$

where $a_1, \ldots, a_n, b_1, \ldots, b_n$ are constants. U_1 and U_2 are said to be orthogonal if $\text{Cov}(U_1, U_2) = 0$.

(a) Show that U_1 and U_2 are orthogonal if and only if $\sum_{i=1}^{n} a_i b_i = 0$.

(b) Suppose, in addition, that $Y_1, \ldots, Y_n$ are normally distributed. Conclude that U_1 and U_2 are independent if they are orthogonal.

6

Functions of Random Variables

6.1 *Introduction*

Each preceding chapter plays a role in the story of statistical inference, but none is so closely related as a study of functions of random variables. This is because all quantities used to estimate population parameters or to make decisions about a population are functions of the n random observations that appear in a sample.

To illustrate, consider the problem of estimating a population mean. Intuitively we draw a random sample of n observations $y_1, y_2, \ldots, y_n$, from the population and employ the sample mean,

$$\bar{y} = \frac{y_1 + y_2 + \cdots + y_n}{n} = \frac{\sum_{i=1}^{n} y_i}{n}$$

as an estimate of μ. How good will this estimate be? The answer depends upon the behavior of the random variables $Y_1, Y_2, \ldots, Y_n$ and their effect on $\bar{Y} = (1/n) \sum_{i=1}^{n} Y_i$.

The measure of goodness of an estimate is the *error of estimation*, the difference between the estimate and the parameter estimated (for our example, $\bar{y}$ and μ). Because $Y_1, Y_2, \ldots, Y_n$ are random variables in repeated sampling, so will be $\bar{Y}$. Hence we cannot be certain that the error of estimation will be less than a specific value, B, but we can frequently state that the probability of an error less than or equal to B is .95 or some other relatively high probability. To do this, we must know the probability distribution of the estimator, $\bar{Y}$, which is indeed a function of the n random variables, $Y_1, Y_2, \ldots, Y_n$.

6.2 Random Sampling

We will make some additional observations on random sampling, a term that was defined in Chapter 2.

The population of interest is defined by the experimenter. This is a very important point to grasp. If we wish to make inferences concerning the efficiencies of a shipment of $N = 10$ jet engines, the population contains only $N = 10$ elements. If the population consists of the yield measurements for a chemical process over a "long" period of time, we might view the population as very large indeed, perhaps infinitely large.

Suppose that $n = 3$ jet engines were selected without replacement from the population of $N = 10$ and their efficiencies, y_1, y_2, and y_3, measured. Intuitively, we would suspect that the probability of observing a particular efficiency on the second test is dependent on the engine selected for the first test. In other words, we imply that sampling without replacement from finite populations gives rise to dependent sample observations.

Example 6.1: *Suppose that a population contains $N = 4$ observations, 7, 2, 1, and 6. A random sample of $n = 2$ observations is drawn from the population. Show that the events "draw a 2" and "draw a 7" are dependent.*

Solution:

$$\binom{4}{2} = \frac{4!}{2!2!} = 6$$

different samples can be drawn from the population of $N = 4$ elements. They are

$$(7, 2), (7, 1), (7, 6), (2, 1), (2, 6), (1, 6).$$

The six combinations associated with the selection of two elements from four identify six simple events, $E_1, E_2, \ldots, E_6$, to which we assign equal probabilities of 1/6 (because the sampling is random).

The probability of observing a 7 is

$$p(7) = P(E_1) + P(E_2) + P(E_3)$$
$$= 3/6 = 1/2.$$

Similarly, the probability of drawing a 2 is 1/2. Observing the six sample points for the experiment, the probability of drawing both *a 7 and a 2 is*

$$p(7, 2) = 1/6.$$

Because $p(7, 2) \neq p(7)p(2)$, the events 7 and 2 are dependent (by Definition 2.13).

If the population is very large, say infinitely large, any two observations in a random sample, say the first and second observations, Y_1 and Y_2, would be independent. That is, drawing $Y_1 = y_1$ on the first observation would not affect the probability of drawing $Y_2 = y_2$ on the second.

The importance of random sampling is that it implies independence of the random variables $Y_1, Y_2, \ldots, Y_n$ when the population is large. Then the multivariate density (or probability function) for $Y_1, Y_2, \ldots, Y_n$ is

$$f(y_1, y_2, \ldots, y_n) = f(y_1)f(y_2)\cdots f(y_n)$$

if Y_i, $i = 1, 2, \ldots, n$, are continuous random variables or

$$p(y_1, y_2, \ldots, y_n) = p(y_1)p(y_2)\cdots p(y_n)$$

if Y_i, $i = 1, 2, \ldots, n$, are discrete random variables. Unless otherwise stated, we shall henceforth assume populations to be infinite. Thus the statement "$Y_1, Y_2, \ldots, Y_n$ is a random sample from $f(y)$" will imply that the random variables are independent with common density function $f(y)$.

6.3 Finding the Probability Distribution of a Function of Random Variables

We shall present four methods for finding the probability distribution for a function of random variables. Any one of these may be employed for a given function, but one method usually leads to a simpler derivation than another. Hence an acquaintance with all four is desirable. Although the four methods for finding the probability distribution of a function of random variables will be discussed separately in the next four sections, we will summarize the methods here.

Consider random variables, $Y_1, Y_2, \ldots, Y_n$ and a function $U(Y_1, Y_2, \ldots, Y_n)$ which we will denote simply as U. Then the four methods for finding the probability distribution of U are

1. Method of Distribution Functions: *Find* $P(U < u) = F(u)$. *This is a probability problem. For every joint event* $(Y_1 = y_1, Y_2 = y_2, \ldots, Y_n = y_n)$, *there corresponds one and only one value of* U. *Thus one must find the region in the* $y_1, y_2, \ldots, y_n$ *space for which* $U < u$ *and then find* $P(U < u)$ *by integrating* $f(y_1, y_2, \ldots, y_n)$ *over this region. A detailed account of this procedure will be presented in Section 6.4.*
2. Method of Moment-Generating Functions: *The moment-generating-function method is based on a uniqueness theorem, Theorem 6.1, which states that if two moment-generating functions are identical, the two random variables possess the same probability distributions. Hence we must find the moment-generating function for* U *and compare it with the moment-generating functions for the common discrete and continuous random variables derived in Chapters 3 and 4. If it is identical to some well-known moment-generating function, the probability distribution of* U *will be identified because of the uniqueness theorem. The applications of moment-generating functions will be presented in Section 6.5. Probability-generating functions can be employed in a similar way. The reader interested in their use is directed to the references at the end of the chapter.*
3. Method of Transformations: *Given the density function of a random variable* Y, *the method of transformations results in a general expression for the density of*

$U = h(Y)$ for an increasing or decreasing function $h(y)$. Then, if Y_1 and Y_2 are jointly distributed random variables, we can use the univariate result explained above to find the joint density of Y_1 and U. By integrating over y_1, we then find the marginal probability density function of U, which is our objective. This method will be illustrated in Section 6.6.

4. Method of Conditioning: *In some cases it may be easy to find the conditional density of U when some of the other variables are held fixed. We then can find the density for U through a relation such as the following, stated for the case $n = 2$,*

$$f_U(u) = \int_{-\infty}^{\infty} f(u|y_2) f(y_2)\, dy_2.$$

This method will be discussed in Section 6.7.

6.4 Method of Distribution Functions

Let Y_1 and Y_2 be random variables with joint density $f(y_1, y_2)$. Let $U = U(Y_1, Y_2)$ be a function of Y_1, Y_2. Then for every point (y_1, y_2) there corresponds one and only one value of U. If we can find the points (y_1, y_2), such that $U < u$, then the integral of the joint density function, $f(y_1, y_2)$, over the region will equal $P(U < u)$ or, equivalently, $F_U(u)$ (by Definition 4.1). Then the density function for U can be obtained by differentiation,

$$\frac{dF_U(u)}{du} = f_U(u).$$

We will illustrate with an example.

Example 6.2: *Let*

$$f(y_1, y_2) = \begin{cases} 1, & 0 \le y_1 \le 1;\ 0 \le y_2 \le 1, \\ 0, & \text{elsewhere.} \end{cases}$$

Find the density function for $U = Y_1 + Y_2$.

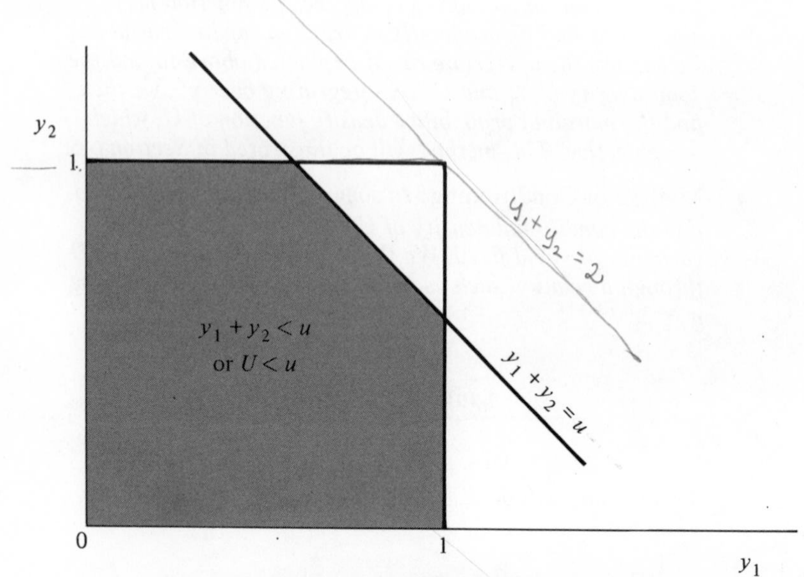

Figure 6.1 The Region of Integration, Example 6.2.

Solution: *The random variables Y_1 and Y_2 are defined over the unit square as shown in Figure 6.1. We wish to find $F_U(u) = P(U \leq u)$.*

The first step is to find the points (y_1, y_2) that imply $y_1 + y_2 < u$. The easiest way to find this region is to locate the points that divide the regions $U \leq u$ and $U > u$. These points lie on the line

$$y_1 + y_2 = u.$$

Graphing this relationship in Figure 6.1 and arbitrarily selecting y_2 as the dependent variable, the line possesses a slope equal to -1 and a y_2 intercept equal to u. The points associated with $U < u$ are either above or below the line and can be obtained by testing. Suppose that $u = 1.5$. Let $y_1 = y_2 = 1/4$; then $y_1 + y_2 = 1/4 + 1/4 = 1/2$ satisfies the inequality $y_1 + y_2 < u$ and falls in the shaded region below the line. The reader can see intuitively that all points such that $y_1 + y_2 < u$ will similarly lie below the line, $y_1 + y_2 = u$. Thus

$$F_U(u) = P(U \leq u) = P(Y_1 + Y_2 \leq u)$$

$$= \iint_{y_1 + y_2 \leq u} f(y_1, y_2)\, dy_1\, dy_2.$$

Note that u can assume any value in the interval, $0 \le u \le 2$, and that the limits of integration depend upon u (where u is the y intercept of the line $y_1 + y_2 = u$). Thus the mathematical expression for $F_U(u)$ changes depending on whether $0 \le u \le 1$ or $1 \le u \le 2$.

The region $y_1 + y_2 \le u$, $0 \le u \le 1$, is the shaded area in Figure 6.2.

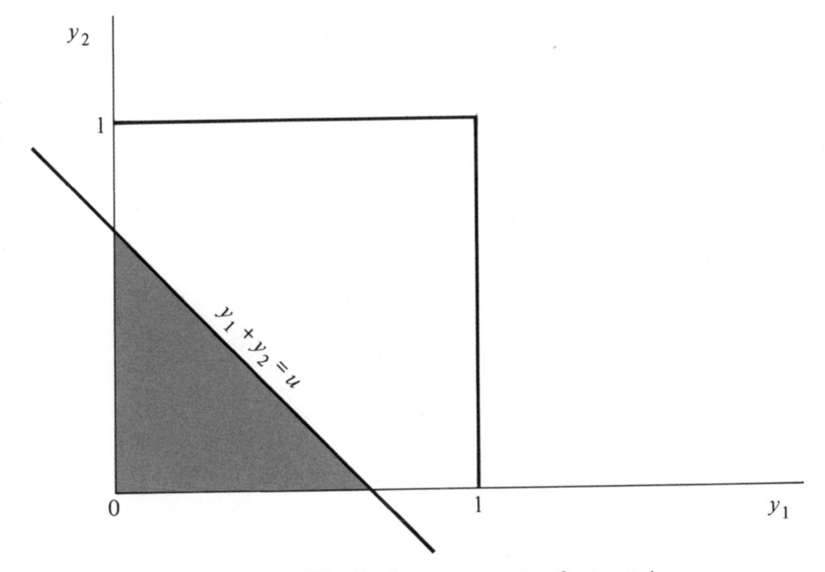

Figure 6.2 The Region $y_1 + y_2 \le u$, $0 \le u \le 1$

Then for $f(y_1, y_2) = 1$,

$$F_U(u) = \int_0^u \int_0^{u-y_2} (1)\, dy_1\, dy_2 = \int_0^u (u - y_2)\, dy_2$$

or

$$F_U(u) = uy_2 - \frac{y_2^2}{2}\Bigg]_0^u = u^2 - \frac{u^2}{2} = \frac{u^2}{2}, \qquad 0 \le u \le 1.$$

The solution, $F_U(u)$, $0 \le u \le 1$, could have been acquired directl; ....ng knowledge of elementary geometry. The bivariate density $f(y_1, y_2) = 1$ is uniform over the unit square, $0 \le y_1 \le 1$, $0 \le y_2 \le 1$. Hence $F_U(u)$ is the volume of a solid with height equal to $f(y_1, y_2) = 1$ and triangular cross section as shown *in Figure 6.2. Hence*

$$F_U(u) = (\text{area of triangle}) \cdot (\text{altitude})$$
$$= \frac{u^2}{2}(1) = \frac{u^2}{2}.$$

The distribution function can be acquired in a similar manner when u is defined over the interval $1 \leq u \leq 2$. Although the geometric solution is easier, we will acquire $F_U(u)$ directly by integration. Thus the region $y_1 + y_2 \leq u$, $1 \leq u \leq 2$, is the shaded area indicated in Figure 6.3.

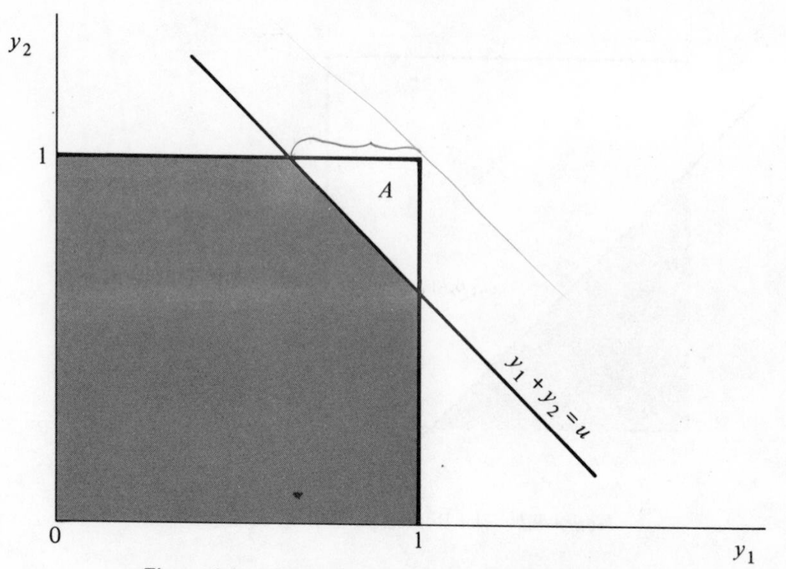

Figure 6.3 The Region $y_1 + y_2 \leq u$, $1 \leq u \leq 2$

The complement of the event, $U \leq u$, is the event that (Y_1, Y_2) falls in the region, A, Figure 6.3. Then

$$F_U(u) = 1 - \int_A\int f(y_1, y_2)\, dy_1\, dy_2, \qquad 1 \leq u \leq 2,$$

$$F_U(u) = 1 - \int_{u-1}^{1}\int_{u-y_2}^{1} (1)\, dy_1\, dy_2 = 1 - \int_{u-1}^{1} y_1\Big]_{u-y_2}^{1} dy_2$$

$$= 1 - \int_{u-1}^{1} (1 - u + y_2)\, dy_2$$

$$= 1 - \left[(1-u)y_2 + \frac{y_2^2}{2}\right]_{u-1}^{1}$$

$$= -\frac{u^2}{2} + 2u - 1, \qquad 1 \leq u \leq 2.$$

The reader will observe that this probability (volume) under $f(y_1, y_2)$ over the region $y_1 + y_2 \leq u$, $1 \leq u \leq 2$, could have been acquired directly and simply using elementary geometry concepts.

To summarize, $F_U(u) = u^2/2$, $0 \leq u \leq 1$, and $F_U(u) = (-u^2/2) + 2u - 1$, $1 \leq u \leq 2$. The distribution function for U is shown in Figure 6.4.

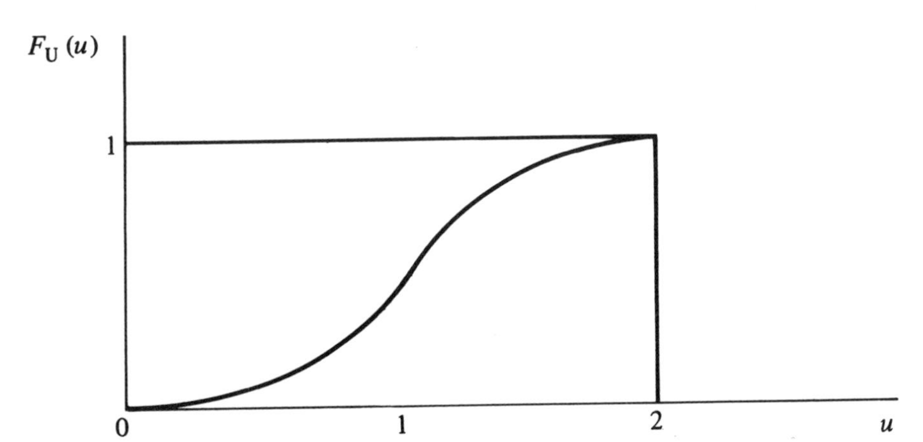

Figure 6.4 Distribution Function, $F_U(u)$, for Example 6.2

The density function, $f_U(u)$, can be obtained by differentiating $F_U(u)$. Thus

$$f_U(u) = \frac{dF_U(u)}{du} = \frac{d(u^2/2)}{du} = u, \qquad 0 \leq u < 1,$$

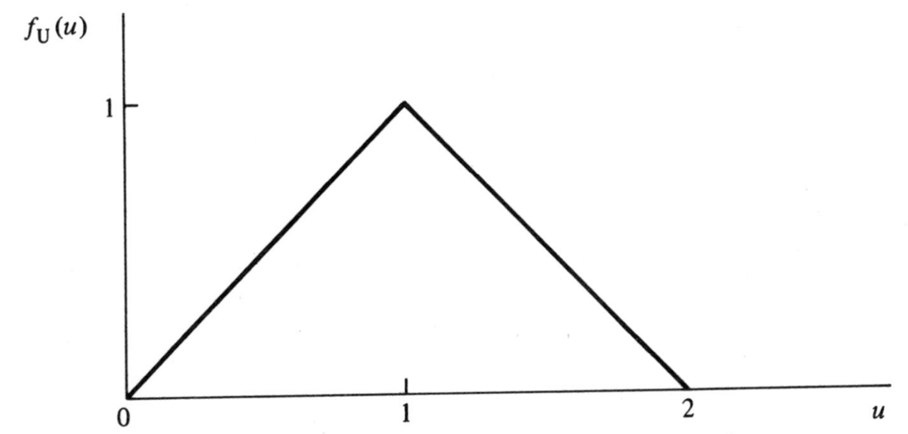

Figure 6.5 The Density Function, $f_U(u)$, for Example 6.2.

and
$$f_U(u) = \frac{d[-(u^2/2) + 2u - 1]}{du} = 2 - u, \qquad 1 \le u \le 2.$$

A graph of $f_U(u)$ is shown in Figure 6.5.

Summary of the Distribution Function Method

Let U be a function of the random variables $Y_1, Y_2, \ldots, Y_n$. Then

1. *Find the region $U = u$ in the $(y_1, y_2, \ldots, y_n)$ space.*
2. *Find the region, $U \le u$.*
3. *Find $F_U(u) = P(U \le u)$ by integrating $f(y_1, y_2, \ldots, y_n)$ over the region $U \le u$.*
4. *Find the density function $f_U(u)$ by differentiating $F_U(u)$. Thus $f_U(u) = dF_U(u)/du$.*

6.5 Method of Moment-Generating Functions

The moment-generating-function method for finding the probability distribution of a function of random variables, $Y_1, Y_2, \ldots, Y_n$, is based on the following uniqueness theorem.

Theorem 6.1: *Suppose that for each of two random variables, X and Y, moment-generating functions exist and are given by $m_X(t)$ and $m_Y(t)$, respectively. If $m_X(t) = m_Y(t)$ for all values of t, then X and Y have the same probability distribution.*

(The proof of Theorem 6.1 is beyond the scope of this text.)

The first step in using Theorem 6.1 is to find the moment-generating function of U where, as before, U is a function of n random variables, $Y_1, Y_2, \ldots, Y_n$. Thus U is a random variable and, by definition, its moment-generating function is

$$m_U(t) = E(e^{tU}).$$

Once the moment-generating function for U has been found, it is compared with the moment-generating functions for other well-known random variables. If $m_U(t)$ is identical to one of these, say the moment-generating function for a random variable V, then by Theorem 6.1, U and V possess identical probability distributions. The density functions, means, variances, and moment-generating functions are presented in Table 1, Appendix III, for some frequently encountered random variables. We shall illustrate the procedure with a few examples.

Example 6.3: *Suppose that Y is a normally distributed random variable with mean μ and variance σ^2. Show that*

$$Z = \frac{Y - \mu}{\sigma}$$

is normally distributed with mean 0 and variance 1.

Solution: *We have seen in Example 4.14 that $Y - \mu$ has moment-generating function $e^{t^2\sigma^2/2}$. Hence*

$$\begin{aligned} m_Z(t) &= E(e^{tZ}) = E[e^{(t/\sigma)(Y-\mu)}] \\ &= m_{(Y-\mu)}(t/\sigma) \\ &= e^{(t/\sigma)^2(\sigma^2/2)} \\ &= e^{t^2/2}. \end{aligned}$$

On comparing $m_Z(t)$ with the moment-generating function of a normal random variable, we see that Z must be normally distributed with $E(Z) = 0$ and $V(Z) = 1$.

Example 6.4: *Let Z be a normally distributed random variable with mean and variance equal to 0 and 1, respectively. Use moment-generating functions to find the probability distribution for Z^2.*

Solution: *The moment-generating function for Z^2 is*

$$m_{Z^2}(t) = E(e^{tZ^2}) = \int_{-\infty}^{\infty} e^{tz^2} f(z)\, dz = \int_{-\infty}^{\infty} e^{tz^2} \frac{e^{-z^2/2}}{\sqrt{2\pi}}\, dz$$

$$= \int_{-\infty}^{\infty} \frac{1}{\sqrt{2\pi}} e^{-(z^2/2)(1-2t)}\, dz.$$

This integral can be evaluated by either consulting a table of integrals or by noting that the integrand

$$\frac{\exp\left[-\left(\frac{z^2}{2}\right)(1-2t)\right]}{\sqrt{2\pi}} = \frac{\exp\left[-\left(\frac{z^2}{2}\right)\Big/(1-2t)^{-1}\right]}{\sqrt{2\pi}}$$

is the major portion of a normal density function for a random variable with mean and variance equal to 0 and $(1-2t)^{-1}$, respectively. To make the integrand a normal density function (so that the definite integral is equal to 1), multiply numerator and denominator by the standard deviation, $(1-2t)^{-1/2}$. Then

$$m_{Z^2}(t) = \frac{1}{(1-2t)^{1/2}} \int_{-\infty}^{\infty} \frac{1}{\sqrt{2\pi}(1-2t)^{-1/2}} \exp\left[-\left(\frac{z^2}{2}\right)\Big/(1-2t)^{-1}\right] dz.$$

Since the integral integrates to 1,

$$m_{Z^2}(t) = \frac{1}{(1-2t)^{1/2}}.$$

A comparison of $m_{Z^2}(t)$ with the moment-generating functions, Table 1, Appendix III, shows $m_{Z^2}(t)$ identical to the moment-generating function for the gamma-type density function with $\alpha = 1/2$ and $\beta = 2$. Thus Z^2 has a chi-square distribution with $\nu =$ one degree of freedom. Substituting these values for the chi-square density function, Section 4.5, the density function for $Y = Z^2$ is

$$f_Y(y) = \begin{cases} \dfrac{y^{-1/2}e^{-y/2}}{\Gamma(1/2)2^{1/2}}, & y \geq 0, \\ 0, & \text{elsewhere.} \end{cases}$$

Another very useful application of moment-generating functions comes when dealing with sums of independent random variables.

> ***Theorem 6.2:*** *Let $Y_1, \ldots, Y_n$ be independent random variables with moment-generating functions $m_{Y_1}(t), \ldots, m_{Y_n}(t)$, respectively. If $U = Y_1 + Y_2 + \cdots + Y_n$, then*
>
> $$m_U(t) = m_{Y_1}(t)m_{Y_2}(t)\cdots m_{Y_n}(t).$$

Proof: *We know that*

$$\begin{aligned} m_U(t) &= E\left[e^{t(Y_1+\cdots+Y_n)}\right] \\ &= E(e^{tY_1}e^{tY_2}\cdots e^{tY_n}) \\ &= E(e^{tY_1})E(e^{tY_2})\cdots E(e^{tY_n}) \end{aligned}$$

since the random variables $Y_1, \ldots, Y_n$ are independent. Thus, by the definition of moment-generating function,

$$m_U(t) = m_{Y_1}(t)m_{Y_2}(t)\cdots m_{Y_n}(t).$$

Example 6.5: *Let $Y_1, \ldots, Y_n$ be independent random variables, each with density*

$$f(y) = \begin{cases} \dfrac{1}{\theta}\, e^{-y/\theta} & y \geq 0, \\ 0,. & \text{elsewhere.} \end{cases}$$

Find the density function of $U = Y_1 + \cdots + Y_n$.

Solution: *To use Theorem 6.2 we must first find $m_{Y_i}(t)$, where*

$$\begin{aligned} m_{Y_i}(t) &= \int_0^\infty \frac{1}{\theta} e^{ty} e^{-y/\theta}\, dy \\ &= \frac{1}{\theta}\int_0^\infty e^{-y[(1-\theta t)/\theta]}\, dy \\ &= (1-\theta t)^{-1}. \end{aligned}$$

Hence

$$\begin{aligned} m_U(t) &= m_{Y_1}(t)\cdots m_{Y_n}(t) \\ &= (1-\theta t)^{-n}, \end{aligned}$$

and it follows that U has a gamma-type density function with $\alpha = n$ and $\beta = \theta$. That is,

$$f_U(u) = \begin{cases} \dfrac{1}{\Gamma(n)\theta^n} u^{n-1} e^{-u/\theta}, & u > 0, \\ 0, & \text{elsewhere.} \end{cases}$$

Summary of the Moment-Generating-Function Method

Let U be a function of the random variables $Y_1, Y_2, \ldots, Y_n$. Then

1. *Find the moment-generating function for U, $m_U(t)$.*
2. *Compare $m_U(t)$ with other well-known moment-generating functions. If $m_U(t) = m_V(t)$, for all values of t, then U and V have identical density functions by Theorem 6.1.*

6.6 Method of Transformations

The transformation method for finding the probability distribution of a function of random variables is simply a generalization of the distribution function method, Section 6.4. Through the distribution function approach we can arrive at a simple method of writing down the density function of $U = h(Y)$

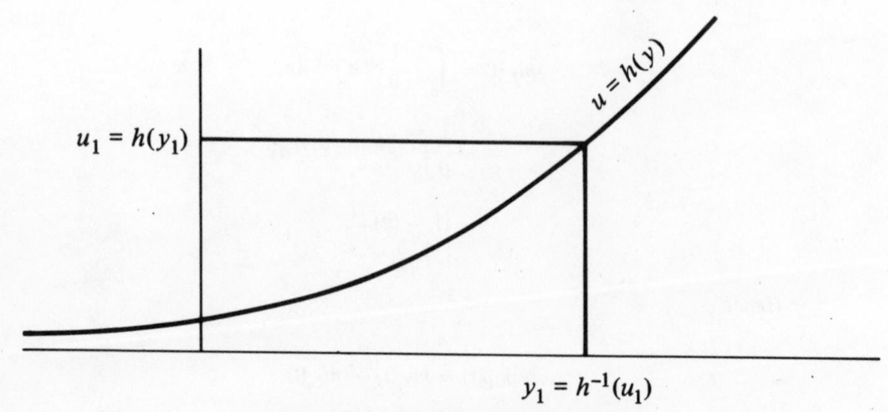

Figure 6.6 An Increasing Function

provided that $h(y)$ is either decreasing or increasing. [By $h(y)$ *increasing* we mean that if $y_1 < y_2$ then $h(y_1) < h(y_2)$ for any real numbers y_1 and y_2.]

Suppose that $h(y)$ is an increasing function of y and suppose $U = h(Y)$, where Y has density function $f_Y(y)$. The symbol $h^{-1}(u)$ denotes the inverse function. That is, if $u = h(y)$, then we can solve for y, obtaining $y = h^{-1}(u)$.

The graph of an increasing function $h(y)$ appears in Figure 6.6. We see from the figure that the set of points y such that $h(y) \leq u_1$ is precisely the same as the set of points y such that $y \leq h^{-1}(u_1)$.

To find the density of $U = h(Y)$ by the distribution function method, we write

$$\begin{aligned} F_U(u) &= P(U \leq u) \\ &= P[h(Y) \leq u] \\ &= P[Y \leq h^{-1}(u)] \\ &= F_Y[h^{-1}(u)], \end{aligned}$$

where $F_Y(y)$ is the distribution function of Y.

To find the density function of U, $f_U(u)$, we must differentiate $F_U(u)$. Since $y = h^{-1}(u)$,

$$F_U(u) = F_Y[h^{-1}(u)] = F_Y(y).$$

Then

$$\begin{aligned} f_U(u) &= \frac{dF_U(u)}{du} \\ &= \frac{dF_Y(y)}{dy}\frac{dy}{du} \\ &= f_Y(y)\frac{dy}{du} \\ &= f_Y[h^{-1}(u)]\frac{dy}{du}. \end{aligned}$$

Note that $dy/du = 1/(du/dy)$.

Example 6.6: *Let Y have the probability density function given by*

$$f_Y(y) = \begin{cases} 2y, & 0 < y < 1, \\ 0, & \text{elsewhere.} \end{cases}$$

Find the density function of $U = 3Y - 1$.

Solution: *The function of interest here is* $h(y) = 3y - 1$, *which is increasing in y. If* $u = 3y - 1$, *then*

$$y = h^{-1}(u) = \frac{u + 1}{3}$$

and

$$\frac{dy}{du} = \frac{1}{3}.$$

Thus

$$f_U(u) = f_Y[h^{-1}(u)]\frac{dy}{du} = 2y\frac{dy}{du}$$

$$= 2 \cdot \frac{u + 1}{3} \cdot \frac{1}{3}$$

$$= \frac{2(u + 1)}{9}, \qquad -1 < u < 2,$$

$$f_U(u) = 0, \qquad \text{elsewhere.}$$

The range over which $f_U(u)$ *is positive is simply the interval* $0 < y < 1$ *transformed to the u axis by the function* $u = 3y - 1$.

If $h(y)$ is a decreasing function, as in Figure 6.7, then the set of points y such that $h(y) \le u_1$ is the same as the set of points y such that $y \ge h^{-1}(u_1)$. If follows that, for $U = h(Y)$,

$$\begin{aligned} F_U(u) &= P(U \le u) \\ &= P[h(Y) \le u] \\ &= P[Y \ge h^{-1}(u)] \\ &= 1 - F_Y[h^{-1}(u)]. \end{aligned}$$

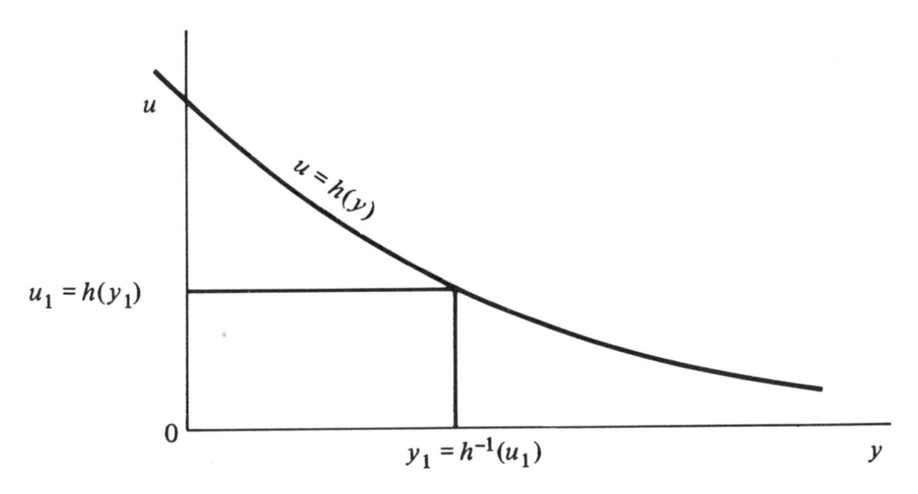

Figure 6.7 A Decreasing Function

On taking derivatives with respect to u, we have

$$f_U(u) = -f_Y[h^{-1}(u)]\frac{dy}{du}.$$

Since dy/du is negative for a decreasing function, the above equation is equivalent to

$$f_U(u) = f_Y[h^{-1}(u)]\left|\frac{dy}{du}\right|.$$

The above results combine into the following statement.

> *Let Y have probability density function $f_Y(y)$. If $h(y)$ is either increasing or decreasing in y, then $U = h(Y)$ has density function given by*
>
> $$f_U(u) = f_Y[h^{-1}(u)]\left|\frac{dy}{du}\right|.$$

Example 6.7: *Let Y have the probability density function given by*

$$f_Y(y) = \begin{cases} 2y, & 0 < y < 1, \\ 0, & \text{elsewhere.} \end{cases}$$

Find the density function of $U = -4Y + 3$.

Solution: *The function of interest, $h(y) = -4y + 3$, is decreasing in y. If $u = -4y + 3$, then*

$$y = h^{-1}(u) = \frac{3-u}{4}$$

and

$$\frac{dy}{du} = -\frac{1}{4}.$$

Thus

$$f_U(u) = f_Y[h^{-1}(u)]\left|\frac{dy}{du}\right| = 2y\left|\frac{dy}{du}\right|$$

$$= 2\cdot\frac{3-u}{4}\cdot\frac{1}{4}$$

$$= \frac{3-u}{8}, \qquad -1 < u < 3,$$

$$f_U(u) = 0, \qquad \text{elsewhere.}$$

The transformation method as outlined above can be readily applied to some functions that are neither increasing nor decreasing. To illustrate, we shall consider the case $U = h(Y) = Y^2$, where Y is still continuous with distribution function $F_Y(y)$ and density function $f_Y(y)$. We then have

$$F_U(u) = P(U \leq u) = P(Y^2 \leq u)$$

$$= P(-\sqrt{u} \leq Y \leq \sqrt{u})$$

$$= F_Y(\sqrt{u}) - F_Y(-\sqrt{u}).$$

On differentiating with respect to u we see that

$$f_U(u) = f_Y(\sqrt{u})\frac{1}{2\sqrt{u}} + f_Y(-\sqrt{u})\frac{1}{2\sqrt{u}}$$

$$= \frac{1}{2\sqrt{u}}[f_Y(\sqrt{u}) + f_Y(-\sqrt{u})].$$

Example 6.8: *Let Y have the probability density function given by*

$$f_Y(y) = \begin{cases} \dfrac{y+1}{2}, & -1 \leq y \leq 1, \\ 0, & \textit{elsewhere.} \end{cases}$$

Find the density function for $U = Y^2$.

Solution: *We know that*

$$f_U(u) = \frac{1}{2\sqrt{u}}[f_Y(\sqrt{u}) + f_Y(-\sqrt{u})]$$

and, on substituting into this equation,

$$f_U(u) = \frac{1}{2\sqrt{u}}\left(\frac{\sqrt{u}+1}{2} + \frac{-\sqrt{u}+1}{2}\right)$$

$$= \frac{1}{2\sqrt{u}}, \qquad 0 \leq u \leq 1,$$

$$f_U(u) = 0, \qquad \textit{elsewhere.}$$

Note that since Y has positive density over the interval $-1 \leq y \leq 1$, $U = Y^2$ *has positive density over the interval* $0 \leq u \leq 1$.

The transformation method can also be used in multivariate situations. The following example gives an illustration for the bivariate case.

Example 6.9: *Let Y_1 and Y_2 have joint density function given by*

$$f(y_1, y_2) = \begin{cases} e^{-(y_1 + y_2)}, & 0 \le y_1;\ 0 \le y_2, \\ 0, & \text{elsewhere.} \end{cases}$$

Find the density function for $U = Y_1 + Y_2$.

Solution: *This problem must be solved in two stages; first we will find the joint density of Y_1 and U and, second, we will find the marginal density of U. The approach is to let one of the original variables, say Y_1, be fixed at a value y_1. Then $U = y_1 + Y_2$ and we can consider the one-dimensional transformation problem in which $U = h(Y_2) = y_1 + Y_2$. Letting $g(y_1, u)$ denote the joint density of Y_1 and U, we have*

$$g(y_1, u) = f[y_1, h^{-1}(u)]\left|\frac{dy_2}{du}\right|$$

$$= e^{-u}(1), \qquad 0 \le u;\ 0 \le y_1 \le u,$$

$$g(y_1, u) = 0, \qquad \text{otherwise.}$$

(Note that $Y_1 \le U$.) The marginal density of U is then given by

$$f_U(u) = \int_{-\infty}^{\infty} g(y_1, u)\, dy_1$$

$$= \int_0^u e^{-u}\, dy_1$$

$$= ue^{-u}, \qquad 0 \le u,$$

$$f_U(u) = 0, \qquad \text{elsewhere.}$$

Summary of the Transformation Method

Let U be an increasing or decreasing function of the random variable Y, say $U = h(Y)$.

1. *Find the inverse function, $Y = h^{-1}(U)$.*
2. *Evaluate dy/du.*
3. *Find $f_U(u)$ by*

$$f_U(u) = f_Y[h^{-1}(u)]\left|\frac{dy}{du}\right|.$$

6.7 Method of Conditioning

Conditional density functions frequently provide a convenient path to follow in finding distributions of functions of random variables. Suppose that Y_1 and Y_2 have a joint density function, and we want to find the density function for $U = h(Y_1, Y_2)$. We note that the density of $U, f_U(u)$, can be written

$$f_U(u) = \int_{-\infty}^{\infty} f(u, y_2)\, dy_2 .$$

Since the conditional density of U given Y_2 is given by

$$f(u|y_2) = \frac{f(u, y_2)}{f_2(y_2)},$$

it follows that

$$f_U(u) = \int_{-\infty}^{\infty} f(u|y_2) f_2(y_2)\, dy_2 .$$

Example 6.10: *Let Y_1 and Y_2 be independent random variables with each having the density function*

$$f_Y(y) = \begin{cases} e^{-y}, & y > 0, \\ 0, & \text{elsewhere.} \end{cases}$$

Find the density function for $U = Y_1/Y_2$.

Solution: *We will first find the conditional density of U given $Y_2 = y_2$, which is obtained by the transformation method. When Y_2 is held fixed at the constant value y_2, U is simply Y_1/y_2, where Y_1 has the exponential density function. Thus we want the density function of $U = Y_1/y_2$. The inverse function is $Y_1 = Uy_2$.*

By the method of Section 6.6,

$$f(u|y_2) = f_{Y_1}(uy_2)\left|\frac{dy_1}{du}\right|$$

$$= e^{-uy_2}y_2, \qquad u > 0,$$

$$f(u|y_2) = 0, \qquad \text{elsewhere.}$$

Now

$$f_U(u) = \int_{-\infty}^{\infty} f(u|y_2)f(y_2)\,dy_2$$

$$= \int_0^{\infty} e^{-uy_2}y_2(e^{-y_2})\,dy_2$$

$$= \int_0^{\infty} y_2 e^{-y_2(u+1)}\,dy_2$$

$$= (u+1)^{-2}, \qquad u > 0,$$

$$f_U(u) = 0, \qquad \text{elsewhere.}$$

The evaluation of the above integral is facilitated by observing that the integrand is a gamma function.

Summary of the Conditioning Method

Let U be a function of the random variables Y_1 and Y_2.

1. *Find the conditional density of U given $Y_2 = y_2$. (This is usually found by transformations.)*
2. *Find $f_U(u)$ from the relation*

$$f_U(u) = \int_{-\infty}^{\infty} f(u|y_2) f(y_2)\, dy_2.$$

6.8 Order Statistics

Many functions of random variables which are of interest in practice depend on the relative magnitudes of the observed variables. For instance, we may be interested in the fastest time in an automobile race or the heaviest mouse among those fed on a certain diet. Thus we often order observed random variables according to their magnitudes. The resulting ordered variables are called *order statistics.*

Formally, let $Y_1, Y_2, \ldots, Y_n$ denote independent continuous random variables with distribution function $F(y)$ and density function $f(y)$. We will denote the ordered random variables, Y_i, by $Y_{(1)}, Y_{(2)}, \ldots, Y_{(n)}$, where $Y_{(1)} \leq Y_{(2)} \leq \cdots \leq Y_{(n)}$. (Since the random variables are continuous, the equality signs can be ignored.) That is,

$$Y_{(1)} = \min(Y_1, \ldots, Y_n),$$

the minimum of the Y_i's, and

$$Y_{(n)} = \max(Y_1, \ldots, Y_n),$$

the maximum of the Y_i's.

The probability density functions for $Y_{(1)}$ and $Y_{(n)}$ are easily found. Looking at $Y_{(n)}$ first, we see that

$$P[Y_{(n)} \leq y] = P(Y_1 \leq y, Y_2 \leq y, \ldots, Y_n \leq y),$$

since $[Y_{(n)} \leq y]$ implies that all the Y_i's must be less than or equal to y, and vice versa. But $Y_1, \ldots, Y_n$ are independent; hence

$$P[Y_{(n)} \leq y] = [F(y)]^n.$$

Letting $g_n(y)$ denote the density function of $Y_{(n)}$, we see that, on taking derivatives on both sides,

$$g_n(y) = n[F(y)]^{n-1}f(y).$$

The density function of $Y_{(1)}$, denoted by $g_1(y)$, can be found by a similar device. We have that

$$\begin{aligned} P[Y_{(1)} \leq y] &= 1 - P[Y_{(1)} > y] \\ &= 1 - P(Y_1 > y, Y_2 > y, \ldots, Y_n > y) \\ &= 1 - [1 - F(y)]^n. \end{aligned}$$

Hence

$$g_1(y) = n[1 - F(y)]^{n-1}f(y).$$

Let us now consider the case $n = 2$ and find the joint density of $Y_{(1)}$ and $Y_{(2)}$. Now the event $[Y_{(1)} \le y_1, Y_{(2)} \le y_2]$ means that either $(Y_1 \le y_1, Y_2 \le y_2)$ or $(Y_2 \le y_1, Y_1 \le y_2)$. [Note that $Y_{(1)}$ could be either Y_1 or Y_2, whichever is smaller.] Thus for $y_1 \le y_2$,

$$
\begin{aligned}
P[Y_{(1)} &\le y_1, Y_{(2)} \le y_2] \\
&= P[(Y_1 \le y_1, Y_2 \le y_2) \cup (Y_2 \le y_1, Y_1 \le y_2)] \\
&= P(Y_1 \le y_1, Y_2 \le y_2) + P(Y_2 \le y_1, Y_1 \le y_2) \\
&\quad - P(Y_1 \le y_1, Y_2 \le y_1)
\end{aligned}
$$

(by the additive laws of probabilities)

$$= 2F(y_1)F(y_2) - [F(y_1)]^2.$$

Letting $g_{12}(y_1, y_2)$ denote the joint density of $Y_{(1)}$ and $Y_{(2)}$, we see that, on differentiating first with respect to y_2 and then with respect to y_1,

$$g_{12}(y_1, y_2) = \begin{cases} 2f(y_1)f(y_2), & y_1 \le y_2, \\ 0, & \text{elsewhere.} \end{cases}$$

The same method can be used to find the joint density of $Y_{(1)}, \ldots, Y_{(n)}$, which turns out to be

$$
\begin{aligned}
&g_{12\cdots n}(y_1, \ldots, y_n) \\
&\quad = \begin{cases} n!f(y_1), \ldots, f(y_n), & y_1 \le y_2 \le \cdots \le y_n, \\ 0, & \text{elsewhere.} \end{cases}
\end{aligned}
$$

The marginal density function for any of the order statistics can be found from this joint density function, but we will not pursue the matter in this text.

Example 6.11: *Electronic components of a certain type have life length, Y, with probability density given by*

$$f(y) = \begin{cases} (1/100)e^{-y/100}, & y > 0y, \\ 0, & \text{elsewhere.} \end{cases}$$

(Life length is measured in hours.) Suppose that two such components operate independently and in series in a certain system (that is, the system fails when either component fails). Find the density function for X, the life length of the system.

Solution: *Since the system fails at the first component failure,* $X = \min(Y_1, Y_2)$, *where* Y_1 *and* Y_2 *are independent random variables with the given density. Then*

$$\begin{aligned} f_X(x) &= g_1(y) = n[1 - F(y)]^{n-1}f(y) \\ &= 2e^{-y/100}(1/100)e^{-y/100} \\ &= 50e^{-y/50}, \qquad y > 0, \\ f_X(x) &= 0, \qquad \text{elsewhere.} \end{aligned}$$

Note that, for this problem, $F(y) = 1 - e^{-y/100}$, $y \geq 0$.

Example 6.12: *Suppose that in Example 6.11 the components operate in parallel (that is, the system does not fail until both components fail). Find the density function for X, the life length of the system.*

Solution: *Now* $X = \max(Y_1, Y_2)$ *and*

$$f_X(x) = g_2(y) = n[F(y)]^{n-1}f(y) = \begin{cases} 2(1 - e^{y/100})e^{-y/100}, & y > 0, \\ 0, & \text{elsewhere.} \end{cases}$$

6.9 Distributions of Some Common Functions of Random Variables

We shall now investigate the probability distributions of some common functions of random variables which will be used throughout Chapters 8, 9, and 10. Theorems 6.3, 6.4, and 6.5 all involve functions of normally distributed random variables.

> ***Theorem 6.3:*** *Let $Y_1, \ldots, Y_n$ be independent normally distributed random variables with $E(Y_i) = \mu_i$ and $V(Y_i) = \sigma_i^2$, $i = 1, \ldots, n$. Define U by*
>
> $$U = \sum_{i=1}^{n} a_i Y_i,$$
>
> *where $a_1, \ldots, a_n$ are constants. Then U is a normally distributed random variable with*
>
> $$E(U) = \sum_{i=1}^{n} a_i \mu_i$$
>
> *and*
>
> $$V(U) = \sum_{i=1}^{n} a_i^2 \sigma_i^2.$$

Proof: *The moment-generating-function method works nicely for proving this theorem: Y_i has moment-generating function*

$$m_{Y_i}(t) = \exp\left(\mu_i t + \frac{\sigma_i^2 t^2}{2}\right).$$

Then $a_i Y_i$ has moment-generating function

$$\begin{aligned} m_{a_i Y_i}(t) &= E(e^{t a_i Y_i}) \\ &= m_{Y_i}(a_i t) \\ &= \exp\left(\mu_i a_i t + \frac{a_i^2 \sigma_i^2 t^2}{2}\right). \end{aligned}$$

It then follows from Theorem 6.2 that

$$m_U(t) = m_{a_1Y_1}(t)m_{a_2Y_2}(t)\cdots m_{a_nY_n}(t)$$
$$= \exp\left(\mu_1 a_1 t + \frac{a_1^2\sigma_1^2 t^2}{2}\right)\ldots\exp\left(\mu_n a_n t + \frac{a_n^2\sigma_n^2 t^2}{2}\right)$$
$$= \exp\left(t\sum_{i=1}^n a_i\mu_i + t^2\sum_{i=1}^n a_i^2\sigma_i^2/2\right)$$

[exp( ) *is simply a more convenient way to write* $e^{(\ )}$ *when the term is long or complex.*] *Thus U is normal with mean* $\sum_{i=1}^n a_i\mu_i$ *and variance* $\sum_{i=1}^n a_i^2\sigma_i^2$.

Theorem 6.4: *Let* $Y_1, \ldots, Y_n$ *be as in Theorem 6.3 and define* Z_i *by*

$$Z_i = \frac{Y_i - \mu_i}{\sigma_i}, \qquad i = 1, \ldots, n.$$

Then $\sum_{i=1}^n Z_i^2$ *has a* χ^2 *distribution with n degrees of freedom.*

Proof: *Recall from Chapter 4 that a* χ^2 *density with n degrees of freedom is a gamma-type density with* $\alpha = n/2$ *and* $\beta = 2$. *Now*

$$m_{Z_i}(t) = E(e^{tZ_i}) = E\left[\exp\left(\frac{tY_i}{\sigma_i} - \frac{t\mu_i}{\sigma_i}\right)\right]$$
$$= e^{-t\mu_i/\sigma_i}E(e^{(t/\sigma_i)Y_i})$$
$$= e^{-t\mu_i/\sigma_i}m_{Y_i}(t/\sigma_i)$$
$$= e^{-t\mu_i/\sigma_i}\exp\left(\frac{t\mu_i}{\sigma_i} + \frac{t^2}{2}\right)$$
$$= e^{t^2/2},$$

and hence Z_i *is normally distributed with mean* 0 *and variance* 1.

From Example 6.4 we then have that Z_i^2 is a χ^2 random variable with one degree of freedom. Thus

$$m_{Z_i^2}(t) = (1 - 2t)^{-1/2},$$

and from Theorem 6.2, with $V = \sum_{i=1}^{n} Z_i^2$,

$$m_V(t) = (1 - 2t)^{-n/2}$$

or V is distributed as a χ^2 random variable with n degrees of freedom.

Theorem 6.5: *If $Y_1, \ldots, Y_n$ are independent normal random variables with common mean μ and common variance σ^2, then*

$$\frac{1}{\sigma^2}\sum_{i=1}^{n}(Y_i - \bar{Y})^2 = \frac{(n-1)S^2}{\sigma^2}$$

has a χ^2 distribution with $(n - 1)$ degrees of freedom. Also, $\bar{Y}$ and S^2 are independent random variables.

The proof of Theorem 6.5 is outlined in Exercise 13.27.

We now define two random variables which play a prominent role in the theory of statistics.

Definition 6.1: *Let U be a standard normal random variable and V a χ^2 random variable with n degrees of freedom. Then, if U and V are independent,*

$$T = \frac{U}{\sqrt{V/n}}$$

is said to have a t distribution *with n degrees of freedom.*

Definition 6.2: *Let V and W be χ^2 random variables with n_1 and n_2 degrees of freedom, respectively. Then, if V and W are independent,*

$$F = \frac{V/n_1}{W/n_2}$$

is said to have an F distribution *with n_1 and n_2 degrees of freedom.*

Hints for the derivation of the density functions for T and F are given in Exercises 6.25 and 6.26. Some of the areas under the T and F densities are given in Tables 4, 6, and 7, Appendix III.

6.10 Summary

This chapter has been concerned with finding probability distributions for functions of random variables. This is an important problem in statistics because estimators of population parameters are functions of random variables. Hence it is necessary to know something about the probability distributions of these functions (or estimators) in order to evaluate the goodness of our statistical procedures. A discussion of estimation will be presented in Chapters 8 and 9.

The methods for finding probability distributions for functions of random variables are the distribution function method, Section 6.4, the moment-generating-function method, Section 6.5, the transformation method, Section 6.6, and the conditioning method, Section 6.7. It should be noted that no one method is always best, as the method of solution depends a great deal on the nature of the function involved. Facility for handling these methods can only be achieved through practice. The exercises at the end of the chapter provide a good starting point.

References

1. Hoel, P. G., *Introduction to Mathematical Statistics*, 3rd ed. New York: John Wiley & Sons, Inc., 1962.
2. Hogg, R. V., and A. T. Craig, *Introduction to Mathematical Statistics*, 3rd ed. New York: The Macmillan Company, 1970.
3. Mood, A. M., and F. A. Graybill, *Introduction to the Theory of Statistics*, 2nd ed. New York: McGraw-Hill Book Company, 1963.
4. Parzen, E., *Modern Probability Theory and Its Applications*. New York: John Wiley & Sons, Inc., 1964.

Exercises

6.1. Let Y be a random variable with probability density function given by

$$f(y) = \begin{cases} 2(1 - y), & 0 \le y \le 1, \\ 0, & \text{elsewhere.} \end{cases}$$

Find the density function of
(a) $U_1 = 2Y - 1$.
(b) $U_2 = 1 - 2Y$.
(c) $U_3 = Y^2$.

6.2. Let Y be a random variable with density function given by

$$f(y) = \begin{cases} (3/2)y^2, & -1 \le y \le 1, \\ 0, & \text{elsewhere.} \end{cases}$$

Find the density function of
(a) $U_1 = 3Y$.
(b) $U_2 = 3 - Y$.
(c) $U_3 = Y^2$.

6.3. Suppose that Y_1 and Y_2 are independent standard normal random variables. Find the density function of

$$U = Y_1^2 + Y_2^2.$$

6.4. Let $Y_1, Y_2, \ldots, Y_n$ be independent normal random variables, each with mean μ and variance σ^2. Let $a_1, a_2, \ldots, a_n$ denote known constants. Find the density function of the linear combination

$$U = \sum_{i=1}^{n} a_i Y_i.$$

6.5. Let Y_1 and Y_2 be independent and uniformly distributed over the interval (0, 1). Find the probability density function of
(a) $U_1 = Y_1^2$.
(b) $U_2 = Y_1/Y_2$.
(c) $U_3 = -\ln(Y_1 Y_2)$.
(d) $U_4 = Y_1 Y_2$.

6.6. Let Y_1 be a binomial random variable with n_1 trials and probability of success given by p. Let Y_2 be another binomial random variable with n_2 trials and probability of success given by p. If Y_1 and Y_2 are independent, find the probability function of $Y_1 + Y_2$.

6.7. Let Y_1 and Y_2 be independent Poisson random variables with mean λ_1 and λ_2, respectively.
(a) Find the probability function of $Y_1 + Y_2$.
(b) Find the conditional probability function of Y_1, given $Y_1 + Y_2 = m$.

6.8. Refer to Exercise 6.1.
(a) Find the expected values of U_1, U_2, and U_3 directly (without using the density functions of U_1, U_2, and U_3).
(b) Find the expected values of U_1, U_2, and U_3 by using the derived density functions for these random variables.

6.9. A parachutist wants to land at a target, T, but finds that he is equally likely to land at any point on a straight line (A, B) of which T is the midpoint. Find the probability density function of the distance between his landing point and the target. (*Hint*: Denote A by -1, B by $+1$, and T by 0. Then the parachutist's landing point has a coordinate, X, which is uniformly distributed between -1 and $+1$. The distance between X and T is $|X|$.)

6.10. Let Y_1 denote the amount of a bulk item stocked by a supplier at the beginning of a day, and let Y_2 denote the amount of that item sold during the day. Suppose that Y_1 and Y_2 have joint density function

$$f(y_1, y_2) = \begin{cases} 2, & 0 \le y_2 \le y_1 \le 1, \\ 0, & \text{elsewhere.} \end{cases}$$

Of interest to this supplier is the random variable $U = Y_1 - Y_2$, which denotes the amount he has left at the end of the day.
(a) Find the probability density function for U.
(b) Find $E(U)$.
(c) Find $V(U)$.

6.11. An efficiency expert takes two independent measurements, Y_1 and Y_2, on the length of time it takes workmen to complete a certain task. Each measurement is assumed to have the density function given by

$$f(y) = \begin{cases} (1/4)ye^{-y/2}, & y > 0, \\ 0, & \text{elsewhere.} \end{cases}$$

Find the density function for the average

$$U = (1/2)(Y_1 + Y_2).$$

(*Hint*: Use moment-generating functions.)

6.12. The opening prices per share of two similar stocks, Y_1 and Y_2, are independent random variables, each with density function

$$f(y) = \begin{cases} (1/2)e^{-(1/2)(y-4)}, & y \geq 4, \\ 0, & \text{elsewhere.} \end{cases}$$

On a given morning Mr. A is going to buy shares of whichever stock is less expensive. Find the probability density function for the price per share that Mr. A will have to pay.

6.13. A certain type of elevator has a maximum weight capacity, Y_1, which is normally distributed with a mean and standard deviation of 5000 and 300 pounds, respectively. For a certain building equipped with this type of elevator, the elevator loading, Y_2, is a normally distributed random variable with a mean and standard deviation of 4000 and 400 pounds, respectively. For any given time that the elevator is in use, find the probability that it will be overloaded, assuming that Y_1 and Y_2 are independent.

6.14. The length of time that a certain machine operates without failure is denoted by Y_1 and the length of time to repair a failure is denoted by Y_2. After a repair is made, the machine is assumed to operate like a new machine. Y_1 and Y_2 are independent and each has the density function

$$f(y) = \begin{cases} e^{-y}, & y > 0, \\ 0, & \text{elsewhere.} \end{cases}$$

Find the probability density function for

$$U = \frac{Y_1}{Y_1 + Y_2},$$

the proportion of time that the machine is in operation during any one operation–repair cycle.

6.15. The Weibull density function is given by

$$f(y) = \begin{cases} \frac{1}{\alpha} m y^{m-1} e^{-y^m/\alpha}, & y > 0, \\ 0, & \text{elsewhere}, \end{cases}$$

where α and m are positive constants.
If Y has the above Weibull density,
(a) Find the density function of $U = Y^m$.
(b) Find $E(Y^k)$ for any positive integer k.

6.16. Let Y_1 and Y_2 be independent and uniformly distributed over the interval (0, 1). Find the probability density function of
(a) $U_1 = \min(Y_1, Y_2)$.
(b) $U_2 = \max(Y_1, Y_2)$.

6.17. Suppose that a unit of mineral ore contains a proportion, Y_1, of metal A and a proportion, Y_2, of metal B. Experience has shown that the joint probability density function of (Y_1, Y_2) is uniform over the region $0 \leq y_1 \leq 1$, $0 \leq y_2 \leq 1$, $0 \leq y_1 + y_2 \leq 1$. Let $U = Y_1 + Y_2$, the proportion of metals A and B per unit.
(a) Find the probability density function for U.
(b) Find $E(U)$ by using the answer to part (a).
(c) Find $E(U)$ by using the marginal densities of Y_1 and Y_2 only.

6.18. Two sentries are sent to patrol a road 1 mile long. The sentries are sent to points chosen independently and at random along the road. Find the probability that the sentries will be less than 1/2 mile apart when they reach their assigned posts.

6.19. Let Y_1 and Y_2 be independent random variables, each having a normal distribution with mean μ and variance 1. Define $U_1 = Y_1 + Y_2$, $U_2 = Y_1 - Y_2$.

(a) Find the joint density of U_1 and U_2.
(b) Are U_1 and U_2 independent?

6.20. Let Y_1 and Y_2 be independent standard normal random variables. Find the probability density function of $U = Y_1/Y_2$.

6.21. If Y is a continuous random variable with distribution function $F(y)$, find the probability density function of $U = F(Y)$.

6.22. Let Y be uniformly distributed over the interval $(-1, 3)$. Find the probability density function of $U = Y^2$.

6.23. If Y denotes the life length of a component and $F(y)$ is the distribution function of Y, then $P(Y > y) = 1 - F(y)$ is called the reliability of the component. Suppose that a system consists of four components with identical reliability functions, $1 - F(y)$, operating as indicated:

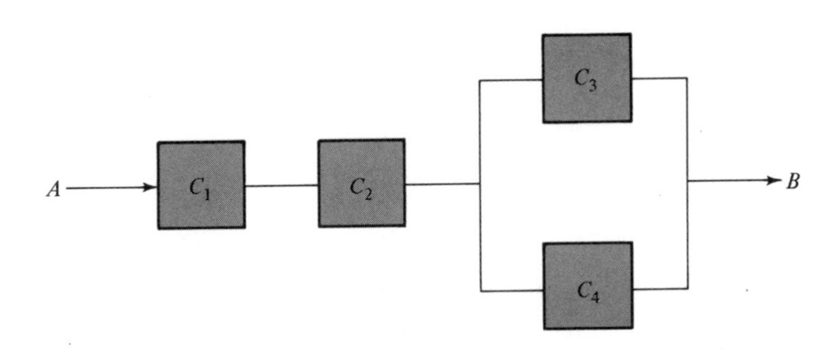

The system operates correctly if an unbroken chain of components is in operation between A and B. If the four components operate independently, find the reliability of the system, in terms of $F(y)$.

6.24. Let $Y_1, Y_2, \ldots, Y_n$ denote a random sample from the uniform distribution, $f(y) = 1, 0 \leq y \leq 1$. Find the probability density function for the range, $R = Y_{(n)} - Y_{(1)}$.

6.25. Suppose that T is defined as in Definition 6.1.
(a) If V is fixed at v, then T is given by U/c, where $c = \sqrt{v/n}$. Use this idea to find the conditional density of T for fixed $V = v$.
(b) Find the joint density of T and $V, f(t, v)$, by using

$$f(t, v) = f(t|v)f(v).$$

(c) Integrate over v to show that

$$f(t) = \frac{\Gamma[(n+1)/2]}{\sqrt{\pi n}\,\Gamma(n/2)}(1 + t^2/n)^{-(n+1)/2}, \qquad -\infty < t < \infty.$$

6.26. Suppose F is defined as in Definition 6.2.

(a) If W is fixed at w, then $F = W/c$, where $c = wn_1/n_2$. Thus find the conditional density of F for fixed $W = w$.

(b) Find the joint density of F and W.

(c) Integrate over w to show that the probability density function of F, $g(f)$, is given by

$$g(f) = \frac{\Gamma[(n_1 + n_2)/2](n_1/n_2)^{n_1/2}}{\Gamma(n_1/2)\Gamma(n_2/2)}(f)^{\left(\frac{n_1}{2} - 1\right)}\left(1 + \frac{n_1 f}{n_2}\right)^{-(n_1+n_2)/2},$$

$$0 < f < \infty.$$

7

Some Approximations to Probability Distributions: Limit Theorems

7.1 *Introduction*

As noted in Chapter 6, we are frequently interested in functions of random variables, such as their average or sum. These quantities will be used to estimate or make decisions about population parameters. To accomplish these inferential goals, we presented methods in Chapter 6 for finding probability distributions for functions of random variables.

Unfortunately, the application of these methods may lead to intractable mathematical problems. Hence we need some simple methods for approximating the probability distributions of functions of random variables.

In Chapter 7 we will discuss some properties of functions of random variables when the number of variables, n, gets large (approaches infinity). We will see, for example, that the distribution for certain functions of random variables can be easily approximated for large n even though the exact distribution for fixed n may be difficult to obtain. Even more important, the approximations are sometimes good for samples of modest size and, in some instances, for samples as small as $n = 5$ or 6. In Section 7.2 we shall present extremely useful theorems, called *limit theorems*, which give properties of random variables as n tends to infinity.

7.2 Convergence in Probability

Suppose that a coin has probability p, $0 \leq p \leq 1$, of coming up heads on a single toss and that we toss the coin n times. What can be said about the fraction of heads observed in the n tosses? Intuition tells us that the sampled fraction of heads provides an estimate of p and we would expect the estimate to fall closer to p for larger sample sizes. That is, we would expect the estimate to fall closer to p as the quantity of information in the sample is increased. Although our supposition and intuition may be correct for many problems of estimation, it is *not* always true that larger sample sizes lead to better estimates. Hence this example gives rise to a question that occurs in all estimation problems. What can be said about the random distance between an estimate and its target parameter?

Notationally, let X denote the number of heads observed in the n tosses. Then $E(X) = np$ and $V(X) = np(1 - p)$. One way to measure the closeness of X/n to p is to examine the probability that the distance, $|(X/n) - p|$, will be less than a preassigned real number ε. This probability,

$$P\left(\left|\frac{X}{n} - p\right| \leq \varepsilon\right),$$

should be close to unity for large n if our intuition is correct.

The following definition formalizes this convergence concept.

Definition 7.1: *The sequence of random variables, $X_1, X_2, \ldots, X_n$, is said to* converge in probability *to the constant c if for every positive number ε,*

$$\lim_{n\to\infty} P(|X_n - c| \leq \varepsilon) = 1.$$

The following theorem often provides a mechanism for proving convergence in probability.

Theorem 7.1: *Let $X_1, \ldots, X_n$ be independent and identically distributed random variables, with $E(X_i) = \mu$ and $V(X_i) = \sigma^2 < \infty$. Let $\overline{X}_n = (1/n) \sum_{i=1}^{n} X_i$. Then, for any positive real number ε,*

$$\lim_{n \to \infty} P(|\overline{X}_n - \mu| > \varepsilon) = 0$$

or

$$\lim_{n \to \infty} P(|\overline{X}_n - \mu| \leq \varepsilon) = 1.$$

That is, $\overline{X}_n$ converges in probability to μ.

Proof: *Note that $E(\overline{X}_n) = \mu$ and $V(\overline{X}_n) = \sigma^2/n$. To prove the theorem, we appeal to Tchebysheff's theorem (see Section 3.11 or Section 4.10), which states that*

$$P(|X - \mu| > k\sigma) < \frac{1}{k^2},$$

where $E(X) = \mu$ and $V(X) = \sigma^2$. In the context of our theorem, X is to be replaced by $\overline{X}_n$ and σ^2 by σ^2/n. It then follows that

$$P\left(|\overline{X}_n - \mu| > k\frac{\sigma}{\sqrt{n}}\right) < \frac{1}{k^2}.$$

Note that k can be any real number, so we will choose

$$k = \frac{\varepsilon}{\sigma}\sqrt{n}.$$

Then

$$P\left(|\overline{X}_n - \mu| > \frac{\varepsilon\sqrt{n}}{\sigma}\frac{\sigma}{\sqrt{n}}\right) < \frac{\sigma^2}{\varepsilon n}$$

or

$$P(|\overline{X}_n - \mu| > \varepsilon) < \frac{\sigma^2}{\varepsilon n}.$$

Now let us take the limit of this expression as n tends to infinity.

Recall that σ^2 is finite and ε is a positive real number. On taking the limit as n tends to infinity, we have

$$\lim_{n\to\infty} \frac{\sigma^2}{n\varepsilon} = \lim_{n\to\infty} P(|\overline{X}_n - \mu| > \varepsilon) = 0.$$

That $\lim_{n\to\infty} P(|\overline{X}_n - \mu| \le \varepsilon) = 1$ follows directly because

$$P(|\overline{X}_n - \mu| > \varepsilon) = 1 - P(|\overline{X}_n - \mu| \le \varepsilon).$$

We will now apply Theorem 7.1 to our coin-tossing example.

Example 7.1: *Let X be a binomial random variable with probability of success p and number of trials n. Show that X/n converges in probability to p.*

Solution: *We have seen that we can write X as $\sum_{i=1}^{n} X_i$, where $X_i = 1$ if the ith trial results in success and $X_i = 0$ otherwise. Then*

$$\frac{X}{n} = \frac{1}{n} \sum_{i=1}^{n} X_i.$$

Also, $E(X_i) = p$ and $V(X_i) = p(1 - p)$. The conditions of Theorem 7.1 are then fulfilled with $\mu = p$ and $\sigma^2 = p(1 - p)$, and we conclude that

$$\lim_{n\to\infty} \frac{X_i}{n} = \lim_{n\to\infty} P\left(\left|\frac{X}{n} - p\right| > \varepsilon\right) = 0$$

for any positive ε. from $\frac{X}{n}$ as X is finite and n gets infinitely large

Theorem 7.1 is sometimes called the *law of large numbers.* It is the theoretical justification for the averaging process employed by many experimenters to obtain precision in measurements. For example, an experimenter may take the average of five measurements of the weight of an animal to obtain a more precise estimate of an animal's weight. His feeling, a feeling born out by Theorem 7.1, is that the average of a number of independently selected weights should be quite close to the true weight, with high probability.

Like the law of large numbers, the theory of convergence in probability has many applications. Theorem 7.2, which we present without proof, points out some properties of the concept of convergence in probability.

Theorem 7.2: *Suppose that X_n converges in probability to μ_1 and Y_n converges in probability to μ_2. Then*
(a) $X_n + Y_n$ converges in probability to $\mu_1 + \mu_2$.
(b) $X_n Y_n$ converges in probability to $\mu_1 \mu_2$.
(c) X_n / Y_n converges in probability to μ_1/μ_2, provided that $\mu_2 \neq 0$.
(d) $\sqrt{X_n}$ converges in probability to $\sqrt{\mu_1}$, provided that $P(X_n \geq 0) = 1$.

Example 7.2: *Suppose that $X_1, X_2, \ldots, X_n$ are independent and identically distributed random variables with $E(X_i) = \mu$, $E(X_i^2) = \mu_2'$, $E(X_i^3) = \mu_3'$, and $E(X_i^4) = \mu_4'$, all assumed finite. Let S'^2 denote the sample variance given by*

$$S'^2 = \frac{1}{n} \sum_{i=1}^{n} (X_i - \overline{X})^2.$$

Show that S'^2 converges in probability to $V(X_i)$.

Solution: *First, note that we can write*

$$S'^2 = \frac{1}{n} \sum_{i=1}^{n} X_i^2 - \overline{X}^2,$$

where

$$\overline{X} = \frac{1}{n} \sum_{i=1}^{n} X_i.$$

To show that S'^2 converges in probability to $V(X_i)$, we will apply both Theorem 7.1 and 7.2. Look at the terms in S'^2. The quantity $(1/n) \sum_{i=1}^{n} X_i^2$ is the average of n independent and identically distributed variables of the form X_i^2, with $E(X_i^2) = \mu_2'$ and $V(X_i^2) = \mu_4' - (\mu_2')^2$. Since $V(X_i^2)$ is assumed to be finite, Theorem 7.1 tells us that $(1/n) \sum_{i=1}^{n} X_i^2$ converges in probability to μ_2'. Now consider the limit of $\overline{X}^2$ as n approaches infinity. Theorem 7.1 tells us that $\overline{X}$ converges in probability to μ, and it follows from Theorem 7.2, part (b), that $\overline{X}^2$ converges in probability to μ^2. This leads to the final step. Having shown that $(1/n) \sum_{i=1}^{n} X_i^2$ and $\overline{X}^2$ converge in probability to μ_2' and μ^2, respectively, it follows from Theorem 7.2 that

$$S'^2 = \frac{1}{n} \sum_{i=1}^{n} X_i^2 - \overline{X}^2$$

converges in probability to $\mu_2' - \mu^2 = V(X_i)$.

This example shows that, for large samples, the sample variance should be close to the population variance with high probability.

7.3 Convergence in Distribution

In Section 7.2 we only dealt with the convergence of certain random variables to constants and we said nothing about the form of the probability distributions. In this section we will look at what happens to the probability distributions of certain types of random variables as n tends to infinity. We need the following definition before proceeding.

Definition 7.2: *Let $X_1, \ldots, X_n$ be random variables and let $g(X_1, \ldots, X_n)$ be a function of $X_1, \ldots, X_n$ with distribution function $F_n(x)$. Let Y be a random variable with distribution function $F(y)$. If*

$$\lim_{n \to \infty} F_n(y) = F(y)$$

at every point y for which $F(y)$ is continuous, then $g(X_1, \ldots, X_n)$ is said to converge in distribution *to Y. $F(y)$ is called the* limiting distribution function *of $g(X_1, \ldots, X_n)$.*

We will illustrate Definition 7.2 with the following example.

Example 7.3: *Let $X_1, \ldots, X_n$ be independent uniform random variables over the interval $(0, \theta)$ for a positive constant θ. Let $Y_n = \max(X_1, \ldots, X_n)$. Find the limiting distribution of Y_n.*

Solution: *The distribution function for the uniform random variable X_i is*

$$F_X(y) = P(X_i \le y) = \begin{cases} 0, & y \le 0, \\ \dfrac{y}{\theta}, & 0 < y < \theta, \\ 1, & y \ge \theta. \end{cases}$$

In Section 6.8 we found that the distribution function for Y_n is

$$G(y) = P(Y_n \leq y) = [F_X(y)]^n$$

where $F_X(y)$ is the distribution function for each X_i. Then

$$\lim_{n\to\infty} G(y) = \begin{cases} 0, & y \leq 0, \\ \lim_{n\to\infty}\left(\dfrac{y}{\theta}\right)^n = 0, & 0 < y < \theta, \\ 1, & y \geq \theta. \end{cases}$$

Thus Y_n converges in distribution to a random variable that has probability 1 at the point θ and probability 0 elsewhere.

It is often easier to find limiting distributions by working with moment-generating functions. The following theorem gives the relationship between convergence of distribution functions and convergence of moment-generating functions.

Theorem 7.3: *Let Y_n and Y be random variables with moment-generating functions $m_n(t)$ and $m(t)$, respectively. If*

$$\lim_{n\to\infty} m_n(t) = m(t)$$

for all real t, then Y_n converges in distribution to Y.

Proof of Theorem 7.3 is beyond the scope of this text.

Example 7.4: *Let X_n be a binomial random variable with n trials and probability p of success on each trial. If n tends to infinity with np remaining fixed, show that X_n converges in distribution to a Poisson random variable.*

Solution: *This problem was solved in Chapter 3 when we derived the Poisson probability distribution. We will now solve it using moment-generating functions and Theorem 7.3.*

We know that the moment-generating function of X_n, $m_n(t)$, is given by

$$m_n(t) = (q + pe^t)^n$$

where $q = 1 - p$. This can be rewritten as

$$m_n(t) = [1 + p(e^t - 1)]^n.$$

Letting $np = \lambda$ and substituting into $m_n(t)$, we obtain

$$m_n(t) = \left[1 + \frac{\lambda}{n}(e^t - 1)\right]^n.$$

Now let us take the limit of this expression as n approaches infinity.

From the calculus you recall that

$$\lim_{n\to\infty}\left(1 + \frac{k}{n}\right)^n = e^k.$$

Letting $k = \lambda(e^t - 1)$, we have

$$\lim_{n\to\infty} m_n(t) = \exp[\lambda(e^t - 1)].$$

We recognize the right-hand expression as the moment-generating function for the Poisson random variable. Hence it follows from Theorem 7.3 that X_n converges in distribution to a Poisson random variable.

Example 7.5: *In the interest of pollution control, an experimenter wants to count the number of bacteria per small volume of water. The sample size in this problem is really the volume of water in which the count is made. We do not have an n as in previous problems. For purposes of approximating the probability distribution of counts, we can think of the volume, and hence the average count per volume, as the quantity that is getting large.*

Let X denote the bacteria count per cubic centimeter of water and assume that X has a Poisson probability distribution with mean λ. We want to approximate the probability distribution of X for large values of λ, and we will do this by showing that

$$Y = \frac{X - \lambda}{\sqrt{\lambda}}$$

converges in distribution to a standard normal random variable as λ tends to infinity.

Specifically, if the allowable pollution in a water supply is a count of 110 *per cubic centimeter, approximate the probability that X will be at most* 110, *assuming that $\lambda = 100$.*

Solution: *We will proceed by taking the limit of the moment-generating function of Y as $\lambda \to \infty$, and then use Theorem 7.3. The moment-generating function of $X, m_X(t)$, is given by*

$$m_X(t) = \theta^{\lambda(e^t - 1)},$$

and hence the moment-generating function for Y, $m_Y(t)$, is

$$m_Y(t) = e^{-\sqrt{\lambda}t} m_x\left(\frac{t}{\sqrt{\lambda}}\right)$$

$$= e^{-\sqrt{\lambda}t} \exp[\lambda(e^{t/\sqrt{\lambda}} - 1)].$$

The term $e^{t/\sqrt{\lambda}} - 1$ can be written as

$$e^{t/\sqrt{\lambda}} - 1 = \frac{t}{\sqrt{\lambda}} + \frac{t^2}{2\lambda} + \frac{t^3}{6\lambda^{3/2}} + \cdots$$

and, on adding exponents,

$$m_Y(t) = \exp\left[-\sqrt{\lambda}t + \lambda\left(\frac{t}{\sqrt{\lambda}} + \frac{t^2}{2\lambda} + \frac{t^3}{6\lambda^{3/2}} + \cdots\right)\right]$$

$$= \exp\left(\frac{t^2}{2} + \frac{t^3}{6\sqrt{\lambda}} + \cdots\right).$$

In the exponent of $m_Y(t)$ the first term, $t^2/2$, is free of λ and the remaining terms all have a λ to some positive power in the denominator. Therefore, as $\lambda \to \infty$, all the terms after the first will tend to zero sufficiently fast to allow

$$\lim_{\lambda \to \infty} m_Y(t) = e^{t^2/2},$$

and the right-hand expression is the moment-generating function for a standard normal random variable.

We now want to approximate $P(X \leq 110)$. Note that

$$P(X \leq 110) = P\left(\frac{X - \lambda}{\sqrt{\lambda}} \leq \frac{110 - \lambda}{\sqrt{\lambda}}\right).$$

We have shown that $Y = (X - \lambda)/\sqrt{\lambda}$ is approximately a standard normal random variable for large λ. Hence, for $\lambda = 100$, we have

$$P\left(Y \leq \frac{110 - 100}{10}\right) = P(Y \leq 1)$$

$$= .8413,$$

from Table 3, Appendix III.

The normal approximation to Poisson probabilities works reasonably well for $\lambda \geq 25$.

7.4 The Central Limit Theorem

Example 7.5 gives a random variable that converges in distribution to the standard normal random variable. That this phenomenon is shared by a large class of random variables is shown by the following theorem.

Theorem 7.4 The Central Limit Theorem: *Let $X_1, \ldots, X_n$ be independent and identically distributed random variables with $E(X_i) = \mu$ and $V(X_i) = \sigma^2 < \infty$. Define Y_n to be*

$$Y_n = \sqrt{n}\frac{\overline{X} - \mu}{\sigma},$$

where

$$\overline{X} = \frac{1}{n}\sum_{i=1}^{n} X_i.$$

Then Y_n converges in distribution to a standard normal random variable.

Proof: *We will sketch a proof for the case in which the moment-generating function for X_i exists. (This is not the most general proof, because moment-generating functions do not always exist.)*

Define a random variable Z_i by

$$Z_i = \frac{X_i - \mu}{\sigma}.$$

Note that $E(Z_i) = 0$ and $V(Z_i) = 1$. The moment-generating function of Z_i, $m_Z(t)$, can then be written as

$$m_Z(t) = 1 + \frac{t^2}{2} + \frac{t^3}{3!}E(Z_i^3) + \cdots.$$

Now

$$\begin{aligned} Y_n &= \sqrt{n}\frac{\bar{X} - \mu}{\sigma} = \frac{1}{\sqrt{n}} \frac{\sum_{i=1}^{n} X_i - n\mu}{\sigma} \\ &= \frac{1}{\sqrt{n}} \sum_{i=1}^{n} Z_i \end{aligned}$$

and the moment-generating function of Y_n, $m_n(t)$, can be written as

$$m_n(t) = \left[m_Z\left(\frac{t}{\sqrt{n}}\right)\right]^n.$$

Recall that the moment-generating function of the sum of independent random variables is the product of their individual moment-generating functions. Hence

$$\begin{aligned} m_n(t) &= \left[m_Z\left(\frac{t}{\sqrt{n}}\right)\right]^n \\ &= \left(1 + \frac{t^2}{2n} + \frac{t^3}{3!n^{3/2}}k + \cdots\right)^n, \end{aligned}$$

where $k = E(Z_i^3)$.

Now take the limit of $m_n(t)$ *as* $n \to \infty$. *One way to evaluate the limit is to consider* $\log m_n(t)$, *where*

$$\log m_n(t) = n \log\left[1 + \left(\frac{t^2}{2n} + \frac{t^3k}{6n^{3/2}} + \cdots\right)\right].$$

A standard series expansion for $\log(1 + x)$ *is*

$$\log(1 + x) = x - \frac{x^2}{2} + \frac{x^3}{3} - \frac{x^4}{4} + \cdots.$$

Letting

$$x = \left(\frac{t^2}{2n} + \frac{t^3k}{6n^{3/2}} + \cdots\right),$$

$$\log m_n(t) = n \log(1 + x) = n\left(x - \frac{x^2}{2} + \cdots\right)$$

$$= n\left[\left(\frac{t^2}{2n} + \frac{t^3k}{6n^{3/2}} + \cdots\right) - \frac{1}{2}\left(\frac{t^2}{2n} + \frac{t^3k}{6n^{3/2}} + \cdots\right)^2 + \cdots\right],$$

where the succeeding terms in the expansion involve x^3, x^4, *and so on. Multiplying through by n we see that the first term,* $t^2/2$, *does not involve n, while all other terms will have n to a positive power in the denominator. Thus it can be shown that*

$$\lim_{n\to\infty} \log m_n(t) = t^2/2$$

or

$$\lim_{n\to\infty} m_n(t) = e^{t^2/2},$$

the moment-generating function for a standard normal random variable. Applying Theorem 7.3, we conclude that Y_n *converges in distribution to a standard normal random variable.*

Another way to say that a random variable converges in distribution to a standard normal is to say that it is asymptotically normal. Thus we would say that $\overline{X}$ is asymptotically normally distributed with mean μ and variance

σ^2/n. We note in passing that Theorem 7.4 is not the most general form of the central limit theorem. Similar theorems exist for some cases in which the X_i's are not identically distributed and in which they are dependent.

The normal approximation to the probability distribution of $\bar{X}$ is reasonably good, in most cases, for sample sizes of 30 or more. However, if the distribution of each X_i is somewhat mound-shaped, the approximation may be accurate for sample sizes as small as 5.

Example 7.6: *Achievement test scores from all high school seniors in a certain state have a mean and variance of 60 and 64, respectively. A specific high school class of $n = 100$ students had a mean score of 58. Is there evidence to suggest that this high school is inferior? (Calculate the probability that the sample mean is at most 58 when $n = 100$.)*

Solution: *Let $\bar{X}$ denote the mean of a random sample of $n = 100$ scores from a population with $\mu = 60$ and $\sigma^2 = 64$. We want to approximate $P(\bar{X} \leq 58)$. We know from Theorem 7.4 that $\sqrt{n}(\bar{X} - \mu)/\sigma$ is approximately a standard normal random variable, which we denote by Z. Hence*

$$P(\bar{X} \leq 58) = P\left(Z \leq \frac{58 - 60}{\sqrt{64/100}}\right)$$

$$= P(Z \leq -2.5)$$

$$= .0062,$$

using Table 3, Appendix III.

Since this probability is so small, it is unlikely that the specific class of interest can be regarded as a random sample from a population with $\mu = 60$ and $\sigma^2 = 64$. There is evidence to suggest that this class could be set aside as inferior.

Example 7.7: *Candidate A believes that he can win a city election if he can poll at least 55 percent of the votes in precinct I. He also believes that about 50 percent of the city's voters favor him. If $n = 100$ voters show up to vote at precinct I, what is the probability that candidate A receives at least 55 percent of the votes?*

Solution: *Let X denote the number of voters at precinct I who vote for candidate A. We must approximate $P(X/n \geq .55)$, when p, the probability that a randomly selected voter favors candidate A, is .5. If we think of the $n = 100$*

voters at precinct I as a random sample from the city, then X has a binomial distribution with $p = .5$. We have seen that the fraction of voters can be written as

$$\frac{X}{n} = \frac{1}{n} \sum_{i=1}^{n} X_i,$$

where $X_i = 1$ if the ith voter favors candidate A and $X_i = 0$ otherwise. Now $E(X_i) = p$ and $V(X_i) = p(1 - p)$. Thus X/n is a sample mean that satisfies the conditions of Theorem 7.4. In other words, X/n is asymptotically normal with mean p and variance $p(1 - p)/n$.

Applying Theorem 7.4, it follows that

$$P(X/n \geq .55) \approx P\left[Z \geq \frac{.55 - .5}{\sqrt{.5(.5)/100}}\right]$$

$$= P(Z \geq 1) = .1587,$$

from Table 3, Appendix III.

7.5 A Combination of Convergence in Probability and Convergence in Distribution

Many times we will be interested in the limiting behavior of the product or quotient of several functions of a set of random variables. The following theorem, which combines convergence in probability with convergence in distribution, applies to the quotient of two functions, X_n and Y_n.

> ***Theorem 7.5:*** *Suppose that X_n converges in distribution to a random variable X, and Y_n converges in probability to unity. Then X_n/Y_n converges in distribution to X.*

The proof of this theorem is beyond the scope of this text, but we will illustrate its usefulness with the following example.

Example 7.8: *Suppose that $X_1, \ldots, X_n$ are independent and identically distributed random variables with $E(X_i) = \mu$ and $V(X_i) = \sigma^2$. Define S'^2 as*

$$S'^2 = \frac{1}{n}\sum_{i=1}^{n}(X_i - \overline{X})^2.$$

Show that

$$\sqrt{n}\frac{\overline{X} - \mu}{S'}$$

converges in distribution to a standard normal random variable.

Solution: *In Example 7.2 we showed that S'^2 converges in probability to σ^2. Hence it follows from Theorem 7.2, parts (c) and (d), that S'^2/σ^2 (and hence S'/σ) converges in probability to 1. We also know from Theorem 7.4 that*

$$\sqrt{n}\frac{\overline{X} - \mu}{\sigma}$$

converges in distribution to a standard normal random variable. Therefore,

$$\sqrt{n}\frac{\overline{X} - \mu}{S'} = \sqrt{n}\frac{\overline{X} - \mu}{\sigma}\Bigg/\frac{S'}{\sigma}$$

converges in distribution to a standard normal random variable by Theorem 7.5.

7.6 Summary

You will recall that the objective of statistics is to make inferences about a population based on information contained in a sample. Specifically, we need to know the probability distributions for certain functions of the sample measurements in order to make inferences about population parameters. The object of this chapter has been to present some methods for approximating the probability distributions of functions of random variables when the exact distributions are difficult, if not impossible, to find. These methods, which are valid for large sample sizes, are based on limit theorems.

This chapter contains a discussion of two types of limit theorems, one type involving convergence in probability and the other, convergence in distribution.

Convergence in probability, as we presented it, provides a probability statement concerning the distance between a random variable and some constant as the sample size gets large. Often the "constant" is the expected value of the random variable. The major result of convergence in probability is the law of large numbers, Theorem 7.1.

Convergence in distribution involves finding the limiting distribution of a random variable as the sample size tends to infinity. The most important theorem based on this concept is the central limit theorem.

Theorem 7.5 gives a useful result for a situation in which both convergence in probability and in distribution are needed to find the limiting distribution function for a random variable.

References

1. Hoel, P. G., *Introduction to Mathematical Statistics*, 3rd ed. New York: John Wiley & Sons, Inc., 1962.
2. Hogg, R. V., and A. T. Craig, *Introduction to Mathematical Statistics*, 3rd ed. New York: The Macmillan Company, 1970.
3. Mood, A. M., and F. A. Graybill, *Introduction to the Theory of Statistics*, 2nd ed. New York: McGraw-Hill Book Company, 1963.
4. Parzen, E., *Modern Probability Theory and Its Applications*. New York: John Wiley & Sons, Inc., 1964.

Exercises

7.1. Let $Y_1, \ldots, Y_n$ be independent random variables, each with probability density function

$$f(y) = \begin{cases} 3y^2, & 0 \le y \le 1, \\ 0, & \text{elsewhere.} \end{cases}$$

Show that

$$\overline{Y}_n = \frac{1}{n} \sum_{i=1}^{n} Y_i$$

converges in probability to a constant as $n \to \infty$, and find the constant.

7.2. Let $Y_1, \ldots, Y_n$ be independent gamma-type random variables with a density function given by

$$f(y) = \begin{cases} \dfrac{1}{b^a \Gamma(a)} y^{a-1} e^{-y/b}, & y \geq 0, \\ 0, & \text{elsewhere.} \end{cases}$$

Show that the mean, $\overline{Y}$, converges in probability to a constant, and find the constant.

7.3. Let $Y_1, \ldots, Y_n$ be independent random variables, each uniformly distributed over the interval $(0, \theta)$.

(a) Show that the mean, $\overline{Y}$, converges in probability to a constant as $n \to \infty$, and find the constant.

(b) Show that $\max(Y_1, \ldots, Y_n)$ converges in probability to θ as $n \to \infty$.

7.4. Let $Y_1, \ldots, Y_n$ be independent random variables, each possessing the density function

$$f(y) = \begin{cases} \dfrac{2}{y^2}, & y \geq 2, \\ 0, & \text{elsewhere.} \end{cases}$$

Does the law of large numbers apply to $\overline{Y}$ in this case? If so, find the limit in probability of $\overline{Y}$.

7.5. An experimenter is counting bacteria of two types, A and B, in a certain liquid. Let X denote the number of type A bacteria per cubic centimeter and let Y denote the number of type B bacteria per cubic centimeter. X and Y are assumed to have Poisson distributions with means λ_1 and λ_2, respectively. The experimenter plans to observe a number of independent values for X and Y. What function of the observed X's and Y's would you suggest as an estimator of the ratio

$$\frac{\lambda_1}{\lambda_1 + \lambda_2}?$$

Why?

7.6. If the probability that a person suffers a bad reaction from an injection of a certain serum is .001, use the Poisson distribution to approximate the probability that, of 1000 persons, 2 or more will suffer a bad reaction.

7.7. The number of accidents per year at a given intersection, Y, is assumed to have a Poisson distribution. Over the past few years, there has been an average of 36 accidents per year at this intersection. If the number of accidents per year is at least 45, an intersection can qualify to be rebuilt under an emergency program set up by the state. Approximate the probability that the intersection in question will come under the emergency program at the end of next year.

7.8. A large industry has an average hourly wage of $4.00 per hour with a standard deviation of $.50. A certain ethnic group consisting of 64 workers has an average wage of $3.90 per hour. Is it reasonable to assume that the ethnic group is a random sample of workers from the industry? (Calculate the probability of obtaining a sample mean less than or equal to $3.90 per hour.)

7.9. An anthropologist wishes to estimate the average height of men for a certain race of people. If the population standard deviation is assumed to be 2.5 inches and if he randomly samples 100 men, find the probability that the difference between the sample mean and the true population mean will not exceed .5 inch.

7.10. Suppose that the anthropologist of Exercise 7.9 wants the difference between the sample mean and the population mean to be less than .4 inch with probability .95. How many men should he sample to achieve this objective?

7.11. A machine is shut down for repairs if a random sample of 100 items selected from the daily output of the machine reveals at least 15 percent defectives. (Assume that the daily output is a large number of items.) If the machine is, in fact, only producing 10 percent defective items, find the probability that it will be shut down on a given day.

7.12. A pollster believes that 20 percent of the voters in a certain area favor a bond issue. If 64 voters are randomly sampled from the large number of voters in this area, approximate the probability that the sampled fraction of voters favoring the bond issue will not differ from the true fraction by more than .06.

7.13. Twenty-five heat lamps are connected in a greenhouse so that when one lamp fails, another takes over immediately. (Only one lamp is turned on at any time.) The lamps operate independently, and each has mean life of 50 hours and standard deviation of 4 hours. If the greenhouse is not checked for 1300 hours after the lamp system is turned on, what is the probability that a lamp will be burning at the end of the 1300-hour period?

7.14. Suppose that $X_1, \ldots, X_n$ are independent random variables, each with mean μ_1 and variance σ_1^2. Suppose, also, that $Y_1, \ldots, Y_n$ are independent random variables, each with mean μ_2 and variance σ_2^2. Show that the random variable

$$\frac{(\bar{X} - \bar{Y}) - (\mu_1 - \mu_2)}{\sqrt{\dfrac{\sigma_1^2 + \sigma_2^2}{n}}}$$

converges in distribution, as $n \to \infty$, to a standard normal random variable.

7.15. An experiment is designed to test whether operator A or operator B gets the job of operating a new machine. Each operator is timed on 50 independent trials involving the performance of a certain task on the machine. If the sample means for the 50 trials differ by more than 1 second, the operator with the smaller mean time gets the job. Otherwise, the experiment is considered to end in a tie. If the standard deviations of times for both operators are assumed to be 2 seconds, what is the probability that operator A gets the job even though both operators have equal ability?

7.16. Let Y have a χ^2 distribution with n degrees of freedom. That is, Y has the density function

$$f(y) = \begin{cases} \dfrac{1}{2^{n/2}\Gamma(n/2)} y^{(n/2)-1} e^{-y/2}, & y \geq 0, \\ 0, & \text{elsewhere.} \end{cases}$$

Show that the random variable

$$\frac{Y - n}{\sqrt{2n}}$$

is asymptotically standard normal in distribution, as $n \to \infty$.

7.17. A machine in a heavy-equipment factory produces steel rods of length Y, where Y is a normal random variable with a mean μ of 6 inches and a variance of .2. The cost, C, of repairing a rod that is not exactly 6 inches in length is proportional to the square of the error, and is given, in dollars, by

$$C = 4(Y - \mu)^2.$$

If 50 rods with independent lengths are produced in a given day, approximate the probability that the total cost for repairs for that day exceeds \$48.

8

Estimation

8.1 Introduction

We now leave the theory of probability and turn to the objective of statistics, making an inference about a population based on information contained in a sample. Since populations are characterized by numerical descriptive measures, the objective of many statistical investigations is to make an inference about one or more population *parameters.*

For example, a manufacturer of washing machines might be interested in estimating the fraction of washers that would fail prior to a 1-year guarantee time. Letting Y equal the length of a washer's life, the parameter of interest is $P(Y \leq 1 \text{ year})$, which is the area under the probability density function over the interval $y \leq 1$. Other important population parameters are the population mean, variance, and standard deviation. For example, we might wish to estimate the mean waiting time, μ, at a supermarket checkout station or the standard deviation of the error of measurement, σ, of an electronic instrument. To simplify our terminology, we will call the parameter of interest to the experiment the *target parameter*.

Methods for making inferences about population parameters fall in one of two categories. We can make *decisions* concerning the value of a parameter or we can *estimate* its value. Methods of estimation, which will occupy our attention in this chapter, can be classified as belonging to one of two types, *point estimation* and *interval estimation.*

Suppose that we wish to estimate the average amount of mercury that can be removed from 1 ounce of ore obtained at a particular geographic location. We could give the estimate as a single number, for instance .13 ounce, or we might estimate the mean to fall in an interval, say .07 to .19 ounce. The first type of estimate is called a *point estimate*, because a single number, representing the estimate, may be associated with a point on a line. The second type, involving two points that define an interval, is called an *interval estimate.*

A point-estimation procedure utilizes information in a sample to arrive at a single number or point which estimates the target parameter. The actual estimation is accomplished by an estimator.

Definition 8.1: *An estimator is a rule that tells how to calculate an estimate based on the measurements contained in a sample.*

Most frequently an estimator is expressed as a formula. For example, the sample mean,

$$\bar{y} = \frac{1}{n} \sum_{i=1}^{n} y_i,$$

is one possible estimator of the population mean, μ. Clearly, the expression for $\bar{y}$ is both a rule and a formula. It tells us to sum the sample observations and divide by the sample size, n.

An experimenter who wants an interval estimate of a parameter must use the sample data to calculate two points. Hopefully, the interval formed by the two points will possess a high probability of including the target parameter. Examples of interval estimators will be given in subsequent sections.

It is possible to obtain many different estimators (rules for estimating) for the same population parameter, and this should not be surprising. Ten engineers, each assigned to estimate the cost of a large construction job, would most likely arrive at different estimates of the total cost. Such engineers, called "estimators" in the construction industry, use certain fixed guidelines plus intuition to achieve their estimates. Each represents a unique human subjective "rule" for obtaining a single estimate. This brings us to a most important point, some estimators are considered *good*, some bad. How would the management of a construction firm define "good" and "bad" as they relate to the estimation of the cost of a job—and how would we establish a criterion of goodness to compare one statistical estimator with another?

8.2 *Some Properties of Estimators*

An investigation of the reasoning used in calculating the goodness of a point estimator is facilitated by considering an analogy. Point estimation is similar, in many respects, to firing a revolver at a target. The estimator, generating estimates, is analogous to the revolver, a particular estimate to the bullet, and the parameter of interest to the bull's-eye. Drawing a sample from the population and estimating the value of the parameter is equivalent to firing a single shot at the target.

Suppose that a man fires a single shot at a target and that the shot pierces the bull's-eye. Do we conclude that he is an excellent shot? Obviously, the answer is no, because not one of us would consent to hold the target while a second shot was fired. On the other hand, if 1 million shots in succession hit the bull's-eye, we might acquire sufficient confidence in the marksman to hold the target for the next shot, if the compensation were adequate. The point we wish to make is certainly clear. We cannot evaluate the goodness of an estimation procedure on the basis of a single estimate; rather, we must observe the results when the estimation procedure is used over and over again, many many times. Since the estimates are numbers, we would evaluate the goodness of the estimator by constructing a frequency distribution of the estimates obtained in repeated sampling and note how closely the distribution centers about the parameter of interest.

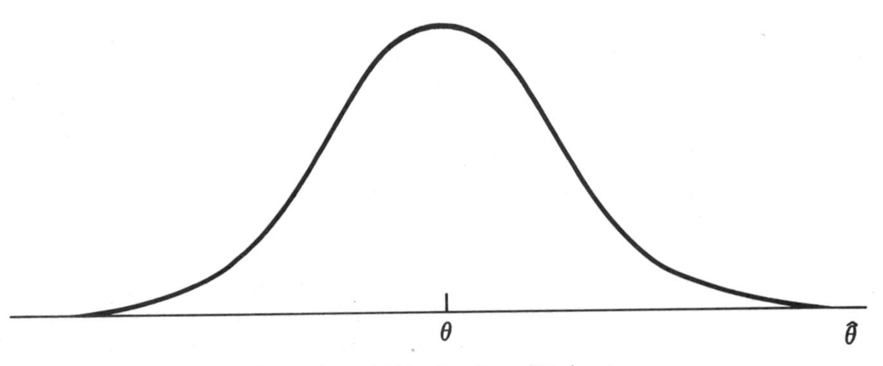

Figure 8.1 A Distribution of Estimates

Suppose that we wish to estimate a population parameter that we will call θ. The estimator of θ will be indicated by the symbol $\hat{\theta}$, where the "hat" indicates that we are estimating the parameter immediately beneath. With the revolver-firing example in mind, we see that the desirable properties of a good estimator are quite clear. We would like the distribution of estimates or, more properly, the probability distribution of the estimator to center about the target parameter as shown in Figure 8.1. In other words, we would like the mean or expected value of the distribution of estimates to equal the parameter estimated.

Definition 8.2: *Let $\hat{\theta}$ be an estimator of a parameter θ. Then $\hat{\theta}$ is an* unbiased estimator *if $E(\hat{\theta}) = \theta$. Otherwise, $\hat{\theta}$ is said to be* biased.

Definition 8.3: *The bias, B, of an estimator $\hat{\theta}$ is given by $B = E(\hat{\theta}) - \theta$.*

The probability distribution for a positively biased estimator, one for which $E(\hat{\theta}) > \theta$, is shown in Figure 8.2.

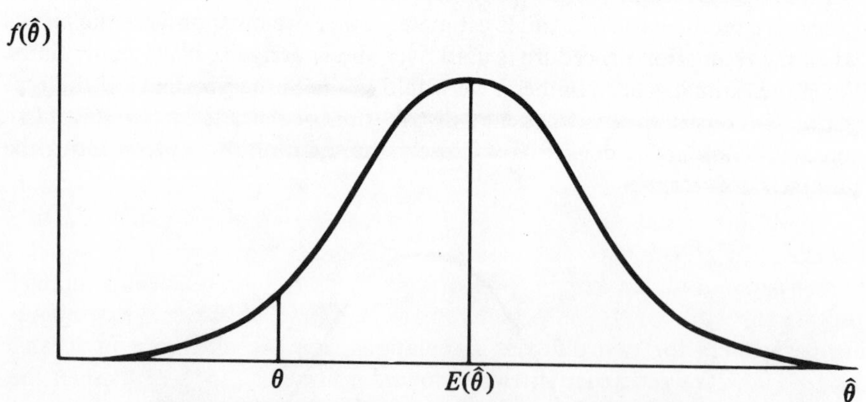

Figure 8.2 Probability Distribution for a Positively Biased Estimator

In addition to unbiasedness, we like the spread of a distribution of estimates to be as small as possible. That is, we want $V(\hat{\theta})$ to be a minimum. Given two unbiased estimators of a parameter θ, and all other things being equal, we would select the estimator with the smaller variance.

Rather than use the bias and variance to characterize the goodness of an estimator, we might look at the expected value of $(\hat{\theta} - \theta)^2$, the square of the distance between $\hat{\theta}$ and its target parameter.

Definition 8.4: The mean-square error *of an estimator $\hat{\theta}$ is defined to be the expected value of $(\hat{\theta} - \theta)^2$.*

The mean-square error of an estimator $\hat{\theta}$, denoted by the symbol $\mathrm{MSE}(\hat{\theta})$, is a function of both its variance and its bias.

It can be shown that

$$\mathrm{MSE}(\hat{\theta}) = V(\hat{\theta}) + B^2.$$

We will leave the proof of this result to Exercise 8.1.

Now let us consider some common and very useful unbiased point estimators.

8.3 Some Common Unbiased Point Estimators

Some estimators for population parameters are selected intuitively. For example, it seems natural to use the sample mean, $\overline{Y}$, to estimate the population mean μ, and the sample proportion, $\hat{p} = Y/n$, to estimate a binomial parameter p. How, then, would you estimate the difference between corresponding parameters for two different populations, say the difference in means, $(\mu_1 - \mu_2)$, or the difference in two binomial parameters, $(p_1 - p_2)$, when the inference is to be based on random samples of n_1 and n_2 observations selected independently from the two populations? Again, our intuition suggests the point estimators, $(\overline{Y}_1 - \overline{Y}_2)$, the difference in the sample means, for estimating $(\mu_1 - \mu_2)$, and $(\hat{p}_1 - \hat{p}_2)$, the difference in the sample proportions, for estimating $(p_1 - p_2)$.

Since the four estimators, $\overline{Y}$, $\hat{p}$, $(\overline{Y}_1 - \overline{Y}_2)$, and $(\hat{p}_1 - \hat{p}_2)$ are functions of the random sample measurements, we could find their expected values and variances using the expectation theorems of Sections 5.6, 5.7, and 5.8. Such an effort would show that all four estimators are unbiased and that they possess the variances shown in Table 8.1, when random sampling has been employed.

Table 8.1 Expected Values and Variances of Some Common Point Estimators

Target Parameter θ	*Sample Size(s)*	*Point Estimator* $\hat{\theta}$	*Expected Value of* $\hat{\theta}$	*Variance of* $\hat{\theta}$
μ	n	$\overline{Y}$	μ	$\frac{\sigma^2}{n}$
p	n	$\hat{p} = \frac{Y}{n}$	p	$\frac{pq}{n}$
$\mu_1 - \mu_2$	n_1 and n_2	$\overline{Y}_1 - \overline{Y}_2$	$\mu_1 - \mu_2$	$\frac{\sigma_1^2}{n_1} + \frac{\sigma_2^2}{n_2}$*
$p_1 - p_2$	n_1 and n_2	$\hat{p}_1 - \hat{p}_2$	$p_1 - p_2$	$\frac{p_1 q_1}{n_1} + \frac{p_2 q_2}{n_2}$

* *Note*: σ_1^2 and σ_2^2 are the variances of populations 1 and 2, respectively.

A survey of Chapter 5 would show that we have done much of the derivation required for Table 8.1. In particular, we found the means and variances of $\overline{Y}$ and $\hat{p}$ in Exercises 5.14 and 5.16, respectively. Using these results and Theorem 5.8, it follows that

$$E(\overline{Y}_1 - \overline{Y}_2) = E(\overline{Y}_1) - E(\overline{Y}_2) = \mu_1 - \mu_2$$

and

$$V(\overline{Y}_1 - \overline{Y}_2) = V(\overline{Y}_1) + V(\overline{Y}_2) = \frac{\sigma_1^2}{n_1} + \frac{\sigma_2^2}{n_2}.$$

The expected value and variance of $(\hat{p}_1 - \hat{p}_2)$, shown in Table 8.1, are acquired in a similar manner.

Not all estimators are unbiased. Example 8.1 shows that the sample variance,

$$S'^2 = \frac{\sum_{i=1}^{n} (Y_i - \overline{Y})^2}{n}$$

is a biased estimator of σ^2 and, particularly, that this bias can be corrected by dividing the sum of squares of deviations of the measurements about $\overline{Y}$ by $(n - 1)$ rather than n. Because it is most often used in practice to estimate σ^2, the unbiased estimator,

$$S^2 = \frac{\sum_{i=1}^{n} (Y_i - \overline{Y})^2}{n - 1},$$

is often called the *sample variance.*

Example 8.1: *Let $Y_1, \ldots, Y_n$ be a random sample with $E(Y_i) = \mu$ and $V(Y_i) = \sigma^2$. Show that*

$$S'^2 = \frac{1}{n} \sum_{i=1}^{n} (Y_i - \overline{Y})^2$$

is a biased estimator of σ^2.

Solution: *It can be shown (see Exercise 1.3) that*

$$\sum_{i=1}^{n} (Y_i - \overline{Y})^2 = \sum_{i=1}^{n} Y_i^2 - n\overline{Y}^2.$$

Hence

$$E\left[\sum_{i=1}^{n} (Y_i - \overline{Y})^2\right] = E\left(\sum_{i=1}^{n} Y_i^2\right) - nE(\overline{Y}^2)$$

$$= \sum_{i=1}^{n} E(Y_i^2) - nE(\overline{Y}^2).$$

Now note that $E(Y_i^2)$ is the same for $i = 1, 2, \ldots, n$ and use the fact that the variance of a random variable is given by $V(Y) = E(Y^2) - \mu^2$. Then $E(Y^2) = V(Y) + \mu^2$ and

$$E\left[\sum_{i=1}^{n} (Y_i - \overline{Y})^2\right] = \sum_{i=1}^{n} (\sigma^2 + \mu^2) - n\left(\frac{\sigma^2}{n} + \mu^2\right) = n(\sigma^2 + \mu^2) - n\left(\frac{\sigma^2}{n} + \mu^2\right)$$

$$= n\sigma^2 - \sigma^2$$

$$= (n - 1)\sigma^2.$$

It follows that

$$E(S'^2) = \frac{n-1}{n}\sigma^2$$

and that

$$S^2 = \frac{1}{n-1}\sum_{i=1}^{n}(Y_i - \bar{Y})^2$$

is an unbiased estimator of σ^2.

Two final comments can be made concerning the point estimators of Table 8.1. First, the expected values and variances shown in the table are valid regardless of the form of the population probability density functions. Second, all four estimators will possess probability distributions that are approximately normal for "large samples." The central limit theorem justifies this statement for $\bar{Y}$ and $\hat{p}$ while Theorem 6.3, which attributes normality to all linear functions of normally distributed random variables, justifies the assertion for $(\bar{Y}_1 - \bar{Y}_2)$ and $(\hat{p}_1 - \hat{p}_2)$. How large is "large"? For most populations, the probability distributions of $\bar{Y}$ will be mound-shaped for relatively small samples, as low as $n = 5$, and will tend rapidly to normality as the sample size approaches $n = 30$ or larger. However, you will sometimes need to select larger samples from binomial populations because the required sample size depends on p. The binomial probability distribution is perfectly symmetrical about its mean when $p = 1/2$ and becomes more and more asymmetric as p tends to 0 or 1. As a rough rule you can assume that the distribution of $\hat{p}$ will be mound-shaped and approaching normality for sample sizes such that $p \pm 2\sqrt{pq/n}$ lies in the interval (0, 1).

Since we know that $\bar{Y}$, $\hat{p}$, $(\bar{Y}_1 - \bar{Y}_2)$, and $(\hat{p}_1 - \hat{p}_2)$ are unbiased with near-normal (at least mound-shaped) probability distributions for moderate-sized samples, let us now see how we can use this information to answer a practical question. If we use an estimator once and acquire a single estimate, how good will this estimate be? How much faith can we place in the validity of our inference?

8.4 Evaluating the Goodness of a Point Estimate

One way to measure the goodness of *any* point estimate is in terms of the distance between the estimate and its target parameter. This quantity, which varies in a random manner in repeated sampling, is called the error of estimation.

> ***Definition 8.5:*** *The error of estimation, ε, is the distance between an estimator and its target parameter. That is, $\varepsilon = |\hat{\theta} - \theta|$.*

Naturally, we would like the error of estimation to be as small as possible.

Since the error of estimation will be a random quantity, we cannot say how large or small it will be for a particular estimate, but we can make probability statements about it. For example, suppose that $\hat{\theta}$, an unbiased estimator of θ, possesses a probability distribution as shown in Figure 8.3. If we select two points, $(\theta - b)$ and $(\theta + b)$, located near the tails of the probability distribution, the probability that the error of estimation, ε, is less than b is the shaded area in Figure 8.3.

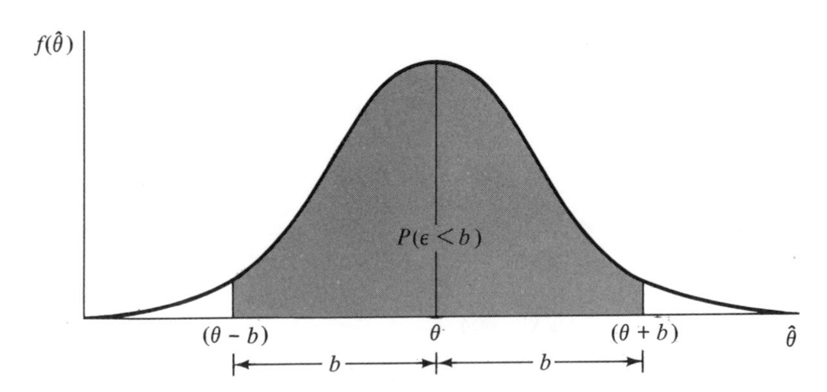

Figure 8.3 The Probability Distribution of an Estimator, $\hat{\theta}$

We can think of b as a probabilistic bound on the error of estimation. Hence we are not certain that a given error will be less than b, but we know that the probability of such an event is high. If b can be regarded as small, from a practical point of view, then $P(\varepsilon < b)$ provides a measure of the goodness of a single estimate.

Finding the value of b for a given estimation problem is easy if you know the probability distribution of $\hat{\theta}$. Suppose that you want ε to be less than b with probability equal to .90. Then you seek a value, b, such that

$$\int_{\theta - b}^{\theta + b} f(\hat{\theta})\, d\hat{\theta} = .90.$$

Whether you know the probability distribution of $\hat{\theta}$ or not, an approximate bound on ε for unbiased estimators can be found by expressing b as a multiple of the standard deviation of $\hat{\theta}$. For example, if we let $b = k\sigma_{\hat{\theta}}, k \geq 1$, then we know from Tchebysheff's theorem that ε will be less than $k\sigma_{\hat{\theta}}$ with probability at least $1 - 1/k^2$. A convenient and often used value of k is $k = 2$. Hence we know that ε will be less than $b = 2\sigma_{\hat{\theta}}$ with probability at least .75. In fact, from the empirical rule (Chapter 1) we would expect the probability to be near .95. We will illustrate with an example.

Example 8.2: *A sample of $n = 1000$ voters, randomly selected from a city, showed $y = 560$ in favor of candidate Jones. Estimate the fraction of voters in the population favoring Jones, p, and place a bound on the error of estimation.*

Solution: *We will use the estimator, $\hat{p} = Y/n$, to estimate p. Hence the estimate of p, the fraction of voters favoring candidate Jones, is*

$$\hat{p} = \frac{y}{n} = \frac{560}{1000} = .56.$$

How much faith can we place in this figure?

The probability distribution of $\hat{p}$ will be very accurately approximated by a normal probability distribution for samples as large as $n = 1000$. Then, letting $b = 2\sigma_{\hat{p}}$, the probability that ε will be less then b is approximately .95.

From Table 8.1, $V(\hat{p}) = pq/n$. Hence

$$b = 2\sigma_{\hat{p}} = 2\sqrt{\frac{pq}{n}}.$$

Unfortunately, we need to know p in order to calculate b, and finding p was the objective of our sampling. This apparent stalemate is not a handicap, because $\sigma_{\hat{p}}$ will vary little for small changes in p. Hence the substitution of the estimate, $\hat{p}$, for p will produce little error in calculating the exact value of $b = 2\sigma_{\hat{p}}$. Then, for our example,

$$b = 2\sigma_{\hat{\theta}} = 2\sqrt{\frac{pq}{n}} \approx 2\sqrt{\frac{(.56)(.44)}{1000}} = .03.$$

Now what is the significance of our calculations? The probability that the error of estimation is less than .03 is approximately .95. Hence we are reasonably confident that our estimate, .56, is within .03 of the population value of p.

Example 8.3: *A comparison of the wearing quality of two types of automobile tires was obtained by road testing samples of $n_1 = n_2 = 100$ tires for each type. The number of miles until wear-out was recorded, where wear-out was defined as a specific amount of tire wear. The test results were as follows:*

$$\bar{y}_1 = 26{,}400 \text{ miles}, \qquad \bar{y}_2 = 25{,}100 \text{ miles};$$

$$s_1^2 = 1{,}440{,}000, \qquad s_2^2 = 1{,}960{,}000.$$

Estimate the difference in mean time to wear-out and place a bound on the error of estimation.

Solution: *The point estimate of $(\mu_1 - \mu_2)$ is*

$$(\bar{y}_1 - \bar{y}_2) = 26{,}400 - 25{,}100 = 1300 \text{ miles}$$

and

$$\sigma_{(\bar{Y}_1 - \bar{Y}_2)} = \sqrt{\frac{\sigma_1^2}{n_1} + \frac{\sigma_2^2}{n_2}}.$$

You will note that we must know σ_1^2 and σ_2^2, or have good approximate values for them, in order to calculate $\sigma_{(\bar{Y}_1 - \bar{Y}_2)}$. Fairly accurate values of σ_1^2 and σ_2^2 can often be calculated from similar experimental data collected at some prior time or they can be obtained from the current sample data using the unbiased estimator

$$\hat{\sigma}_i^2 = S_i^2 = \frac{\sum_{j=1}^{n} (Y_{ij} - \bar{Y}_i)^2}{n_i - 1}, \qquad i = 1, 2.$$

These estimates will be adequate if the sample sizes are reasonably large, say $n_i \geq 30$, $i = 1, 2$. The calculated values of S_1^2 and S_2^2, based on the two wear tests, are $s_1^2 = 1{,}440{,}000$ and $s_2^2 = 1{,}960{,}000$. Substituting these values for σ_1^2 and σ_2^2 in the formula for $\sigma_{(\bar{Y}_1 - \bar{Y}_2)}$, we have

$$\sigma_{(\bar{Y}_1 - \bar{Y}_2)} \approx \sqrt{\frac{s_1^2}{n_1} + \frac{s_2^2}{n_2}} = \sqrt{\frac{1{,}440{,}000}{100} + \frac{1{,}960{,}000}{100}}$$

$$= \sqrt{34{,}000} = 184 \text{ miles}.$$

Consequently, we estimate the difference in mean wear to be 1300 *miles and we expect the error of estimation to be less than* $2\sigma_{(\bar{Y}_1 - \bar{Y}_2)}$ *or* 368 *miles.*

Whether you believe it or not, you will find that most random variables observed in nature lie within two standard deviations of their mean, with a probability in the vicinity of .95. The probability that Y lies in the interval $(\mu \pm 2\sigma)$ is shown in Table 8.2 for the normal, uniform, and exponential probability distributions. The point we make, of course, is that $B = 2\sigma_{\hat{\theta}}$ is a good approximate bound on the error of estimation in a practical situation. The probability that the error of estimation will be less than this bound will be near .95.

Table 8.2 Probability That $(\mu - 2\sigma) < Y < (\mu + 2\sigma)$

Distribution	*Probability*
Normal	.9544
Uniform	1.0000
Exponential	.9502

8.5 *Confidence Intervals*

You will recall that an interval estimator for a parameter requires the construction of two points that are functions of the sample measurements. For a given sample, we wish to calculate two points so that the resulting interval will contain the target parameter, θ. Note that one or both of the end points of the interval, being functions of the sample measurements, will vary in a random manner from sample to sample, and hence the length and location of the interval are random quantities. This being the case, the objective is to find an interval estimator that generates narrow intervals which enclose θ with a high probability.

Interval estimators are commonly called *confidence intervals*. The upper and lower endpoints of a confidence interval are called the *upper* and *lower confidence limits*, respectively. The probability that a confidence interval will enclose θ is called the *confidence coefficient*. The confidence coefficient provides a measure of the confidence we can place in a given interval estimate. If we know that the confidence coefficient associated with our estimator is high, we will be highly confident that a particular confidence interval, constructed in a real-life sampling situation, will enclose θ.

For example, suppose that $\hat{\theta}_L$ and $\hat{\theta}_U$ are the lower and upper confidence limits, respectively, for a parameter θ. Then if

$$P(\hat{\theta}_L < \theta < \hat{\theta}_U) = 1 - \alpha,$$

the probability, $(1 - \alpha)$, is the confidence coefficient. The resulting random interval defined by $\hat{\theta}_L$ to $\hat{\theta}_U$ is called a *two-sided confidence interval.*

It is also possible to form a *one-sided confidence* interval, such that

$$P(\hat{\theta}_L < \theta) = 1 - \alpha.$$

Although only one point is random in this case, the confidence interval is $(\hat{\theta}_L, \infty)$. Similarly, we could have an upper one-sided confidence interval such that

$$P(\theta < \hat{\theta}_U) = 1 - \alpha.$$

The implied confidence interval is $(-\infty, \hat{\theta}_U)$.

A useful method for finding confidence intervals, called the *pivotal method,* is demonstrated in Section 8.6 and is employed to find large-sample confidence intervals for μ, p, $(\mu_1 - \mu_2)$, and $(p_1 - p_2)$, the parameters for which we found point estimators in Section 8.3. The same method can be used to acquire the confidence intervals discussed in the remaining sections of this chapter.

8.6 A Large-Sample Confidence Interval

The method for finding confidence intervals in this and subsequent sections requires that you find a statistic, called a *pivotal statistic,* that possesses two characteristics: (1) it is a function of θ and the sample measurements, and

(2) it possesses a probability distribution that is independent of θ. This procedure, called the *pivotal method*, is illustrated by the following example.

Example 8.4: *Let $\hat{\theta}$ be a statistic that is normally distributed with expected value and variance equal to θ and $\sigma_{\hat{\theta}}^2$, respectively. Find a confidence interval for θ that possesses a confidence coefficient equal to $(1 - \alpha)$.*

Solution: *The statistic*

$$Z = \frac{\hat{\theta} - \theta}{\sigma_{\hat{\theta}}}$$

has a standard normal distribution. Now select two tail-end values of this distribution, $z_{\alpha/2}$ and $-z_{\alpha/2}$, such that

$$P(-z_{\alpha/2} < Z < z_{\alpha/2}) = 1 - \alpha.$$

See Figure 8.4.

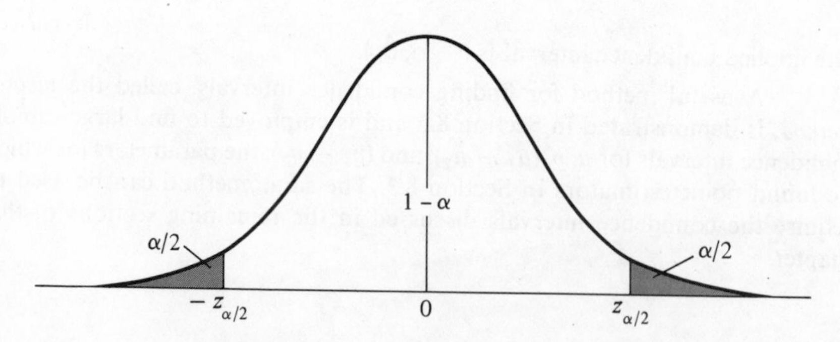

Figure 8.4 Location of $z_{\alpha/2}$ and $-z_{\alpha/2}$

From this point on we employ the following logic. If Y is a random variable, c is a constant ($c > 0$), and

$$P(a < Y < b) = .7,$$

then certainly

$$P(ac < cY < bc) = .7.$$

Similarly, $P(a + c < Y + c < b + c) = .7$. *That is, the probability that* $a < Y < b$ *is unaffected by a change of scale or translation of* Y. *Now let us apply this information to our example. Substituting for* Z *in the probability statement,*

$$P\left(-z_{\alpha/2} < \frac{\hat{\theta} - \theta}{\sigma_{\hat{\theta}}} < z_{\alpha/2}\right) = 1 - \alpha.$$

Multiplying by $\sigma_{\hat{\theta}}$,

$$P(-z_{\alpha/2}\sigma_{\hat{\theta}} < \hat{\theta} - \theta < z_{\alpha/2}\sigma_{\hat{\theta}}) = 1 - \alpha,$$

and subtracting $\hat{\theta}$ *from each term of the inequality,*

$$P(-\hat{\theta} - z_{\alpha/2}\sigma_{\hat{\theta}} < -\theta < -\hat{\theta} + z_{\alpha/2}\sigma_{\hat{\theta}}) = 1 - \alpha.$$

Finally, multiplying each term by -1 *and, consequently, changing the direction of the inequalities, we have*

$$P(\hat{\theta} - z_{\alpha/2}\sigma_{\hat{\theta}} < \theta < \hat{\theta} + z_{\alpha/2}\sigma_{\hat{\theta}}) = 1 - \alpha.$$

Thus the lower and upper confidence limits for θ *are*

$$\text{lower confidence limit (LCL)} = \hat{\theta} - z_{\alpha/2}\sigma_{\hat{\theta}},$$

$$\text{upper confidence limit (UCL)} = \hat{\theta} + z_{\alpha/2}\sigma_{\hat{\theta}}.$$

Example 8.4 can be used to find large-sample confidence intervals for μ, p, $(\mu_1 - \mu_2)$, and $(p_1 - p_2)$, the parameters estimated under the conditions described in Section 8.3. We will let $\hat{\theta}$ represent any one of the point estimators of Section 8.3. The point estimator, $\hat{\theta}$, will be approximately normally distributed for the reasons given in Section 8.3 and hence will satisfy the assumptions of Example 8.4.

Example 8.5: A random sample of $n = 64$ customers at a local supermarket showed that the average shopping time was 33 minutes, with a sample variance of 256. Estimate the true average shopping time per customer, μ, with a confidence coefficient of .90.

Solution: We have $\bar{y} = 33$ and $s^2 = 256$ for the sample of $n = 64$ customers. The point estimator of μ is $\bar{Y}$, and $V(\bar{Y}) = \sigma^2/n$. Since σ^2 is unknown, we will use s^2 as its estimated value.

From Table 3, Appendix III, $z_{\alpha/2} = z_{.05} = 1.645$, and hence the confidence limits are given by

$$\bar{y} - z_{\alpha/2}\frac{s}{\sqrt{n}} = 33 - 1.645\frac{16}{8} = 29.71$$

and

$$\bar{y} + z_{\alpha/2}\frac{s}{\sqrt{n}} = 33 + 1.645\frac{16}{8} = 36.29.$$

Thus our confidence interval for μ is (29.71, 36.29). The chances are good that this interval does include μ because, in repeated sampling, approximately 90 percent of all intervals of the form $\bar{Y} \pm 1.645(\sigma/\sqrt{n})$ do include $\mu = E(\bar{Y})$.

Example 8.6: Two brands of refrigerators, denoted by A and B, are each guaranteed for 1 year. In a random sample of 50 refrigerators of brand A, 12 where observed to fail before the guarantee period ended. A random sample of 60 brand B refrigerators also revealed 12 failures during the guarantee period. Estimate the true difference between proportions of failures during the guarantee period, $(p_1 - p_2)$, with confidence coefficient .98.

Solution: The confidence interval

$$\hat{\theta} \pm z_{\alpha/2}\sigma_{\hat{\theta}}$$

now has the form

$$(\hat{p}_1 - \hat{p}_2) \pm z_{\alpha/2}\sqrt{\frac{p_1q_1}{n_1} + \frac{p_2q_2}{n_2}}.$$

For this example, $\hat{p}_1 = .24$, $\hat{p}_2 = .20$, *and* $z_{.01} = 2.33$. *The confidence interval then becomes*

$$(.24 - .20) \pm 2.33\sqrt{\frac{(.24)(.76)}{50} + \frac{(.20)(.80)}{60}}$$

or

$$.04 \pm .1857.$$

Note that this confidence interval overlaps zero, so the true difference, $(p_1 - p_2)$, *could quite possibly be positive or negative.*

8.7 Selecting the Sample Size

The design of an experiment is essentially a plan for purchasing a quantity of information which, like any other commodity, may be acquired at varying prices depending upon the manner in which the data are obtained. Some measurements contain a large amount of information concerning the parameter of interest, whereas others may contain little or none. Since the sole product of research is information, it behooves us to make its purchase at minimum cost.

The sampling procedure, or "experimental design" as it is usually called, affects the quantity of information per measurement. This, together with the sample size, n, controls the total amount of relevant information in a sample. At this point in our study we will be concerned with the simplest sampling situation—random sampling from a relatively large population—and will devote our attention to the selection of the sample size, n.

The researcher makes little progress in planning an experiment before encountering the problem of selecting the sample size. Indeed, perhaps one of the most frequent questions asked of the statistician is: How many measurements should be included in the sample? Unfortunately, the statistician cannot answer this question without knowing how much information the experimenter wishes to buy. Certainly, the total amount of information in the sample will affect the measure of goodness of the method of inference and must be specified by the experimenter. Referring specifically to estimation, we would like to know how accurate the experimenter wishes his estimate to be. This may be stated by specifying a bound on the error of estimation.

For instance, suppose that we wish to estimate the average daily yield of a chemical, μ, and we wish the error of estimation to be less than 5 tons with

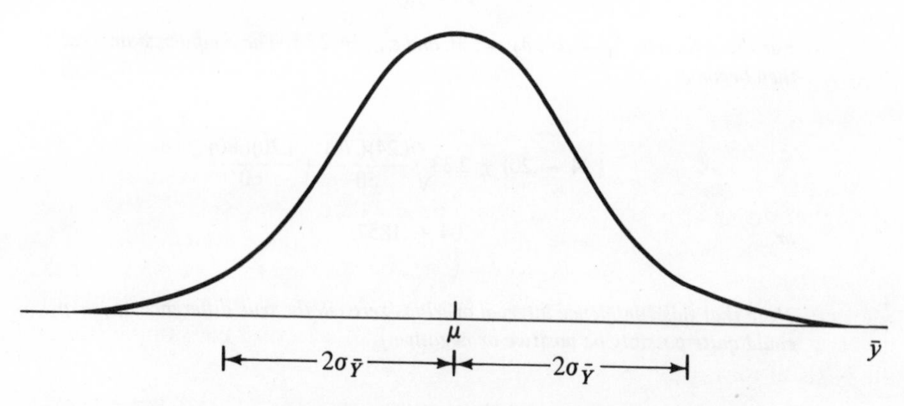

Figure 8.5 The Approximate Distribution of $\bar{Y}$ for Large Samples

a probability of .95. Since approximately 95 percent of the sample means will lie within $2\sigma_{\bar{Y}}$ of μ in repeated sampling, we are asking that $2\sigma_{\bar{Y}}$ equal 5 tons (see Figure 8.5). Then

$$\frac{2\sigma}{\sqrt{n}} = 5.$$

Solving for n we obtain

$$n = \frac{4\sigma^2}{25}.$$

The reader will quickly note that we cannot obtain a numerical value for n unless the population standard deviation, σ, is known. And this is exactly what we would expect, because the variability of $\bar{Y}$ depends upon the variability of the population from which the sample was drawn.

Lacking an exact value for σ, we would use the best approximation available, such as an estimate, s, obtained from a previous sample or knowledge of the range in which the measurements will fall. Since the range is approximately equal to 4σ (the empirical rule), one-fourth of the range will provide an approximate value for σ. For our example, suppose we know that the range of the daily yields is approximately 84 tons. Then $\sigma \approx 21$ and

$$n = \frac{4\sigma^2}{25} \approx \frac{(4)(21)^2}{25} = 70.56$$

or

$$n = 71.$$

Using a sample size $n = 71$, we would be reasonably certain (with probability approximately equal to .95) that our estimate will lie within $2\sigma_{\bar{Y}} = 5$ tons of the true average daily yield.

Actually we would expect the error of estimation to be much less than 5 tons. According to the empirical rule, the probability is approximately equal to .68 that the error of estimation would be less than $\sigma_{\bar{Y}} = 2.5$ tons. The reader will note that the probabilities, .95 and .68, used in these statements will be inexact because σ was approximated. Although this method of choosing the sample size is only approximate for a specified desired accuracy of estimation, it is the best available and is certainly better than selecting the sample size on the basis of our intuition.

The method of choosing the sample size for all the large-sample estimation procedures outlined in Table 8.1 is identical to that described above. The experimenter must specify a desired bound on the error of estimation and an associated confidence level, $1 - \alpha$. For example, if the parameter is θ and the desired bound is B, we would equate

$$z_{\alpha/2}\sigma_{\hat{\theta}} = B,$$

where $z_{\alpha/2}$ is the z value defined in Section 8.6; that is,

$$P(Z > z_{\alpha/2}) = \frac{\alpha}{2}.$$

We will illustrate with examples.

Example 8.7: *The reaction of an individual to a stimulus in a psychological experiment may take one of two forms, A or B. If an experimenter wishes to estimate the probability, p, that a person will react in favor of A, how many people must be included in the experiment? Assume that he will be satisfied if the error of estimation is less than .04 with probability equal to .90. Assume also that he expects p to lie somewhere in the neighborhood of .6.*

Solution: *Since the confidence coefficient is $1 - \alpha = .90$, α must equal .10 and $\alpha/2 = .05$. The z value corresponding to an area equal to .05 in the upper tail of the z distribution is $z_{\alpha/2} = 1.645$. We then require that*

$$1.645\sigma_{\hat{p}} = .04$$

or

$$1.645\sqrt{\frac{pq}{n}} = .04.$$

Since the variability of $\hat{p}$ is dependent upon p, which is unknown, we must use the guessed value of $p = .6$ provided by the experimenter as an approximation. Then

$$1.645\sqrt{\frac{(.6)(.4)}{n}} = .04$$

or

$$n = 406.$$

(Note: If we did not know that $p \approx .6$, we would use $p = .5$, which would yield that maximum value for n, $n = 423$.)

Example 8.8: *An experimenter wishes to compare the effectiveness of two methods of training industrial employees to perform a certain assembly operation. A number of employees is to be divided into two equal groups, the first receiving training method 1 and the second training method 2. Each will perform the assembly operation, and the length of assembly time will be recorded. It is expected that the measurements for both groups will have a range of approximately 8 minutes. If the estimate of the difference in mean time to assemble is desired correct to within 1 minute with probability equal to .95, how many workers must be included in each training group?*

Solution: *Equating $2\sigma_{(\bar{Y}_1 - \bar{Y}_2)}$ to 1 minute, we obtain*

$$2\sqrt{\frac{\sigma_1^2}{n_1} + \frac{\sigma_2^2}{n_2}} = 1.$$

Or, since we desire n_1 to equal n_2, we may let $n_1 = n_2 = n$ and obtain the equation

$$2\sqrt{\frac{\sigma_1^2}{n} + \frac{\sigma_2^2}{n}} = 1.$$

As noted above, the variability of each method of assembly is approximately the same, and hence $\sigma_1^2 = \sigma_2^2 = \sigma^2$. Since the range, equal to 8 minutes, is approximately equal to 4σ, then

$$4\sigma \approx 8$$

and

$$\sigma \approx 2.$$

Substituting this value for σ_1 and σ_2 in the above equation, we obtain

$$2\sqrt{\frac{(2)^2}{n} + \frac{(2)^2}{n}} = 1.$$

Solving, we have $n = 32$, so each group should contain $n = 32$ members.

8.8 Small-Sample Confidence Intervals for μ and $\mu_1 - \mu_2$

The following confidence interval is based on the assumption that the experimenter's sample has been randomly selected from a normal population. It is appropriate for samples of any size and works satisfactorily even when the population is nonnormal, as long as the departure from normality is not excessive. That is, we rarely know the form of the population frequency distribution before we sample. So if a confidence interval is to be of any value, it must "work" reasonably well even when the population is nonnormal. Working "well" means that the confidence coefficient should not be affected by modest departures from normality. Experimental studies indicate that this particular confidence interval will maintain a confidence coefficient close to the experimenter's specified value for most mound-shaped probability distributions.

We assume that $Y_1, Y_2, \ldots, Y_n$ represents a random sample selected from a normal population and let $\bar{Y}$ and S^2 represent the sample mean and variance, respectively. We would like to construct a confidence interval for the population mean when $V(Y_i) = \sigma^2$ is unknown. A pivotal statistic for this situation can be formed using Student's t statistic (Definition 6.1).

Recall from Theorem 6.5 that $\bar{Y}$ and $S^2 = [1/(n-1)] \sum_{i=1}^{n} (Y_i - \bar{Y})^2$ are independent, and $(n-1)S^2/\sigma^2$ has a χ^2 distribution with $(n-1)$ degrees of freedom. Then

$$T = \frac{Z}{\sqrt{\frac{\chi^2}{\nu}}}, \qquad \text{where } Z = \frac{\bar{Y} - \mu}{\sigma/\sqrt{n}} \text{ and } \chi^2 = \frac{(n-1)S^2}{\sigma^2}$$

with $\nu = (n-1)$ degrees of freedom. Substituting for Z, χ^2, and ν, this quantity reduces to

$$T = \frac{\sqrt{n}(\bar{Y}-\mu)/\sigma}{\sqrt{\dfrac{(n-1)(S^2/\sigma^2)}{n-1}}} = \frac{\bar{Y}-\mu}{S/\sqrt{n}},$$

which has a t distribution with $(n-1)$ degrees of freedom.

From Table 4, Appendix III, we can find values $t_{\alpha/2}$ and $-t_{\alpha/2}$ so that

$$P(-t_{\alpha/2} \leq T \leq t_{\alpha/2}) = 1 - \alpha.$$

The t distribution has a density function very much like the standard normal except that the tails are thicker, as illustrated in Figure 8.6.

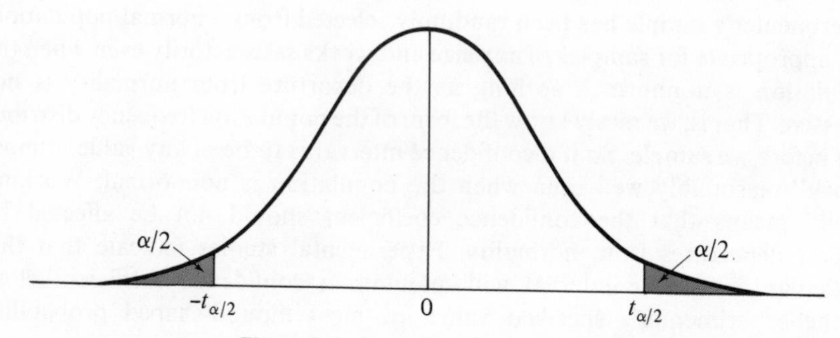

Figure 8.6 Location of $t_{\alpha/2}$ and $-t_{\alpha/2}$

The confidence interval for μ is developed just as in Example 8.4. We thus have a confidence interval for μ of the form

$$\bar{y} \pm t_{\alpha/2}\frac{s}{\sqrt{n}},$$

which means that $\bar{y} - t_{\alpha/2}(s/\sqrt{n})$ is the lower confidence limit and $\bar{y} + t_{\alpha/2}(s/\sqrt{n})$ is the upper confidence limit.

Note that the values of $t_{\alpha/2}$ depend upon the degrees of freedom, $(n-1)$, as well as the confidence coefficient, $(1-\alpha)$.

Example 8.9: *A manufacturer of gunpowder has developed a new powder, which was tested in eight shells. The resulting muzzle velocities, in feet per second, were as follows:*

3005	2925	2935	2965
2995	3005	2935	2905.

Find a confidence interval for the true average velocity, μ, for shells of this type with confidence coefficient .95.

Solution: *If we assume the velocities, Y_i, to be normally distributed, the confidence interval for μ is*

$$\bar{y} \pm t_{\alpha/2}\frac{s}{\sqrt{n}}.$$

For the above data, $\bar{y} = 2959$ and $s = 39.4$. From Table 4, Appendix III, $t_{\alpha/2} = t_{.025} = 2.365$ since there are $(n - 1) = 7$ degrees of freedom associated with this sample. Thus we obtain

$$2959 \pm 2.365\frac{39.4}{\sqrt{8}}$$

or

$$2959 \pm 33$$

as the confidence interval for μ. Note that this interval is longer than the corresponding large-sample confidence interval, for which $z_{\alpha/2} = 1.96$.

Suppose that we are interested in comparing means from two normal populations, one with mean μ_1 and variance σ_1^2 and the other with mean μ_2 and variance σ_2^2. A confidence interval for $\mu_1 - \mu_2$ based on a T random variable can be constructed if we assume that $\sigma_1^2 = \sigma_2^2 = \sigma^2$.

The large-sample confidence interval for $(\mu_1 - \mu_2)$ is developed from the random variable

$$Z = \frac{(\bar{Y}_1 - \bar{Y}_2) - (\mu_1 - \mu_2)}{\sqrt{\dfrac{\sigma_1^2}{n_1} + \dfrac{\sigma_2^2}{n_2}}}$$

which has approximately a standard normal distribution, $\bar{Y}_1$ and $\bar{Y}_2$ are the respective sample means obtained from random sampling. Under the assumption $\sigma_1^2 = \sigma_2^2 = \sigma^2$, the above ratio becomes

$$Z = \frac{(\bar{Y}_1 - \bar{Y}_2) - (\mu_1 - \mu_2)}{\sigma\sqrt{\dfrac{1}{n_1} + \dfrac{1}{n_2}}}.$$

Now we need an estimator of the common variance, σ^2, in order to construct a t statistic.

Let $Y_{11}, Y_{12}, \ldots, Y_{1n_1}$ denote the random sample of size n_1 from the first population and $Y_{21}, Y_{22}, \ldots, Y_{2n_2}$ the random sample from the second. Then

$$\bar{Y}_1 = \frac{1}{n_1}\sum_{i=1}^{n_1} Y_{1i} \quad \text{and} \quad \bar{Y}_2 = \frac{1}{n_2}\sum_{i=1}^{n_2} Y_{2i}.$$

The usual unbiased estimator of the common variance, σ^2, is obtained by pooling the sample data to obtain

$$\begin{aligned} S^2 &= \frac{\sum_{i=1}^{n_1} (Y_{1i} - \bar{Y}_1)^2 + \sum_{i=1}^{n_2} (Y_{2i} - \bar{Y}_2)^2}{n_1 + n_2 - 2} \\ &= \frac{(n_1 - 1)S_1^2 + (n_2 - 1)S_2^2}{n_1 + n_2 - 2}, \end{aligned}$$

where S_i^2 is the sample variance from the ith sample, $i = 1, 2$. Note that

$$\frac{(n_1 + n_2 - 2)S^2}{\sigma^2} = \frac{\sum_{i=1}^{n_1} (Y_{1i} - \bar{Y}_1)^2}{\sigma^2} + \frac{\sum_{i=1}^{n_2} (Y_{2i} - \bar{Y}_2)^2}{\sigma^2}$$

is the sum of two independent χ^2 random variables with $(n_1 - 1)$ and $(n_2 - 1)$ degrees of freedom, respectively. Thus $(n_1 + n_2 - 2)S^2/\sigma^2$ has a χ^2 distribution with $\nu = (n_1 + n_2 - 2)$ degrees of freedom. (See Theorems 6.4 and 6.5.) We now utilize this χ^2 variable and the Z defined in the previous paragraph to form a pivotal T statistic. That is,

$$T = \frac{Z}{\sqrt{\dfrac{\chi^2}{\nu}}} = \frac{(\bar{Y}_1 - \bar{Y}_2) - (\mu_1 - \mu_2)}{\sigma\sqrt{\dfrac{1}{n_1} + \dfrac{1}{n_2}}} \frac{1}{\sqrt{\dfrac{(n_1 + n_2 - 2)S^2}{\sigma^2(n_1 + n_2 - 2)}}}$$

$$= \frac{(\bar{Y}_1 - \bar{Y}_2) - (\mu_1 - \mu_2)}{S\sqrt{\dfrac{1}{n_1} + \dfrac{1}{n_2}}}$$

has a t distribution with $(n_1 + n_2 - 2)$ degrees of freedom.

The confidence interval for $(\mu_1 - \mu_2)$ then has the form

$$(\bar{y}_1 - \bar{y}_2) \pm t_{\alpha/2} s\sqrt{\frac{1}{n_1} + \frac{1}{n_2}}$$

where $t_{\alpha/2}$ comes from the t distribution with $(n_1 + n_2 - 2)$ degrees of freedom.

Example 8.10: *An assembly operation in a manufacturing plant requires approximately a 1-month training period for a new employee to reach maximum efficiency. A new method of training was suggested and a test was conducted to compare the new method with the standard procedure. Two groups of nine new employees were trained for a period of 3 weeks, one group using the new method and the other following standard training procedure. The length of time in minutes required for each employee to assemble the device was recorded at the end of the 3-week period. The measurements are as follows:*

Standard Procedure	32	37	35	28	41	44	35	31	34
New Procedure	35	31	29	25	34	40	27	32	31

Estimate the true mean difference, $(\mu_1 - \mu_2)$, with confidence coefficient .95.

Solution: *For the data given above, with sample 1 denoting the standard procedure, we have*

$$\bar{y}_1 = 35.22, \qquad \bar{y}_2 = 31.56$$

$$\sum_{i=1}^{9} (y_{1i} - \bar{y}_1)^2 = 195.56, \qquad \sum_{i=1}^{9} (y_{2i} - \bar{y}_2)^2 = 160.22.$$

Hence

$$s^2 = \frac{195.56 + 160.22}{9 + 9 - 2} = 22.24$$

and $$s = 4.71.$$

Also, $t_{.025} = 2.120$ *for* $(n_1 + n_2 - 2) = 16$ *degrees of freedom. The confidence interval is then*

$$(\bar{y}_1 - \bar{y}_2) \pm t_{\alpha/2} s \sqrt{\frac{1}{n_1} + \frac{1}{n_2}},$$

$$(35.22 - 31.56) \pm (2.120)(4.71)\sqrt{\frac{1}{9} + \frac{1}{9}},$$

or $$3.66 \pm 4.68.$$

As the sample size, n, gets large, the T random variable converges in distribution to the standard normal (see Example 7.8). Thus the small-sample confidence intervals of this section are equivalent to the large-sample confidence intervals of Section 8.6 for large n (or large n_1 and n_2). The intervals are nearly equivalent when $(n_1 + n_2 - 2) \geq 30$.

8.9 A Confidence Interval for σ^2

Again assume that we have a random sample, $Y_1, \ldots, Y_n$, from a normal distribution with mean μ (assumed unknown) and variance σ^2. It is now of interest to estimate σ^2. We know from Theorem 6.5 that

$$\frac{\sum_{i=1}^{n} (Y_i - \bar{Y})^2}{\sigma^2} = \frac{(n-1)S^2}{\sigma^2}$$

has a χ^2 distribution with $(n - 1)$ degrees of freedom. We can then proceed, by the pivotal method, to find two numbers, χ_L^2 and χ_U^2, such that

$$P\left[\chi_L^2 \leq \frac{(n-1)S^2}{\sigma^2} \leq \chi_U^2\right] = 1 - \alpha$$

for any confidence coefficient, $(1 - \alpha)$. (The subscripts L and U stand for "lower" and "upper," respectively.) The χ^2 density function is not symmetric, and so there is some freedom in the choice of χ_L^2 and χ_U^2. We would like to find the shortest interval that includes probability $(1 - \alpha)$, but this is generally difficult. We compromise by choosing points that cut off equal tail areas, as indicated in Figure 8.7.

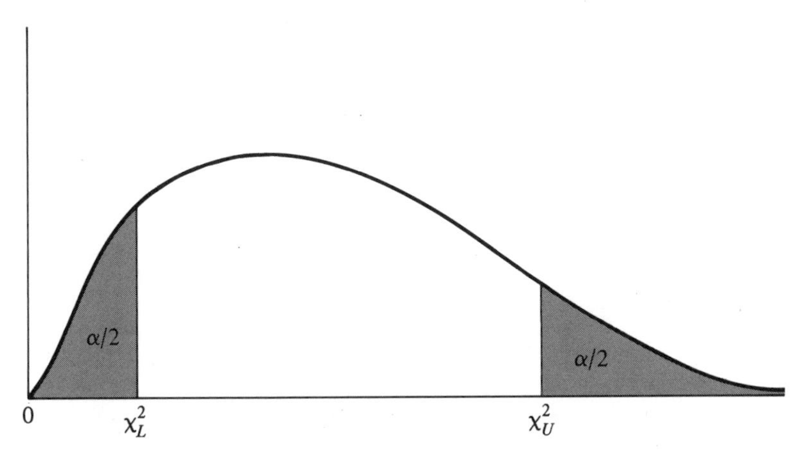

Figure 8.7 Location of χ_L^2 and χ_U^2

A reordering of the inequality in the above probability statement gives

$$P\left[\frac{(n-1)S^2}{\chi_U^2} \le \sigma^2 \le \frac{(n-1)S^2}{\chi_L^2}\right] = 1 - \alpha,$$

and hence the confidence interval for σ^2 is given by

$$\left(\frac{(n-1)s^2}{\chi_U^2}, \frac{(n-1)s^2}{\chi_L^2}\right).$$

Example 8.11: *An experimenter wanted to check the variability of equipment designed to measure the volume of an audio source. Three independent measurements recorded by this equipment for the same sound were* 4.1, 5.2, *and* 10.2. *Estimate* σ^2 *with confidence coefficient*, .90.

Solution: *Assuming normality of the measurements recorded by this equipment, the confidence interval developed above applies. For the data given,* $s^2 = 10.57$. *With* $\alpha/2 = .05$ *and* $(n - 1) = 2$ *degrees of freedom, Table 5, Appendix III, gives* $\chi_L^2 = .103$ *and* $\chi_U^2 = 5.991$. *Thus the confidence interval for* σ^2 *is*

$$\left(\frac{(n-1)s^2}{\chi_U^2}, \frac{(n-1)s^2}{\chi_L^2}\right),$$

$$\left(\frac{(2)(10.57)}{5.991}, \frac{(2)(10.57)}{.103}\right),$$

or

$$(3.53, 205.24).$$

Note that this interval for σ^2 *is very large, primarily because n is quite small.*

8.10 Summary

The objective of many statistical investigations is to make inferences about population parameters based on sample data. Often these inferences are in the form of estimates, either point estimates or interval estimates.

In this chapter we have discussed the fact that we like to have unbiased estimators with small variance. The goodness of an estimator $\hat{\theta}$ can be measured by $\sigma_{\hat{\theta}}$ because the error of estimation will generally be smaller than $2\sigma_{\hat{\theta}}$ with high probability.

Interval estimates of many parameters, such as μ and p, can be derived from the normal distribution for large sample sizes because of the central limit theorem. If sample sizes are small, the normality of the population must be assumed and the t distribution is used in deriving confidence intervals.

If sample measurements have been selected from a normal distribution, a confidence interval for σ^2 can be developed through use of the χ^2 distribution.

References

1. Hoel, P. G., *Introduction to Mathematical Statistics*, 3rd ed. New York: John Wiley & Sons, Inc., 1962.

2. Hogg, R. V., and A. T. Craig, *Introduction to Mathematical Statistics*, 3rd ed. New York: The Macmillan Company, 1970.

3. Mood, A. M., and F. A. Graybill, *Introduction to the Theory of Statistics*, 2nd ed. New York: McGraw-Hill Book Company, 1963.

Exercises

8.1. Using the identity

$$(\hat{\theta} - \theta) = [\hat{\theta} - E(\hat{\theta})] + [E(\hat{\theta}) - \theta] = [\hat{\theta} - E(\hat{\theta})] + B,$$

show that

$$\text{MSE}(\hat{\theta}) = E[(\hat{\theta} - \theta)^2] = V(\hat{\theta}) + B^2.$$

8.2. Suppose that $E(\hat{\theta}_1) = E(\hat{\theta}_2) = \theta$, $V(\hat{\theta}_1) = \sigma_1^2$, and $V(\hat{\theta}_2) = \sigma_2^2$. A new unbiased estimator, $\hat{\theta}_3$, is to be formed by

$$\hat{\theta}_3 = a\hat{\theta}_1 + (1 - a)\hat{\theta}_2.$$

How should the constant, a, be chosen so as to minimize the variance of $\hat{\theta}_3$? Assume that $\hat{\theta}_1$ and $\hat{\theta}_2$ are independent.

8.3. The mean and standard deviation for the life of a random sample of 100 light bulbs were calculated to be 1280 and 142 hours, respectively. Estimate the mean life of the population of light bulbs from which the sample was drawn and place bounds on the error of estimation.

8.4. A new type of photoflash bulb was tested to estimate the probability, p, that the new bulb would produce the required light output at the appropriate time. A sample of 1000 bulbs was tested and 920 were observed to function according to specifications. Estimate p and place bounds on the error of estimation.

8.5. Using a confidence coefficient equal to .90, place a confidence interval on the mean life of the light bulbs of Exercise 8.3.

8.6. A random sample of 400 radio tubes was tested and 40 tubes were found to be defective. With confidence coefficient equal to .90, estimate the interval within which the true fraction defective lies.

8.7. A hospital wished to estimate the average number of days required for treatment of patients between the ages of 25 and 34. A random sample of 500 hospital patients between these ages produced a mean and standard deviation equal to 5.4 and 3.1 days, respectively. Estimate the mean length of stay for the population of patients from which the sample was drawn. Place bounds on the error of estimation.

8.8. An experiment was conducted to compare the depth of penetration for two hydraulic mining nozzles. The rock structure and the length of drilling time were the same for both nozzles. With nozzle A the average penetration was 10.8 inches with a standard deviation of 1.2 inches for a sample of 50 holes. With nozzle B the average and standard deviation of the penetration measurements were 9.1 and 1.6 inches, respectively, for a sample of 80 holes. Estimate the difference in mean penetration rate and place bounds on the error of estimation.

8.9. Construct a confidence interval for the difference between the population means of Exercise 8.8, using a confidence coefficient equal to .90.

8.10. The percentage of D's and F's awarded to students by two college history professors was duly noted by the Dean. Professor A achieved a rate equal to 32 percent as opposed to 21 percent for professor B, based upon 200 and 180 students, respectively. Estimate the difference in the percentage of D's and F's awarded by the professors. Place bounds on the error of estimation.

8.11. A chemist has prepared a product designed to kill 60 percent of a particular type of insect. How large a sample should be used if he desires to be 95 percent confident that he is within .02 of the true fraction of insects killed?

8.12. Past experience shows that the standard deviation of the yearly income of textile workers in a certain state is \$400. How large a sample of textile workers would one need if one wished to estimate the population mean to within \$50.00, with a probability of .95 of being correct?

8.13. How many voters must be included in a sample collected to estimate the fraction of the popular vote favorable to a presidential candidate in a national election if the estimate is desired correct to within .005? Assume that the true fraction will lie somewhere in the neighborhood of .5. Use a confidence coefficient of approximately .95.

8.14. In a poll taken among college students, 300 of 500 fraternity men favored a certain proposition, whereas 64 of 100 nonfraternity men favored it.

Estimate the difference in the fractions favoring the proposition and place a bound upon the error of estimation.

8.15. Refer to Exercise 8.14. How many fraternity and nonfraternity students must be included in a poll if we wish to estimate the difference in the fractions correct to within .05? Assume that the groups will be of equal size and that $p = .6$ will suffice as an approximation to both fractions.

8.16. From each of two normal populations with identical means and with standard deviations of 6.40 and 7.20, independent random samples of 64 observations are drawn. Find the probability that the difference between the means of the samples exceeds 0.60 in absolute value.

8.17. We wish to use the sample mean, $\bar{y}$, to estimate the mean of a normally distributed population with an error of less than .5 with probability .9. If it is known that the variance of the population is equal to 4, how large should the sample be to achieve the accuracy stated above?

8.18. We wish to estimate the difference in grade-point average between two groups of college students accurate to within .2 grade point. If the standard deviation of the grade-point measurements is approximately equal to .6, how many students must we include in each group? (Assume that the groups will be of equal size.)

8.19. A chemical process has produced, on the average, 800 tons of chemical per day. The daily yields for the past week are 785, 805, 790, 793, and 802 tons. Estimate the mean daily yield, with confidence coefficient .90, from the above data.

8.20. The mean and standard deviation for a sample of 19 measurements were found to equal 24.7 and 1.8, respectively. Find a 98 percent confidence interval for the mean of the population.

8.21. The main-stem growth, measured for a sample of 17 4-year-old red pine trees, produced a mean and standard deviation equal to 11.3 and 3.4 inches, respectively. Find a 90 percent confidence interval for the mean growth of a population of 4-year-old red pine trees subjected to similar environmental conditions.

8.22. Owing to the variability of trade-in allowance, the profit per new car sold by an automobile dealer varies from car to car. The profits per sale (in hundreds of dollars), tabulated for the past week, were

$$2.1, \quad 3.0, \quad 1.2, \quad 6.2, \quad 4.5, \quad 5.1.$$

Find a 90 percent confidence interval for the mean profit per sale.

8.23. Two random samples, each containing 11 measurements, were drawn from normal populations possessing means, μ_1 and μ_2, respectively, and a common variance, σ^2. The sample means and variances are as follows:

Population I	*Population II*
$\bar{y}_1 = 60.4$	$\bar{y}_2 = 65.3$
$s_1^2 = 31.40$	$s_2^2 = 44.82$

Find a 90 percent confidence interval for the difference between the population means.

8.24. Two methods for teaching reading were applied to two randomly selected groups of elementary school children and compared on the basis of a reading comprehension test given at the end of the learning period. The sample means and variances computed from the test scores are as follows:

	Method 1	*Method 2*
No. children in group	11	14
$\bar{y}$	64	69
s^2	52	71

Find a 95 percent confidence interval for $(\mu_1 - \mu_2)$.

8.25. A comparison of reaction times for two different stimuli in a psychological word-association experiment produced the following results (in seconds) when applied to a random sample of 16 people:

Stimulus 1	*Stimulus 2*
1	4
3	2
2	3
1	3
2	1
1	2
3	3
2	3

Obtain a 90 percent confidence interval for $(\mu_1 - \mu_2)$.

8.26. The following data give readings (in foot-pounds) on the impact strength of two types of packaging material. Find a 98 percent confidence interval for $(\mu_1 - \mu_2)$. What assumptions are necessary?

	A	B
	1.25	.89
	1.16	1.01
	1.33	.97
	1.15	.95
	1.23	.94
	1.20	1.02
	1.32	.98
	1.28	1.06
	1.21	.98
$\sum y =$	11.13	8.80
$\bar{y} =$	1.237	.978
$\sum y_i^2 =$	13.7973	8.6240

8.27. Refer to Exercise 8.19. Find a 90 percent confidence interval for σ^2, the variance of the population of daily yields.

8.28. A precision instrument is guaranteed to read accurate to within 2 units. A sample of four instrument readings on the same object yielded the measurements 353, 351, 351, and 355. Find a 90 percent confidence interval for the population variance. Does the guarantee seem reasonable?

8.29. Let $Y_1, Y_2, \ldots, Y_n$ be a random sample of n observations from a normal population. Use the fact that $[(n-1)S^2]/\sigma^2 = \chi^2$ with $\nu = (n-1)$ degrees of freedom to show that S^2 is an unbiased estimator of σ^2 and that $V(S^2) = 2\sigma^4/(n-1)$.

8.30. (Refer to Exercise 8.29.) We noted in Section 8.3 that

$$S'^2 = \frac{\sum_{i=1}^{n} (Y_i - \bar{Y})^2}{n}$$

is a biased estimator of σ^2 and that

$$S^2 = \frac{\sum_{i=1}^{n} (Y_i - \bar{Y})^2}{n-1}$$

is an unbiased estimator of the same parameter.

(a) Find $V(S'^2)$.
(b) Show that $V(S^2)$ exceeds $V(S'^2)$.

8.31. Exercise 8.30 suggests that S^2 is superior to S'^2 in regard to bias and that S'^2 is superior to S^2 because it possesses smaller variance. Which is the better estimator? Compare the mean-square errors.

8.32. Suppose that two independent random samples of n_1 and n_2 observations, respectively, are selected from normal populations. Further, assume that the populations possess a common variance σ^2. Let

$$S_i^2 = \frac{\sum_{j=1}^{n} (Y_{ij} - \bar{Y}_i)}{n_i - 1}, \qquad i = 1, 2.$$

(a) Show that the pooled estimator of σ^2,

$$S^2 = \frac{(n_1 - 1)S_1^2 + (n_2 - 1)S_2^2}{n_1 + n_2 - 2},$$

is unbiased.
(b) Find $V(S^2)$.

8.33. Suppose that two independent random samples of n_1 and n_2 observations, respectively, are selected from normal populations with means μ_i and variances σ_i^2, $i = 1, 2$. We wish to construct a confidence interval for the variance ratio σ_1^2/σ_2^2. Let S_i^2, $i = 1, 2$, be as defined in Exercise 8.32. Use the fact that

$$\frac{(n_i - 1)S_i^2}{\sigma_i^2}, \qquad i = 1, 2,$$

has a χ^2 distribution with $(n_i - 1) = \nu_i$ degrees of freedom and the definition of the F-random variable, Definition 6.2, to construct a pivotal statistic. Then, find a confidence interval for σ_1^2/σ_2^2 with confidence coefficient $(1 - \alpha)$.

8.34. Note that the small-sample confidence interval for μ, based on Student's t (Section 8.8), possesses a random width [in contrast to the large-sample confidence interval (Section 8.6), where the width is nonrandom]. Find the expected value of the interval width in the small-sample case.

8.35. A confidence interval is *unbiased* if the expected value of the interval midpoint is equal to the estimated parameter. The expected value of the midpoint of the large-sample confidence interval (Section 8.6) is equal to the estimated parameter and the same is true for the small-sample confidence intervals for μ and $(\mu_1 - \mu_2)$ (Section 8.8). For example, the midpoint of the interval,

$$\bar{y} \pm ts/\sqrt{n},$$

is $\bar{y}$ and $E(\bar{Y}) = \mu$. Now consider the confidence interval for σ^2. Show that the expected value of the midpoint of this confidence interval is not equal to σ^2.

8.36. As we have noted, the sample mean, $\bar{Y}$, is a good point estimator of the population mean, μ. It can also be used to predict a future value of Y independently selected from the population. Assume that you have a sample mean and variance, $\bar{Y}$ and S^2, based on a random sample of n measurements from a normal population. Use Student's t to form a pivotal statistic to find a prediction interval for some new value of Y, say Y_p, to be observed in the future. (*Hint:* Start with the quantity $Y_p - \bar{Y}$.) Note the terminology: parameters are *estimated*; values of random variables are *predicted.*

9

Methods of Estimation and Properties of Point Estimators

9.1 Introduction

Chapter 8 contains a somewhat intuitive discussion of the process of finding and using some common estimators, along with a brief mention of the basic properties of these estimators. Chapter 9 will present a more unified and theoretical approach to finding estimators and a more detailed look at some properties of point estimators.

We will first present two methods for finding estimators, the method of moments and the method of maximum likelihood. This will be followed by a discussion of three important properties of estimators: efficiency, consistency, and sufficiency. Finally, a decision theoretic approach to the problem of estimation will be discussed briefly.

9.2 The Method of Moments

The method of moments is a very simple procedure for finding an estimator for one or more population parameters. You recall that the kth moment of a random variable, taken about the origin, is

$$\mu'_k = E(Y^k).$$

The corresponding kth sample moment is the average,

$$m'_k = \frac{1}{n}\sum_{i=1}^{n} Y_i^k.$$

The method of moments is based on the assumption that sample moments should provide good estimates of the corresponding population moments. That is, m'_k should be a good estimator of μ'_k, $k = 1, 2, \ldots$. Then, since the population moments, $\mu'_1, \mu'_2, \ldots, \mu'_k$, will be functions of the population parameters, we can equate corresponding population and sample moments and solve for the desired parameters. Hence the method of moments can be stated as follows.

> ***Method of Moments:*** *Choose as estimates those values of the parameters that are solutions of the equations $\mu'_k = m'_k$, $k = 1, 2, \ldots, p$, where p equals the number of parameters.*

Example 9.1: *A random sample of n observations, $y_1, y_2, \ldots, y_n$, is selected from a population in which Y_i, $i = 1, 2, \ldots, n$, possesses a uniform probability density function, $f(y) = 1/\theta$, $0 \le y \le \theta$. Use the method of moments to estimate the unknown parameter, θ.*

Solution: *The value of μ'_1 for a uniform random variable is*

$$\mu'_1 = \mu = \frac{\theta}{2}.$$

(This is easy to derive; or see Table 1, Appendix III.) The corresponding first sample moment is

$$m'_1 = \frac{\sum_{i=1}^{n} y_i}{n} = \bar{y}.$$

Equating corresponding moments and solving for the unknown parameter, θ, we have

$$\mu_1' = \frac{\theta}{2} = \bar{y}$$

or

$$\hat{\theta} = 2\bar{y}.$$

(Note: The hat symbol over a parameter is to be read "estimator of.") Thus $2\bar{Y}$ is the moment estimator of θ.

Example 9.2: *A random sample of n observations, $y_1, y_2, \ldots, y_n$, is selected from a population where Y_i, $i = 1, 2, \ldots, n$, possesses a gamma probability density function with parameters α and β (see Section 4.5 for the gamma probability density function). Find moment estimators for the unknown parameters, α and β.*

Solution: *Since we seek estimators for* two parameters, *α and β, it will be necessary to equate two pairs of population and sample moments.*

You can verify (either by deriving or consulting Table 1, Appendix III) that the first two moments of the gamma distribution are

$$\mu_1' = \mu = \alpha\beta,$$

$$\mu_2' = \sigma^2 + \mu^2 = \alpha\beta^2 + \alpha^2\beta^2.$$

Now equate these quantities to their corresponding sample moments and solve for α and β. Thus

$$\mu_1' = \alpha\beta = m_1' = \bar{y},$$

$$\mu_2' = \alpha\beta^2 + \alpha^2\beta^2 = m_2' = \frac{\sum_{i=1}^{n} y_i^2}{n}.$$

From the first equation $\beta = m_1'/\alpha$. Substituting into the second equation and solving for α, we obtain

$$\alpha = \frac{m_1'^2}{m_2' - m_1'^2} = \frac{\bar{y}^2}{\frac{\sum y_i^2}{n} - \bar{y}^2} = \frac{n\bar{y}^2}{\sum_{i=1}^{n} (y_i - \bar{y})^2}.$$

Substituting α into the first equation, we obtain

$$\beta = \frac{m_1'}{\alpha} = \frac{\bar{y}}{\alpha} = \frac{\sum_{i=1}^{n} (y_i - \bar{y})^2}{n\bar{y}}.$$

Hence the two moment estimators of the parameters α and β are

$$\hat{\alpha} = \frac{n\bar{Y}^2}{\sum_{i=1}^{n} (Y_i - \bar{Y})^2}$$

and

$$\hat{\beta} = \frac{\sum_{i=1}^{n} (Y_i - \bar{Y})^2}{n\bar{Y}}.$$

To summarize, the method of moments finds estimates of unknown population parameters by equating corresponding population and sample moments.

9.3 The Method of Maximum Likelihood

We shall use an example to illustrate the logic upon which the method of maximum likelihood is based. Suppose that we are confronted with a box which contains three balls. We know that some of the balls are white and some are red, but we do not know the number of either color. However, we are allowed to randomly sample two of the balls. If our random sample yields two red balls, what would be a good estimate of the total number of red balls in the box? Obviously, the number of red balls in the box must be two or three. If there are two red balls and one white ball in the box, the probability of sampling two red balls is $\binom{2}{2}\binom{1}{0}\Big/\binom{3}{2} = 1/3$. On the other hand, if there are three red balls in the box, the probability of sampling two red balls is $\binom{3}{2}\Big/\binom{3}{2} = 1$. It should seem reasonable to choose three as the estimate of the number of red balls in the box, because this estimate maximizes the probability of the observed sample. Of course it is possible for the box to contain only two red balls, but the observed outcome gives more credence to there being three red balls in the box.

The above example illustrates a method for finding an estimator that can be applied to any situation. The technique, called the *method of maximum likelihood*, selects as estimates those values of the parameters that maximize the probability (or joint density) of the observed sample. The following definitions provide a precise statement of these ideas.

Definition 9.1: *Let $y_1, y_2, \ldots, y_n$ be sample observations taken on corresponding random variables, $Y_1, Y_2, \ldots, Y_n$. Then if $Y_1, Y_2, \ldots, Y_n$ are discrete random variables, the* likelihood of the sample, *L, is defined to be the joint probability of $y_1, y_2, \ldots, y_n$. If $Y_1, Y_2, \ldots, Y_n$ are continuous random variables, the likelihood, L, is defined to be the joint density evaluated at $y_1, y_2, \ldots, y_n$.*

Method of Maximum Likelihood: *Choose as estimates those values of the parameters that maximize the likelihood, L.*

We will illustrate the method with an example.

Example 9.3: *A binomial experiment consisting of n trials resulted in observations $y_1, y_2, \ldots, y_n$, where $y_i = 1$ if the ith trial was a success, $y_i = 0$ otherwise. Find the maximum likelihood estimator of p, the probability of a success.*

Solution: *The likelihood of the observed sample is the probability of observing $y_1, y_2, \ldots, y_n$. Hence*

$$L = p^y(1-p)^{n-y},$$

where

$$y = \sum_{i=1}^{n} y_i.$$

We now wish to find the value of p that maximizes L. This can be done by setting the derivative, dL/dp, equal to zero and solving for p.

You will note that $\ln L$ is a monotonically increasing function of L and hence both $\ln L$ and L will be maximized for the same value of p. Since L is a product of functions of p, and finding the derivative of products is tedious, it is easier to find the value of p that maximizes $\ln L$. Then

$$\ln L = y \ln p + (n-y)\ln(1-p).$$

Taking the derivative of $\ln L$ with respect to p,

$$\frac{d \ln L}{dp} = y\frac{1}{p} + (n-y)\frac{-1}{1-p}.$$

Then the values of p that maximize (or minimize) $\ln L$ are the solutions of the equation

$$\frac{y}{\hat{p}} - \frac{n-y}{1-\hat{p}} = 0.$$

Solving, we obtain the estimator, $\hat{p} = Y/n$. You can easily verify that this solution occurs when $\ln L$ (and hence L) achieves a maximum.

Note that the estimator, $\hat{p} = Y/n$, is the fraction of successes in the total number of trials, n. Hence the maximum likelihood estimator of p is one that is intuitively appealing.

Example 9.4: *Let $Y_1, Y_2, \ldots, Y_n$ be a random sample from a normal distribution with mean μ and variance σ^2. Find the maximum likelihood estimators of μ and σ^2.*

Solution: *Since $Y_1, Y_2, \ldots, Y_n$ are continuous random variables, L is the joint density of the sample. Thus $L = f(y_1, y_2, \ldots, y_n)$. Further, since the sample was selected in a random manner, $Y_1, Y_2, \ldots, Y_n$ are mutually independent random variables (explained in Section 5.4). Hence*

$$L = f(y_1, y_2, \ldots, y_n) = f(y_1)f(y_2)\cdots f(y_n)$$

$$= \frac{\exp\left[\dfrac{-(y_1-\mu)^2}{2\sigma^2}\right]}{\sigma\sqrt{2\pi}} \frac{\exp\left[\dfrac{-(y_2-\mu)^2}{2\sigma^2}\right]}{\sigma\sqrt{2\pi}} \cdots \frac{\exp\left[\dfrac{-(y_n-\mu)^2}{2\sigma^2}\right]}{\sigma\sqrt{2\pi}}$$

$$= \frac{1}{\sigma^n(2\pi)^{n/2}} \exp\left[-\sum_{i=1}^{n}(y_i-\mu)^2/2\sigma^2\right]$$

[recall that exp( ) is just another way of writing $e^{(\ )}$] and

$$\ln L = -\frac{n}{2}\ln\sigma^2 - \frac{n}{2}\ln 2\pi - \frac{\sum_{i=1}^{n}(y_i-\mu)^2}{2\sigma^2}.$$

The maximum likelihood estimators of μ and σ^2 are those values that make $\ln L$ a maximum. Taking derivatives with respect to μ and σ^2, we obtain

$$\frac{d(\ln L)}{d\mu} = \frac{\sum_{i=1}^{n}(y_i-\mu)}{\sigma^2}$$

and

$$\frac{d\ln L}{d\sigma^2} = -\frac{n}{2}\frac{1}{\sigma^2} + \frac{\sum_{i=1}^{n}(y_i-\mu)^2}{2\sigma^4}.$$

Setting the derivatives equal to zero and solving simultaneously, from the first equation,

$$\frac{\sum_{i=1}^{n}(y_i-\hat{\mu})}{\hat{\sigma}^2} = 0,$$

$$\sum_{i=1}^{n} y_i - n\hat{\mu} = 0,$$

or

$$\hat{\mu} = \frac{\sum_{i=1}^{n} y_i}{n} = \bar{y}.$$

Substituting $\bar{y}$ for $\hat{\mu}$ in the second equation and solving for $\hat{\sigma}^2$,

$$\frac{-n}{\hat{\sigma}^2} + \frac{\sum_{i=1}^{n}(y_i-\bar{y})^2}{\hat{\sigma}^4} = 0$$

or

$$\hat{\sigma}^2 = \frac{\sum_{i=1}^{n}(y_i-\bar{y})^2}{n} = s'^2.$$

Example 9.5: *Let $y_1, y_2, \ldots, y_n$ be a random sample of observations from a uniform distribution with probability density function, $f(y_i) = 1/\theta$, $0 \le y_i \le \theta$, $i = 1, 2, \ldots, n$. Find the maximum likelihood estimator of θ.*

Solution: *Since $y_1, y_2, \ldots, y_n$ are measurements obtained by random sampling, $Y_1, Y_2, \ldots, Y_n$ are independent random variables and*

$$\begin{aligned} L = f(y_1, y_2, \ldots, y_n) &= f(y_1)f(y_2)\cdots f(y_n) \\ &= \frac{1}{\theta}\frac{1}{\theta}\cdots\frac{1}{\theta} \\ &= \frac{1}{\theta^n}. \end{aligned}$$

You will note that L is a monotonically decreasing function of θ and hence nowhere in the interval $0 < \theta < \infty$ is $dL/d\theta$ equal to zero. However, note that L increases as θ decreases and that θ must be equal to or greater than the maximum observation in the set, $y_1, y_2, \ldots, y_n$. Hence the value of θ that maximizes L is the largest observation in the sample. That is, $\hat{\theta} = y_{(n)} = \max(y_1, \ldots, y_n)$.

9.4 Properties of Estimators

Two desirable properties of estimators, unbiasedness and minimum variance, were briefly discussed in Section 8.2. This section, along with the next three sections, will present a more detailed look at properties of point estimators.

Since estimators are functions of the random variables that appear in a sample, they possess probability distributions. That is, if we were to repeatedly sample from a population, always selecting the same number of observations in each sample, the estimates of a parameter, say θ, would generate a probability distribution. For this reason, the probability distributions of estimators are called *sampling distributions*. Sampling distributions can often be derived theoretically by using the methods of Chapter 6. Even if the sampling distribution for a certain estimator cannot be derived, some properties of the estimator can be found through knowledge of its mean and variance.

Recall that one of the basic desirable properties for an estimator is that of unbiasedness. By Definition 8.2, an estimator, $\hat{\theta}$, is unbiased for θ if $E(\hat{\theta}) = \theta$. Examples of unbiased estimators are given in Chapter 8.

We also mentioned in Chapter 8 that we would like estimators to have small variances. The problem of comparing variances of estimators will be discussed in Section 9.5 in terms of efficiency.

9.5 Efficiency

Unbiased estimators are frequently compared using the ratio of their variances. This quantity, called *efficiency*, is defined as follows:

Definition 9.2: *Given two unbiased estimators, $\hat{\theta}_1$ and $\hat{\theta}_2$, of a parameter θ, with variances $V(\hat{\theta}_1)$ and $V(\hat{\theta}_2)$, respectively, then the efficiency of $\hat{\theta}_1$ relative to $\hat{\theta}_2$ is defined to be the ratio*

$$\text{efficiency} = \frac{V(\hat{\theta}_2)}{V(\hat{\theta}_1)}.$$

For example, you can construct many estimators for a population mean. Suppose that we wish to estimate the mean of a normal population. Let $\hat{\theta}_1$ be the sample median, the middle observation when the sample measurements are ordered according to magnitude (n odd) or the average of the two middle observations (n even). Let $\hat{\theta}_2$ be the sample mean. Although proof is omitted, it can be shown that the variance of the sample median is, for large n, $V(\hat{\theta}_1) = (1.2533)^2(\sigma^2/n)$. Then the efficiency of the sample median relative to the sample mean is

$$\text{efficiency} = \frac{V(\hat{\theta}_2)}{V(\hat{\theta}_1)} = \frac{\sigma^2/n}{(1.2533)^2\sigma^2/n} = \frac{1}{(1.2533)^2}$$

$$= .6366.$$

Example 9.6: *Let $Y_1, \ldots, Y_n$ denote a random sample from a uniform distribution on the interval $(0, \theta)$. The moment estimator of θ, denoted by $\hat{\theta}_1$, is given in Example 9.1 to be $\hat{\theta}_1 = 2\bar{Y}$. The maximum likelihood estimator of θ, given in Example 9.5, is $Y_{(n)} = \max(Y_1, \ldots, Y_n)$. The maximum likelihood estimator adjusted to be unbiased is given by $\hat{\theta}_2 = [(n+1)/n]Y_{(n)}$. Find the efficiency of $\hat{\theta}_1$ relative to $\hat{\theta}_2$.*

Solution: *Observe that $\hat{\theta}_1$ is unbiased and*

$$V(\hat{\theta}_1) = 4V(\bar{Y}) = \frac{4}{n}V(Y_i)$$

$$= \frac{4}{n}\frac{\theta^2}{12} = \frac{\theta^2}{3n},$$

since $V(Y_i) = \theta^2/12$ in this uniform case.

To find the mean and variance of $\hat{\theta}_2$ we must first recall that the density function of $Y_{(n)}$ is given by

$$g_n(y) = n[F_Y(y)]^{n-1}f_Y(y)$$

$$= n\left(\frac{y}{\theta}\right)^{n-1}\frac{1}{\theta}$$

$$= \frac{ny^{n-1}}{\theta^n}, \qquad 0 \le y \le \theta,$$

$$g_n(y) = 0, \qquad \text{elsewhere},$$

since

$$P(Y_i \leq y) = F_Y(y) = \frac{y}{\theta}, \qquad 0 \leq y \leq \theta.$$

Thus

$$E[Y_{(n)}] = \frac{n}{\theta^n}\int_0^\theta y^n \, dy = \frac{n}{n+1}\theta$$

and

$$E[Y_{(n)}^2] = \frac{n}{\theta^n}\int_0^\theta y^{n+1} \, dy$$

$$= \frac{n}{n+2}\theta^2.$$

If follows that

$$V[Y_{(n)}] = E[Y_{(n)}^2] - \{E[Y_{(n)}]\}^2$$

$$= \left[\frac{n}{n+2} - \left(\frac{n}{n+1}\right)^2\right]\theta^2,$$

and

$$V(\hat{\theta}_2) = V\left[\frac{n+1}{n}Y_{(n)}\right]$$

$$= \left(\frac{n+1}{n}\right)^2 V[Y_{(n)}]$$

$$= \left[\frac{(n+1)^2}{n(n+2)} - 1\right]\theta^2.$$

$$= \frac{\theta^2}{n(n+2)}.$$

Therefore, the efficiency of $\hat{\theta}_1$ relative to $\hat{\theta}_2$ is given by

$$\text{efficiency} = \frac{V(\hat{\theta}_2)}{V(\hat{\theta}_1)} = \frac{\theta^2/n(n+2)}{\theta^2/3n}$$

$$= \frac{3}{n+2}.$$

Note that this efficiency is less than unity for $n > 1$. That is, $\hat{\theta}_2$ has smaller variance than $\hat{\theta}_1$, and therefore $\hat{\theta}_2$ would generally be preferred over $\hat{\theta}_1$ as an estimator of θ.

9.6 Consistency

We mentioned earlier that convergence in probability concerns the convergence of a random variable to a constant. Hence it is not surprising that we should use convergence in probability as a desirable criterion for an estimator.

> ***Definition 9.3:*** *Let $\hat{\theta}$ be an estimator of θ. Then $\hat{\theta}$ is said to be a* consistent estimator *if it converges in probability to θ.*

Theorem 7.1 shows that the sample mean, $\overline{Y}$, of n independent random variables, $Y_1, Y_2, \ldots, Y_n$, with common mean μ is a consistent estimator of μ. Example 7.1 shows that the sample fraction of successes is a consistent estimator of the binomial parameter, p.

Consistency is a property possessed by many estimators. In fact, most moment estimators and maximum likelihood estimators are consistent. Sufficient conditions for consistency are given by the following theorem.

> ***Theorem 9.1:*** *An unbiased estimator, $\hat{\theta}$, of θ is consistent if*
>
> $$\lim_{n\to\infty} V(\hat{\theta}) = 0.$$

Proof of Theorem 9.1 follows directly from Tchebysheff's Theorem, as in the proof of Theorem 7.1.

9.7 Sufficient Statistics

Before proceeding, let us define precisely what we mean by the term "statistic."

> ***Definition 9.4:*** *A* statistic *is a function of observable random variables which contains no unknown parameters.*

In a sense, a sufficient statistic for a parameter θ is a function of the sampled random variables, $Y_1, \ldots, Y_n$, which utilizes all the sample information pertinent to θ. We will explain. Suppose that we wish to estimate θ from a random sample of n measurements, $y_1, \ldots, y_n$, and suppose that U is a statistic (that is, a function of $Y_1, \ldots, Y_n$). Now, assume that the likelihood of the sample, L, can be factored into two functions,

$$L = g(u, \theta)h(y_1, \ldots, y_n),$$

where $g(u, \theta)$ is a function of only u and θ and $h(y_1, \ldots, y_n)$ does not depend on θ. Then

$$\ln L = \ln g(u, \theta) + \ln h(y_1, \ldots, y_n).$$

Taking the derivative with respect to θ, we note that derivative of the second term on the right is equal to zero and the equation, $d \ln L/d\theta = 0$, is a function only of θ and u. Hence the maximum likelihood estimator can only be a function of the statistic, U. That is, from the maximum likelihood point of view, all the information pertinent to θ is contained in the statistic U.

> ***Definition 9.5:*** *Let U be a statistic based on the random sample, $Y_1, Y_2, \ldots, Y_n$. Then U is a* sufficient statistic *for the estimation of a parameter θ if the likelihood, L, can be factored into two nonnegative functions,*
>
> $$L = g(u, \theta)h(y_1, y_2, \ldots, y_n),$$
>
> *where $g(u, \theta)$ is a function only of u and θ and $h(y_1, y_2, \ldots, y_n)$ is not a function of θ.*

Example 9.7: *Let $Y_1, Y_2, \ldots, Y_n$ be a random sample in which Y_i possesses the probability density function, $f(y_i) = (1/\alpha)e^{-y_i/\alpha}, 0 \le y_i \le \infty, i = 1, 2, \ldots, n$. Show that $\overline{Y}$ is a sufficient statistic for the estimation of α.*

Solution: *The likelihood of the sample, L, is the joint density,*

$$L = f(y_1, y_2, \ldots, y_n) = f(y_1)f(y_2)\cdots f(y_n)$$

$$= \frac{e^{-y_1/\alpha}}{\alpha}\frac{e^{-y_2/\alpha}}{\alpha}\cdots\frac{e^{-y_n/\alpha}}{\alpha}$$

$$= \frac{e^{-\Sigma y_i/\alpha}}{\alpha^n} = \frac{e^{-n\bar{y}/\alpha}}{\alpha^n}.$$

Note that L is a function only of α and $\bar{y}$ and that

$$g(\bar{y}, \alpha) = \frac{e^{-n\bar{y}/\alpha}}{\alpha^n}$$

and

$$h(y_1, y_2, \ldots, y_n) = 1.$$

Hence $\overline{Y}$ is a sufficient statistic for the estimation of α.

Sufficient statistics play a prominent role in the theory of statistics. For almost any probability distribution that is likely to come up in practice, the unique minimum variance unbiased estimator of an unknown parameter θ, if such an estimator exists, will be found as a function of the sufficient statistic for θ. This concept is particularly useful when related to maximum likelihood estimation. The method of maximum likelihood will always produce the sufficient statistic, if one exists. It then remains to adjust the statistic to make it unbiased for θ. The result is generally a minimum variance unbiased estimator for θ. (Henceforth we shall denote "minimum variance unbiased estimator" by MVUE.)

Example 9.8: *Let $Y_1, \ldots, Y_n$ be a random sample from a normal distribution with mean μ and variance σ^2. Find the MVUE for σ^2.*

Solution: *Example 9.4 shows the maximum likelihood estimator of σ^2 to be*

$$S'^2 = \frac{1}{n}\sum_{i=1}^{n}(Y_i - \overline{Y})^2.$$

That S'^2 is not unbiased for σ^2 is shown in Example 8.1. The sufficient statistic for σ^2, with μ unknown, is

$$\sum_{i=1}^{n} (Y_i - \overline{Y})^2.$$

This statistic can be made into an unbiased estimator for σ^2 by dividing by $n - 1$. Hence

$$S^2 = \frac{1}{n-1} \sum_{i=1}^{n} (Y_i - \overline{Y})^2$$

is the MVUE of σ^2, if μ is unknown.

Sufficient statistics are also useful in finding the MVUE of a function of a parameter θ, say $h(\theta)$. If a certain statistic, say U, is sufficient for θ, then U will be sufficient for $h(\theta)$. For uncomplicated functions $h(\theta)$, the MVUE can usually be found by first finding the sufficient statistic, U, and then adjusting it to be unbiased for $h(\theta)$.

Example 9.9: *Let $Y_1, \ldots, Y_n$ be a random sample from the exponential density given by*

$$f(y) = \begin{cases} \frac{1}{\theta} e^{-y/\theta}, & y > 0, \\ 0, & \text{elsewhere.} \end{cases}$$

Find the MVUE for $V(Y_i)$.

Solution: *Methods of Chapter 4 are used to derive the fact that $V(Y_i) = \theta^2$. In this case, the likelihood, L, is given by*

$$L = \frac{1}{\theta^n} \exp\left(- \sum_{i=1}^{n} \frac{y_i}{\theta} \right)$$

$$= \frac{1}{\theta^n} e^{-n\bar{y}/\theta}.$$

Hence $\overline{Y}$ is a sufficient statistic for θ and it follows that $\overline{Y}$ is also sufficient for θ^2. Intuition tells us that $\overline{Y}^2$ might be a reasonable function to investigate in looking for an estimator of θ^2. Now

$$\begin{aligned} E(\overline{Y}^2) &= V(\overline{Y}) + [E(\overline{Y})]^2 \\ &= \frac{1}{n}V(Y_i) + [E(\overline{Y})]^2 \\ &= \frac{1}{n}\theta^2 + \theta^2 = \theta^2\frac{n+1}{n}, \end{aligned}$$

and hence

$$\frac{n}{n+1}\overline{Y}^2$$

is an unbiased estimator of θ^2. Since this estimator is a function of $\overline{Y}$, the sufficient statistic for θ, it is the MVUE for $\theta^2 = V(Y_i)$.

9.8 Other Methods for Finding Estimators

Many other methods for finding estimators can be devised, such as the method of least squares, which will be discussed in Chapter 11. One of the most general methods is based on decision theory. Suppose that $\hat{\theta}$ is an estimator of a parameter θ. When a given sample is observed, a value of $\hat{\theta}$ is calculated and compared with θ. Being "wrong" is usually costly. That is, the farther the observed value of $\hat{\theta}$ lies from θ, the greater the loss you might incur in basing any decision on $\hat{\theta}$. In fact, it is conceivable that positive errors might be more expensive than negative errors (or vice versa) and the most ideal situation would occur when $\hat{\theta} = \theta$. Then you would estimate with perfect accuracy and no loss would occur. The decision theoretic approach defines a loss function, call it $L(\hat{\theta}, \theta)$, which gives the loss associated with every value that θ might assume. Hence the loss is a function of the sample values, $y_1, y_2, \ldots, y_n$, and θ. Then you select the estimator, $\hat{\theta}$, so as to minimize the expected loss.

The decision theoretic method for finding estimators is theoretically appealing but possesses certain practical shortcomings. It is generally impossible to find a single estimator, $\hat{\theta}$, which minimizes the expected loss for all possible values of θ without some restrictions on the nature of $\hat{\theta}$ or $L(\hat{\theta}, \theta)$. Note that if $\hat{\theta}$ is restricted to be unbiased for θ and $L(\hat{\theta}, \theta) = (\hat{\theta} - \theta)^2$, then $E[L(\hat{\theta}, \theta)] = V(\hat{\theta})$ and the $\hat{\theta}$ which minimizes the expected loss is the MVUE for θ, as discussed in previous sections.

Another criterion that is often used for finding an estimator, $\hat{\theta}$, for a given loss function, $L(\hat{\theta}, \theta)$, is the *minimax* criterion. That is, we choose as the estimator of θ that function $\hat{\theta}$ which minimizes the maximum expected loss over all possible values of θ. The next example illustrates the method.

Example 9.10: *A box contains four identical replacement parts for an engine. Suppose that θ denotes the number of good parts among the four in the box, and it is known that θ equals 2 or 3. (One or two parts may be defective.) A mechanic randomly chooses one part from the box and tests it to see if it is good. Let Y denote the number of good parts in the sample of size one (that is, $Y = 0$ or $Y = 1$). The mechanic assumes that his loss function, $L(\hat{\theta}, \theta)$, is as follows (in dollars):*

$$L(2,2) = 0, \qquad L(3,2) = 4,$$

$$L(2,3) = 1, \qquad L(3,3) = 0.$$

Note that the loss for estimating θ to be 3 when it is, in fact, 2 is quite large relative to the others. Practically speaking, this means that it is costly to overestimate the number of good parts on hand. Some estimators of θ are $\hat{\theta}_1 = Y + 2$, $\hat{\theta}_2 = 3 - Y$, $\hat{\theta}_3 = 2$, and $\hat{\theta}_4 = 3$. $\hat{\theta}_4 = 3$ means that we always estimate θ to be 3 regardless of the value of Y. Find the minimax estimator of θ in this set.

Solution: *We must first compute the expected loss for each estimator and each value of θ:*

$$E[L(\hat{\theta}_1, 2)] = L(2,2)P(Y=0) + L(3,2)P(Y=1)$$

$$= 0(1/2) + 4(1/2) = 2,$$

$$E[L(\hat{\theta}_1, 3)] = L(2,3)P(Y=0) + L(3,3)P(Y=1)$$

$$= 1(1/4) + 0(3/4) = 1/4,$$

$$E[L(\hat{\theta}_2, 2)] = L(3,2)P(Y=0) + L(2,2)P(Y=1)$$

$$= 4(1/2) + 0(1/2) = 2,$$

$$E[L(\hat{\theta}_2, 3)] = L(3,3)P(Y=0) + L(2,3)P(Y=1)$$

$$= 0(1/4) + 1(3/4) = 3/4,$$

$$E[L(\hat{\theta}_3, 2)] = L(2, 2) = 0,$$

$$E[L(\hat{\theta}_3, 3)] = L(2, 3) = 1,$$

$$E[L(\hat{\theta}_4, 2)] = L(3, 2) = 4,$$

$$E[L(\hat{\theta}_4, 3)] = L(3, 3) = 0.$$

The maximum expected losses for $\hat{\theta}_1, \hat{\theta}_2, \hat{\theta}_3$, and $\hat{\theta}_4$, respectively, are 2, 2, 1, and 4. Since $\hat{\theta}_3$ has the minimum of the maximum expected losses, $\hat{\theta}_3$ is the minimax estimator of θ. The cost of overestimating θ is so large that we will always estimate θ to be 2, thus guarding against incurring large losses. The interested reader can observe that slight changes in the loss function will produce different minimax estimators.

Sometimes it is realistic to regard the unknown parameter θ as a random variable with some density function, say $g(\theta)$. We then assume that we know the conditional density (or probability function) of the observable random variables, $Y_1, \ldots, Y_n$, for a fixed value of θ. We denote the density of Y for a given θ by $f(y|\theta)$. If $Y_1, \ldots, Y_n$ are independent random variables, the joint conditional density of $Y_1, \ldots, Y_n$ given θ is

$$f(y_1, y_2, \ldots, y_n|\theta) = f(y_1|\theta)f(y_2|\theta)\cdots f(y_n|\theta).$$

From this conditional density and $g(\theta)$ it is possible to find the conditional density of θ given values $y_1, \ldots, y_n$ for the random sample. Notationally, the conditional density of θ given $y_1, \ldots, y_n$ is

$$f(\theta|y_1, \ldots, y_n) = \frac{f(y_1, \ldots, y_n, \theta)}{f(y_1, \ldots, y_n)}$$

$$= \frac{f(y_1, \ldots, y_n|\theta)g(\theta)}{\int_{-\infty}^{\infty} f(y_1, \ldots, y_n|\theta)g(\theta)\, d\theta}$$

since the integral in the denominator represents the marginal density of $Y_1, \ldots, Y_n$ (not conditioned on θ).

The density functions $g(\theta)$ and $f(\theta|y_1, \ldots, y_n)$ are called the *prior* and *posterior* densities of θ, respectively.

After the values $y_1, \ldots, y_n$ have been observed in a sample, $L(\hat{\theta}, \theta)$ is simply a function of θ if the estimator $\hat{\theta}$ is specified. Thus we can take the expectation of $L(\hat{\theta}, \theta)$ with respect to the conditional density $f(\theta|y_1, \ldots, y_n)$. The resulting expectation is called the *posterior risk*. The Bayes estimator of θ is that function $\hat{\theta}$ which minimizes the posterior risk. The following theorem gives one case in which Bayes estimators are easily found (proof is omitted).

Theorem 9.2: *If $L(\hat{\theta}, \theta) = (\hat{\theta} - \theta)^2$, a squared-error loss, then the function $\hat{\theta}$ which minimizes the posterior risk is the mean of the posterior density of θ.*

Example 9.11: *Let Y denote the number of defectives observed in a random sample of n items produced by a given machine in one day. The proportion, p, of defectives produced by this machine varies from day to day, and is assumed to have probability density function*

$$g(p) = \begin{cases} 1, & 0 \le p \le 1, \\ 0, & \text{elsewhere.} \end{cases}$$

If a single Y is observed and if the loss function is $L(\hat{p}, p) = (\hat{p} - p)^2$, find the Bayes estimator of p.

Solution: *Y is a binomial random variable for fixed p, and hence*

$$f(y|p) = \binom{n}{y} p^y (1 - p)^{n-y}, \quad y = 0, 1, \ldots, n.$$

Then

$$\begin{aligned} f(y, p) &= f(y|p)g(p) \\ &= \binom{n}{y} p^y (1 - p)^{n-y}, \ y = 0, 1, \ldots, n; \quad 0 \le p \le 1, \end{aligned}$$

and

$$f(y) = \int_0^1 f(y, p)\, dp = \int_0^1 \binom{n}{y} p^y (1 - p)^{n-y}\, dp.$$

By referring to the beta density function in Chapter 4, we see that

$$f(y) = \binom{n}{y}\frac{y!(n-y)!}{(n+1)!}$$

$$= \frac{1}{n+1}, \qquad y = 0, 1, \ldots, n.$$

Note that $f(y)$ says that each possible value of Y has the same unconditional probability, since nothing is known about p other than $0 \leq p \leq 1$. It follows that

$$f(p|y) = \frac{f(y, p)}{f(y)}$$

$$= (n+1)\binom{n}{y}p^y(1-p)^{n-y}$$

$$= \frac{(n+1)!}{y!(n-y)!}p^y(1-p)^{n-y}, \qquad 0 \leq p \leq 1,$$

another beta density function. The Bayes estimator is the mean of this beta density, $f(p|y)$, and so we choose as the estimate of p

$$\hat{p} = \int_0^1 pf(p|y)\,dp = \frac{y+1}{n+2}.$$

9.9 Summary

In this chapter, a continuation of Chapter 8, we have extended our discussion of estimation. Many methods are available for finding estimators. We have presented the method of moments, the method of maximum likelihood, and a decision theoretic approach emphasizing minimax and Bayes estimators.

Given a set of estimators for a parameter, we need a way to determine which estimator is best. For this purpose, a number of different desirable characteristics of estimators have been proposed. These include the notions of unbiasedness, minimum variance, consistency, and sufficiency. For our purposes, the most desirable estimators are those that are unbiased and possess minimum variance.

References

1. Cramér, H., *Mathematical Statistics.* Princeton, N.J.: Princeton University Press, 1946.
2. Hogg, R. V., and A. T. Craig, *Introduction to Mathematical Statistics*, 3rd ed. New York: The Macmillan Company, 1970.
3. Mood, A. M., and F. A. Graybill, *Introduction to the Theory of Statistics*, 2nd ed. New York: McGraw-Hill Book Company, 1963.
4. Wilks, S. S., *Mathematical Statistics.* New York: John Wiley & Sons, Inc., 1962.

Exercises

9.1. Suppose that a coin has probability p of coming up heads, and that the coin is tossed until the first head appears. If the first head appears on trial Y (Y is a random variable) find the maximum likelihood estimator of p. Note that Y has the geometric probability distribution.

9.2. Find an estimator for the p of Exercise 9.1 by the method of moments.

9.3. Let $Y_1, \ldots, Y_n$ denote a random sample from the density function given by

$$f(y) = \begin{cases} \dfrac{1}{\Gamma(\alpha)\theta^\alpha} y^{\alpha-1} e^{-y/\theta}, & y > 0, \\ 0, & \text{elsewhere.} \end{cases}$$

(a) Find the maximum likelihood estimator, $\hat{\theta}$, of θ if α is known.
(b) Find the expected value and variance of $\hat{\theta}$.

(c) Show that $\hat{\theta}$ is consistent for θ.
(d) What is the sufficient statistic for θ in this problem?
(e) Suppose that $n = 5$, $\alpha = 2$. Use the sufficient statistic to construct a 90 percent confidence interval for θ. (*Hint*: Transform to a χ^2 distribution.)

9.4. Suppose $Y_1, \ldots, Y_n$ denotes a random sample from the Poisson distribution with mean λ.
(a) Find the maximum likelihood estimator, $\hat{\lambda}$, for λ.
(b) Find the expected value and variance of $\hat{\lambda}$.
(c) Show that the estimator of part (a) is consistent for λ.
(d) Suppose you have a sample of $n = 20$ observations and $\hat{\lambda} = 12$. Find an approximate bound on the error of estimating λ. Since usually λ will be unknown, you can use $\hat{\lambda}$ to approximate λ in the expression for $\sigma_{\hat{\lambda}}$.

9.5. Let $Y_1, \ldots, Y_n$ denote a random sample from the density function given by

$$f(y) = \begin{cases} \dfrac{1}{\theta} r y^{r-1} e^{-y^r/\theta}, & \theta > 0,\ y > 0, \\ 0, & \text{elsewhere,} \end{cases}$$

where r is a known positive constant.
(a) Find a sufficient statistic for θ.
(b) Find the maximum likelihood estimator of θ.
(c) Is the estimator in part (b) the MVUE for θ?

9.6. A random sample of size n is taken from the probability density function

$$f(y) = \begin{cases} (\theta + 1)y^{\theta}, & 0 < y < 1, \\ 0, & \text{elsewhere,} \end{cases}$$

where θ is an unknown, positive constant.
(a) Find an estimator, $\hat{\theta}_1$, for θ by the method of moments.
(b) Find an estimator, $\hat{\theta}_2$, for θ by the method of maximum likelihood.

9.7. A random sample of 100 voters selected from a large population revealed 30 favoring candidate A, 38 favoring candidate B, and 32 favoring candidate C. Find maximum likelihood estimates for the proportions of voters, in the population, favoring candidates A, B, and C, respectively. Estimate the difference between the fractions favoring A and B and place a bound on the error of estimation.

9.8. Suppose that $Y_1, \ldots, Y_n$ is a random sample from the density function

$$f(y) = \begin{cases} e^{-(y-\theta)}, & y > \theta, \\ 0, & \text{elsewhere,} \end{cases}$$

where θ is an unknown, positive constant.
(a) Find an estimator, $\hat{\theta}_1$, for θ by the method of moments.
(b) Find an estimator, $\hat{\theta}_2$, for θ by the method of maximum likelihood.
(c) Adjust $\hat{\theta}_1$ and $\hat{\theta}_2$ so that they are unbiased. Find the efficiency of the adjusted $\hat{\theta}_1$ relative to the adjusted $\hat{\theta}_2$.

9.9. If $Y_1, \ldots, Y_n$ is a random sample from the probability density function given by

$$f(y) = \begin{cases} \dfrac{2}{\theta^2}(\theta - y), & 0 < y < \theta, \\ 0, & \text{elsewhere,} \end{cases}$$

find an estimator for θ by the method of moments. Is this estimator a function of a sufficient statistic?

9.10. It is observed that the number of breakdowns per day, Y, for a certain machine is a Poisson random variable with mean λ. The cost per day of repairing these breakdowns is given by $C = 3Y^2$. If $Y_1, \ldots, Y_n$, denote the observed number of breakdowns for n independently selected days, find the MVUE for $E(C)$.

9.11. A certain type of electronic component has a lifetime, Y (in hours), with probability density function given by

$$f(y) = \begin{cases} \dfrac{1}{\theta^2} y e^{-y/\theta}, & y > 0, \\ 0, & \text{otherwise.} \end{cases}$$

Let $\hat{\theta}$ denote the maximum likelihood estimator of θ. If three such components, tested independently, gave lifetimes of 120, 130, and 128 hours,

(a) Find the maximum likelihood estimate of θ.

(b) Find $E(\hat{\theta})$ and $V(\hat{\theta})$.

(c) Suppose that θ actually equals 130. Give an approximate bound that you might expect for the error of estimation.

9.12. Suppose that $X_1, \ldots, X_m$, representing yields per acre for corn variety A, is a random sample from a normal distribution with mean μ_1 and variance σ^2. Also, $Y_1, \ldots, Y_n$, representing yields for corn variety B, is a random sample from a normal distribution with mean μ_2 and variance σ^2. If the X's and Y's are independent, find the maximum likelihood estimator for the common variance, σ^2. Assume that μ_1 and μ_2 are unknown.

9.13. $Y_1, Y_2, \ldots, Y_n$ is a random sample from a population with mean μ and variance σ^2. Consider the following three estimators for μ:

(1)
$$\hat{\mu}_1 = \frac{Y_1 + Y_2}{2}.$$

(2)
$$\hat{\mu}_2 = \frac{1}{4}Y_1 + \frac{Y_2 + \cdots + Y_{n-1}}{2(n-2)} + \frac{1}{4}Y_n.$$

(3)
$$\hat{\mu}_3 = \frac{Y_1 + \cdots + Y_n}{n}.$$

(a) Show that each of the three estimators is unbiased.

(b) Show that only (3) is consistent.

(c) By comparing variances show that (3) is the most efficient.

9.14. $Y_1, Y_2, \ldots, Y_n$ is a random sample from a uniform distribution with probability density function

$$f(y) = \begin{cases} \dfrac{1}{2\theta + 1}, & 0 \le y \le 2\theta + 1, \\ 0, & \text{otherwise.} \end{cases}$$

Obtain the maximum likelihood estimator of θ.

9.15. Use the result of Exercise 9.14 to obtain an unbiased estimator for θ. Compute the relative efficiency of the resulting unbiased estimator relative to the unbiased estimator obtained by the method of moments.

9.16. In Example 9.10, assume that the loss function is $L(\hat{\theta}, \theta) = |\theta - \hat{\theta}|$ (that is, loss due to overestimation = loss due to underestimation). Show that the minimax estimator for θ is the intuitive estimator $\hat{\theta}(Y) = 2$ if $Y = 0$, $\hat{\theta}(Y) = 3$ if $Y = 1$. Will your solution change if $L(\hat{\theta}, \theta) = (\hat{\theta} - \theta)^2$?

9.17. Let Y denote the number of defects per yard for a certain type of cloth. For a given mean, λ, Y has a Poisson distribution. But λ varies from yard to yard according to the density function

$$f(\lambda) = \begin{cases} e^{-\lambda}, & \lambda > 0, \\ 0, & \text{elsewhere.} \end{cases}$$

If the loss function is given by $L(\hat{\lambda}, \lambda) = (\hat{\lambda} - \lambda)^2$, find the Bayes estimator of λ based on a single observation of Y.

9.18. In Example 9.11, suppose that

$$g(p) = \begin{cases} 2, & 0 \le p \le 1/2, \\ 0, & \text{otherwise.} \end{cases}$$

If two items are produced in a given day and one is defective, find the Bayes estimate of p.

9.19. Suppose that n integers are drawn at random and *with replacement* from the integers $1, 2, \ldots, N$. That is, each sampled integer has probability $1/N$ of taking on any of the values $1, 2, \ldots, N$, and the sampled values are independent.
(a) Find the moment estimator, $\hat{N}_1$, of N.
(b) Find $E(\hat{N}_1)$ and $V(\hat{N}_1)$.

9.20. Refer to Exercise 9.19.
(a) Find the maximum likelihood estimator, $\hat{N}_2$, of N.
(b) Show that $E(\hat{N}_2)$ is approximately $[n/(n + 1)]N$. Adjust $\hat{N}_2$ to form an estimator $\hat{N}_3$ which is approximately unbiased for N.
(c) Find an approximate variance for $\hat{N}_3$ using the fact that, for large N, the variance of the largest sampled integer is approximately

$$\frac{nN^2}{(n + 1)^2(n + 2)}.$$

(d) Show that, for large N and $n > 1$, $V(\hat{N}_3) < V(\hat{N}_1)$.

9.21. Suppose that enemy tanks have serial numbers $1, 2, \ldots, N$. A spy randomly observes five tanks (with replacement) with serial numbers 97, 64, 118, 210, and 57. Estimate N and place a bound on the error of estimation.

9.22. Suppose that $\hat{\theta}$ is the maximum likelihood estimator for a parameter θ. Let $g(\theta)$ be a function of θ which possesses a unique inverse. Show that $g(\hat{\theta})$ is the maximum likelihood estimator of $g(\theta)$.

10

Hypothesis Testing

10.1 Introduction

Recall that the objective of statistics is to make inferences about unknown population parameters based on information contained in sample data. These inferences are phrased in one of two ways, as estimates of the respective parameters or as tests of hypotheses about their values. Chapters 8 and 9 dealt with estimation; now we turn to hypothesis testing.

Hypothesis testing arises as a natural consequence of the scientific method. The scientist observes nature, formulates a theory, and then tests his theory against observation. In our context, the scientist poses a theory concerning one or more population parameters—that they equal specified values. He then samples the population and compares observation with theory. If the observations disagree with the theory, he rejects his hypothesis. If not, he concludes that either his theory is true or that his sample did not detect the difference between the real and hypothesized values of the population parameters.

For example, a medical researcher may hypothesize that a new drug is more effective than another in combating a disease. To test his hypothesis he randomly selects a number of patients infected with the disease and randomly divides them into two groups. The new drug, A, is applied to the first group of patients and the other drug, B, is applied to the second. Then, based on the number of patients in each group that recover from the disease, the researcher must conclude whether or not the new drug is more effective than the old.

Hypothesis tests are conducted in all fields in which theory can be tested against observation. A quality-control engineer may hypothesize that a new assembly method produces only 5 percent defective items. An educator may claim that two methods of reading are equally effective, and a political candidate may claim that a plurality of voters favor his election. All these hypotheses can be subjected to statistical verification by comparing the hypotheses with observed sample data.

What is the role of statistics in testing hypothesis? Putting it more bluntly, of what value is statistics in this hypothesis-testing procedure? Note that testing a hypothesis requires a decision when comparing the observed sample with theory. How do we decide whether the sample disagrees with the scientist's hypothesis? When should we reject the hypothesis, when should we accept it, and when should we withhold judgment? What is the probability that we will make the wrong decision and be led to a consequential loss? And particularly, what function of the sample measurements should be employed to reach a decision? The answers to these questions are contained in a study of statistical hypothesis testing.

10.2 Elements of a Statistical Test

The logic employed in a statistical test is best illustrated by an example. A political candidate, Jones, claims that he will gain more than 50 percent of the votes in a city election and thereby emerge as the winner. To test his claim, $n = 15$ voters are randomly selected from the city and the number favoring Jones, Y, is recorded. If none in the sample favor Jones ($Y = 0$), what do you conclude concerning his claim? Without hesitation we conclude that Jones' claim is false, not because it is impossible to draw $Y = 0$ in a sample of $n = 15$, assuming Jones is a winner, but because it is highly improbable. Thus any "small" value of Y would have led us to the same conclusion.

An examination of the mental process that produced our rapid intuitive decision explains the reasoning employed in a statistical test and identifies its elements. If Jones truly possesses a majority of votes, then p, the probability of selecting a voter favoring Jones, is greater than or equal to .5. If his claim is false, p is less than .5. Hence we mentally hypothesized that $p = .5$, the minimum value needed for a plurality, and compared the observed value of Y with our hypothesis. Knowing that the probability of observing $Y = 0$ was extremely small assuming that $p = .5$, we rejected this hypothesis in favor of the alternative, $p < .5$.

Any statistical test of a hypothesis works in exactly the same way and is composed of the same essential elements.

The Elements of a Statistical Test

1. *Null hypothesis, H_0*
2. *Alternative hypothesis, H_a*
3. *Test statistic*
4. *Rejection region*

For our example, the hypothesis to be tested, called the *null hypothesis* and denoted as H_0, is $p = .5$. The *alternative hypothesis*, denoted as H_a, is the hypothesis to be accepted in case H_0 is rejected; thus H_a is $p < .5$. The functioning parts of a statistical test are the test statistic and its associated rejection region. The *test statistic* is a function of the sample measurements (Y for our example) upon which the statistical decision will be based. To make it function, the entire set of values that the test statistic can assume is divided into two sets, a *rejection* or *critical region*, denoted by the symbol C, and its complement, $\bar{C}$. If for a particular sample, the computed value of the test statistic falls in the rejection region, we reject the null hypothesis and accept H_a. If it falls in $\bar{C}$, we accept H_0. We could have selected the rejection region as $C = \{y : y \leq 1\}$ for our particular test.

Finding a good rejection region for a statistical test is an interesting problem and one that bears further attention. It is clear that small values of Y, say $y \leq k$ (see Figure 10.1), are contradictory to the hypothesis, $H_0: p = .5$, but favorable to the alternative, $H_a: p < .5$. So we intuitively choose the rejection region as $C = \{y : y \leq k\}$. But what value should we choose for k?

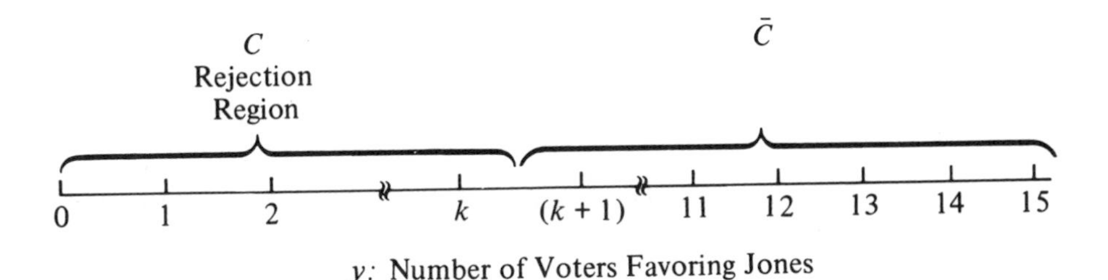

Figure 10.1 The Rejection Region, C, for a Test of the Hypothesis, $H_0: p = .5$, Against the Alternative, $H_a: p < .5$

More generally, how can we evaluate the goodness of a test so that we can compare one test with another? A measure of goodness would provide an objective (rather than intuitive) procedure for choosing a rejection region.

Note that, for any fixed rejection region (determined by a particular value of k), two types of errors can be made in reaching a decision. We can decide in favor of H_a when H_0 is true (called a type I error) or we can favor H_0 when H_a is true (called a type II error).

> ***Definition 10.1:*** *A type I error is the rejection of H_0 when H_0 is true. The probability of a type I error is denoted by α.*
>
> *A type II error is the acceptance of H_0 when H_a is true. The probability of a type II error is denoted by β.*

For Jones' political poll, a type I error, rejecting $H_0: p = .5$ when in fact H_0 is true, means that we conclude that Jones will lose when, in fact, he is going to win. In contrast, making a type II error means that we accept $H_0: p = .5$ when $p < .5$ and conclude that Jones will win when, in fact, he will lose. For most real situations, incorrect decisions cost money, prestige, and so on, and imply a loss. Thus α and β, the probabilities of making these two types of errors, measure the risks associated with the two possible erroneous decisions that might result from a statistical test. As such, they provide a very practical way to measure the goodness of a test. We will illustrate with the following example.

Example 10.1: *For Jones' political poll, calculate α for the test $H_0: p = .5$ against the alternative, $H_a: p < .5$, if we select $C = \{y: y \leq 2\}$ as the rejection region for the test.*

Solution: *By definition,*

$$\alpha = P(\text{rejecting } H_0 \text{ when } H_0 \text{ is true})$$

$$= P(Y \in C \text{ when } p = .5) = P(Y \leq 2 \text{ when } p = .5).$$

Noting that Y is a binomial random variable with $n = 15$ and $p = .5$, it follows that

$$\alpha = \sum_{y=0}^{2} \binom{15}{y} (.5)^{15} = \binom{15}{0} (.5)^{15} + \binom{15}{1} (.5)^{15} + \binom{15}{2} (.5)^{15}.$$

Using Table 1, Appendix III, to circumvent this computation, we find

$$\alpha = .004.$$

Thus we note that we subject ourselves to a very small risk of concluding that Jones will lose when he is, in fact, a winner.

Example 10.2: *Is our test equally good in protecting us from concluding that Jones is a winner when, in fact, he will lose? Suppose that he really only will win 30 percent of the vote ($p = .30$). What is the probability, β, that the sample will erroneously lead us to conclude that H_0 is true and that Jones is going to win?*

Solution: *By definition,*

$$\beta = P(Y \in \bar{C} \text{ when } H_a \text{ is true and } p = .30)$$

$$= P(Y > 2 \text{ when } p = .30)$$

$$= \sum_{y=3}^{15} \binom{15}{y} (.30)^y (.70)^{15-y}.$$

Again, consulting Table 1, Appendix III, we find that

$$\beta = .873.$$

In other words, our test will almost always lead us to conclude that Jones is a winner (with probability, $\beta = .873$), even if p is as low as $p = .3$. Hence its use is particularly suitable if we wish to spare Jones the discontent associated with pre-election uncertainty. We will almost always tell him what he wants to hear.

Note that β depends on the true value of p. The larger the difference between p and $H_0: p = .5$, the less likely we will fail to reject the null hypothesis.

Example 10.3: *Calculate the value of β assuming that Jones really has only 10 percent of the votes ($p = .1$).*

Solution: *Now*

$$\beta = P(Y \in \bar{C} \text{ when } H_a \text{ is true and } p = .1)$$

$$= P(Y > 2 \text{ when } p = .1)$$

$$= \sum_{y=3}^{15} \binom{15}{y} (.1)^y (.9)^{15-y} = .184.$$

We note that our test is so poor that we face a fair probability of claiming Jones a winner when, in fact, he will draw only 10 percent of the votes.

How can we improve our test? One way is to balance α and β by increasing or decreasing the size of the rejection region. If we enlarge C to a new set, C' (that is, $C \subset C'$), then it is clear that α will increase. That is, if $C \subset C'$, then

$$P(Y \in C') \geq P(Y \in C).$$

Likewise, enlarging C will reduce $\bar{C}$, so β will decrease as α increases. Hence α and β are inversely related; as one increases, the other decreases. We will illustrate with an example.

Example 10.4: *Refer to the test, Exercise* 10.1, *but assume that* $C = \{y : y \leq 5\}$. *Calculate* α. *Calculate* β *when* $p = .3$. *Compare with the results of* 10.1 *and* 10.2.

Solution: *In this case,*

$$\alpha = P(Y \in C \text{ when } H_0 \text{ is true})$$

$$= P(Y \leq 5 \text{ when } p = .5) = \sum_{y=0}^{5} \binom{15}{y} (.5)^{15}$$

$$= .151.$$

Calculating β *for an alternative,* $p = .3$, *we see that*

$$\beta = P(Y \in \bar{C} \text{ when } H_a \text{ is true and } p = .3)$$

$$= P(Y > 5 \text{ when } p = .3) = \sum_{y=6}^{15} \binom{15}{y} (.3)^y (.7)^{15-y}$$

$$= .278.$$

A comparison of the α and β calculated for Example 10.4 with the results of Examples 10.1 and 10.2 show that enlarging C from $C = \{y : y \leq 2\}$ to $C = \{y : y \leq 5\}$ increased α and decreased β (see Table 10.1).

Table 10.1 A Comparison of α and β for Two Different Rejection Regions

	C	
	(0, 1, 2)	(0, 1, 2, ..., 5)
α	.004	.151
β for $p = .3$	.873	.278

Hence we have achieved a better balance between the risks of type I and type II errors, but both α and β are uncomfortably large. *How then can we reduce both α and β?* The answer is intuitively clear—shed more light on the true nature of the population by increasing the sample size. Thus for almost all statistical tests, α and β will both decrease as the sample size increases.

As you will subsequently see, it is sometimes difficult to determine β for a test. It may be difficult to calculate β or it may be difficult to specify a particular meaningful alternative to the null hypothesis. When this occurs (β is unknown), we are reluctant to accept the null hypothesis when the test statistic falls in the acceptance region because the risk of a type II error is unknown. Our usual solution to this difficulty is to "withhold judgment." That is, we seek additional data before committing ourselves to acceptance of H_0. For this reason, we most often choose H_0 and H_a so that the point we want to "prove" corresponds to the alternative hypothesis. Then if the test statistic falls in the rejection region and supports H_a, we will know α, the risk of making a type I error. Hence for Jones' poll, if we wish to show that $p < .5$, we select $H_0: p = .5$ and hope that the data will lead to its rejection. If successful, we have demonstrated clear evidence favoring our theory ($p < .5$) and we know α, the risk of being led to an erroneous conclusion.

The probability of making a type I error, α, is often called the *significance level* of the test. This is because, if you reject H_0, the smaller the value of α, the greater will be the weight of evidence favoring H_a.

Now let us turn to some common statistical tests and see how they are employed in practice.

10.3 Common Large-Sample Tests

Suppose that we want to test a hypothesis concerning a parameter θ based on a random sample $Y_1, \ldots, Y_n$. In this section, we assume that we have an estimator, $\hat{\theta}$, of θ where $\hat{\theta}$ is asymptotically normal with mean θ and variance $\sigma_{\hat{\theta}}^2$, and that n is large enough for the approximate normality of $\hat{\theta}$ to hold. The large-sample estimators of Chapter 8 (Table 8.1), such as $\bar{Y}$ and $\hat{p}$ used for estimating a population mean μ and proportion p, respectively, satisfy these requirements. So also do the estimators for the comparison of two means, $(\mu_1 - \mu_2)$, and the comparison of two binomial parameters, $(p_1 - p_2)$.

We desire to test the null hypothesis, $H_0: \theta = \theta_0$, against the alternative hypothesis, $H_a: \theta > \theta_0$, where θ_0 is a specified value of θ. If $\hat{\theta}$ is a good estimator of θ, then $\hat{\theta}$ should serve as a good test statistic. Intuitively, it should seem reasonable to accept H_0 if $\hat{\theta}$ is "close" to θ_0 and to reject H_0 in favor of H_a if $\hat{\theta}$ is "much larger" than θ_0. Thus the rejection region, C, is of the form

$$C = \{\hat{\theta}: \hat{\theta} > k\},$$

the set of all values of $\hat{\theta}$ larger than some constant k. The actual size of C is again determined by fixing the type I error probability, α, at a small level and choosing k accordingly.

In calculating α we make use of the fact that $(\hat{\theta} - \theta_0)/\sigma_{\hat{\theta}}$ is approximately standard normal in distribution when H_0 is true. Since θ_0 and $\sigma_{\hat{\theta}}$ are constants, we can write C as

$$C = \left\{\hat{\theta}: \frac{\hat{\theta} - \theta_0}{\sigma_{\hat{\theta}}} > z_\alpha\right\}$$

where z_α is the constant which cuts off an area of size α in the right-hand tail of the standard normal density function (see Figure 10.2). If we choose C as

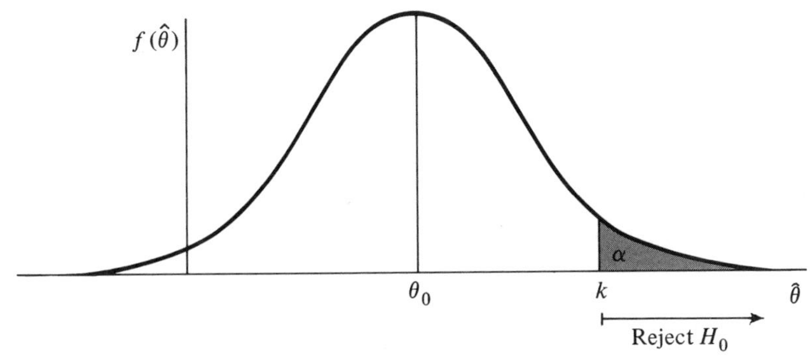

$\alpha = P[\hat{\theta} > k \text{ when } E(\hat{\theta}) = \theta_0]$ where $k = \theta_0 + z_\alpha \sigma_{\hat{\theta}}$

Figure 10.2 A Large-Sample Test [$f(\hat{\theta})$ represents the density function for $\hat{\theta}$]

above and define

$$Z = \frac{\hat{\theta} - \theta_0}{\sigma_{\hat{\theta}}},$$

then the probability of making a type I error is

$$\alpha = P(Z > z_\alpha \text{ when } \theta = \theta_0).$$

For any fixed α, z_α can be obtained from Table 3, Appendix III.

The above discussion is summarized as follows:

A Large-Sample Test

$$H_0: \theta = \theta_0.$$

$$H_a: \theta > \theta_0.$$

Test Statistic: $Z = \dfrac{\hat{\theta} - \theta_0}{\sigma_{\hat{\theta}}}.$

Rejection Region: $z > z_\alpha$, *for fixed* α.

Note that Z measures the number of standard deviations between $\hat{\theta}$ and θ_0. Hence we reject H_0 in favor of H_a when $\hat{\theta}$ is "too many" standard deviations away from θ_0.

Example 10.5: *A vice president in charge of sales for a large corporation claims that the salesmen are only averaging 15 sales contacts per week. (He would like to increase this figure.) To check his claim, $n = 36$ salesmen are selected at random and the number of contacts is recorded for a single randomly selected week. The sample reveals a mean of 17 contacts and a variance of 9. Does the evidence contradict the vice president's claim at the 5 percent level of significance?*

Solution: *We are interested in testing the mean number of sales per week, μ. Specifically, we are testing*

$$H_0\colon \mu = 15$$

against

$$H_a\colon \mu > 15.$$

We know that the sample mean, $\overline{Y}$, is a point estimator of μ that satisfies the assumptions described above. Hence our test statistic is

$$Z = \frac{\overline{Y} - \mu}{\sigma_{\overline{Y}}} = \frac{\overline{Y} - \mu}{\sigma/\sqrt{n}}.$$

The rejection region, with $\alpha = .05$, is given by $z > 1.645$ (see Table 3, Appendix III).

The population variance, σ^2, is not known, but it can be estimated very accurately (because $n = 36$ is quite large) by the sample variance, $s^2 = 9$. Thus the observed value of the test statistic is approximately

$$z = \frac{\bar{y} - \mu}{s/\sqrt{n}}$$

$$= \frac{17 - 15}{3/\sqrt{36}} = 4.$$

Since the observed value of Z lies in the rejection region (exceeds $z_\alpha = 1.645$) we reject $H_0: \mu = 15$. Thus it appears that the vice president's claim is incorrect and that the average number of sales contacts per week exceeds 15.

Example 10.6: *A machine in a certain factory must be repaired if it produces more than 10 percent defectives among the large lot of items it produces in a day. A random sample of 100 items from the day's production contains 15 defectives, and the foreman says that the machine must be repaired. Does the sample evidence support his decision? Use $\alpha = .01$.*

Solution: *If Y denotes the number of observed defectives, then Y is a binomial random variable with p denoting the probability that a randomly selected item is defective. Hence we want to test the null hypothesis,*

$$H_0: p = .10$$

against the alternative

$$H_a: p > .10.$$

The test statistic is based on $\hat{p} = Y/n$, the unbiased point estimator of p, and is given by

$$Z = \frac{\hat{p} - p_0}{\sigma_{\hat{p}}} = \frac{\hat{p} - p_0}{\sqrt{\dfrac{p_0(1 - p_0)}{n}}}$$

From Table 3, Appendix III, we see that $P(Z > 2.33) = .01$. Hence we take $z > 2.33$ as the rejection region. The observed value of the test statistic is given by

$$z = \frac{\hat{p} - p_0}{\sqrt{\dfrac{p_0(1 - p_0)}{n}}}$$

$$= \frac{.15 - .10}{\sqrt{\dfrac{(.1)(.9)}{100}}} = \frac{5}{3}.$$

Since the observed value of Z is not in the rejection region, we conclude that the evidence does not support the foreman's decision. Note that we did not accept H_0

and would not unless we had calculated β for some value of p that differs (some difference of practical significance to the manufacturer) from $p = .10$. This was not required because all we wished to do was to check the foreman's claim.

An additional comment is required to explain why we selected $\alpha = .01$ (rather than .05) for Example 10.6. People familiar with manufacturing concerns know that production is terminated only if a major fault is found in the process. Hence we wished to protect ourselves against the possibility of claiming that p had increased when, in fact, it was still $p = .10$. This was done by reducing α, the probability of rejecting H_0 when it is true.

The rejection regions for the preceding examples were constructed for the one-sided alternative $\theta > \theta_0$. Now, let us consider two other possible alternatives, $\theta < \theta_0$ and $\theta \neq \theta_0$.

A test of $H_0: \theta = \theta_0$ against $H_a: \theta < \theta_0$ would be carried out in an analogous manner, except that we now reject H_0 for values of $\hat{\theta}$ which are much smaller than θ_0. The test statistic remains as

$$Z = \frac{\hat{\theta} - \theta_0}{\sigma_{\hat{\theta}}},$$

but, for fixed α, we reject the null hypothesis when $z < -z_\alpha$.

If we wish to test $H_0: \theta = \theta_0$ against $H_a: \theta \neq \theta_0$, we would want to reject H_0 if $\hat{\theta}$ is either much smaller or much larger than θ_0. The test statistic remains at Z, as above, but the rejection region is located symmetrically in the two tails of the probability distribution for Z. Thus we reject H_0 if either $z < -z_{\alpha/2}$ or $z > z_{\alpha/2}$. Equivalently, we reject H_0 if $|z| > z_{\alpha/2}$. This test is called a *two-tailed test*, as opposed to the *one-tailed tests* used for the alternatives $\theta < \theta_0, \theta > \theta_0$. The rejection regions for the one-sided alternative, $H_a: \theta < \theta_0$, and the two-sided alternative, $H_a: \theta \neq \theta_0$, are displayed in Figure 10.3.

How do you decide which alternative you should use for a test? The answer depends on what you are trying to prove. If you are only interested in detecting an increase in the percentage of defectives (Example 10.6), you should locate the rejection region in the upper tail of the Z distribution. On the other hand, if it is practically important for you to detect a change in p either above or below $p = .10$, then locate the rejection region in both tails of the Z distribution and employ a two-tailed test.

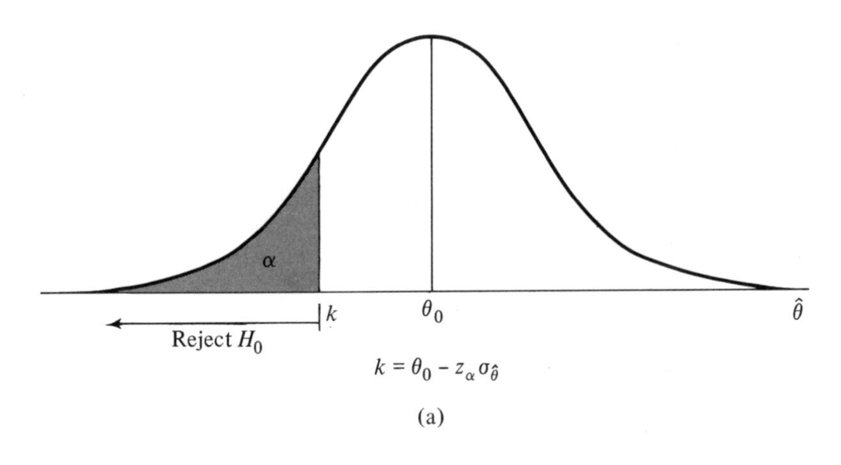

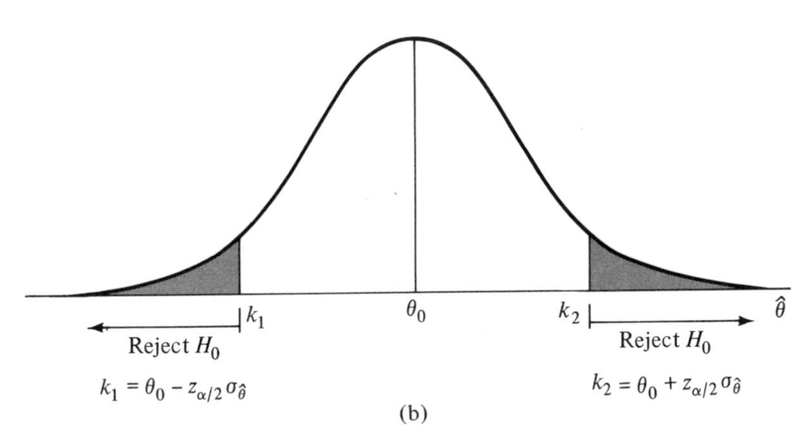

Figure 10.3 The Rejection Regions for the Alternative Hypothesis: (a) $H_0: \theta = \theta_0$, $H_a: \theta < \theta_0$; (b) $H_0: \theta = \theta_0$, $H_a: \theta \neq \theta_0$

Example 10.7: *A psychological study was conducted to compare the reaction times of men and women to a certain stimulus. Independent random samples of 50 men and 50 women were employed in the experiment and the results were as follows:*

Men	*Women*
$n_1 = 50$	$n_2 = 50$
$\bar{y}_1 = 42$ seconds	$\bar{y}_2 = 38$ seconds
$s_1^2 = 18$	$s_2^2 = 14$

Do the data present sufficient evidence to suggest a difference between true mean reaction times for men and women? Use $\alpha = .05$.

Solution: *Let μ_1 and μ_2 denote the true mean reaction times for men and women, respectively. Then if we wish to test the hypothesis that the means are equal, we will test $H_0: (\mu_1 - \mu_2) = 0$ against $H_a: (\mu_1 - \mu_2) \neq 0$. Note that we use the two-sided alternative to detect either the case $\mu_1 > \mu_2$, or the reverse, $\mu_2 > \mu_1$, in case H_0 is false.*

The point estimator of $(\mu_1 - \mu_2)$, $(\bar{Y}_1 - \bar{Y}_2)$, satisfies the assumptions of our large-sample test. Hence the test statistic is given by

$$Z = \frac{(\bar{Y}_1 - \bar{Y}_2) - (\mu_1 - \mu_2)}{\sqrt{\dfrac{\sigma_1^2}{n_1} + \dfrac{\sigma_2^2}{n_2}}},$$

where σ_1^2 and σ_2^2 are the respective population variances. For $\alpha = .05$, equally divided between the two tails of the Z distribution, we reject H_0 for $|z| > z_{\alpha/2} = z_{.025} = 1.96$.

For large samples (say $n > 30$), the sample variances provide good estimates of their corresponding population variances. Substituting these values, along with $\bar{y}_1$, $\bar{y}_2$, n_1, and n_2, into the formula for the test statistic, we have

$$z = \frac{\bar{y}_1 - \bar{y}_2}{\sqrt{\dfrac{\sigma_1^2}{n_1} + \dfrac{\sigma_2^2}{n_2}}} = \frac{42 - 38}{\sqrt{\dfrac{18}{50} + \dfrac{14}{50}}} = 5.$$

This value exceeds $z_{\alpha/2} = 1.96$ and therefore falls in the rejection region. Hence we reject the hypothesis of no difference in mean reaction times for men and women.

10.4 Calculating Type II Error Probabilities and Finding the Sample Size for the Z Test

Calculating β can be very difficult for some statistical tests, but it is easy for the test of Section 10.3. Consequently, we can use the Z test to demonstrate both the calculation of β and the logic employed in selecting the sample size for a test.

In testing $H_0: \theta = \theta_0$ against $H_a: \theta > \theta_0$ it is only possible to calculate type II error probabilities for specific points in H_a. Suppose that the experimenter has a specific alternative, say $\theta = \theta_a (\theta_a > \theta_0)$, in mind. Since the rejection region, C, is of the form

$$C = \{\hat{\theta} : \hat{\theta} > k\},$$

the probability of a type II error, β, is

$$\begin{aligned}\beta &= P(\hat{\theta} \in \bar{C} \text{ when } H_a \text{ is true})\\ &= P(\hat{\theta} \le k \text{ when } \theta = \theta_a)\\ &= P\left(\frac{\hat{\theta} - \theta_a}{\sigma_{\hat{\theta}}} \le \frac{k - \theta_a}{\sigma_{\hat{\theta}}} \text{ when } \theta = \theta_a\right).\end{aligned}$$

If θ_a is the true value of $\theta = E(\hat{\theta})$, then $(\hat{\theta} - \theta_a)/\sigma_{\hat{\theta}}$ is a standard normal random variable, and the probability, β, will be an area under the standard normal curve.

The size of β will depend on the distance between θ_a and θ_0 for fixed n. If θ_a is close to θ_0, it will be difficult to detect a difference between them, and the probability of accepting H_0 when H_a is true will tend to be large. If θ_a is far from θ_0, it will be relatively easy to see which is the true value, and β will be considerably smaller. As we saw in Section 10.2, α and β can both be made small by choosing a large sample size, n.

Example 10.8: *Suppose that the vice president of Example 10.5 wants to be able to detect a difference equal to one call in the mean number of customer calls per week. That is, he is interested in testing $H_0: \mu = 15$ against $H_a: \mu = 16$. With the data as given in Example 10.5, find β for this test.*

Solution: *In Example 10.5 we had $n = 36$, $\bar{y} = 17$, and $s^2 = 9$. The rejection region was given by ($\alpha = .05$)*

$$z = \frac{\bar{y} - \mu_0}{\sigma/\sqrt{n}} > 1.645,$$

which is equivalent to

$$\bar{y} - \mu_0 > 1.645\frac{\sigma}{\sqrt{n}}$$

or

$$\bar{y} > \mu_0 + 1.645\frac{\sigma}{\sqrt{n}}.$$

Substituting $n = 36$ *and using* s *to approximate* σ, *we find the rejection region to be*

$$\bar{y} > 15 + 1.645\frac{3}{\sqrt{36}}$$

or

$$\bar{y} > 15.8225.$$

This rejection region is shown in Figure 10.4. Then, by definition, $\beta = P(\bar{Y} \leq 15.8225$ *when* $\mu = 16)$ *and is the shaded area under the dashed curve to the left of* k *in Figure 10.4. Thus*

$$\beta = P\left(\frac{\bar{Y} - \mu_a}{\sigma/\sqrt{n}} \leq \frac{15.8225 - 16}{3/\sqrt{36}}\right)$$

$$= P(Z \leq -.36)$$

$$= .3594,$$

which is a fairly large error probability.

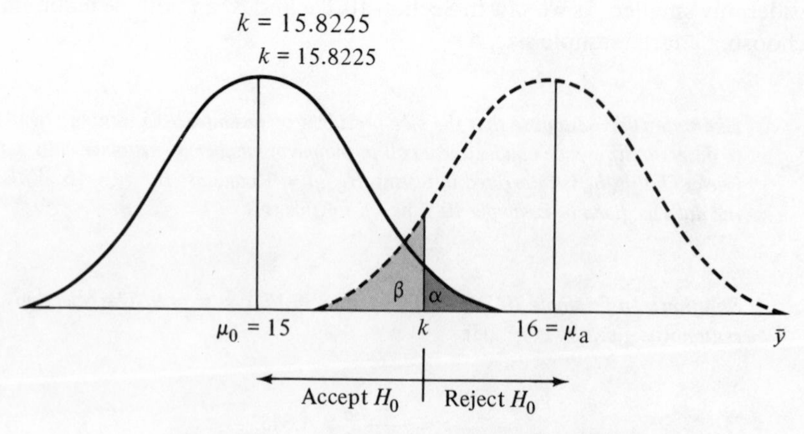

Figure 10.4 Rejection Region for Example 10.8

The large value of β tells us that samples of size $n = 36$ frequently will not detect a difference of one unit from the hypothesized population mean. This situation can be corrected by increasing the sample size.

The procedure for calculating β was demonstrated for the alternative $\theta_a > \theta_0$, but the method is similar for the alternatives $\theta_a < \theta_0$ and $\theta_a \neq \theta_0$. The procedure would also be the same for any large-sample test satisfying the conditions of Section 10.3. For the general method, just substitute $\hat{\theta}$ for $\bar{y}$ and σ_{θ} for $\sigma/\sqrt{n}$ in Example 10.8.

To conclude this section, we present a detailed discussion of the method for choosing the sample size when testing $H_0: \mu = \mu_0$ against the one-sided alternative $H_a: \mu = \mu_a$, where $\mu_a > \mu_0$, with α and β both specified. The method employed would be identical for any estimator satisfying the assumptions of Section 10.3.

After suitable error probabilities, α and β, are specified by the experimenter, the test has two remaining quantities which must be determined. These are n, the sample size, and k, the point at which the rejection region begins. Since α and β can be written as probabilities involving n and k, we have two equations in two unknowns, which can be solved simultaneously for n.

Formally, we have

$$\alpha = P(\bar{Y} > k \text{ when } \mu = \mu_0)$$

$$= P\left(\frac{\bar{Y} - \mu_0}{\sigma/\sqrt{n}} > \frac{k - \mu_0}{\sigma/\sqrt{n}} \text{ when } \mu = \mu_0\right)$$

$$= P(Z > z_\alpha)$$

and

$$\beta = P(\bar{Y} \leq k \text{ when } \mu = \mu_a)$$

$$= P\left(\frac{\bar{Y} - \mu_a}{\sigma/\sqrt{n}} \leq \frac{k - \mu_a}{\sigma/\sqrt{n}} \text{ when } \mu = \mu_a\right)$$

$$= P(Z \leq -z_\beta).$$

(See Figure 10.4.)

From the equation for α we have

$$\frac{k - \mu_0}{\sigma/\sqrt{n}} = z_\alpha$$

and, from the equation for β,

$$\frac{k - \mu_a}{\sigma/\sqrt{n}} = -z_\beta.$$

(Note that k is always between μ_0 and μ_a, so $k - \mu_a$ is negative.) Eliminating k from these two equations gives

$$\mu_0 + z_\alpha \frac{\sigma}{\sqrt{n}} = \mu_a - z_\beta \frac{\sigma}{\sqrt{n}}.$$

Thus

$$(z_\alpha + z_\beta)\frac{\sigma}{\sqrt{n}} = \mu_a - \mu_0,$$

$$\sqrt{n} = \frac{(z_\alpha + z_\beta)\sigma}{(\mu_a - \mu_0)},$$

or

$$n = \frac{(z_\alpha + z_\beta)^2\sigma^2}{(\mu_a - \mu_0)^2}.$$

Exactly the same solution would be obtained for the alternative $H_a\colon \mu = \mu_a$ with $\mu_a < \mu_0$. The method employed above can be used to develop a similar formula for the sample size for any one-tailed hypothesis-testing problem satisfying the conditions of Section 10.3.

Example 10.9: *Suppose that the vice president of Example 10.5 wants to test $H_0: \mu = 15$ against $H_a: \mu = 16$ with $\alpha = \beta = .05$. Find the sample size that will ensure this accuracy. Assume that σ^2 is approximately 9.*

Solution: *Since $\alpha = \beta = .05$, it follows that $z_\alpha = z_\beta = z_{.05} = 1.645$. Then*

$$\begin{aligned} n &= \frac{(z_\alpha + z_\beta)^2\sigma^2}{(\mu_a - \mu_0)^2} \\ &= \frac{(1.645 + 1.645)^2(9)}{(15 - 16)^2} \\ &= 97.4. \end{aligned}$$

Hence $n = 98$ observations should be used to ensure that $\alpha = \beta = .05$ for this test.

10.5 Two Tests Based on Student's t Statistic

The statistical tests of Section 10.3 are useful but are limited to the analysis of data based on large samples. This section concerns the use of Student's t as a test statistic, particularly its application to small-sample tests for a single population mean and the comparison of two means.

First we must make an assumption concerning the probability distribution of the sample measurements, because with small samples we can no longer appeal to the central limit theorem to justify normality of the distribution of the test statistic. We will assume that we have a random sample of n observations from a normal population with mean and variance μ and σ^2, respectively, and that $\bar{y}$ and s^2 are the corresponding quantities computed from the sample.

The test statistic that we seek needs to be a function of the sample measurements and possess a probability distribution that is dependent solely on μ. These requirements lead us to the Student's t statistic of Section 8.8, given by

$$T = \frac{\bar{Y} - \mu}{S/\sqrt{n}}.$$

You will note that this t statistic is identical to the large-sample test statistic of Section 10.3, except that we assume that s is calculated from a small sample and hence that it is not a good approximation to σ. Since the t distribution is symmetric and mound-shaped, it is clear that the rejection region for a small-sample test of the hypothesis, $H_0: \mu = \mu_0$, would be located in the tails of the t distribution and would be determined in exactly the same way as for the large-sample z statistic. Thus for the one-sided alternative, $H_a: \mu > \mu_0$, the rejection region, C, is determined by α, where

$$\alpha = P(T > t_\alpha \text{ when } \mu = \mu_0)$$

and T is based on $v = (n - 1)$ degrees of freedom. The constant t_α, the value that cuts off a right-hand tail area of α, is tabulated in Table 4, Appendix III, for various values of v and $\alpha = .100, .050, .025, .010$, and $.005$.

A Small-Sample Test on μ

Assumptions:

$Y_1, \ldots, Y_n$ *is a random sample from a normal distribution with* $E(Y_i) = \mu$.

$H_0: \mu = \mu_0$.

$H_a: \mu > \mu_0$.

Test Statistic: $T = \dfrac{\bar{Y} - \mu_0}{S/\sqrt{n}}$.

Rejection Region: $t > t_\alpha$ *for specified* α. *(See Table 4, Appendix III, for values of* t_α.*)*

Example 10.10: *Example 8.9 gives muzzle velocities of eight shells tested with a new gunpowder along with the sample mean and standard deviation* $\bar{y} = 2959$ *and* $s = 39.4$. *The manufacturer claims that the new gunpowder produces an average velocity of no less than 3000 feet per second. Do the sample data provide sufficient evidence to contradict the manufacturer's claim at the .025 level of significance?*

Solution: *Assuming that the observed velocities come from a normal distribution, the test outlined above can be employed. We want to test $H_0: \mu = 3000$ versus the alternative, $H_a: \mu < 3000$. The rejection region is given by $t < -t_{.025} = -2.365$, where t possesses $\nu = (n - 1) = 7$ degrees of freedom. Computing, we find that the observed value of the test statistic,*

$$t = \frac{\bar{y} - \mu_0}{s/\sqrt{n}} = \frac{2959 - 3000}{39.4/\sqrt{8}} = -2.943,$$

falls in the rejection region (that is, t is less than -2.365), and hence the null hypothesis is rejected. The sample data do not support the manufacturer's claim.

Remember that the test of Example 10.10 is based on the assumption that the muzzle-velocity measurements have been randomly selected from a normal population and that, in most cases, it is impossible to verify this assumption. We might ask how this predicament affects the validity of our conclusions.

Empirical studies of the test statistic,

$$\frac{\bar{Y} - \mu}{S/\sqrt{n}},$$

have been conducted for sampling from certain nonnormal populations. Such investigations have shown that moderate departures from normality in the probability distribution for Y have little effect on the probability distribution of the test statistic. This result, coupled with the common occurrence of near-normal distributions of data in nature, make the t test of a population mean extremely useful. Statistical tests that lack sensitivity to departures from the assumptions upon which they are based possess wide applicability. Because of their insensitivity to assumptions, they have been called *robust statistical tests.*

A second application of the t statistic is its use in constructing a small sample test to compare the means of two normal populations that possess equal variances.

Suppose that independent random samples are selected from each of two normal populations, $Y_{11}, Y_{12}, \ldots, Y_{1n_1}$ from the first and $Y_{21}, Y_{22}, \ldots, Y_{2n_2}$ from the second, where the mean and variance of the ith population are μ_i and σ^2, $i = 1, 2$. Further, assume that $\bar{Y}_i$ and S_i^2, $i = 1, 2$, are the corresponding sample means and variances. Then a test statistic for the hypothesis,

$H_0: (\mu_1 - \mu_2) = 0$, is readily provided by the t statistic of Section 8.8. Thus

$$T = \frac{(\bar{Y}_1 - \bar{Y}_2) - (\mu_1 - \mu_2)}{S\sqrt{1/n_1 + 1/n_2}},$$

where

$$S^2 = \frac{(n_1 - 1)S_1^2 + (n_2 - 1)S_2^2}{n_1 + n_2 - 2}$$

$$= \frac{\sum_{i=1}^{n_1} (Y_{1i} - \bar{Y}_1)^2 + \sum_{i=1}^{n_2} (Y_{2i} - \bar{Y}_2)^2}{n_1 + n_2 - 2}$$

is the pooled estimator of σ^2 and T has $(n_1 + n_2 - 2)$ degrees of freedom. Note that this small-sample test statistic is similar in form to its large-sample counterpoint, the Z statistic of Section 10.3.

A Small-Sample Test for Comparing Two Population Means

$H_0: (\mu_1 - \mu_2) = 0.$

$H_a: (\mu_1 - \mu_2) > 0.$

(Note: We give a one-sided alternative as an example.)

Test Statistic: $T = \dfrac{\bar{Y}_1 - \bar{Y}_2}{S\sqrt{\dfrac{1}{n_1} + \dfrac{1}{n_2}}}.$

Rejection Region: $t > t_\alpha$ *(for the one-sided alternative) where* $P(T > t_\alpha) = \alpha$. *(See Table 4, Appendix III, for values of* t_α*.)*

Tests of the hypothesis, $H_0: (\mu_1 - \mu_2) = 0$, are conducted in the same manner as for the large-sample test except that we employ the t statistic. We illustrate with an example.

Example 10.11: *Example 8.10 gives data on length of time to complete an assembly procedure for two different training methods. The sample data are as follows:*

Standard Procedure	New Procedure
$n_1 = 9$	$n_2 = 9$
$\bar{y}_1 = 35.22$ seconds	$\bar{y}_2 = 31.56$ seconds
$\sum_{i=1}^{9} (y_{1i} - \bar{y}_1)^2 = 195.50$	$\sum_{i=1}^{9} (y_{2i} - \bar{y}_2)^2 = 160.22$

Is there sufficient evidence to indicate a difference in true mean times for the two methods? Test at the $\alpha = .05$ level of significance.

Solution: *We are testing $H_0: (\mu_1 - \mu_2) = 0$ against the alternative, $H_a: (\mu_1 - \mu_2) \neq 0$. Consequently, we will require a two-tailed test. The test statistic is*

$$T = \frac{(\bar{Y}_1 - \bar{Y}_2) - (\mu_1 - \mu_2)}{S\sqrt{\frac{1}{n_1} + \frac{1}{n_2}}},$$

and the rejection region for $\alpha = .05$ is $|t| > t_{\alpha/2}$, where

$$P[|T| > t_{\alpha/2} \text{ when } (\mu_1 - \mu_2) = 0] = .05.$$

In this case $t_{.025} = 2.120$ since t is based on $(n_1 + n_2 - 2) = 16$ degrees of freedom.

The observed value of the test statistic is found by first computing

$$s^2 = \frac{195.56 + 160.22}{9 + 9 - 2} = 22.24.$$

Then

$$t = \frac{\bar{y}_1 - \bar{y}_2}{s\sqrt{\frac{1}{n_1} + \frac{1}{n_2}}} = \frac{35.22 - 31.56}{4.71\sqrt{\frac{1}{9} + \frac{1}{9}}}$$

$$= 1.65.$$

This value does not fall in the rejection region ($|t| > 2.120$), and hence the null hypothesis is not rejected. There is not sufficient evidence to indicate a difference in the mean assembly times.

Like the t test for a single mean, this test is robust to the assumption of normality. It is also robust to the assumption that $\sigma_1^2 = \sigma_2^2$ when n_1 and n_2 are equal (or nearly equal).

10.6 Testing Hypotheses Concerning Variances

We again assume that our random sample, $Y_1, \ldots, Y_n$, comes from a normal distribution with mean μ and variance σ^2. However, we now assume that μ is unknown and we wish to test hypotheses concerning σ^2.

A test statistic for testing the hypothesis $H_0: \sigma^2 = \sigma_0^2$ is readily provided by the chi-square statistic of Section 8.9. Thus $\chi^2 = (n - 1)S^2/\sigma^2$ has a chi-square probability distribution with $(n - 1)$ degrees of freedom. Note that for a given σ^2, χ^2 will increase (or decrease) as S^2 increases (or decreases). Hence if $\sigma^2 > \sigma_0^2$, both S^2 and χ^2 would be larger than expected. The larger the value, the greater the disagreement with H_0. Thus for the alternative, $H_a: \sigma^2 > \sigma_0^2$, we select the rejection region, C, as

$$C = \{\chi^2 : \chi^2 > \chi_\alpha^2\}$$

where $P(\chi^2 > \chi_\alpha^2) = \alpha$. Values of χ_α^2 can be found in Table 5, Appendix III.

A similar rejection region, located in the *lower* tail of the chi-square distribution, is appropriate for the alternative, $H_a: \sigma^2 < \sigma_0^2$. Or, we could run a two-sided test necessary for the alternative, $H_a: \sigma^2 \neq \sigma_0^2$.

Test of a Hypothesis Concerning a Population Variance

$H_0: \sigma^2 = \sigma_0^2$.
$H_a: \sigma^2 > \sigma_0^2$ (*a one-sided alternative is given as an example*).

Test Statistic: $\chi^2 = \dfrac{(n-1)S^2}{\sigma_0^2}$.

Rejection Region: $\chi^2 > \chi_\alpha^2$, *where* χ_α^2 *is chosen so that* $P(\chi^2 > \chi_\alpha^2) = \alpha$. (*See Table 5, Appendix III.*)

We will illustrate the test, given above, with an example.

Example 10.12: *A machined engine part produced by a company is claimed to have diameter variance no larger than .0002 (diameters measured in inches). A random sample of 10 parts gave a sample variance of .0003. Test, at the 5 percent level,* $H_0: \sigma^2 = .0002$ *against* $H_a: \sigma^2 > .0002$.

Solution: *We must assume that the measured diameters are normally distributed. Then the test statistic is* $\chi^2 = (n-1)S^2/\sigma_0^2$, *and we reject* H_0 *for values of this statistic larger than* $\chi_{.05}^2 = 16.919$ *(based on 9 degrees of freedom). The observed value of the test statistic is*

$$\frac{(n-1)s^2}{\sigma_0^2} = \frac{(9)(.0003)}{.0002} = 13.5.$$

Thus H_0 *is not rejected. There is not sufficient evidence to indicate that* σ^2 *exceeds .0002.*

Sometimes we wish to compare the variances of two normal distributions, particularly by testing to determine whether or not they are equal. These problems arise in comparing the precision of two measuring instruments, the variation in quality characteristics of a manufactured product or the variation in scores for two testing procedures. For example, suppose that $Y_{11}, \ldots, Y_{1n_1}$ and $Y_{21}, \ldots, Y_{2n_2}$ are independent random samples from normal distributions with unknown means, that $V(Y_{1i}) = \sigma_1^2$ and $V(Y_{2i}) = \sigma_2^2$, and that we want to test the null hypothesis, $H_0: \sigma_1^2 = \sigma_2^2$, against the alternative, $H_a: \sigma_1^2 > \sigma_2^2$.

It seems reasonable that we would utilize the ratio or difference of S_1^2 and S_2^2, the two sample variances for our test statistic and would reject H_0 in favor of H_a if S_1^2 is much larger than S_2^2. We will employ the ratio S_1^2/S_2^2 as our test statistic and select the rejection region C of the form

$$C = \left\{\frac{s_1^2}{s_2^2} : \frac{s_1^2}{s_2^2} > k'\right\}.$$

That is, we reject H_0 if S_1^2/S_2^2 is very large.

The next step is to find the probability distribution for S_1^2/S_2^2. Note that $(n_1 - 1)S_1^2/\sigma_1^2$ and $(n_2 - 1)S_2^2/\sigma_2^2$ are independent chi-square random variables, and from Definition 6.2 it follows that

$$F = \frac{(n_1 - 1)S_1^2}{\sigma_1^2(n_1 - 1)} \bigg/ \frac{(n_2 - 1)S_2^2}{\sigma_2^2(n_2 - 1)}$$

$$= \frac{S_1^2\sigma_2^2}{S_2^2\sigma_1^2}$$

has an F distribution with $(n_1 - 1)$ and $(n_2 - 1)$ degrees of freedom. Thus C could be reformulated as

$$C = \left\{\frac{s_1^2}{s_2^2} : \frac{s_1^2\sigma_2^2}{s_2^2\sigma_1^2} > k\right\}$$

or, equivalently, $C = \{F : F > k\} = \{F : F > F_\alpha\}$, where $k = F_\alpha$ is the value of the F distribution with $\nu_1 = (n_1 - 1)$ and $\nu_2 = (n_2 - 1)$ such that $P(F > F_\alpha) = \alpha$. Values of F_α are given in Tables 6 and 7, Appendix III.

Test of a Hypothesis, $\sigma_1^2 = \sigma_2^2$

$H_0: \sigma_1^2 = \sigma_2^2$.
$H_a: \sigma_1^2 > \sigma_2^2$ *(a one-sided alternative is given as an example).*

Test Statistic: $F = \dfrac{S_1^2}{S_2^2}$.

Rejection Region: $F > F_\alpha$ *when* F_α *is chosen so that* $P(F > F_\alpha) = \alpha$. *(See Tables 6 and 7, Appendix III.)*

Example 10.13: *Suppose that we wish to compare the variation in diameters for the company of Example 10.12 with the variation of its competitor. Recall that the sample variance for our company, based on $n = 10$ diameters, was $s_1^2 = .0003$. In contrast, the variance of the diameter measurements for a sample of 20 of the competitor's parts was $s_2^2 = .0001$. Do the data provide sufficient information to indicate a smaller variation in diameters for the competitor? Test using $\alpha = .05$.*

Solution: *We are testing $H_0: \sigma_1^2 = \sigma_2^2$ against the alternative, $H_a: \sigma_1^2 > \sigma_2^2$. The test statistic, $F = (S_1^2/S_2^2)$, is based on $\nu_1 = 9$ and $\nu_2 = 19$ degrees of freedom and we reject H_0 for values of F larger than $F_{.05} = 2.42$. (See Table 6, Appendix III, for $F_{.05}$ values.) Since the observed value of the test statistic is*

$$\frac{s_1^2}{s_2^2} = \frac{.0003}{.0001} = 3,$$

$F > F_\alpha$, and we reject $H_0: \sigma_1^2 = \sigma_2^2$. It appears that the variation of the parts diameters is less for the new company.

Suppose that, for Example 10.13, we wished to detect either $\sigma_1^2 > \sigma_2^2$ or $\sigma_2^2 > \sigma_1^2$. Then we would locate $\alpha/2$ in each tail of the F distribution. To avoid the necessity of finding the critical value in the lower tail, always place the larger sample variance in the numerator of the F statistic. Then you would reject $H_0: \sigma_1^2 = \sigma_2^2$, against the alternative, $H_a: \sigma_1^2 \neq \sigma_2^2$, if $F > F_{\alpha/2}$.

10.7 *Power of Tests: The Neyman–Pearson Lemma*

Now we move from practical examples of statistical tests to a theoretical discussion of their properties. We have suggested specific tests for a number of practical hypothesis-testing situations, but the alert reader may be asking why we chose those particular tests. How did we decide on the test statistics that were presented and how did we know that we had selected the best rejection regions?

Recall that the goodness of a test is measured by α and β, the probability of the type I and type II errors. A related but more useful concept is that of *power*.

Definition 10.2: *Suppose that $\hat{\theta}$ is the test statistic and C is the rejection region for a test of a hypothesis concerning a parameter θ. Then the power of the test at $\theta = \theta_a$ is the probability of rejecting $H_0: \theta = \theta_0$ when, in fact, $H_a: \theta = \theta_a$ is true. In general, the power of the test for a given θ is*

$$P(\hat{\theta} \in C|\theta).$$

Note that the power of a test measures the ability of the test to detect that the null hypothesis is false when $\theta = \theta_a$. Then, for $\theta = \theta_a$,

$$\text{power} = P[\text{rejecting } H_0 \text{ given } \theta = \theta_a]$$

and

$$\beta = P[\text{accepting } H_0 \text{ given } \theta = \theta_a].$$

It follows that the power of a test at θ_a and the probability of a type II error are related as follows:

$$\textit{power} = 1 - \beta.$$

Particularly note that the power of the test for $\theta = \theta_0$ is equal to the probability of rejecting H_0 when H_0 is true, namely, α. A typical *power curve*, a graph of power versus θ, is shown in Figure 10.5.

How should we select the rejection region C for the test $H_0: \theta = \theta_0$ against the alternative $H_a: \theta = \theta_a$. Intuitively, C should be chosen so that the power is small (close to 0) at θ_0 and large (close to 1) at θ_a. Since, for a fixed sample size, α and β cannot both be made arbitrarily small, we adopted the procedure of fixing α, selecting a rejection region C, and letting β fall where it may. Now let us optimize over all possible rejection regions that might have been selected for a given α. That is, we will choose C so as to maximize the power (minimize β) at $\theta = \theta_a$. This can be done using a theorem, known as the Neyman–Pearson lemma, that tells us how to make this choice when the density function (or probability function) of the observable random variables is completely defined except for the value of θ.

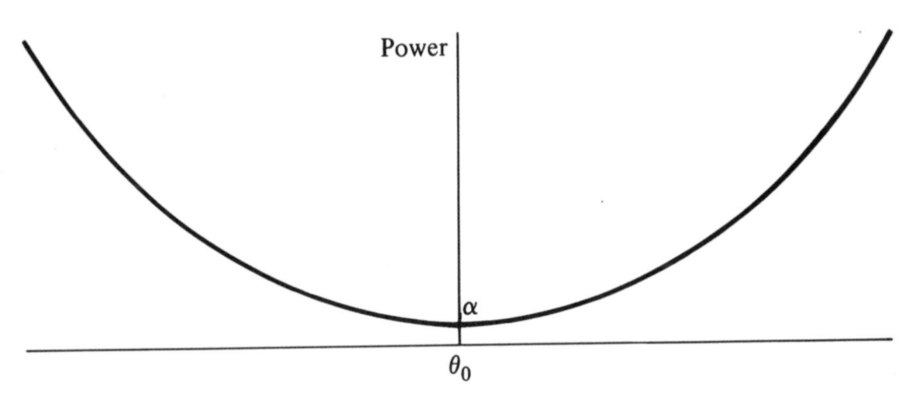

Figure 10.5 A Typical Power Curve for the Test $H_0: \theta = \theta_0$ Against the Alternative, $H_a: \theta \neq \theta_0$

To understand the Neyman–Pearson lemma (Theorem 10.1), you will need to recall the definition of likelihood given in Definition 9.1.

Theorem 10.1: *Suppose that we wish to test $H_0: \theta = \theta_0$ against the alternative, $H_a: \theta = \theta_a$, based on a random sample, $Y_1, Y_2, \ldots, Y_n$, from the probability density function $f(y; \theta)$. Let $L(\theta)$ denote the likelihood of the sample at the point θ.*

Then, for a given α, the test that maximizes the power at θ_a has a rejection region C determined by

$$\frac{L(\theta_0)}{L(\theta_a)} < k.$$

Such a test will be called a most powerful *test.*

The proof of Theorem 10.1 will not be given here but it can be found in most of the references given at the end of the chapter. We illustrate the application of the theorem with the following example.

Example 10.14: *Suppose that $Y_1, \ldots, Y_n$ is a random sample from a normal distribution with mean μ and variance σ^2 (where σ^2 is a known constant) and we wish to test $H_0: \mu = \mu_0$ against the alternative, $H_a: \mu = \mu_1$, where $\mu_1 > \mu_0$. Find the most powerful test for a fixed α.*

Solution: *We are given*

$$f(y;\mu) = \frac{1}{\sigma\sqrt{2\pi}} \exp\left[\frac{-(y-\mu)^2}{2\sigma^2}\right], \qquad \infty < y < \infty,$$

and hence

$$L(\mu) = f(y_1;\mu)f(y_2;\mu)\cdots f(y_n;\mu)$$
$$= \left(\frac{1}{\sigma\sqrt{2\pi}}\right)^n \exp\left[-\sum_{i=1}^{n}\frac{(y_i-\mu)^2}{2\sigma^2}\right]$$

[recall that exp( ) *is simply* $e^{(\,)}$ *in another form]. Appealing to Theorem 10.1, we see that the most powerful test is given by*

$$\frac{L(\mu_0)}{L(\mu_1)} < k,$$

which in this case is

$$\frac{\left(\frac{1}{\sqrt{2\pi}\sigma}\right)^n \exp\left[-\sum_{i=1}^{n}\frac{(y_i-\mu_0)^2}{2\sigma^2}\right]}{\left(\frac{1}{\sqrt{2\pi}\sigma}\right)^n \exp\left[-\sum_{i=1}^{n}\frac{(y_i-\mu_1)^2}{2\sigma^2}\right]} < k.$$

This inequality can be rearranged as follows:

$$\exp\left\{-\frac{1}{2\sigma^2}\left[\sum_{i=1}^{n}(y_i-\mu_0)^2 - \sum_{i=1}^{n}(y_i-\mu_1)^2\right]\right\} < k.$$

Taking natural logarithms, and simplifying, we have

$$-\frac{1}{2\sigma^2}\left[\sum_{i=1}^{n}(y_i-\mu_0)^2 - \sum_{i=1}^{n}(y_i-\mu_1)^2\right] < \ln k,$$

$$\sum_{i=1}^{n}(y_i-\mu_0)^2 - \sum_{i=1}^{n}(y_i-\mu_1)^2 > -2\sigma^2 \ln k,$$

$$\sum_{i=1}^{n} y_i^2 - 2n\bar{y}\mu_0 + n\mu_0^2 - \sum y_i^2 + 2n\bar{y}\mu_1 - n\mu_1^2 > -2\sigma^2 \ln k,$$

$$\bar{y}(\mu_1 - \mu_0) > \frac{-2\sigma^2 \ln k - n\mu_0^2 + n\mu_1^2}{2n},$$

or

$$\bar{y} > \frac{-2\sigma^2 \ln k - n\mu_0^2 + n\mu_1^2}{2n(\mu_1 - \mu_0)}.$$

Since the quantity on the right-hand side of this inequality is a constant (call it k'), it follows that $\bar{y}$ is the test statistic that yields the most powerful test and that the rejection region is $\bar{y} > k'$. That is,

$$C = \{\bar{y} : \bar{y} > k'\}.$$

The precise value of k' is determined by fixing α and noting that

$$\alpha = P(\bar{Y} \in C \text{ when } \mu = \mu_0).$$

Hypotheses that specify a single value of θ, say $H_a: \theta = \theta_a$, are called *simple*; those that permit θ to assume more than one value, say $H_a: \theta > \theta_a$, are said to be *composite*. For most (but not all) practical problems, for example, the tests of Sections 10.3 through 10.5, the null hypotheses are simple and the alternative hypotheses are composite.

Unfortunately, there is no general theorem, like Theorem 10.1, that applies to tests where the alternative is expressed as a composite hypothesis. However, in many situations the form of the rejection region does not depend on the specific value under H_a. For instance, in Example 10.14 the form of C, $C = \{\bar{y} : \bar{y} > k'\}$, only depends on the fact that $\mu_1 > \mu_0$, not on any specific value of μ_1. When a test obtained by Theorem 10.1 actually maximizes the power for any value of θ greater than θ_0, it will be called a *uniformly most powerful* test of $H_0: \theta = \theta_0$ versus $H_a: \theta > \theta_0$. An analogous situation holds for testing $H_0: \theta = \theta_0$ against the alternative, $H_a: \theta < \theta_0$.

Now consider the problem of selecting the rejection region if both the null and alternative hypotheses are composite, say $H_0: \theta \le \theta_0$ versus $H_a: \theta > \theta_0$. The solution for most tests is provided by examination of the typical power curve of Figure 10.5. If you select your rejection region to obtain adequate power to detect a particular value of θ_a for $H_0: \theta = \theta_0$, you will possess even greater power for $H_0: \theta < \theta_1$ where θ_1 is the same value less than θ. This is because the difference between the null value of θ and the alternative, θ_a, is larger if $\theta < \theta_0$ than if $\theta = \theta_0$. Consequently, if we find a good test for $H_0: \theta = \theta_0$, it will most likely be better for $H_0: \theta < \theta_0$. For this reason, composite null hypotheses do not play a major role in the most widely employed statistical tests.

10.8 Likelihood Ratio Tests

Theorem 10.1 shows how to construct most powerful tests for simple hypotheses when the density function under consideration is completely known except for one unknown parameter, θ. Although few tests fall into this category, the method employed in the theorem suggests a very general procedure for producing test statistics.

Our method for finding a test statistic requires that we calculate the likelihood of the observed sample under the null and alternative hypotheses. To accomplish this, we need to define the following parameter spaces.

Suppose that we have a random sample of size n from a density function, $f(y;\theta)$, which may involve other unknown parameters besides θ. The null hypothesis specifies that θ lies in a set of possible values, say Ω_0; the alternative specifies that θ lies in another, nonoverlapping set, say Ω_a. For example, we may select $\Omega_0 = \{\theta : \theta = \theta_0\}$ and $\Omega_a = \{\theta : \theta > \theta_0\}$, as we have already seen. Denote the union of these two sets, Ω_0 and Ω_a, by Ω, that is, $\Omega = \Omega_0 \cup \Omega_a$. We cannot completely specify the likelihood of the sample under either hypothesis, because the hypothesis may be composite or other unknown parameters may be present. However, we can employ the observed data to estimate all unknown parameters by the maximum likelihood method, under either hypothesis. Under H_0, this estimation would be accomplished by using the methods of Section 9.3 to find the maximum likelihood estimators of all unknown parameters, subject to the restriction that $\theta \in \Omega_0$.

Let $L(\hat{\Omega}_0)$ denote the likelihood function with all unknown parameters replaced by their maximum likelihood estimators, subject to the restriction that $\theta \in \Omega_0$. Similarly, let $L(\hat{\Omega})$ be obtained the same way, but with the restriction that $\theta \in \Omega$. A likelihood ratio test is then based on the ratio $L(\hat{\Omega}_0)/L(\hat{\Omega})$.

A Likelihood Ratio Test:

Define λ by

$$\lambda = \frac{L(\hat{\Omega}_0)}{L(\hat{\Omega})}.$$

A likelihood ratio test of $H_0 : \theta \in \Omega_0$ versus $H_a : \theta \in \Omega_a$ employs λ as a test statistic, and the rejection region is determined by $\lambda \le k$.

It can easily be shown that $0 \leq \lambda \leq 1$. A value of λ close to zero indicates that the likelihood of the sample appears to be very small under H_0 in comparison with its value under H_a. Equivalently, the data suggest favoring H_a over H_0. The actual value of k is chosen so that α remains at a predetermined level.

We illustrate the mechanics of this method with the following example.

Example 10.15: *Suppose that $Y_1, \ldots, Y_n$ is a random sample from a normal distribution with unknown mean, μ, and variance, σ^2. We want to test $H_0: \mu = \mu_0$ versus $H_a: \mu > \mu_0$. Find the appropriate likelihood ratio test.*

Solution: *Note that Ω_0 is the point μ_0, $\Omega_a = \{\mu: \mu > \mu_0\}$, and hence that $\Omega = \Omega_0 \cup \Omega_a = \{\mu: \mu \geq \mu_0\}$. The constant σ^2 is completely unspecified. We must now find $L(\hat{\Omega}_0)$ and $L(\hat{\Omega})$.*

Restricting μ to Ω_0 implies that $\mu = \mu_0$, and hence it need not be estimated. However, we must estimate σ^2 for the case where $\mu = \mu_0$. From Example 9.4 we see that the maximum likelihood estimate of σ^2 is

$$\hat{\sigma}_0^2 = \frac{\sum_{i=1}^{n} (Y_i - \mu_0)^2}{n}$$

when $\mu = \mu_0$.

The unrestricted maximum likelihood estimator of μ (see Example 9.4) is $\bar{Y}$. Therefore, for μ restricted to Ω, the maximum likelihood estimator of μ is $\hat{\mu} = \max(\bar{Y}, \mu_0)$. (If the actual maximum of L is outside of the region Ω, the maximum within Ω occurs at the boundary point, μ_0.) Just as above, the maximum likelihood estimator of σ^2 in Ω is

$$\hat{\sigma}^2 = \frac{\sum_{i=1}^{n} (Y_i - \hat{\mu})^2}{n}.$$

For the normal density function we have

$$L = \left(\frac{1}{\sqrt{2\pi}}\right)^n \left(\frac{1}{\sigma^2}\right)^{n/2} \exp\left[-\sum_{i=1}^{n} \frac{(y_i - \mu)^2}{2\sigma^2}\right].$$

$L(\hat{\Omega}_0)$ is obtained by replacing μ by μ_0 and σ^2 by $\hat{\sigma}_0^2$, which gives

$$L(\hat{\Omega}_0) = \left(\frac{1}{\sqrt{2\pi}}\right)^n \left(\frac{1}{\hat{\sigma}_0^2}\right)^{n/2} e^{-n/2}.$$

$L(\hat{\Omega})$ is obtained by replacing μ by $\hat{\mu}$ and σ^2 by $\hat{\sigma}^2$, which yields

$$L(\hat{\Omega}) = \left(\frac{1}{\sqrt{2\pi}}\right)^n \left(\frac{1}{\hat{\sigma}^2}\right)^{n/2} e^{-n/2}.$$

Thus

$$\lambda = \frac{L(\hat{\Omega}_0)}{L(\hat{\Omega})} = \left(\frac{\hat{\sigma}^2}{\hat{\sigma}_0^2}\right)^{n/2}$$

$$= \left[\frac{\sum_{i=1}^{n} (Y_i - \overline{Y})^2}{\sum_{i=1}^{n} (Y_i - \mu_0)^2}\right]^{n/2} \qquad \text{if } \overline{Y} > \mu_0,$$

$$= 1 \qquad \text{if } \overline{Y} \le \mu_0.$$

Now note that

$$\sum_{i=1}^{n} (Y_i - \mu_0)^2 = \sum_{i=1}^{n} [(Y_i - \overline{Y}) + (\overline{Y} - \mu_0)]^2$$

$$= \sum_{i=1}^{n} (Y_i - \overline{Y})^2 + n(\overline{Y} - \mu_0)^2.$$

Then the rejection region, $\lambda \le k$, is equivalent to

$$\frac{\sum_{i=1}^{n} (Y_i - \overline{Y})^2}{\sum_{i=1}^{n} (Y_i - \mu_0)^2} < k^{2/n} = k',$$

$$\frac{\sum_{i=1}^{n} (Y_i - \overline{Y})^2}{\sum_{i=1}^{n} (Y_i - \overline{Y})^2 + n(\overline{Y} - \mu_0)^2} < k',$$

or

$$\frac{1}{1 + \dfrac{n(\overline{Y} - \mu_0)^2}{\sum_{i=1}^{n} (Y_i - \overline{Y})^2}} < k'.$$

This inequality, in turn, is equivalent to

$$\frac{n(\overline{Y} - \mu_0)^2}{\sum_{i=1}^{n} (Y_i - \overline{Y})^2} > \frac{1}{k'} - 1 = k'',$$

$$\frac{n(\overline{Y} - \mu_0)^2}{\sum_{i=1}^{n} \frac{(Y_i - \overline{Y})^2}{n-1}} > (n-1)k'',$$

or

$$\frac{\sqrt{n}(\overline{Y} - \mu_0)}{S} > \sqrt{(n-1)k''},$$

where

$$S^2 = \frac{\sum_{i=1}^{n} (Y_i - \overline{Y})^2}{n-1}.$$

The last inequality follows because $\overline{Y} > \mu_0$ in the cases for which $\lambda \leq k < 1$. Note that $\sqrt{n}(\overline{Y} - \mu_0)/S$ is simply the t statistic employed in previous sections. The likelihood ratio test is equivalent to the t test of Section 10.5.

Situations in which the likelihood ratio test assumes a well-known form are not uncommon. In fact, all the tests of Sections 10.5 and 10.6 can be obtained by the likelihood ratio method. For most practical problems, the likelihood ratio method produces the best possible test, in the sense of power.

Unfortunately, the likelihood ratio method does not always produce a test statistic with a known probability distribution, such as the t statistic of Example 10.15. However, if the sample size is large, an approximation to the distribution of λ is available.

> ***Theorem 10.2:*** *Under certain regularity conditions on $f(y;\theta)$, $-2 \ln \lambda$ converges in distribution, as $n \to \infty$, to a χ^2 random variable with degrees of freedom equal to the number of parameters or functions of parameters assigned specific numerical values under H_0.*

Theorem 10.2 allows us to use the χ^2 table for finding rejection regions with fixed α, when n is large. Suitable values of n vary from problem to problem.

The necessary regularity conditions will not be discussed, but the theorem holds for most practical distributions.

Example 10.16: *Suppose that an engineer wishes to compare the number of complaints per week filed by union stewards for two different shifts at a manufacturing plant. One hundred independent observations on the number of complaints gave means $\bar{x} = 20$ for shift 1 and $\bar{y} = 22$ for shift 2. Assume that the number of complaints per week on the ith shift has a Poisson distribution with mean θ_i, $i = 1, 2$. Test $H_0: \theta_1 = \theta_2$ versus $H_a: \theta_1 \neq \theta_2$ by the likelihood ratio method, with $\alpha = .01$.*

Solution: *The likelihood of the sample is now the joint density of all X's and Y's and is given by*

$$L = \frac{1}{k} \theta_1^{\sum_{i=1}^n x_i} e^{-n\theta_1} \theta_2^{\sum_{i=1}^n y_i} e^{-n\theta_2},$$

where $k = x_1!, \ldots, x_n!, y_1!, \ldots, y_n!$, and $n = 100$. Now Ω_0 is the set of parameter values in which $\theta_1 = \theta_2 = \theta$. Hence

$$L = \frac{1}{k} \theta^{\sum_{i=1}^n x_i + \sum_{i=1}^n y_i} e^{-2n\theta}.$$

Solving for the maximum likelihood estimator of θ in Ω_0, we find

$$\hat{\theta} = \frac{1}{2n}\left(\sum_{i=1}^n X_i + \sum_{i=1}^n Y_i\right) = \tfrac{1}{2}(\bar{X} + \bar{Y}).$$

Separate maximum likelihood estimators are needed for θ_1 and θ_2 in Ω. Using the likelihood for $\theta_1 \neq \theta_2$, and solving, the estimators are $\hat{\theta}_1 = \bar{X}$ and $\hat{\theta}_2 = \bar{Y}$. Thus

$$\lambda = \frac{L(\hat{\Omega}_0)}{L(\hat{\Omega})}$$

$$= \frac{(\hat{\theta})^{n\bar{x} + n\bar{y}} e^{-2n\hat{\theta}}}{(\hat{\theta}_1)^{n\bar{x}} (\hat{\theta}_2)^{n\bar{y}} e^{-n\hat{\theta}_1 - n\hat{\theta}_2}}$$

$$= \frac{(\hat{\theta})^{n\bar{x} + n\bar{y}}}{(\bar{x})^{n\bar{x}} (\bar{y})^{n\bar{y}}}.$$

The observed value of $\hat{\theta}$ *is* $(1/2)(\bar{x} + \bar{y}) = (1/2)(20 + 22) = 21$. *The observed value of* λ *is*

$$\lambda = \frac{21^{(100)(20+22)}}{20^{(100)(20)}22^{(100)(22)}},$$

and hence

$$\begin{aligned} -2 \ln \lambda &= -(2)[4200 \ln(21) - 2000 \ln(20) - 2200 \ln(22)] \\ &= 14.8. \end{aligned}$$

The degrees of freedom associated with $-2 \ln \lambda$ *is the number of parameters, or functions of parameters, assigned a specific value under* H_0. *The null hypothesis* $H_0\colon \theta_1 = \theta_2$ *could be stated* $H_0\colon (\theta_1 - \theta_2) = 0$ *and thus only one function of the parameters,* $(\theta_1 - \theta_2)$, *has a specified value. From Theorem 10.2, we have that* $-2 \ln \lambda$ *is approximately a* χ^2 *random variable with one degree of freedom. Small values of* λ *correspond to large values of* $-2 \ln \lambda$, *and so the rejection region contains those values of* $-2 \ln \lambda$ *larger than* $\chi^2_{.01} = 6.635$, *the value that cuts off an area of* .01 *in the right-hand tail of a* χ^2 *density with one degree of freedom.*

Since the observed value of $-2 \ln \lambda$ *is larger than* $\chi^2_{.01}$ *in this case, we reject* $H_0\colon \theta_1 = \theta_2$. *It appears that the mean numbers of complaints filed by the union stewards do differ.*

10.9 Minimax and Bayes Tests

The minimax and Bayes methods for choosing estimators were discussed in Section 9.8. We now extend that discussion to include tests of hypotheses. Only the case of simple hypotheses, $H_0\colon \theta = \theta_0$ versus $H_a\colon \theta = \theta_a$, will be considered here. For a more thorough discussion of these procedures, see the references at the end of the chapter.

For testing $H_0\colon \theta = \theta_0$ versus $H_a\colon \theta = \theta_a$ we assume that we have n independent observations on a random variable Y upon which we must base a decision to either accept or reject H_0. For simplicity of presentation, assume that Y is a discrete random variable so that we need consider only a finite number of possible decisions, denoted as $d_i, i = 0, 1, 2, \ldots, r$. Then, as in all decision theoretic problems, we must select a loss function, $L(d_i, \theta)$, which represents the loss incurred if decision d_i is employed when θ is the true parameter value. The *minimax* procedure chooses the decision, d_i, which minimizes the maximum expected loss over both values of θ. We illustrate with an example.

Example 10.17: *The following simplified example illustrates an important application of a minimax test. A supplier of transistors ships two large lots of transistors, one containing* 10 *percent defectives, the other* 20 *percent. The fraction defectives in the lot, say θ, is very important to purchaser A because it is uneconomical for him to operate his manufacturing process if $\theta = .2$. In fact, his manufacturing costs are three times as high for $\theta = .2$ as they are for $\theta = .1$. If he decides that $\theta = .2$, he will send the lot back to the supplier; if he decides that $\theta = .1$, he will operate as usual. Assume that the purchaser will decide which of the two lots he possesses, $\theta = .2$ or $\theta = .1$, based on the number of defectives in a random sample of $n = 2$. Find the purchaser's minimax decision.*

Solution: *We are testing $H_0: \theta = .1$ versus $H_a: \theta = .2$ based on a single observation of the number of defectives, Y, a binomial variable with $n = 2$. Since we would be inclined to reject H_0 for large values of Y, the only four reasonable decisions to consider are*

$$d_0: \text{reject } H_0 \text{ when } Y \geq 0,$$

$$d_1: \text{reject } H_0 \text{ when } Y \geq 1,$$

$$d_2: \text{reject } H_0 \text{ when } Y = 2,$$

and

$$d_3: \text{never reject } H_0.$$

We will display some of the loss functions below, assuming that the cost of overestimating θ is 1 *unit and the cost of underestimating θ is* 3 *units:*

$$L(d_0, \theta_0) = 1, \qquad L(d_0, \theta_a) = 0,$$

$$L(d_1, \theta_0) = \begin{cases} 0 & \text{if } Y = 0, \\ 1 & \text{if } Y \geq 1, \end{cases} \qquad L(d_1, \theta_a) = \begin{cases} 3 & \text{if } Y = 0, \\ 0 & \text{if } Y \geq 1, \end{cases}$$

$$L(d_3, \theta_0) = 0, \qquad L(d_3, \theta_a) = 3.$$

We will define the risk associated with any pair (d_i, θ) as the expected loss for decision d_i when that particular value of θ occurs. That is,

$$R(d_i, \theta_0) = E[L(d_i, \theta_0)]$$

and

$$R(d_i, \theta_a) = E[L(d_i, \theta_a)].$$

It now remains to compute the risk functions. Since Y is binomial with $n = 2$, we have

$$R(d_0, \theta_0) = E[L(d_0, \theta_0)]$$
$$= 1P(Y \geq 0) = 1,$$
$$R(d_0, \theta_a) = E[L(d_0, \theta_a)] = 0,$$
$$R(d_1, \theta_0) = 0P(Y = 0 \text{ when } \theta = .1)$$
$$+ 1P(Y \geq 1 \text{ when } \theta = .1)$$
$$= 0 + .19$$
$$= .19,$$

and

$$R(d_1, \theta_a) = 3P(Y = 0 \text{ when } \theta = .2)$$
$$+ 0P(Y \geq 1 \text{ when } \theta = .2)$$
$$= (3)(.64) + 0$$
$$= 1.92.$$

Similarly,

$$R(d_2, \theta_0) = .01,$$
$$R(d_2, \theta_a) = 2.88$$

and

$$R(d_3, \theta_0) = 0,$$
$$R(d_3, \theta_a) = 3.$$

We see that the maximum values of the risks associated with d_0, d_1, d_2, and d_3 are, respectively, 1, 1.92, 2.88, and 3. Thus d_0 has the smallest maximum risk, and is therefore the minimax decision. Under d_0, the purchaser always rejects H_0, regardless of the value of Y. The solution turns out this way because it is so costly to underestimate θ.

The Bayes procedure assumes that θ is itself a random variable with a prior probability distribution. It is then possible to take the expectation of the risk function, for a given decision, with respect to the distribution on θ, and the result is called the Bayes risk. Denoting the Bayes risk by $B(d_i)$ we have

$$B(d_i) = E[R(d_i, \theta)]$$

$$= R(d_i, \theta_0)p(\theta_0) + R(d_i, \theta_a)p(\theta_a),$$

where $p(\theta_0)$ and $p(\theta_a)$ are the prior probabilities assigned to θ_0 and θ_a, respectively. The Bayes decision is the rule, d_i, which minimizes the Bayes risk.

Example 10.18: *Suppose that in Example 10.17 the supplier is fairly certain that the purchaser got the better lot (the one with $\theta = .1$). In fact, he is willing to say that $p(\theta_0) = .8$ and $p(\theta_a) = .2$. (He is actually saying that the chances are 4 to 1 that the purchaser got the better lot.) Find the Bayes decision for the purchaser.*

Solution: *Using the risks found in Example 10.17, we see that*

$$B(d_0) = R(d_0, \theta_0)p(\theta_0) + R(d_0, \theta_a)p(\theta_a)$$

$$= (1)(.8) + (0)(.2)$$

$$= .8,$$

$$B(d_1) = .536,$$

$$B(d_2) = .584,$$

and $$B(d_3) = .6.$$

Thus d_1, which says "reject H_0 when $Y \geq 1$," is the Bayes decision. Note that this decision-making procedure requires the purchaser to take the supplier's judgment into account. Frequently, this sort of thing is done informally anyway, and the Bayes procedure simply provides a formal mechanism for working personal judgments into the decision-making process.

10.10 Summary

In Chapters 8, 9, and 10 we have presented the basic concepts associated with the two methods for making inferences, estimation, and tests of hypotheses. Philosophically, estimation (Chapters 8 and 9) answers the question: What is the numerical value of θ? In contrast, a test of a hypothesis asks the question: Is θ equal to a specific numerical value, θ_0? Which method you employ for a given situation often depends on how you, the experimenter, like to phrase your inference, but sometimes this decision is taken out of your hands. That is, the practical question clearly implies either estimation or a statistical test of hypothesis. For example, the acceptance or rejection of incoming supplies or outgoing products in a manufacturing process clearly requires a decision, or statistical test.

Associated with both methods for making inferences are measures of their goodness. Thus the expected width of a confidence interval and the confidence coefficient measure the goodness of the estimation procedure. Likewise, the goodness of a statistical test is measured by the probabilities, α and β, for type I and type II errors. These measures of goodness enable us to compare one statistical test with another and to develop a theory for acquiring statistical tests with desirable properties. Being able to evaluate the goodness of an inference is one of the major contributions of statistics to the analysis of experimental data. Of what value is an inference if you have no measure of its validity?

In this chapter we have presented the elements of a statistical test and explained how a test works. Some useful tests are given to show how they can be used in practical situations and you will see other interesting applications in the chapters that follow. Although many test statistics are suggested by parameter estimators, the likelihood ratio procedure provides a general method for acquiring a statistical test—and one that very often has good properties.

References

1. Hoel, P. G., *Introduction to Mathematical Statistics*, 3rd ed. New York: John Wiley & Sons, Inc., 1962.
2. Hogg, R. V., and A. T. Craig, *Introduction to Mathematical Statistics*, 3rd ed. New York: The Macmillan Company, 1970.
3. Lehmann, E. L., *Testing Statistical Hypotheses*. New York: John Wiley & Sons, Inc., 1959.
4. Mood, A. M., and F. A. Graybill, *Introduction to the Theory of Statistics*, 2nd ed. New York: McGraw-Hill Book Company, 1963.
5. Raiffa, H., and R. Schlaifer, *Applied Statistical Decision Theory*. Boston: Harvard Graduate School of Business Administration, 1961.
6. Wilks, S. S., *Mathematical Statistics*. New York: John Wiley & Sons, Inc., 1962.

Exercises

10.1. Define α and β for a statistical test of a hypothesis.

10.2. What is the level of significance of a statistical test of a hypothesis?

10.3. The daily wages in a particular industry are normally distributed with a mean of \$13.20 and a standard deviation of \$2.50. If a company in this industry employing 40 workers pays these workers on the average \$12.20, can this company be accused of paying inferior wages at the 1 percent level of significance?

10.4. Two sets of 50 elementary school children were taught to read by two different methods. At the conclusion of the instructional period, a reading test gave the results $\bar{y}_1 = 74$, $\bar{y}_2 = 71$, $s_1 = 9$, $s_2 = 10$. Test to see if there is evidence of a real difference between the two population means. (Use $\alpha = .10$.)

10.5. A manufacturer of automatic washers provides a particular model in one of three colors, *A*, *B*, or *C*. Of the first 1000 washers sold, it is noticed that 400 of the washers were of color *A*. Would you conclude that customers have a preference for color *A*? Justify your answer.

10.6. A manufacturer claimed that at least 20 percent of the public preferred his product. A sample of 100 persons is taken to check his claim. With $\alpha = .05$, how small would the sample percentage need to be before the claim could be rightfully refuted? (Note that this would require a one-tailed test of a hypothesis.)

10.7. What conditions must be met in order that the z test may be used to test a hypothesis concerning a population mean, μ?

10.8. A manufacturer claimed that at least 95 percent of the equipment which he supplied to a factory conformed to specification. An examination of a sample of 700 pieces of equipment revealed that 53 were faulty. Test his claim at a significance level of .05.

10.9. Mr. Sands believes that the fraction p_1 of Republicans in favor of the death penalty is greater than the fraction p_2 of Democrats in favor of the death penalty. He acquired an independent random sample of 200 Republicans and 200 Democrats, respectively, and found 46 Republicans and 34 Democrats favoring the death penalty. Does this evidence provide statistical support at the .05 level of significance for Mr. Sands' belief?

10.10. Refer to Exercise 10.9. Mr. Sands should have put some thought into designing a test for which β is tolerably low when p_1 exceeds p_2 by an important amount. For example, show Mr. Sands how to find a common sample size n for a test with $\alpha = .05$ and $\beta \leq .20$ when, in fact, p_1 exceeds p_2 by .1. [*Hint*: The maximum value of $p(1 - p)$ is .25.]

10.11. Presently 20 percent of potential customers buy a certain brand of soap, say brand A. In order to increase sales, an extensive advertising campaign will be conducted. At the end of the campaign, a sample of 400 potential customers will be interviewed to determine if the campaign was successful.

(a) State H_0 and H_a in terms of p, the probability that a customer prefers soap brand A.

(b) It is decided to conclude that the advertising campaign was a success if at least 92 of the 400 customers interviewed prefer brand A. (Use the normal approximation to the binomial distribution to evaluate the desired probability.) Find α.

10.12. In the past, a chemical plant has produced an average of 1100 pounds of chemical per day. The records for the past year, based on 260 operating days, show the following:

$$\bar{y} = 1060 \text{ pounds/day},$$

$$s = 340 \text{ pounds/day}.$$

We wish to test whether or not the average daily production has dropped significantly over the past year.

(a) Give the appropriate null and alternative hypotheses.

(b) If z is used as a test statistic, determine the rejection region corresponding to a level of significance $\alpha = .05$.

(c) Do the data provide sufficient evidence to indicate a drop in average daily production?

10.13. The braking ability was compared for two types of 1972 automobiles. Random samples of 64 automobiles were tested for each type. The recorded measurement was the distance required to stop when the brakes were applied at 40 miles per hour. The computed sample means and variances were

$$\bar{y}_1 = 118, \qquad \bar{y}_2 = 109;$$

$$s_1^2 = 102, \qquad s_2^2 = 87.$$

Do the data provide sufficient evidence to indicate a difference in the mean stopping distance for the two types of automobiles?

10.14. Refer to Exercise 10.6 and calculate the value of β for an alternative, $p_a = .15$.

10.15. Why is the z test usually inappropriate as a test statistic when the sample size is small?

10.16. What assumptions are made when Student's t test is employed to test a hypothesis concerning a population mean?

10.17. A chemical process has produced, on the average, 800 tons of chemical per day. The daily yields for the past week are 785, 805, 790, 793, and 802 tons. Do these data indicate that the average yield is less than 800 tons and hence that something is wrong with the process? Test at the 5 percent level of significance.

10.18. A coin-operated soft-drink machine was designed to discharge, on the average, 7 ounces of beverage per cup. To test the machine, ten cupfuls of beverage were drawn from the machine and measured. The mean and standard deviation of the ten measurements were 7.1 ounces and .12 ounce, respectively. Do these data present sufficient evidence to indicate that the mean discharge differs from 7 ounces? Test at the 10 percent level of significance.

10.19. What assumptions are made about the populations from which independent random samples are obtained when utilizing the t distribution in making small-sample inferences concerning the differences in population means?

10.20. Two methods for teaching reading were applied to two randomly selected groups of elementary school children and compared on the basis of a reading comprehension test given at the end of the learning period. The sample means and variances computed from the test scores are shown below.

	Method 1	*Method 2*
No. children in group	11	14
$\bar{y}$	64	69
s^2	52	71

Do the data present sufficient evidence to indicate a difference in the mean scores for the populations associated with the two teaching methods? Test at an $\alpha = .05$ level of significance.

10.21. A comparison of reaction times for two different stimuli in a psychological word-association experiment produced the following results when applied to a random sample of 16 people:

Stimulus 1	*Stimulus 2*
1	4
3	2
2	3
1	3
2	1
1	2
3	3
2	3

Do the data present sufficient evidence to indicate a difference in mean reaction time for the two stimuli? Test at the $\alpha = .05$ level of significance.

10.22. A manufacturer of a machine to package soap powder claimed that his machine could load cartons at a given weight with a range of no more than 2/5 ounce. The mean and variance of a sample of eight 3-pound boxes were found to equal 3.1 and .018, respectively. Test the hypothesis that the variance of the population of weight measurements is $\sigma^2 = .01$ against the alternative, $\sigma^2 > .01$. Use an $\alpha = .05$ level of significance.

10.23. Under what assumptions may the F distribution be used in making inferences about the ratio of population variances?

10.24. A dairy is in the market for a new bottle-filling machine and is considering models A and B manufactured by companies X and Y, respectively. If ruggedness, cost, and convenience are comparable in the two models, the deciding factor is the variability of fills (the model producing fills with the smaller variance being preferred). Let σ_1^2 and σ_2^2 be the fill variances for models A and B, respectively, and consider various tests of the null hypothesis $H_0: \sigma_1^2 = \sigma_2^2$. Obtaining samples of fills from the two machines and utilizing the test statistic s_1^2/s_2^2, one could set up as the rejection regions an upper-tail area, a lower-tail area, or a two-tailed area of the F distribution depending on his point of view. Which type of rejection region would be most favored by the following persons:

(a) The manager of the dairy? Why?
(b) A salesman for company X? Why?
(c) A salesman for company Y? Why?

10.25. The closing prices of two common stocks were recorded for a period of 15 days. The means and variances were

$$\bar{y}_1 = 40.33, \qquad \bar{y}_2 = 42.54;$$

$$s_1^2 = 1.54, \qquad s_2^2 = 2.96.$$

Do these data present sufficient evidence to indicate a difference in variability of the two stocks for the populations associated with the two samples?

10.26. A precision instrument is guaranteed to be accurate to within 2 units. A sample of four instrument readings on the same object yielded the measurements 353, 351, 351, and 355. Test the null hypothesis that $\sigma = .7$ against the alternative $\sigma > .7$. Conduct the test at the $\alpha = .05$ level of significance.

10.27. For a normal distribution with mean μ and variance $\sigma^2 = 25$, one wishes to test $H_0: \mu = 10$ versus $H_a: \mu = 5$. Find the sample size, n, so that the most powerful test will have $\alpha = \beta = .025$.

10.28. Let $Y_1, \ldots, Y_n$ denote a random sample from a normal distribution with mean μ (unknown) and variance σ^2. For testing $H_0: \sigma^2 = \sigma_0^2$ against $H_a: \sigma^2 > \sigma_0^2$, show that the likelihood ratio test is equivalent to the χ^2 test given in Section 10.6.

10.29. A survey of voter sentiment was conducted in four midcity political wards to compare the fraction of voters favoring candidate A. Random samples of 200 voters were polled in each of the four wards with the following results:

	Ward				
	1	*2*	*3*	*4*	*Total*
Favor A	76	53	59	48	236
Do not favor A	124	147	141	152	564
Total	200	200	200	200	800

The numbers of voters favoring A in the four samples can be regarded as four independent binomial random variables. Construct a likelihood ratio test of the hypothesis that the fractions of voters favoring candidate A are the same in all four wards. Use $\alpha = .05$.

10.30. Suppose that $X_1, \ldots, X_{n_1}$, $Y_1, \ldots, Y_{n_2}$, and $W_1, \ldots, W_{n_3}$ are independent random samples from normal distributions with respective means μ_1, μ_2, and μ_3 and variances σ_1^2, σ_2^2, and σ_3^2.

(a) Find the likelihood ratio test for $H_0: \sigma_1^2 = \sigma_2^2 = \sigma_3^2$ against the alternative of at least one inequality, with μ_1, μ_2, and μ_3 unknown in both cases.

(b) Find an approximate critical region for the test in (a) if n_1, n_2, and n_3 are large and $\alpha = .05$.

10.31. Let $X_1, \ldots, X_m$ denote a random sample from the exponential density with mean θ_1 and let $Y_1, \ldots, Y_n$ denote an independent random sample from an exponential density with mean θ_2. The exponential density is given by

$$f(x) = \frac{1}{\theta_i} e^{-x/\theta_i}, \qquad \theta_i > 0; \quad x > 0.$$

(a) Find the likelihood ratio criterion for testing

$$H_0: \theta_1 = \theta_2 \text{ versus } H_a: \theta_1 \neq \theta_2.$$

(b) Show that the test in (a) is equivalent to an exact F test.

10.32. A merchant figures his weekly profit to be a function of three variables, retail sales, denoted by X; wholesale sales, denoted by Y; and overhead costs, denoted by W. X, Y, and W are regarded as independent normally distributed random variables with means μ_1, μ_2, and μ_3 and variances $\sigma^2, a\sigma^2$, and $b\sigma^2$, respectively, for known constants a and b but unknown σ^2. The merchant's expected profit per week is $\mu_1 + \mu_2 - \mu_3$. If the merchant has independent observations on X, Y, and W for the past n weeks, construct a test of $H_0: \mu_1 + \mu_2 - \mu_3 = k$ against the alternative $H_a: \mu_1 + \mu_2 - \mu_3 \neq k$, for a given constant, k. You may specify $\alpha = .05$.

10.33. A reading exam is given to the sixth grades at three large elementary schools. The scores on the exam at each school are regarded as having normal distributions with unknown means μ_1, μ_2, and μ_3, respectively, and unknown common variance σ^2 ($\sigma_1^2 = \sigma_2^2 = \sigma_3^2 = \sigma^2$). Using the following data on independent random samples from each school, test to see if there is evidence of a difference between μ_1 and μ_2. Use $\alpha = .05$.

School I	*School II*	*School III*
$n_1 = 10$	$n_2 = 10$	$n_3 = 10$
$\sum x_i^2 = 36{,}950$	$\sum y_i^2 = 25{,}850$	$\sum w_i^2 = 49{,}900$
$\bar{x} = 60$	$\bar{y} = 50$	$\bar{w} = 70$

10.34. Let S_1^2 and S_2^2 denote, respectively, the variances of independent random samples of sizes n and m selected from normal distributions with means μ_1 and μ_2 and common variance σ^2. If μ_1 and μ_2 are unknown, construct a test of $H_0: \sigma^2 = \sigma_0^2$ against $H_1: \sigma^2 = \sigma_1^2$, assuming that $\sigma_1^2 > \sigma_0^2$.

10.35. Suppose we have a random sample of four observations from the density function

$$f(y) = \begin{cases} \dfrac{1}{2\beta^3} y^2 e^{-y/\beta}, & y > 0, \\ 0, & \text{elsewhere.} \end{cases}$$

(a) Find the most powerful critical region for testing $H_0: \beta = \beta_0$ against $H_1: \beta = \beta_1$, assuming that $\beta_1 > \beta_0$. (*Hint*: Make use of the χ^2 distribution.)

(b) Is the test given in part (a) uniformly most powerful for the alternative $\beta > \beta_0$?

10.36. Let $Y_1, \ldots, Y_n$ be a random sample from the probability density function given by

$$f(y) = \begin{cases} \dfrac{1}{\theta} m y^{m-1} e^{-y^m/\theta}, & y > 0, \\ 0, & \text{elsewhere,} \end{cases}$$

with m denoting a known constant.

(a) Find the uniformly most powerful test for testing $H_0: \theta = \theta_0$ against $H_a: \theta > \theta_0$.

(b) If the test in part (a) is to have $\theta_0 = 100$, $\theta_a = 400$, and $\alpha = \beta = .05$, find the appropriate sample size and critical region.

10.37. Let $Y_1, \ldots, Y_n$ denote a random sample from a uniform distribution over the interval $(0, \theta)$.

(a) Find a most powerful level α test for testing $H_0: \theta = \theta_0$ against $H_a: \theta = \theta_a$, where $\theta_a < \theta_0$.

(b) Is the test in part (a) uniformly most powerful for testing $H_0: \theta = \theta_0$ against $H_a: \theta < \theta_0$?

10.38. Refer to the random sample of Exercise 10.37.

(a) Find a most powerful level α test for testing $H_0: \theta = \theta_0$ against $H_a: \theta = \theta_a$ where $\theta_a > \theta_0$.

(b) Is the test in part (a) uniformly most powerful for testing $H_0: \theta = \theta_0$ against $H_a: \theta > \theta_0$?

(c) Is the most powerful level α test found in part (a) unique?

10.39. A certain type of fuse is sold in packages of five for $5.00. To ensure high quality, one fuse is randomly selected from each package and checked before the package leaves the factory. (The checking is not destructive.) After the inspection, the package is either sold as is or returned to the factory for complete checking and repackaging. Letting θ denote the number of defectives in a given package, the manufacturer figures his loss function (in dollars) to be $2\theta - 5$ if he sells the package and $5 - \theta$ if he returns it to the factory for complete checking. (That is, he loses $2.00 for each defective that he sells due to guarantees and customer good-will. Also, if he unnecessarily rechecks a good fuse, he loses $1.00.) Let $Y = 1$ if the sample fuse is defective and $Y = 0$ otherwise. The four possible decisions for each package are:

$$d_1 = (\text{sell if } Y = 0, \text{ sell if } Y = 1),$$

$$d_2 = (\text{sell if } Y = 0, \text{ return if } Y = 1),$$

$$d_3 = (\text{return if } Y = 0, \text{ sell if } Y = 1),$$

$$d_4 = (\text{return if } Y = 0, \text{ return if } Y = 1).$$

(a) Find the risk for each decision as a function of θ.

(b) Find the minimax deision if θ is allowed to range over 0, 1, 2, 3, 4, and 5.

(c) Find the minimax decision if it is known that $\theta = 0$ or 1.

(d) Find the minimax decision if it is known that $\theta = 4$ or 5.

10.40. In Exercise 10.39, suppose each possible value of θ is equally likely to occur. That is, $p(\theta) = 1/6$ for $\theta = 0, 1, \ldots, 5$. Find the Bayes decision.

10.41. In Exercise 10.39, suppose that values of θ equal to 3, 4, and 5 never occur, and the other possible values of θ are equally likely. Find the Bayes decision for this case.

11

Linear Models and Estimation by Least Squares

11.1 Introduction

Two methods for finding parameter estimators, the methods of moments and maximum likelihood, were presented in Chapter 9. A third technique, the method of least squares, is the topic of this chapter.

Because of its use in connection with linear models, the method of least squares follows most appropriately a general treatment of estimation and tests of hypotheses. Based on the results of the least-squares procedure, confidence intervals and tests of hypotheses will be constructed for parameters, and functions of parameters, of a linear model. The methods of Chapters 8 and 10 will be employed.

The method of least squares is employed, almost exclusively, in studying the relation between a set of independent variables, $x_1, x_2, \ldots, x_k$, and a response, y. Thus we think of y as a function of $x_1, x_2, \ldots, x_k$, where, using standard mathematical terminology, y is called the *dependent variable* and $x_1, x_2, \ldots, x_k$ are called *independent variables*. For example, an experimenter might wish to relate and eventually predict the height of high tides, y, at a coastal location as a function of the independent variables, wind velocity, wind direction, time, barometric pressure, and so on. Similarly, a biologist would like to relate the potency, y, of an antibiotic produced in a manufacturing process to such independent variables as the age of the culture and the temperature of the vat in which it is produced.

Many different types of mathematical functions can be used to model a response that is a function of one or more independent variables. These can be classified into two categories, *deterministic* and *probabilistic*. For example, suppose that one is interested in relating a response, y, to a variable x and that knowledge of the scientific field suggests that y and x are linearly related,

$$y = \beta_0 + \beta_1 x$$

(where β_0 and β_1 are unknown parameters). This is called a deterministic mathematical model because it does not allow for any error in predicting y. We imply that y always takes the value $\beta_0 + \beta_1(20)$ whenever $x = 20$.

Now suppose that we collect a sample of n values of y corresponding to n different settings of the independent variable, x, and that a plot of the data appears as shown in Figure 11.1. It is quite clear from the figure that the expected value of y may increase as a linear function of x but that a deterministic model is far from an adequate description of reality. Repeated experiments when $x = 20$ would find y bobbing about in a random manner. This tells us that the deterministic model is not an exact representation of the relationship between the two variables. Further, if the model were used to predict y when $x = 20$, the prediction would be subject to some unknown error. This, of course, leads us to statistics. Predicting y for a given value of x is an inferential process and we need to know the properties of the error of prediction if the prediction is to be of value in real life.

In contrast to the deterministic model, statisticians use *probabilistic* models. For example, we might represent the response of Figure 11.1 by the expression

$$Y = \beta_0 + \beta_1 x + \varepsilon,$$

where ε is a random variable with a specified probability distribution. This model accounts for the random behavior of Y, exhibited in Figure 11.1, and provides a more accurate description of reality than does the deterministic model. Further, these assumptions can be used to derive the properties of the error of prediction for Y, properties that will depend on how we estimate the unknown parameters β_0 and β_1.

Scientific and mathematical textbooks are filled with deterministic models of reality. Indeed, the mathematical functions that appear in your

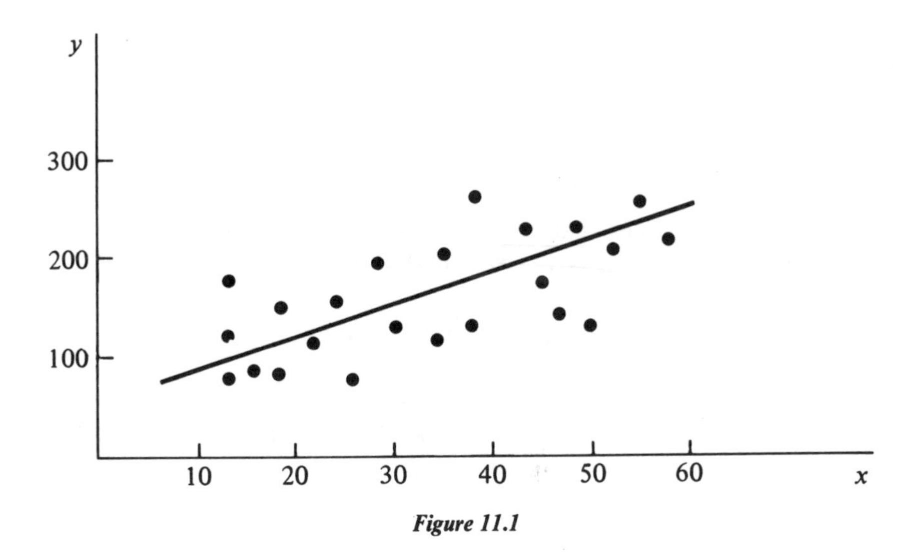

Figure 11.1

calculus book are often used as deterministic mathematical models of nature. For example, Newton's law relating the force of a moving body to its mass and acceleration,

$$F = ma,$$

is a deterministic model that predicts with little error of prediction for practical purposes. In contrast, many other models, functions graphically represented in scientific journals and texts, are very poor indeed. The spatter of points that would give graphic evidence of their inadequacies, similar to the random behavior of the points in Figure 11.1, have been neatly erased. This slight of hand leads many budding scientists to accept the "laws" and theories encountered in their science as an exact description of nature.

So much for deterministic models of reality. If they can be used to predict with negligible error for practical purposes, we use them. If not, we seek a probabilistic model, which will not be an exact characterization of nature but which will enable us to assess the validity of our inferences.

11.2 Linear Statistical Models

A study of probabilistic models and the analyses of data generated by associated experiments is a separate college course in itself. Yet, the recorded response, y, for many experiments can be adequately characterized by a single probabilistic model. Consequently, coverage of this particular model is both instructive and useful.

Definition 11.1: *A* linear statistical *model relating a response, y, to a set of independent variables,* $x_1, x_2, \ldots, x_k$, *is of the form*

$$Y = \beta_0 + \beta_1 x_1 + \beta_2 x_2 + \cdots + \beta_k x_k + \varepsilon,$$

where $\beta_0, \beta_1, \ldots, \beta_k$ *are unknown parameters,* ε *is a random variable, and* $x_1, x_2, \ldots, x_k$ *are recorded without error. Without loss of generality, we will assume that* $E(\varepsilon) = 0$ *and hence that*

$$E(Y) = \beta_0 + \beta_1 x_1 + \beta_2 x_2 + \beta_k x_k.$$

Note the physical interpretation of the linear model, Y. It says that Y is equal to an expected value, $\beta_0 + \beta_1 x_1 + \beta_2 x_2 + \cdots + \beta_k x_k$, a function of the independent variables $x_1, x_2, \ldots, x_k$, plus a random error, ε. From a practical point of view, ε acknowledges our inability to provide an exact model for nature. In repeated experimentation, Y bobs about $E(Y)$ in a random manner because we have failed to include in our model all the many variables that may affect Y. Fortunately, the net effect of these unmeasured, and most often unknown, variables is to cause Y to vary in a manner that is adequately approximated by an assumption of random behavior.

When we say that y is a *linear* statistical model, we mean that $E(Y)$ is a linear function of the unknown parameters, $\beta_0, \beta_1, \ldots, \beta_k$, not a linear function of $x_1, x_2, \ldots, x_k$. Thus $x_1, x_2, \ldots, x_k$ are regarded as known constants, since they are assumed to be measured without error in an experiment. If you think that the yield, y, is a function of the variable T, the temperature of a chemical process, you might let $x_1 = T$ and $x_2 = e^T$ and model $E(Y)$ as $E(Y) = \beta_0 + \beta_1 x_1 + \beta_2 x_2$, or, equivalently, $E(Y) = \beta_0 + \beta_1 T + \beta_2 e^T$. Or, if y is a function of two variables, x_1 and x_2, you might choose a planar approximation to the true mean response using the linear model, $E(Y) = \beta_0 + \beta_1 x_1 + \beta_2 x_2$. Thus $E(Y)$ is a linear function of β_0, β_1, and β_2 and represents a plane in the y, x_1, x_2 space. Similarly,

$$E(Y) = \beta_0 + \beta_1 x + \beta_2 x^2$$

is a linear statistical model where $E(Y)$ is a second-order polynomial function of the independent variable, x, with $x_1 = x$ and $x_2 = x^2$. This model would be appropriate for a response that traces a segment of a parabola over the experimental region.

The expected percentage of water, $E(Y)$, in paper during its manufacture could be represented as a second-order function of the temperature of the dryer, x_1, and the speed of the paper machine, x_2. Thus

$$E(Y) = \beta_0 + \beta_1 x_1 + \beta_2 x_2 + \beta_3 x_1 x_2 + \beta_4 x_1^2 + \beta_5 x_2^2,$$

where $\beta_0, \beta_1, \ldots, \beta_5$ are unknown parameters in the model. Geometrically, $E(Y)$ traces a second-order (conic) surface over the x_1, x_2 plane.

The object of this chapter is to estimate the unknown parameters, $\beta_0, \beta_1, \ldots, \beta_k$, and thereby obtain a model that provides a good fit to the geometric curve (surface) traced by $E(Y)$. In addition, we may wish to estimate $E(Y)$ or to predict some future value of Y for a given setting of $x_1, x_2, \ldots, x_k$. Or, we may wish to test an hypothesis concerning the value of one or more parameters of the model or of $E(Y)$. All these practical inferential problems start with the estimation of $\beta_0, \beta_1, \ldots, \beta_k$, or, as we say in statistics, finding the "best-fitting model" for an observed set of data. To do this, we will use an estimation procedure known as the method of least squares.

11.3 The Method of Least Squares

A procedure for estimating the parameters of any linear model, the method of least squares, can be illustrated simply by employing it to fit a straight line to a set of data points. Suppose that we wish to fit the line

$$E(Y) = \beta_0 + \beta_1 x$$

to the set of data points shown in Figure 11.2. That is, we postulate, $Y = \beta_0 + \beta_1 x + \varepsilon$, where ε possesses some probability distribution with $E(\varepsilon) = 0$. To complete our notation, let $\hat{Y} = \hat{\beta}_0 + \hat{\beta}_1 x$ be the equation of the fitted line that we seek, where $\hat{\beta}_0$ and $\hat{\beta}_1$ are estimators of their respective parameters and $\hat{Y}$ is clearly an estimator of $E(Y)$.

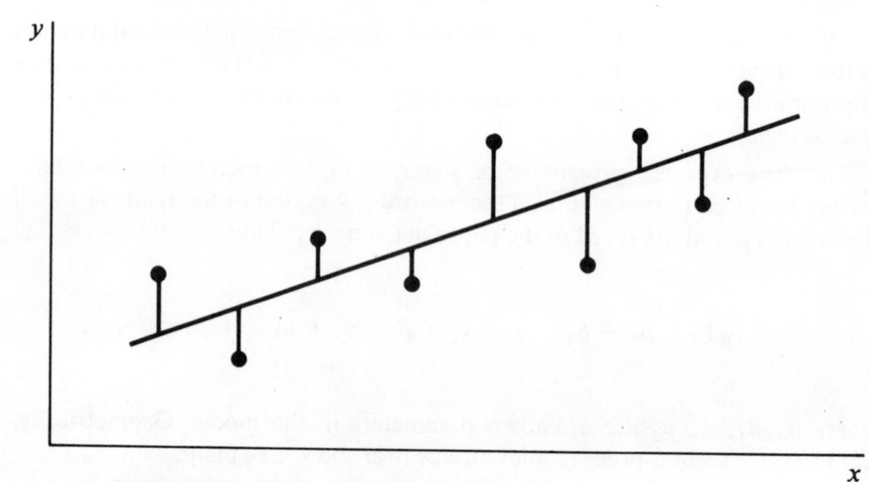

Figure 11.2 Fitting a Straight Line Through a Set of Data Points

The least-squares procedure for fitting a line through a set of n data points is similar to the method we might use if we fit a line by eye; that is, we want the deviations to be "small" in some sense. A convenient way to accomplish this, and one that yields estimators with good properties, is to minimize the sum of squares of the vertical deviations from the fitted line (see the deviations indicated in Figure 11.2). Thus if

$$\hat{y}_i = \hat{\beta}_0 + \hat{\beta}_1 x_i$$

is the predicted value of the ith point (when $x = x_i$), then the deviation of the observed value of y from the $\hat{y}$ line (sometimes called the "error") is

$$y_i - \hat{y}_i$$

and the sum of squares of deviations to be minimized is

$$\text{SSE} = \sum_{i=1}^{n} (y_i - \hat{y}_i)^2 = \sum_{i=1}^{n} [y_i - (\hat{\beta}_0 + \hat{\beta}_1 x_i)]^2.$$

The quantity SSE is also called the *sum of squares for error*, for reasons that will become apparent subsequently.

If SSE possesses a minimum, it will occur for values of β_0 and β_1 that satisfy the equations, $\partial\text{SSE}/\partial\hat{\beta}_0 = 0$ and $\partial\text{SSE}/\partial\hat{\beta}_1 = 0$. Taking the partial derivatives of SSE with respect to $\hat{\beta}_0$ and $\hat{\beta}_1$, and setting them equal to zero, we have

$$\begin{aligned}\frac{\partial\text{SSE}}{\partial\hat{\beta}_0} &= \frac{\partial\left\{\sum_{i=1}^{n}[y_i - (\hat{\beta}_0 + \hat{\beta}_1 x_i)]^2\right\}}{\partial\hat{\beta}_0}\\ &= -\sum_{i=1}^{n} 2[y_i - (\hat{\beta}_0 + \hat{\beta}_1 x_i)]\\ &= -2\left(\sum_{i=1}^{n} y_i - n\hat{\beta}_0 - \hat{\beta}_1\sum_{i=1}^{n} x_i\right) = 0\end{aligned}$$

and

$$\begin{aligned}\frac{\partial\text{SSE}}{\partial\hat{\beta}_1} &= \frac{\partial\left\{\sum_{i=1}^{n}[y_i - (\hat{\beta}_0 + \hat{\beta}_1 x_i)]^2\right\}}{\partial\hat{\beta}_1}\\ &= -\sum_{i=1}^{n} 2[y_i - (\hat{\beta}_0 + \hat{\beta}_1 x_i)]x_i\\ &= -2\left(\sum_{i=1}^{n} x_i y_i - \hat{\beta}_0\sum_{i=1}^{n} x_i - \hat{\beta}_1\sum_{i=1}^{n} x_i^2\right) = 0.\end{aligned}$$

The equations $\partial\text{SSE}/\hat{\beta}_0 = 0$ and $\partial\text{SSE}/\hat{\beta}_1 = 0$ are called the *least-squares equations* for estimating the parameters of a line.

Note that the least-squares equations are linear in $\hat{\beta}_0$ and $\hat{\beta}_1$ and hence are easy to solve simultaneously. You can verify that the solutions are

$$\hat{\beta}_1 = \frac{\sum_{i=1}^{n}(x_i - \bar{x})(y_i - \bar{y})}{\sum_{i=1}^{n}(x_i - \bar{x})^2} = \frac{n\sum_{i=1}^{n} x_i y_i - \sum_{i=1}^{n} x_i \sum_{i=1}^{n} y_i}{n\sum_{i=1}^{n} x_i^2 - \left(\sum_{i=1}^{n} x_i\right)^2}$$

and

$$\hat{\beta}_0 = \bar{y} - \hat{\beta}_1\bar{x}.$$

It can be shown that the simultaneous solution for the two *least-squares equations* yields values of $\hat{\beta}_0$ and $\hat{\beta}_1$ that minimize SSE. We leave this proof to the reader.

We will illustrate the use of the above equations with an example.

Example 11.1: *Use the method of least squares to fit a line to the following $n = 5$ data points:*

x	y
−2	0
−1	0
0	1
1	1
2	3

Solution: *We commence by constructing Table 11.1 to compute the coefficients in the least-squares equations. Then*

Table 11.1

x_i	y_i	$x_i y_i$	x_i^2
−2	0	0	4
−1	0	0	1
0	1	0	0
1	1	1	1
2	3	6	4
$\sum_{i=1}^{n} x_i = 0$	$\sum_{i=1}^{n} y_i = 5$	$\sum_{i=1}^{n} x_i y_i = 7$	$\sum_{i=1}^{n} x_i^2 = 10$

$$\hat{\beta}_1 = \frac{n \sum_{i=1}^{n} x_i y_i - \sum_{i=1}^{n} x_i \sum_{i=1}^{n} y_i}{n \sum_{i=1}^{n} x_i^2 - \left(\sum_{i=1}^{n} x_i\right)^2} = \frac{(5)(7) - (0)(5)}{(5)(10) - (0)^2} = .7,$$

$$\hat{\beta}_0 = \bar{y} - \hat{\beta}_1\bar{x} = \frac{5}{5} - (.7)(0) = 1$$

and the fitted line is

$$\hat{y} = 1 + .7x.$$

The five points and the fitted line are shown in Figure 11.3.

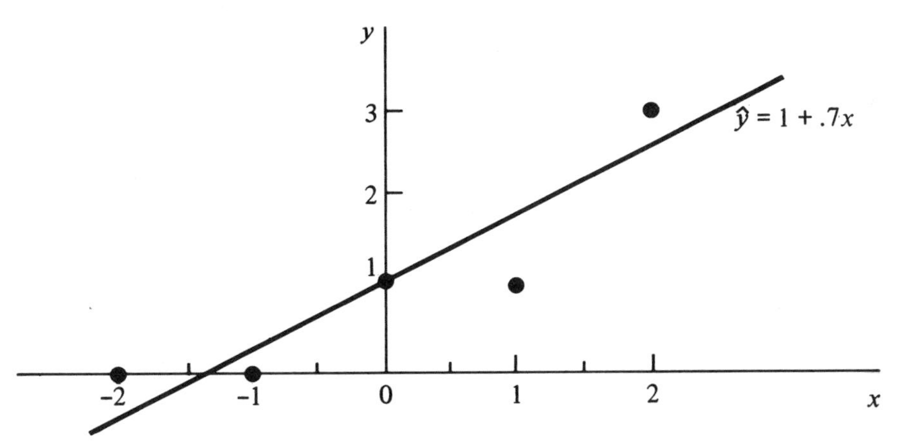

Figure 11.3 Plot of Data Points and Least-Squares Line for Example 11.1

11.4 Fitting the Linear Model Using Matrices

A convenient way to manipulate the linear equations introduced in the previous sections is through matrices. Suppose that we have the linear model

$$Y = \beta_0 + \beta_1 x_1 + \cdots + \beta_k x_k + \varepsilon$$

and we make n independent observations, $y_1, \ldots, y_n$, on Y. We can write the observation y_i as

$$y_i = \beta_0 + \beta_1 x_{i1} + \beta_2 x_{i2} + \cdots + \beta_k x_{ik} + \varepsilon_i,$$

where x_{ij} is the setting of the jth independent variable for the ith observation, $i = 1, \ldots, n$. We now define the following matrices, with $x_0 = 1$:

$$\mathbf{Y} = \begin{bmatrix} y_1 \\ y_2 \\ \vdots \\ y_n \end{bmatrix}, \quad \mathbf{X} = \begin{bmatrix} x_0 & x_{11} & x_{12} & \cdots & x_{1k} \\ x_0 & x_{21} & x_{22} & & x_{2k} \\ \vdots & \vdots & \vdots & & \vdots \\ x_0 & x_{n1} & x_{n2} & & x_{nk} \end{bmatrix},$$

$$\boldsymbol{\beta} = \begin{bmatrix} \beta_0 \\ \beta_1 \\ \vdots \\ \beta_k \end{bmatrix}, \quad \boldsymbol{\varepsilon} = \begin{bmatrix} \varepsilon_1 \\ \varepsilon_2 \\ \vdots \\ \varepsilon_n \end{bmatrix}.$$

Thus the n equations representing y_i as a function of the x's, β's, and ε's can be written simultaneously as

$$\mathbf{Y} = \mathbf{X}\boldsymbol{\beta} + \boldsymbol{\varepsilon}.$$

(See Appendix I for a discussion of matrix operations.)

For n observations from a sample linear model of the form

$$Y = \beta_0 + \beta_1 x + \varepsilon,$$

we have

$$\mathbf{Y} = \begin{bmatrix} y_1 \\ y_2 \\ \vdots \\ y_n \end{bmatrix}, \quad \mathbf{X} = \begin{bmatrix} 1 & x_1 \\ 1 & x_2 \\ \vdots & \vdots \\ 1 & x_n \end{bmatrix},$$

$$\boldsymbol{\varepsilon} = \begin{bmatrix} \varepsilon_1 \\ \varepsilon_2 \\ \vdots \\ \varepsilon_n \end{bmatrix}, \qquad \boldsymbol{\beta} = \begin{bmatrix} \beta_0 \\ \beta_1 \end{bmatrix}.$$

(We suppress the first subscript on x since there is only one x variable involved.) The least-squares equations for β_0 and β_1 were given in Section 11.3 as

$$n\hat{\beta}_0 + \hat{\beta}_1 \sum_{i=1}^{n} x_i = \sum_{i=1}^{n} y_i,$$

$$\hat{\beta}_0 \sum_{i=1}^{n} x_i + \hat{\beta}_1 \sum_{i=1}^{n} x_i^2 = \sum_{i=1}^{n} x_i y_i.$$

Observing that

$$\mathbf{X}'\mathbf{X} = \begin{bmatrix} 1 & 1 & \cdots & 1 \\ x_1 & x_2 & & x_n \end{bmatrix} \begin{bmatrix} 1 & x_1 \\ 1 & x_2 \\ \vdots & \vdots \\ 1 & x_n \end{bmatrix}$$

$$= \begin{bmatrix} n & \sum_{i=1}^{n} x_i \\ \sum_{i=1}^{n} x_i & \sum_{i=1}^{n} x_i^2 \end{bmatrix}$$

and

$$\mathbf{X}'\mathbf{Y} = \begin{bmatrix} \sum_{i=1}^{n} y_i \\ \sum_{i=1}^{n} x_i y_i \end{bmatrix},$$

we see that the least-squares equations are given by

$$(\mathbf{X}'\mathbf{X})\hat{\boldsymbol{\beta}} = \mathbf{X}'\mathbf{Y},$$

where

$$\hat{\boldsymbol{\beta}} = \begin{bmatrix} \hat{\beta}_0 \\ \hat{\beta}_1 \end{bmatrix}.$$

Hence

$$\hat{\boldsymbol{\beta}} = (\mathbf{X}'\mathbf{X})^{-1}\mathbf{X}'\mathbf{Y}.$$

Although we have only shown that this result holds for a simple case, it can be shown that in general the least-squares equations and solutions presented in matrix notation are:

Least-Squares Equations and Solutions for a General Linear Model

1. *Equations*: $(\mathbf{X}'\mathbf{X})\hat{\boldsymbol{\beta}} = \mathbf{X}'\mathbf{Y}$.
2. *Solutions*: $\hat{\boldsymbol{\beta}} = (\mathbf{X}'\mathbf{X})^{-1}\mathbf{X}'\mathbf{Y}$.

Example 11.2: *Solve Example 11.1 using matrix operations.*

Solution: *From the data given in Example 11.1 we see that*

$$\mathbf{Y} = \begin{bmatrix} 0 \\ 0 \\ 1 \\ 1 \\ 3 \end{bmatrix} \quad \textit{and} \quad \mathbf{X} = \begin{matrix} x_0 & x_1 \\ \begin{bmatrix} 1 & -2 \\ 1 & -1 \\ 1 & 0 \\ 1 & 1 \\ 1 & 2 \end{bmatrix} \end{matrix}.$$

It follows that

$$\mathbf{X'X} = \begin{bmatrix} 5 & 0 \\ 0 & 10 \end{bmatrix},$$

$$\mathbf{X'Y} = \begin{bmatrix} 5 \\ 7 \end{bmatrix},$$

and

$$(\mathbf{X'X})^{-1} = \begin{bmatrix} 1/5 & 0 \\ 0 & 1/10 \end{bmatrix}.$$

Thus

$$\hat{\boldsymbol{\beta}} = (\mathbf{X'X})^{-1}\mathbf{X'Y}$$

$$= \begin{bmatrix} 1/5 & 0 \\ 0 & 1/10 \end{bmatrix} \begin{bmatrix} 5 \\ 7 \end{bmatrix} = \begin{bmatrix} 1 \\ .7 \end{bmatrix},$$

or $\hat{\beta}_0 = 1$ *and* $\hat{\beta}_1 = .7$. *Thus*

$$\hat{y} = 1 + .7x,$$

just as in Example 11.1.

Example 11.3: *Fit a parabola to the data of Example 11.1 using the model*

$$Y = \beta_0 + \beta_1 x + \beta_2 x^2 + \varepsilon.$$

Solution: *The* **X** *matrix for this example will differ from that of Example 11.1 only by the addition of a third column corresponding to* x^2. *(Note that* $x_1 = x$, $x_2 = x^2$, *and* $k = 2$ *in the notation of the general linear model.) Thus*

$$\mathbf{Y} = \begin{bmatrix} 0 \\ 0 \\ 1 \\ 1 \\ 3 \end{bmatrix}, \qquad \mathbf{X} = \begin{array}{c} \begin{matrix} x_0 & x & x^2 \end{matrix} \\ \begin{bmatrix} 1 & -2 & 4 \\ 1 & -1 & 1 \\ 1 & 0 & 0 \\ 1 & 1 & 1 \\ 1 & 2 & 4 \end{bmatrix} \end{array}.$$

The three variables, x_0, x, and x^2, are shown above their respective columns in the **X** *matrix. Thus for the first measurement, $y = 0$, $x_0 = 1$, $x = -2$, and $x^2 = 4$. For the second measurement, $y = 0$, $x_0 = 1$, $x = -1$, and $x^2 = 1$. Succeeding rows of the* **Y** *and* **X** *matrices are obtained in a similar manner.*

The matrix products, **X′X** *and* **X′Y**, *are*

$$(\mathbf{X'X}) = \begin{bmatrix} 1 & 1 & 1 & 1 & 1 \\ -2 & -1 & 0 & 1 & 2 \\ 4 & 1 & 0 & 1 & 4 \end{bmatrix} \begin{bmatrix} 1 & -2 & 4 \\ 1 & -1 & 1 \\ 1 & 0 & 0 \\ 1 & 1 & 1 \\ 1 & 2 & 4 \end{bmatrix}$$

$$= \begin{bmatrix} 5 & 0 & 10 \\ 0 & 10 & 0 \\ 10 & 0 & 34 \end{bmatrix},$$

$$(\mathbf{X'Y}) = \begin{bmatrix} 1 & 1 & 1 & 1 & 1 \\ -2 & -1 & 0 & 1 & 2 \\ 4 & 1 & 0 & 1 & 4 \end{bmatrix} \begin{bmatrix} 0 \\ 0 \\ 1 \\ 1 \\ 3 \end{bmatrix} = \begin{bmatrix} 5 \\ 7 \\ 13 \end{bmatrix}.$$

We omit the process of inverting **(X′X)** *and simply state that it is equal to*

$$(\mathbf{X'X})^{-1} = \begin{bmatrix} 17/35 & 0 & -1/7 \\ 0 & 1/10 & 0 \\ -1/7 & 0 & 1/14 \end{bmatrix}.$$

The reader may verify that $(\mathbf{X'X})^{-1}(\mathbf{X'X}) = \mathbf{I}$.

Finally,

$$\hat{\boldsymbol{\beta}} = (\mathbf{X'X})^{-1}\mathbf{X'Y}$$

$$= \begin{bmatrix} 17/35 & 0 & -1/7 \\ 0 & 1/10 & 0 \\ -1/7 & 0 & 1/14 \end{bmatrix} \begin{bmatrix} 5 \\ 7 \\ 13 \end{bmatrix} = \begin{bmatrix} 4/7 \\ 7/10 \\ 3/14 \end{bmatrix} \approx \begin{bmatrix} .57 \\ .7 \\ .214 \end{bmatrix}.$$

Hence $\hat{\beta}_0 = .57$, $\hat{\beta}_1 = .7$, *and* $\hat{\beta}_2 = .214$, *and the prediction equation is*

$$\hat{y} = .57 + .7x + .214x^2.$$

A graph of this parabola on Figure 11.3 will indicate a good fit to the data points.

11.5 Properties of the Least-Squares Estimators for the Model $Y = \beta_0 + \beta_1 x + \varepsilon$

We will now derive means, variances, and probability distributions for least-squares estimators of the parameters of the simple linear model of Section 11.4,

$$Y = \beta_0 + \beta_1 x + \varepsilon.$$

Results applicable to the general linear model will be presented without proof in Section 11.6.

Recall that ε was previously assumed to be a random variable with $E(\varepsilon) = 0$. We now add the assumption that ε is normally distributed with $V(\varepsilon) = \sigma^2$. That is, we are assuming that the difference between the random variable Y and $E(Y) = \beta_0 + \beta_1 x$ is normally distributed about zero, with a variance that does not depend on x. Note that $V(Y) = V(\varepsilon) = \sigma^2$, since the other terms in the linear model are constants.

Assume that n independent observations are to be made on the above model so that, before sampling, we have n independent random variables of the form

$$Y_i = \beta_0 + \beta_1 x_i + \varepsilon_i.$$

From Section 11.3 we know that

$$\hat{\beta}_1 = \frac{\sum_{i=1}^{n} (x_i - \bar{x})(Y_i - \bar{Y})}{\sum_{i=1}^{n} (x_i - \bar{x})^2},$$

which can be written

$$\hat{\beta}_1 = \frac{\sum_{i=1}^{n} (x_i - \bar{x})Y_i - \bar{Y} \sum_{i=1}^{n} (x_i - \bar{x})}{\sum_{i=1}^{n} (x_i - \bar{x})^2}.$$

Then, noting that $\sum_{i=1}^{n} (x_i - \bar{x}) = 0$, we have

$$\hat{\beta}_1 = \frac{\sum_{i=1}^{n} (x_i - \bar{x})Y_i}{\sum_{i=1}^{n} (x_i - \bar{x})^2}.$$

Because all summations in the following discussion will be summed from $i = 1$ to n, we will simplify our notation by omitting the variable of summation and its index. Now let us find the expected value and variance of $\hat{\beta}_1$.

From the expectation theorems, Section 5.7, we have

$$E(\hat{\beta}_1) = E\left[\frac{\sum (x_i - \bar{x})Y_i}{\sum (x_i - \bar{x})^2}\right]$$

$$= \frac{\sum (x_i - \bar{x})E(Y_i)}{\sum (x_i - \bar{x})^2}$$

$$= \frac{\sum (x_i - \bar{x})(\beta_0 + \beta_1 x_i)}{\sum (x_i - \bar{x})^2}$$

$$= \beta_0 \frac{\sum (x_i - \bar{x})}{\sum (x_i - \bar{x})^2} + \beta_1 \frac{\sum (x_i - \bar{x})x_i}{\sum (x_i - \bar{x})^2}.$$

Then, since $\sum (x_i - \bar{x}) = 0$ and $\sum (x_i - \bar{x})^2 = \sum (x_i - \bar{x})x_i$,

$$E(\hat{\beta}_1) = 0 + \beta_1 \frac{\sum (x_i - \bar{x})^2}{\sum (x_i - \bar{x})^2}$$

$$= \beta_1,$$

which shows that $\hat{\beta}_1$ is an unbiased estimator of β_1.

To find $V(\hat{\beta}_1)$, we use Theorem 5.8. Then

$$V(\hat{\beta}_1) = V\left[\frac{\sum (x_i - \bar{x})Y_i}{\sum (x_i - \bar{x})^2}\right]$$

$$= \left[\frac{1}{\sum (x_i - \bar{x})^2}\right]^2 \sum V[(x_i - \bar{x})Y_i]$$

$$= \left[\frac{1}{\sum (x_i - \bar{x})^2}\right]^2 \sum (x_i - \bar{x})^2 V(Y_i).$$

Since $V(Y_i) = \sigma^2$, $i = 1, 2, \ldots, n$,

$$V(\hat{\beta}_1) = \frac{\sigma^2}{\sum (x_i - \bar{x})^2}.$$

Now let us find the expected value and variance of $\hat{\beta}_0$, where $\hat{\beta}_0 = \bar{Y} - \hat{\beta}_1\bar{x}$. From Theorem 5.8,

$$V(\hat{\beta}_0) = V(\bar{Y}) + \bar{x}^2 V(\hat{\beta}_1) - 2\bar{x}\,\text{Cov}(\bar{Y}, \hat{\beta}_1).$$

Consequently, we must find $V(\bar{Y})$ and $\text{Cov}(\bar{Y}, \hat{\beta}_1)$ in order to obtain $V(\hat{\beta}_0)$.

Since $Y_i = \beta_0 + \beta_1 x_i + \varepsilon_i$, we see that

$$\bar{Y} = \frac{1}{n}\sum Y_i = \beta_0 + \beta_1\bar{x} + \bar{\varepsilon}.$$

Thus

$$E(\bar{Y}) = \beta_0 + \beta_1\bar{x} + E(\bar{\varepsilon}) = \beta_0 + \beta_1\bar{x}$$

and

$$V(\bar{Y}) = V(\bar{\varepsilon}) = \frac{1}{n}V(\varepsilon_i) = \frac{\sigma^2}{n}.$$

To find $\text{Cov}(\overline{Y}, \hat{\beta}_1)$, rewrite the expression for $\hat{\beta}_1$ as

$$\hat{\beta}_1 = \sum c_i Y_i,$$

where

$$c_i = \frac{x_i - \bar{x}}{\sum (x_i - \bar{x})^2}.$$

Note that $\sum c_i = 0$.

Now, we have

$$\text{Cov}(\overline{Y}, \hat{\beta}_1) = \text{Cov}\left(\sum \frac{1}{n} Y_i, \sum c_i Y_i\right)$$

and, using Theorem 5.8, this becomes

$$\text{Cov}(\overline{Y}, \hat{\beta}_1) = \sum \frac{c_i}{n} V(Y_i) + 2 \sum_{i<j}\sum \frac{c_j}{n} \text{Cov}(Y_i, Y_j).$$

Since Y_i and Y_j, $i \neq j$, are independent, $\text{Cov}(Y_i, Y_j) = 0$. Also $V(Y_i) = \sigma^2$, and hence

$$\text{Cov}(\overline{Y}, \hat{\beta}_1) = \frac{\sigma^2}{n} \sum c_i = 0.$$

Returning to our original task of finding the expected value and variance of

$$\hat{\beta}_0 = \overline{Y} - \hat{\beta}_1 \bar{x},$$

we apply expectation theorems to obtain

$$\begin{aligned} E(\hat{\beta}_0) &= E(\overline{Y}) - E(\hat{\beta}_1)\bar{x} \\ &= \beta_0 + \beta_1\bar{x} - \beta_1\bar{x} = \beta_0. \end{aligned}$$

We have shown that both $\hat{\beta}_0$ and $\hat{\beta}_1$ are unbiased estimators of their respective parameters.

Since we now have $V(\overline{Y})$, $V(\hat{\beta}_1)$, and $\text{Cov}(\overline{Y}, \hat{\beta}_1)$, we are ready to find $V(\hat{\beta}_0)$. Thus

$$V(\hat{\beta}_0) = V(\overline{Y}) + \bar{x}^2 V(\hat{\beta}_1) - 2\bar{x}\,\text{Cov}(\overline{Y}, \hat{\beta}_1).$$

Substituting into the expression for $V(\overline{Y})$, $V(\hat{\beta}_1)$, and $\text{Cov}(\overline{Y}, \hat{\beta}_1)$, we obtain

$$\begin{aligned} V(\hat{\beta}_0) &= \frac{\sigma^2}{n} + \bar{x}^2 \frac{\sigma^2}{\sum (x_i - \bar{x})^2} + 0 \\ &= \sigma^2\left[\frac{1}{n} + \frac{\bar{x}^2}{\sum (x_i - \bar{x})^2}\right] = \frac{\sigma^2 \sum x_i^2}{n \sum (x_i - \bar{x})^2}. \end{aligned}$$

Now let us summarize the results of this tedious algebra. We have shown that $\hat{\beta}_0$ and $\hat{\beta}_1$ are unbiased estimators of β_0 and β_1, respectively, and have derived their variances. We will now find their probability distributions. If ε_i is normal, then Y_i must be normal, $i = 1, \ldots, n$. Since $\hat{\beta}_0$ and $\hat{\beta}_1$ are each linear functions of $Y_1, \ldots, Y_n$, it follows from Theorem 6.3 that they will have normal distributions.

Note, however, that $\hat{\beta}_0$ and $\hat{\beta}_1$ are not independent except in the case $\bar{x} = 0$. We leave this as an exercise for the reader.

We saw in Section 11.4 that $\mathbf{X}'\mathbf{X}$ for the linear model $Y = \beta_0 + \beta_1 x + \varepsilon$ is given by

$$\mathbf{X}'\mathbf{X} = \begin{bmatrix} n & \sum x_i \\ \sum x_i & \sum x_i^2 \end{bmatrix}.$$

It can easily be shown that

$$(\mathbf{X}'\mathbf{X})^{-1} = \begin{bmatrix} \dfrac{\sum x_i^2}{n\sum (x_i - \bar{x})^2} & -\dfrac{\sum x_i}{n\sum (x_i - \bar{x})^2} \\ -\dfrac{\sum x_i}{n\sum (x_i - \bar{x})^2} & \dfrac{1}{\sum (x_i - \bar{x})^2} \end{bmatrix}$$

$$= \begin{bmatrix} c_{00} & c_{01} \\ c_{10} & c_{11} \end{bmatrix}.$$

By checking the variances and covariances derived above, you can see that

$$V(\hat{\beta}_i) = c_{ii}\sigma^2, \qquad i = 0, 1,$$

and

$$\text{Cov}(\hat{\beta}_0, \hat{\beta}_1) = c_{01}\sigma^2 = c_{10}\sigma^2.$$

Example 11.4: *Find the variances of the estimators $\hat{\beta}_0$ and $\hat{\beta}_1$ for Example 11.2.*

Solution: *In Example 11.2,*

$$(\mathbf{X}'\mathbf{X})^{-1} = \begin{bmatrix} 1/5 & 0 \\ 0 & 1/10 \end{bmatrix}.$$

Hence

$$V(\hat{\beta}_0) = c_{00}\sigma^2 = (1/5)\sigma^2$$

and

$$V(\hat{\beta}_1) = c_{11}\sigma^2 = (1/10)\sigma^2.$$

Note that the $\hat{\beta}_0$ and $\hat{\beta}_1$ are independent for this problem, since $\sum x_i = 0$. (Recall that two normally distributed random variables are independent if their covariance equals zero.)

The above expressions give variances for the least-squares estimators in terms of σ^2, but, in order for these expressions to be of practical value, σ^2 must be estimated. We will now find the maximum likelihood estimator of σ^2.

The random variables $Y_1, \ldots, Y_n$ are independent, and normally distributed with

$$\mu_i = E(Y_i) = \beta_0 + \beta_1 x_i \qquad \text{and} \qquad V(Y_i) = \sigma^2.$$

Thus, the likelihood of the sample is given by

$$\begin{aligned} L &= f(y_1)f(y_2)\cdots f(y_n) \\ &= \frac{1}{\sigma\sqrt{2\pi}} \exp\left[\frac{-(y_1 - \mu_1)^2}{2\sigma^2}\right] \cdots \frac{1}{\sigma\sqrt{2\pi}} \exp\left[\frac{-(y_n - \mu_n)^2}{2\sigma^2}\right] \\ &= \left(\frac{1}{\sigma\sqrt{2\pi}}\right)^n \exp\left[\frac{-\sum (y_i - \mu_i)^2}{2\sigma^2}\right]. \end{aligned}$$

We observe that maximizing L with respect to β_0 and β_1 is the same as minimizing

$$\sum (y_i - \mu_i)^2 = \sum [y_i - (\beta_0 + \beta_1 x_i)]^2$$

with respect to β_0 and β_1. But the least-squares procedure minimizes the same quantity, and therefore the least-squares estimators, $\hat{\beta}_0$ and $\hat{\beta}_1$, are the same as the maximum likelihood estimators.

The procedure for finding the maximum likelihood estimator of σ^2 is just as explained in Example 9.4, and the estimator is

$$\hat{\sigma}^2 = \frac{1}{n} \sum (Y_i - \hat{Y}_i)^2,$$

where

$$\hat{Y}_i = \hat{\beta}_0 + \hat{\beta}_1 x_i, \qquad i = 1, \ldots, n.$$

(Note that $\hat{Y}_i$ is the maximum likelihood estimator of μ_i.)

The estimator $\hat{\sigma}^2$ is not quite an unbiased estimator of σ^2, but it can easily be adjusted to be so. In order to find the necessary adjustment we find $E(\hat{\sigma}^2)$. Now

$$
\begin{aligned}
E\left[\sum (Y_i - \hat{Y}_i)^2\right] &= E\left[\sum (Y_i - \hat{\beta}_0 - \hat{\beta}_1 x_i)^2\right] \\
&= E\left[\sum (Y_i - \overline{Y} + \hat{\beta}_1 \bar{x} - \hat{\beta}_1 x_i)^2\right] \\
&= E\left[\sum [(Y_i - \overline{Y}) - \hat{\beta}_1 (x_i - \bar{x})]^2\right] \\
&= E\Big[\sum (Y_i - \overline{Y})^2 + \hat{\beta}_1^2 \sum (x_i - \bar{x})^2 \\
&\qquad - 2\hat{\beta}_1 \sum (x_i - \bar{x})(Y_i - \overline{Y})\Big].
\end{aligned}
$$

Observe that $\sum (x_i - \bar{x})(Y_i - \overline{Y}) = \sum (x_i - \bar{x})^2 \hat{\beta}_1$, and hence the last two terms in the expectation collapse to $-\hat{\beta}_1^2 \sum (x_i - \bar{x})^2$. Also, recall that

$$\sum (Y_i - \overline{Y})^2 = \sum Y_i^2 - n\overline{Y}^2.$$

Thus

$$
\begin{aligned}
E\left[\sum (Y_i - \hat{Y}_i)^2\right] &= E\left[\sum Y_i^2 - n\overline{Y}^2 - \hat{\beta}_1^2 \sum (x_i - \bar{x})^2\right] \\
&= \sum E(Y_i^2) - nE(\overline{Y}^2) - \sum (x_i - \bar{x})^2 E(\hat{\beta}_1^2).
\end{aligned}
$$

Noting that for any random variable $U, E(U^2) = V(U) + [E(U)]^2$, we see that

$$
\begin{aligned}
E\left[\sum (Y_i - \hat{Y}_i)^2\right] &= \sum \{V(Y_i) + [E(Y_i)]^2\} \\
&\quad - n\{V(\overline{Y}) + [E(\overline{Y})]^2\} \\
&\quad - \sum (x_i - \bar{x})^2 \{V(\hat{\beta}_1) + [E(\hat{\beta}_1)]^2\} \\
&= n\sigma^2 + \sum (\beta_0 + \beta_1 x_i)^2
\end{aligned}
$$

$$- n\left[\frac{\sigma^2}{n} + (\beta_0 + \beta_1\bar{x})^2\right]$$

$$- \sum (x_i - \bar{x})^2\left[\frac{\sigma^2}{\sum (x_i - \bar{x})^2} + \beta_1^2\right].$$

This expression simplifies to $(n - 2)\sigma^2$. Thus we find an unbiased estimator of σ^2 is given by

$$S^2 = \frac{1}{n-2}\sum (Y_i - \hat{Y}_i)^2 = \frac{1}{n-2}\text{SSE}.$$

Note that the 2 occurring in the denominator of S^2 corresponds to the number of β's to be estimated in the model.

One task remains, finding an easy way to calculate $\sum (y_i - \hat{y}_i)^2 = \text{SSE}$. A bit of matrix algebra will show that

$$\text{SSE} = \mathbf{Y}'\mathbf{Y} - \hat{\boldsymbol{\beta}}'\mathbf{X}'\mathbf{Y}.$$

Example 11.5: *Estimate σ^2 from the data given in Example 11.1.*

Solution: *For this data*

$$\mathbf{Y} = \begin{bmatrix} 0 \\ 0 \\ 1 \\ 1 \\ 3 \end{bmatrix}, \quad \mathbf{X} = \begin{bmatrix} 1 & -2 \\ 1 & -1 \\ 1 & 0 \\ 1 & 1 \\ 1 & 2 \end{bmatrix}, \quad \hat{\boldsymbol{\beta}} = \begin{bmatrix} 1 \\ .7 \end{bmatrix}.$$

Hence

$$\text{SSE} = \mathbf{Y}'\mathbf{Y} - \hat{\boldsymbol{\beta}}'\mathbf{X}'\mathbf{Y}$$

$$= [0 \quad 0 \quad 1 \quad 1 \quad 3] \begin{bmatrix} 0 \\ 0 \\ 1 \\ 1 \\ 3 \end{bmatrix}$$

$$- [1 \quad .7] \begin{bmatrix} 1 & 1 & 1 & 1 & 1 \\ -2 & -1 & 0 & 1 & 2 \end{bmatrix} \begin{bmatrix} 0 \\ 0 \\ 1 \\ 1 \\ 3 \end{bmatrix}$$

$$= 11 - [1 \quad .7] \begin{bmatrix} 5 \\ 7 \end{bmatrix}$$

$$= 11 - 9.9 = 1.1.$$

Then

$$s^2 = \frac{\text{SSE}}{n - 2} = \frac{1.1}{5 - 2} = \frac{1.1}{3} = .367.$$

11.6 Properties of the Least-Squares Estimators for the General Linear Model

All the theoretical results of the previous section can be extended to the general linear model,

$$Y = \beta_0 + \beta_1 x_1 + \cdots + \beta_k x_k + \varepsilon.$$

Suppose that $\varepsilon_1, \ldots, \varepsilon_n$ are independent, normally distributed random variables with $E(\varepsilon_i) = 0$ and $V(\varepsilon_i) = \sigma^2$. Then the least-squares estimators are given by

$$\hat{\boldsymbol{\beta}} = (\mathbf{X}'\mathbf{X})^{-1}\mathbf{X}'\mathbf{Y},$$

provided that $(\mathbf{X}'\mathbf{X})^{-1}$ exists. The properties of these estimators are as follows:

Properties of the Least-Squares Estimators

1. $E(\hat{\beta}_i) = \beta_i, i = 0, 1, \ldots, k$.
2. $V(\hat{\beta}_i) = c_{ii}\sigma^2$, *where* c_{ij} *is the element in row i and column j of* $(\mathbf{X}'\mathbf{X})^{-1}$. *(Recall that this matrix has a row and column numbered* 0.)
3. $\text{Cov}(\hat{\beta}_i, \hat{\beta}_j) = c_{ij}\sigma^2$.
4. *Each* $\hat{\beta}_i$ *is normally distributed.*
5. *An unbiased estimator of* σ^2 *is* $S^2 = \text{SSE}/[n - (k + 1)]$, *where* $\text{SSE} = \mathbf{Y}'\mathbf{Y} - \hat{\boldsymbol{\beta}}'\mathbf{X}'\mathbf{Y}$. *(Note that there are* $k + 1$ *unknown* β_i*'s in the model.)*
6. *The random variable*

$$\frac{[n - (k + 1)]S^2}{\sigma^2}$$

has a χ^2 *distribution with* $n - (k + 1)$ *degrees of freedom. Furthermore,* S^2 *and* $\hat{\beta}_i, i = 0, 1, \ldots, k$, *are independent.*

11.7 Inferences Concerning the Parameters β_i

Under the assumption that the random error, ε, is a normally distributed random variable, we have established that $\hat{\beta}_i$ is an unbiased, normally distributed estimator of β_i with $V(\hat{\beta}_i) = c_{ii}\sigma^2$. Using this information, we can construct a large-sample test of the hypothesis $H_0: \beta_i = \beta_{i0}$ (β_{i0} a specified value of β_i), using the test statistic

$$Z = \frac{\hat{\beta}_i - \beta_{i0}}{\sigma\sqrt{c_{ii}}}.$$

The rejection region for a two-tailed test would be

$$|z| \geq z_{\alpha/2}.$$

As in the case of the simple normal-deviate tests studied in Chapter 10 one must either know σ or possess a good estimate based upon an adequate number of degrees of freedom. (What would be adequate is a debatable point. We suggest that the estimate be based upon 30 or more degrees of freedom.) When this estimate is unavailable (which is usually the case), an estimate of σ may be calculated from the experimental data (in accordance with the procedure of Section 11.5) and substituted for σ in the z statistic. The resulting quantity,

$$T = \frac{\hat{\beta}_i - \beta_i}{S\sqrt{c_{ii}}},$$

can be shown to possess a Student's t distribution with $[n - (k + 1)]$ degrees of freedom. Thus a test of a hypothesis concerning a parameter, β_i, reverts to the familiar t test of Chapter 10.

A Test for β_i

$H_0: \beta_i = \beta_{i0}$.

$H_a: \beta_i \neq \beta_{i0}$.

Test Statistic: $T = \dfrac{\hat{\beta}_i - \beta_{i0}}{S\sqrt{c_{ii}}}$.

Rejection Region: $|t| \geq t_{\alpha/2}$ *for specified* α, *where* $t_{\alpha/2}$ *is based upon* $[n - (k + 1)]$ *degrees of freedom.*

Example 11.6: *Do the data of Example 11.3 present sufficient evidence to indicate curvature in the response function?*

Solution: *The verbal question stated above assumes that the probabilistic model is a realistic description of the true response and implies a test of the hypothesis* $\beta_2 = 0$ *in the linear model* $Y = \beta_0 + \beta_1 x + \beta_2 x^2 + \varepsilon$. *(If* $\beta_2 = 0$, *the quadratic term will not appear and the expected value of* Y *will represent a straight-line function of* x.*) The first step in the solution is the calculation of* SSE *and* s^2.

$$\text{SSE} = \mathbf{Y}'\mathbf{Y} - \hat{\boldsymbol{\beta}}'\mathbf{X}'\mathbf{Y}$$

$$= 11 - [4/7 \quad 7/10 \quad 3/14]\begin{bmatrix} 5 \\ 7 \\ 13 \end{bmatrix} = .457.$$

Then

$$s^2 = \frac{\text{SSE}}{n-3} = \frac{.457}{2} = .229 \qquad \textit{and} \qquad s = .48.$$

[Note: *The model contains three parameters and hence* SSE *is based upon* $n - 3 = 2$ *degrees of freedom.*]

The estimate of β_2 *obtained from Example 11.3 was* $\hat{\beta}_2 = 3/14 \approx .214$. *Then*

$$t = \frac{\hat{\beta}_2 - 0}{s\sqrt{c_{22}}} = \frac{.214}{.48\sqrt{1/14}} = 1.67.$$

If we take $\alpha = .05$, *the value of* $t_{\alpha/2} = t_{.025}$ *for two degrees of freedom* is 4.303, *and the rejection region would be*

$$\textit{reject if } |t| \geq 4.303.$$

Since the calculated value of t is less than 4.303, *we cannot reject the null hypothesis that* $\beta_2 = 0$. *Note that we* do not accept $H_0: \beta_2 = 0$. *We would have to know the probability of making a type II error—that is, the probability of falsely accepting* H_0 *for a specified alternative value of* β_2*—before we would accept. Or we could look at the width of the confidence interval for* β_2 *to see whether it is sufficiently small to detect a departure from zero that would be of practical significance. We will show that the confidence interval for* β_2 *is quite large, suggesting that the experimenter collect more data before reaching a decision.*

Based on the t statistic given above, we can follow the procedures of Chapter 10 to show that a confidence interval for β_i, with confidence coefficient $1 - \alpha$, is given by

A (1 − α) Confidence Interval for β

$$\hat{\beta}_i \pm t_{\alpha/2} s\sqrt{c_{ii}}.$$

Example 11.7: *Calculate a 95 percent confidence interval for the parameter* β_2 *of Example 11.6.*

Solution: *The tabulated value for* $t_{.025}$, *based upon two degrees of freedom, is* 4.303.

Then the 95 percent confidence interval for β_2 *is*

$$\hat{\beta}_2 \pm t_{.025} s \sqrt{c_{22}}.$$

Substituting, we get

$$.214 \pm (4.303)(.48)\sqrt{\frac{1}{14}} \qquad or \qquad .214 \pm .552.$$

For example, if we wish to estimate β_2 *correct to within .15 unit, it is obvious that the confidence interval is too wide and that the sample size must be increased.*

11.8 Inferences Concerning Linear Functions of the Model Parameters

In addition to inferences about a single β_i, we are frequently interested in making inferences concerning linear functions of the model parameters, $\beta_0, \beta_1, \ldots, \beta_k$. For example, we might wish to estimate $E(Y)$, given by

$$E(Y) = \beta_0 + \beta_1 x_1 + \cdots + \beta_k x_k,$$

where $E(Y)$ might represent the mean yield of a chemical process for settings of controlled process variables, $x_1, x_2, \ldots, x_k$, or the mean profit of a corporation for various investment expenditures, $x_1, x_2, \ldots, x_k$. Properties of estimators of such linear functions will be given in this section.

Suppose that we wish to make an inference about the linear function

$$a_0\beta_0 + a_1\beta_1 + a_2\beta_2 + \cdots + a_k\beta_k,$$

where $a_0, a_1, a_2, \ldots, a_k$ are constants (some of which may equal zero). Then it is easy to see that the same linear function of the parameter estimators,

$$U = a_0\hat{\beta}_0 + a_1\hat{\beta}_1 + a_2\hat{\beta}_2 + \cdots + a_k\hat{\beta}_k,$$

possesses an expected value (by Theorem 5.8),

$$\begin{aligned} E(U) &= a_0E(\hat{\beta}_0) + a_1E(\hat{\beta}_1) + \cdots + a_kE(\hat{\beta}_k) \\ &= a_0\beta_0 + a_1\beta_1 + a_2\beta_2 + \cdots + a_k\beta_k, \end{aligned}$$

and is therefore an unbiased estimator of

$$a_0\beta_0 + a_1\beta_1 + a_2\beta_2 + \cdots + a_k\beta_k.$$

Applying the same theorem, we find the variance of U,

$$\begin{aligned} V(U) = {} & a_0^2V(\hat{\beta}_0) + a_1^2V(\hat{\beta}_1) + a_2^2V(\hat{\beta}_2) + \cdots + a_k^2V(\hat{\beta}_k) \\ & + 2a_0a_1\operatorname{Cov}(\hat{\beta}_0, \hat{\beta}_1) + 2a_0a_2\operatorname{Cov}(\hat{\beta}_0, \hat{\beta}_2) \\ & + \cdots + 2a_1a_2\operatorname{Cov}(\hat{\beta}_1, \hat{\beta}_2) \\ & + \cdots + 2a_{k-1}a_k\operatorname{Cov}(\hat{\beta}_{k-1}, \hat{\beta}_k), \end{aligned}$$

where $V(\hat{\beta}_i) = c_{ii}\sigma^2$ and $\operatorname{Cov}(\hat{\beta}_i, \hat{\beta}_j) = c_{ij}\sigma^2$. Defining the $(k + 1) \times 1$ matrix,

$$\mathbf{a} = \begin{bmatrix} a_0 \\ a_1 \\ a_2 \\ \vdots \\ a_k \end{bmatrix},$$

the reader may verify that $V(U)$ is equal to

$$V(U) = [\mathbf{a}'(\mathbf{X}'\mathbf{X})^{-1}\mathbf{a}]\sigma^2.$$

Finally, when we note that $\hat{\beta}_0, \hat{\beta}_1, \hat{\beta}_2, \ldots, \hat{\beta}_k$ will be normally distributed in repeated sampling (Section 11.5), it is clear that U is a linear function of normally distributed random variables, and hence will itself be normally distributed in repeated sampling.

Knowing that U is normally distributed with

$$E(U) = \beta_0 a + \beta_1 a_1 + \beta_2 a_2 + \cdots + \beta_k a_k$$

and $V(U) = [\mathbf{a}'(\mathbf{X}'\mathbf{X})^{-1}\mathbf{a}]\sigma^2$, we conclude that

$$Z = \frac{U - E(U)}{\sqrt{V(U)}} = \frac{U - E(U)}{\sigma\sqrt{\mathbf{a}'(\mathbf{X}'\mathbf{X})^{-1}\mathbf{a}}}$$

is a standard normal deviate that could be employed to test a hypothesis,

$$H_0\colon E(U) = E_0(U),$$

when $E_0(U)$ is some specified value. Likewise, a large-sample $1 - \alpha$ confidence interval for $E(U)$ would be

$$u \pm z_{\alpha/2}\sigma\sqrt{\mathbf{a}'(\mathbf{X}'\mathbf{X})^{-1}\mathbf{a}}.$$

Furthermore, as we might suspect, if one substitutes S for σ, the quantity

$$T = \frac{U - E(U)}{S\sqrt{\mathbf{a}'(\mathbf{X}'\mathbf{X})^{-1}\mathbf{a}}}$$

possesses a Student's t distribution in repeated sampling with $[n - (k + 1)]$ degrees of freedom and hence provides a test statistic to test the hypothesis

$$H_0: E(U) = E_0(U).$$

For a two-tailed test and a given α, the null hypothesis would be rejected if $|t| \geq t_{\alpha/2}$.

A Test for E(U)

$H_0: E(U) = E_0(U)$.

$H_a: E(U) \neq E_0(u)$.

Test Statistic: $T = \dfrac{U - E_0(U)}{S\sqrt{\mathbf{a}'(\mathbf{X}'\mathbf{X})^{-1}\mathbf{a}}}$.

Rejection Region: $|t| \geq t_{\alpha/2}$, *for specified* α, *where* $t_{\alpha/2}$ *is based upon* $[n - (k + 1)]$ *degrees of freedom.*

The corresponding $(1 - \alpha)$ confidence interval for $E(U)$ is

A (1 − α) Confidence Interval for E(U)

$$u \pm t_{\alpha/2} s\sqrt{\mathbf{a}'(\mathbf{X}'\mathbf{X})^{-1}\mathbf{a}}.$$

In this formula the tabulated value of t, $t_{\alpha/2}$, is based upon $[n - (k + 1)]$ degrees of freedom.

Example 11.8: *For the data of Example 11.1, find a* 90 *percent confidence interval for* $E(Y)$ *when* $x = 1$.

Solution: *For the model of Example 11.1,*

$$E(Y) = \beta_0 + \beta_1 x.$$

To estimate E(Y) we use the unbiased estimator, $\hat{Y}$. Then,

$$u = \hat{y} = \hat{\beta}_0 + \hat{\beta}_1 x = 1 + .7x.$$

Hence

$$\mathbf{a} = \begin{bmatrix} a_0 \\ a_1 \end{bmatrix} = \begin{bmatrix} 1 \\ x \end{bmatrix},$$

and for the case x = 1,

$$\mathbf{a} = \begin{bmatrix} 1 \\ 1 \end{bmatrix}.$$

Now,

$$\begin{aligned} V(\hat{Y}) &= \mathbf{a}'(\mathbf{X}'\mathbf{X})^{-1}\mathbf{a}\sigma^2 \\ &= [1 \quad 1]\begin{bmatrix} 1/5 & 0 \\ 0 & 1/10 \end{bmatrix}\begin{bmatrix} 1 \\ 1 \end{bmatrix}\sigma^2 \\ &= .3\sigma^2. \end{aligned}$$

In Example 11.5 we found s^2 to be .367 or $s = .606$ for this data. The value of $t_{.05}$ with $n - 2 = 3$ degrees of freedom is 2.353.

The confidence interval for E(Y) is then

$$\hat{y} \pm t_{\alpha/2} s \sqrt{\mathbf{a}'(\mathbf{X}'\mathbf{X})^{-1}\mathbf{a}}$$

$$[1 + (.7)(1)] \pm (2.353)(.606)\sqrt{.3}$$

or

$$1.7 \pm .793.$$

11.9 Predicting a Particular Value of Y

Suppose that the yield, Y, for a chemical experiment is a function of the temperature, x_1, and the pressure, x_2, at which the experiment is run. Assume that a linear model of the form

$$Y = \beta_0 + \beta_1 x_1 + \beta_2 x_2 + \varepsilon$$

is an adequate representation of the response surface traced by Y over the experimental region of interest. In Section 11.8 we discussed methods for estimating $E(Y)$ for a given temperature, say x_{10}, and pressure, say x_{20}. That is, we now know how to estimate the mean yield of the process, $E(Y)$, at the settings x_{10} and x_{20}.

Now consider a different problem. Instead of estimating the mean yield at (x_{10}, x_{20}), we wish to *predict* the particular response, Y, that we will observe if the experiment is run at some time in the future (like next Monday). This situation would occur if, for some reason, the response next Monday held a special significance to us. Prediction problems frequently occur in business in which we may be interested in a particular gain associated with the investment we intend to make next month, rather than the mean gain over a long series of investments.

Note that Y is a random variable, not a parameter, and predicting its value therefore presents a departure from the stated objective of making inferences concerning population parameters. Yet, if the distribution of a random variable, Y, is known and a single value of Y is selected at random from the population, what would you predict for the observed value? We contend that one would select a value of Y near the *center* of the distribution, in particular, the expected value of Y. We could employ $\hat{Y}$ as a predictor of a particular value of Y as well as of $E(Y)$.

The error of predicting a particular value of Y, using $\hat{Y}$ as the predictor, will be the difference between the observed value of Y and the predicted:

$$\text{error} = Y - \hat{Y}.$$

Let us now investigate the properties of this error in repeated sampling.

First note that both Y and $\hat{Y}$ are random variables and that the error is a linear function of Y and $\hat{Y}$. Then we conclude that the error is normally distributed because it is a linear function of normally distributed random variables.

Applying Theorem 5.8, which gives the formulas for the expected value and variance of a linear function of random variables, we obtain

$$E(\text{error}) = E(Y - \hat{Y}) = E(Y) - E(\hat{Y}),$$

and since $E(\hat{Y}) = E(Y)$,

$$E(\text{error}) = 0.$$

Likewise,

$$V(\text{error}) = V(Y - \hat{Y}) = V(Y) + V(\hat{Y}) - 2\,\text{Cov}(Y, \hat{Y}).$$

Since we assume that the predicted value of Y was not employed in the calculation of $\hat{Y}$ and that, in fact, it was randomly selected and hence is independent of $\hat{Y}$, it follows that the covariance of Y and $\hat{Y}$ is equal to zero. Then

$$\begin{aligned} V(\text{error}) &= V(Y) + V(\hat{Y}) \\ &= \sigma^2 + [\mathbf{a}'(\mathbf{X}'\mathbf{X})^{-1}\mathbf{a}]\sigma^2 \\ &= \sigma^2[1 + \mathbf{a}'(\mathbf{X}'\mathbf{X})^{-1}\mathbf{a}]. \end{aligned}$$

We have now shown that the error of predicting a particular value of Y is normally distributed with mean and variance equal to zero and $\sigma^2[1 + \mathbf{a}'(\mathbf{X}'\mathbf{X})^{-1}\mathbf{a}]$, respectively. It follows that

$$Z = \frac{\text{error}}{\sigma\sqrt{1 + \mathbf{a}'(\mathbf{X}'\mathbf{X})^{-1}\mathbf{a}}} = \frac{Y - \hat{Y}}{\sigma\sqrt{1 + \mathbf{a}'(\mathbf{X}'\mathbf{X})^{-1}\mathbf{a}}}$$

follows a standard normal distribution. Furthermore, substituting S for σ, it can be shown that

$$T = \frac{Y - \hat{Y}}{S\sqrt{1 + \mathbf{a}'(\mathbf{X}'\mathbf{X})^{-1}\mathbf{a}}}$$

follows a Student's t distribution with $[n - (k + 1)]$ degrees of freedom. We will use this result to place a bound on the error of prediction and, in doing so, will construct a prediction interval for the random variable, Y. The procedure employed will be similar to that used to construct all the confidence intervals presented in the preceding chapters.

We begin by placing a $(1 - \alpha)$ probability statement on T:

$$P(-t_{\alpha/2} < T < t_{\alpha/2}) = 1 - \alpha.$$

Substituting for T,

$$P\left[-t_{\alpha/2} < \frac{Y - \hat{Y}}{S\sqrt{1 + \mathbf{a}'(\mathbf{X}'\mathbf{X})^{-1}\mathbf{a}}} < t_{\alpha/2}\right] = 1 - \alpha.$$

In other words, in repeated sampling, the inequality within the brackets will hold with probability equal to $(1 - \alpha)$. Furthermore, the inequality will continue to hold with the same probability if each term is multiplied by the same factor or if the same quantity is added to each term of the inequality. Hence multiply each term by $S\sqrt{1 + \mathbf{a}'(\mathbf{X}'\mathbf{X})^{-1}\mathbf{a}}$ and then add $\hat{Y}$ to each. The result,

$$P[\hat{Y} - t_{\alpha/2}S\sqrt{1 + \mathbf{a}'(\mathbf{X}'\mathbf{X})^{-1}\mathbf{a}} < Y < \hat{Y} + t_{\alpha/2}S\sqrt{1 + \mathbf{a}'(\mathbf{X}'\mathbf{X})^{-1}\mathbf{a}}] = 1 - \alpha$$

places an interval about Y that will hold with probability $(1 - \alpha)$. The result is a $(1 - \alpha)$ prediction interval for Y,

A (1 − α) Prediction Interval for Y

$$\hat{y} \pm t_{\alpha/2}s\sqrt{1 + \mathbf{a}'(\mathbf{X}'\mathbf{X})^{-1}\mathbf{a}}.$$

Thinking in terms of a bound on the error of predicting Y, we would expect the error to be less in absolute value than

$$t_{\alpha/2}s\sqrt{1 + \mathbf{a}'(\mathbf{X}'\mathbf{X})^{-1}\mathbf{a}}$$

with probability equal to $(1 - \alpha)$. We will illustrate with an example.

Example 11.9: *Suppose the experiment which generated the data of Example 11.1 is to be run again with $x = 2$. Predict the particular value of Y with $1 - \alpha = .90$.*

Solution: *From Example 11.1 we have that*

$$\hat{y} = 1 + .7x$$

and the predicted value of Y with $x = 2$ is then

$$\hat{y} = 1 + (.7)(2) = 2.4.$$

In this case $\mathbf{a} = \begin{bmatrix} 1 \\ 2 \end{bmatrix}$ *and so*

$$\mathbf{a}'(\mathbf{X}'\mathbf{X})^{-1}\mathbf{a} = [1 \quad 2]\begin{bmatrix} 1/5 & 0 \\ 0 & 1/10 \end{bmatrix}\begin{bmatrix} 1 \\ 2 \end{bmatrix}$$

$$= .6.$$

From Example 11.5, $s = .606$. The $t_{.05}$ value with three degrees of freedom is 2.353. Thus the prediction interval becomes

$$\hat{y} \pm t_{\alpha/2} s\sqrt{1 + \mathbf{a}'(\mathbf{X}'\mathbf{X})^{-1}\mathbf{a}},$$

$$2.4 \pm (2.353)(.606)\sqrt{1 + .6},$$

or

$$2.4 \pm 1.804.$$

11.10 A Test Statistic to Test $H_0: \beta_{g+1} = \beta_{g+2} = \ldots = \beta_k = 0$

In seeking an intuitively appealing test statistic to test a hypothesis concerning a set of parameters of the linear model, we are led to a consideration of the sum of squares of deviations, SSE. Suppose, for example, that we were to fit a model,

$$\text{Model 1}: Y = \beta_0 + \beta_1 x_1 + \beta_2 x_2 + \cdots + \beta_g x_g + \varepsilon,$$

$$\text{where } g < k,$$

to the data and then calculate the sum of squares of deviations of the observed values of Y, SSE_1. Having done this, we then fit the linear model,

$$\text{Model 2}: Y = \beta_0 + \beta_1 x_1 + \beta_2 x_2 + \cdots + \beta_g x_g$$

$$+ \beta_{g+1} x_{g+1} + \cdots + \beta_k x_k + \varepsilon,$$

which contains all of the terms of model 1 plus the terms involving x_{g+1}, $x_{g+2}, \ldots, x_k$. Then we calculate the sum of squares of deviations for this model, SSE_2. Finally, let us suppose that $x_{g+1}, x_{g+2}, \ldots, x_k$ really contribute a substantial quantity of information for the prediction of Y not contained in the variables, $x_1, x_2, \ldots, x_g$ (that is, at least one of the parameters $\beta_{g+1}, \beta_{g+2}, \ldots, \beta_k$ actually differs from zero); what would be the relation between SSE_1 and SSE_2? Intuitively, we see that if $x_{g+1}, x_{g+2}, \ldots, x_k$ are important information-contributing variables, model 2, which contains all the variables of model 1 plus the additions $x_{g+1}, x_{g+2}, \ldots, x_k$, should predict with a *smaller* error of prediction than model 1 and hence SSE_2 should be less than SSE_1. The greater the difference, $(\text{SSE}_1 - \text{SSE}_2)$, the stronger the evidence to support the alternative hypothesis that $x_{g+1}, x_{g+2}, \ldots, x_k$ contribute information for the prediction of Y and to reject the null hypothesis,

$$H_0: \beta_{g+1} = \beta_{g+2} = \cdots = \beta_k = 0.$$

Models 1 and 2 are known as the *reduced* and *complete* models, respectively, for a test of the null hypothesis indicated above. The drop in the sum of squares of deviations, $(\text{SSE}_1 - \text{SSE}_2)$, is called the *sum of squares associated with the variables* $x_{g+1}, x_{g+2}, \ldots, x_k$, *adjusted for the variables* $x_1, x_2, x_3, \ldots, x_g$.

How large is "large"? Although the drop in SSE measures the weight of evidence favoring a rejection of the hypothesis,

$$H_0: \beta_{g+1} = \beta_{g+2} = \cdots = \beta_k = 0,$$

we must now use it to acquire a suitable test statistic whose probability distribution, when the null hypothesis is true, is known. Then we will know whether the observed drop is large enough (and sufficiently improbable) to support the rejection of H_0.

To acquire this test statistic, let us *assume* that the null hypothesis is true and then examine the quantities that we have calculated. Particularly, note that

$$\text{SSE}_1 = \text{SSE}_2 + (\text{SSE}_1 - \text{SSE}_2).$$

In other words, we have partitioned SSE_1 into two parts, SSE_2 and the drop $(\text{SSE}_1 - \text{SSE}_2)$, as indicated in Figure 11.4. Further, assuming H_0 to be true

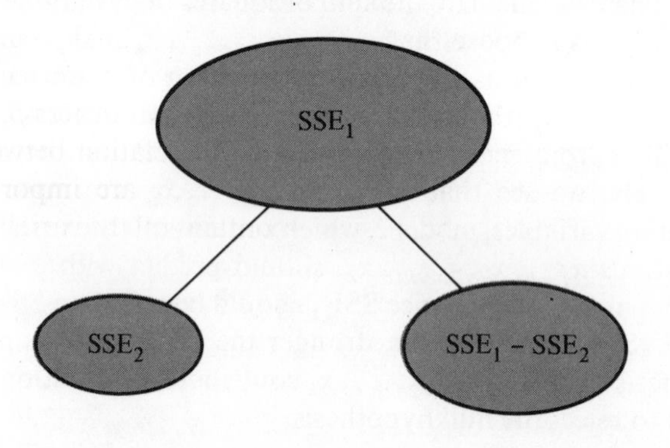

Figure 11.4

$(\beta_{g+1} = \beta_{g+2} = \cdots = \beta_k = 0)$, it follows that

$$S_1^2 = \frac{\text{SSE}_1}{n - (g + 1)}$$

is an unbiased estimator of σ^2, the variance of the random error ε. Keeping in mind that we are primarily concerned with the drop $(\text{SSE}_1 - \text{SSE}_2)$, we now wonder whether the two components of SSE_1, SSE_2 and $(\text{SSE}_1 - \text{SSE}_2)$, might also provide unbiased estimators of σ^2 when divided by appropriate divisors. And, indeed, this can be shown to be true. Both

$$S_2^2 = \frac{\text{SSE}_2}{n - (k + 1)} \quad \text{and} \quad S_3^2 = \frac{\text{SSE}_1 - \text{SSE}_2}{k - g}$$

are unbiased estimators of σ^2 and can be shown to be statistically independent.

Hence a good decision maker for testing the hypothesis

$$H_0: \beta_{g+1} = \beta_{g+2} = \cdots = \beta_k = 0$$

would be the ratio

$$F = \frac{S_3^2}{S_2^2}.$$

If H_0 is true, S_3^2 and S_2^2 will provide independent unbiased estimates of σ^2 and will be of relatively the same magnitude and F, the ratio, will assume a value near 1. (It can be proved that the expected value of F is not equal to 1.) On the contrary, when H_0 is false, S_2^2 will still provide an unbiased estimator of σ^2 but S_3^2, calculated from the drop $(\text{SSE}_1 - \text{SSE}_2)$, will be inflated and possess an expected value equal to σ^2 plus a positive quantity involving the sums of squares of the parameters, $\beta_{g+1}, \beta_{g+2}, \ldots, \beta_k$. The larger the drop, the greater will be the amount of evidence favoring a rejection of H_0 and the more inflated S_3^2 will be in relation to S_2^2. Thus large values of the F statistic will favor a rejection of H_0.

The F random variable is defined in Definition 6.2. Although we omit the proof, it can be shown (assuming H_0 true) that

$$\chi_3^2 = \frac{[n - (g + 1)]S_1^2}{\sigma^2} = \frac{\text{SSE}_1}{\sigma^2},$$

$$\chi_2^2 = \frac{[n - (k + 1)]S_2^2}{\sigma^2} = \frac{\text{SSE}_2}{\sigma^2},$$

and

$$\chi_1^2 = \frac{(k - g)S_3^2}{\sigma^2} = \frac{\text{SSE}_1 - \text{SSE}_2}{\sigma^2}$$

possess chi-square probability distributions in repeated sampling with $[n - (g + 1)]$, $[n - (k + 1)]$, and $(k - g)$ degrees of freedom, respectively. Further, it can be shown that χ_2^2 and χ_1^2 are statistically independent. The resulting ratio,

$$\frac{S_3^2}{S_2^2} = \frac{\chi_1^2[n - (k + 1)]}{\chi_2^2(k - g)},$$

possesses the well-known F distribution with $\nu_1 = (k - g)$ and $\nu_2 = [n - (k + 1)]$ degrees of freedom.

The density function for the F distribution depends upon the two parameters, $\nu_1 = k - g$ and $\nu_2 = n - (k + 1)$, which represent the degrees of freedom associated with the numerator and denominator sums of squares of deviations, $(\text{SSE}_1 - \text{SSE}_2)$ and SSE_2, respectively. To locate the rejection region for the F test, we seek a point F_α such that

$$P(F > F_\alpha) = \alpha.$$

Note that this F test always utilizes a single-tailed rejection region. The values of F for $\alpha = .05$ and $.01$ for specific values of ν_1 and ν_2 are presented in Tables 6 and 7, respectively, of Appendix III.

Example 11.10: *An experiment was conducted to compare the effect of four chemicals, A, B, C, and D, in producing water resistance in textiles. Three types of material, I, II, and III, were used, with each chemical treatment being applied to one piece of each type of material. The data are as follows:*

	Treatment			
Material	*A*	*B*	*C*	*D*
I	10.1	11.4	9.9	12.1
II	12.2	12.9	12.3	13.4
III	11.9	12.7	11.4	12.9

The model for this experiment is

$$Y = \beta_0 + \beta_1 x_1 + \beta_2 x_2 + \beta_3 x_3 + \beta_4 x_4 + \beta_5 x_5 + \varepsilon,$$

where

$$x_1 = \begin{cases} 1 & \text{if material II is used,} \\ 0 & \text{otherwise,} \end{cases}$$

$$x_2 = \begin{cases} 1 & \text{if material III is used,} \\ 0 & \text{otherwise,} \end{cases}$$

$$x_3 = \begin{cases} 1 & \text{if treatment B is used,} \\ 0 & \text{otherwise,} \end{cases}$$

$$x_4 = \begin{cases} 1 & \text{if treatment C is used,} \\ 0 & \text{otherwise,} \end{cases}$$

$$x_5 = \begin{cases} 1 & \text{if treatment D is used,} \\ 0 & \text{otherwise.} \end{cases}$$

Test the hypothesis that there are no treatment differences. That is, test

$$H_0: \beta_3 = \beta_4 = \beta_5 = 0.$$

Use $\alpha = .05$.

Solution: *For the complete model, it can be determined that*

$$\begin{aligned} \text{SSE}_2 &= \mathbf{Y}'\mathbf{Y} - \hat{\boldsymbol{\beta}}'\mathbf{X}'\mathbf{Y} \\ &= 1721.760 - 1721.225 \\ &= .535. \end{aligned}$$

For the reduced model

$$Y = \beta_0 + \beta_1 x_1 + \beta_2 x_2 + \varepsilon,$$

we have

$$\mathbf{Y} = \begin{bmatrix} 10.1 \\ 12.2 \\ 11.9 \\ 11.4 \\ 12.9 \\ 12.7 \\ 9.9 \\ 12.3 \\ 11.4 \\ 12.1 \\ 13.4 \\ 12.9 \end{bmatrix} \quad \text{and} \quad \mathbf{X} = \begin{bmatrix} 1 & 0 & 0 \\ 1 & 1 & 0 \\ 1 & 0 & 1 \\ 1 & 0 & 0 \\ 1 & 1 & 0 \\ 1 & 0 & 1 \\ 1 & 0 & 0 \\ 1 & 1 & 0 \\ 1 & 0 & 1 \\ 1 & 0 & 0 \\ 1 & 1 & 0 \\ 1 & 0 & 1 \end{bmatrix}$$

Thus

$$\hat{\boldsymbol{\beta}} = (\mathbf{X}'\mathbf{X})^{-1}(\mathbf{X}'\mathbf{Y}) = \begin{bmatrix} 10.875 \\ 1.825 \\ 1.350 \end{bmatrix}$$

and

$$\begin{aligned} \text{SSE}_1 &= \mathbf{Y}'\mathbf{Y} - \hat{\boldsymbol{\beta}}'\mathbf{X}'\mathbf{Y} \\ &= 1721.76 - 1716.025 \\ &= 5.735, \end{aligned}$$

where the **X** *matrix is the one given above for the reduced model. It follows that*

$$s_2^2 = \frac{\text{SSE}_2}{n - (k+1)} = \frac{.535}{12 - (6)} = .0891,$$

and

$$s_3^2 = \frac{\text{SSE}_1 - \text{SSE}_2}{k - g} = \frac{5.735 - .535}{3}$$

$$= 1.733.$$

Finally,

$$F = \frac{s_3^2}{s_2^2} = \frac{1.733}{.0891} = 19.4.$$

The tabulated F value for $\alpha = .05$, $\nu_1 = 3$, and $\nu_2 = 6$ is 4.76. Hence the observed value of the test statistic falls in the rejection region, and we conclude that the data present sufficient evidence to indicate that differences among treatments do exist.

11.11 An Example Involving Coding

For many practical problems, the values of the independent variables, $x_1, x_2, \ldots, x_k$, may be large and therefore inconvenient for calculation purposes. This difficulty can frequently be overcome by coding the x's, translating or changing their scale. The concept of coding is illustrated in the following example.

Example 11.11: *An experiment was conducted to determine the effect of pressure and temperature on the yield of a chemical. Two levels of pressure and three of temperature were used:*

Pressure (psi)	*Temperature* (°F)
50	100
80	200
	300

One run of the experiment at each temperature–pressure combination gave the following data:

Yield	*Pressure* (psi)	*Temperature* (°F)
21	50	100
23	50	200
26	50	300
22	80	100
23	80	200
28	80	300

(*a*) *Fit the model* $Y = \beta_0 + \beta_1 x_1 + \beta_2 x_2 + \beta_3 x_2^2 + \varepsilon$, *where* x_1 *refers to pressure and* x_2 *to temperature.*

(*b*) *Test to see if* β_3 *is significantly different from zero, with* $\alpha = .05$.

Solution: *Computing* $\mathbf{X'X}$ *and* $(\mathbf{X'X})^{-1}$ *would be very tedious if we were to use the raw, uncoded pressure and temperature measurements. We will find the computations are much simpler if we code so that the* x_{ij}*'s are small numbers and* $\sum_{j=1}^{n} x_{ij} = 0$. *To accomplish this, we will translate the pressure axis to a new origin at* $P = 65$ *and change the pressure units to 15 pounds per square inch, using*

$$x_1 = \frac{P - 65}{15},$$

where P is the actual pressure employed. Thus $P = 50$ *corresponds to* $x_1 = -1$ *and* $P = 80$ *to* $x_1 = 1$.

Similarly, for temperature settings we use the transformation

$$x_2 = \frac{T - 200}{100}$$

where T denotes the actual temperature in °F. Thus $T = 100, 200,$ *and* 300 *corresponds to* $x_2 = -1, 0,$ *and* 1, *respectively.*

In matrix notation, we now have

$$\mathbf{Y} = \begin{bmatrix} 21 \\ 23 \\ 26 \\ 22 \\ 23 \\ 28 \end{bmatrix} \quad \text{and} \quad \mathbf{X} = \begin{array}{c} \begin{matrix} x_0 & x_1 & x_2 & x_2^2 \end{matrix} \\ \begin{bmatrix} 1 & -1 & -1 & 1 \\ 1 & -1 & 0 & 0 \\ 1 & -1 & 1 & 1 \\ 1 & 1 & -1 & 1 \\ 1 & 1 & 0 & 0 \\ 1 & 1 & 1 & 1 \end{bmatrix} \end{array}.$$

(*a*) *The least-squares estimators involve the matrices*

$$\mathbf{X'X} = \begin{bmatrix} 6 & 0 & 0 & 4 \\ 0 & 6 & 0 & 0 \\ 0 & 0 & 4 & 0 \\ 4 & 0 & 0 & 4 \end{bmatrix},$$

$$(\mathbf{X'X})^{-1} = \begin{bmatrix} 1/2 & 0 & 0 & -1/2 \\ 0 & 1/6 & 0 & 0 \\ 0 & 0 & 1/4 & 0 \\ -1/2 & 0 & 0 & 3/4 \end{bmatrix},$$

and

$$\mathbf{X'Y} = \begin{bmatrix} 143 \\ 3 \\ 11 \\ 97 \end{bmatrix}.$$

Then

$$\hat{\boldsymbol{\beta}} = (\mathbf{X'X})^{-1}\mathbf{X'Y} = \begin{bmatrix} 23 \\ .5 \\ 2.75 \\ 1.25 \end{bmatrix}$$

or

$$\hat{y} = 23 + .5x_1 + 2.75x_2 + 1.25x_2^2.$$

Note that $\hat{y}$ is given as a function of the coded pressure and temperature measurements. Hence if you wish to predict y for some setting of pressure and temperature, you must first compute the coded pressure and temperature readings, x_1 and x_2.

(b) The test statistic for $H_0: \beta_3 = 0$ is

$$T = \frac{\hat{\beta}_3 - 0}{S\sqrt{c_{33}}}.$$

Now

$$\text{SSE} = \mathbf{Y'Y} - \hat{\boldsymbol{\beta}}'\mathbf{X'Y}$$

$$= 3443 - 3442 = 1$$

and

$$s^2 = \frac{\text{SSE}}{n-4} = \frac{1}{6-4} = .5.$$

Thus

$$t = \frac{1.25}{\sqrt{.5}\sqrt{3/4}} = 2.042.$$

Since $t_{.025} = 4.303$, with 2 degrees of freedom, we would not reject $H_0: \beta_3 = 0$. You might conclude that the x_2^2 term could be eliminated from the model, but keep in mind that the test based on 2 degrees of freedom is not very sensitive. The half-width of the confidence interval for estimating β_3,

$$t_{\alpha/2} s\sqrt{c_{33}} = (4.303)(\sqrt{.5})(\sqrt{3/4}) = 2.635,$$

is quite large, so we would expect the power of the test, the probability of detecting a value of β_3 as large as 2, to be quite small. Actually, more observations should be taken before testing hypotheses.

11.12 Correlation

The previous sections of this chapter dealt with modeling a response, Y, as a linear function of a nonrandom variable, x, so that appropriate inferences could be made concerning the expected value of Y, or a future value of Y, for a given value of x. We now want to point out that these models arise in two quite different practical situations.

First, the variable x may be completely controlled by the experimenter. This occurs, for example, if x is the temperature setting and Y the yield in a chemical experiment. Then x is merely the point at which the temperature dial is set when the experiment is run. Of course, x could vary from experiment to experiment, but it is under the complete control, practically speaking, of the experimenter. The linear model

$$Y = \beta_0 + \beta_1 x + \varepsilon$$

then implies that

$$E(Y) = \beta_0 + \beta_1 x,$$

or the average yield is a linear function of the temperature setting.

Second, the variable x may be an observed value of a random variable, X. For example, we may want to relate the volume of usable timber in a tree, Y, to the circumference of the base, X. If a functional relationship could be established, then, in the future, we could predict the amount of timber in any tree by simply measuring the circumference of the base. For this situation, we use the model

$$Y = \beta_0 + \beta_1 x + \varepsilon$$

to imply that

$$E(Y|X = x) = \beta_0 + \beta_1 x.$$

That is, we are assuming that the conditional expectation of Y for a fixed value of X is a linear function of the x value. We generally assume that the vector random variable, (X, Y), has a bivariate normal distribution (see Section 5.9), in which case it can be shown that

$$E(Y|X = x) = \beta_0 + \beta_1 x.$$

The statistical theory for making inferences about the parameters β_0 and β_1 is exactly the same for both of these cases, but the differences in model interpretation should be borne in mind.

For the case where (X, Y) has a bivariate distribution, the experimenter may not always be interested in the linear relationship defining $E(Y|X)$. He may only want to know whether or not X and Y are *independent* random variables. Assuming that (X, Y) has a bivariate normal distribution, testing for independence is equivalent to testing that the correlation coefficient, ρ, is equal to zero. Recall from Section 5.7 that ρ is positive if X and Y tend to increase together and ρ is negative if Y decreases as X increases.

Let (X_1, Y_1), $(X_2, Y_2) \cdots (X_n, Y_n)$ denote a random sample from a bivariate normal distribution. The maximum likelihood estimator of ρ is given by the sample correlation coefficient

$$r = \frac{\sum_{i=1}^{n} (X_i - \overline{X})(Y_i - \overline{Y})}{\sqrt{\sum_{i=1}^{n} (X_i - \overline{X})^2 \sum (Y_i - \overline{Y})^2}}$$

or an equivalent expression,

$$r = \frac{n \sum_{i=1}^{n} X_i Y_i - \sum_{i=1}^{n} X_i \sum_{i=1}^{n} Y_i}{\sqrt{\left[n \sum_{i=1}^{n} X_i^2 - \left(\sum_{i=1}^{n} X_i\right)^2\right]\left[n \sum_{i=1}^{n} Y_i^2 - \left(\sum_{i=1}^{n} Y_i\right)^2\right]}}.$$

It would seem natural to use r as a test statistic to test hypotheses about ρ, but difficulties arise because the probability distribution for r is difficult to obtain. This difficulty can be overcome, for moderately large samples, by using the fact that $(1/2)\ln[(1 + r)/(1 - r)]$ is approximately normally distributed with mean, $(1/2)\ln[(1 + \rho)/(1 - \rho)]$, and variance, $1/(n - 3)$. Thus for testing the hypothesis $H_0\colon \rho = \rho_0$, we can employ a z test in which

$$Z = \frac{(1/2)\ln\left(\frac{1 + r}{1 - r}\right) - (1/2)\ln\left(\frac{1 + \rho_0}{1 - \rho_0}\right)}{\frac{1}{\sqrt{n - 3}}}.$$

H_0 will be rejected for $|z| > z_{\alpha/2}$, for any specified type I error probability α. We illustrate with an example.

Example 11.12: *The data in Table 11.2 represent a sample of mathematics achievement test scores and calculus grades for ten independently selected college freshmen.*

Table 11.2

Student	*Mathematics Achievement Test Score*	*Final Calculus Grade*
1	39	65
2	43	78
3	21	52
4	64	82
5	57	92
6	47	89
7	28	73
8	75	98
9	34	56
10	52	75

Based on this evidence, would you say that achievement test scores and calculus grades are independent? Use $\alpha = .05$.

Solution: *Denoting achievement test scores by x and calculus grades by y, we calculate*

$$\sum_{i=1}^{10} x_i = 460, \qquad \sum_{i=1}^{10} y_i = 760,$$

$$\sum_{i=1}^{10} x_i^2 = 23{,}634, \qquad \sum_{i=1}^{n} y_i^2 = 59{,}816,$$

$$\sum_{i=1}^{10} x_i y_i = 36{,}854.$$

Thus

$$r = \frac{(10)(36{,}854) - (460)(760)}{\sqrt{[(10)(23{,}639) - (460)^2][(10)(59{,}816) - (760)^2]}}$$

$$= 0.84.$$

We state as the null hypothesis that X and Y are independent, or, assuming (X, Y) has a bivariate normal distribution, that $\rho = 0$. *The test statistic is observed to be*

$$z = \frac{(1/2)\ln\left(\frac{1+r}{1-r}\right) - (1/2)\ln\left(\frac{1+\rho_0}{1-\rho_0}\right)}{1/\sqrt{n-3}}$$

$$= \frac{(1/2)\ln\left(\frac{1+.84}{1-.84}\right) - 0}{1/\sqrt{7}}$$

$$= 3.231.$$

Since $z_{\alpha/2} = z_{.025} = 1.96$, *our observed value of the test statistic lies in the rejection region. Thus the evidence strongly suggests that achievement test scores and calculus grades are dependent. Note that* $\alpha = .05$ *is the probability that our test statistic will fall in the rejection region when* H_0 *is true. Hence we are fairly confident that we have made a correct decision.*

11.13 Summary

In this chapter we have been concerned with use of the method of least squares to fit a linear model to an experimental response. We assume the expected value of Y to be a function of a set of variables, $x_1, x_2, \ldots, x_k$, where the function is linear in a set of unknown parameters. We used the expression

$$Y = \beta_0 + \beta_1 x_1 + \beta_2 x_2 + \cdots + \beta_k x_k + \varepsilon$$

to denote a linear statistical model.

Inferential problems associated with the linear statistical model are estimation and tests of hypotheses concerning the model parameters, β_0, $\beta_1, \ldots, \beta_k$ and even more important, estimation of $E(Y)$, the expected response for a particular setting, and the prediction of some future value of Y. Experiments for which the least-squares theory is appropriate include both controlled experiments as well as those where $x_1, x_2, \ldots, x_k$ are observed values of random variables.

Why use the method of least squares to fit a linear model to a set of data? Where the assumptions on the random errors, ε, hold [normality, independence, $V(\varepsilon) = \sigma^2$ for all values of $x_1, x_2, \ldots, x_k$], it can be shown that the least-squares procedure gives the best *linear* unbiased estimators for $\beta_0, \beta_1, \ldots,$ β_k. That is, if we estimate the parameters, $\beta_0, \beta_1, \ldots, \beta_k$, using linear functions of $y_1, y_2, \ldots, y_k$, the least-squares estimators have minimum variance. Some other nonlinear estimators for the parameters may possess a smaller variance than the least-squares estimators, but, if such estimators exist, they are not known at this time. Again, why use least-squares estimators? They are easy to use, we know their properties, and we know that they possess very good properties for many situations.

As you might imagine, the methodology presented in this chapter is widely employed in business and all the sciences for exploring the relationship between a response and a set of independent variables. Estimation of $E(Y)$ or prediction of Y is usually the experimental objective.

References

1. Graybill, F., *An Introduction to Linear Statistical Models*, Vol. 1. New York: McGraw-Hill Book Company, 1961.

2. Li, J. C. R., *Introduction to Statistical Inference*. Ann Arbor, Mich.: J. W. Edwards, Publisher, Inc., 1961.

3. Mendenhall, W., *An Introduction to Linear Models and the Design and Analysis of Experiments*. North Scituate, Mass.: Duxbury Press, 1967.

Exercises

11.1. (a) Fit a straight line through the following five data points. Give the estimates of β_0 and β_1.

y	3	2	1	1	.5
x	-2	-1	0	1	2

Plot the points and sketch the fitted line as a check on the calculations.

(b) Calculate SSE and s^2 for the data. How many degrees of freedom are associated with s^2?

(c) Do the data present sufficient evidence to indicate that the slope β_1 differs from zero? (Test the null hypothesis, $\beta_1 = 0$, using $\alpha = .05$.)

(d) Find a 95 percent confidence interval for the slope, β_1.

11.2. Fit a straight line to the following data, plot the points, and then sketch the fitted line as a check on the calculations:

y	3	2	1	1	.5
x	-1	0	1	2	3

Note that the data points are the same as for Exercise 11.1 except that they are translated one unit in the positive direction along the x axis. What effect does symmetric spacing of the x values about $x = 0$ have on the form of the $(\mathbf{X}'\mathbf{X})$ matrix and the resulting calculations?

11.3. (a) Fit a parabola through the following seven data points by estimating the model parameters, β_0, β_1, and β_2:

y	1	0	0	-1	-1	0	0
x	-3	-2	-1	0	1	2	3

Plot the points and sketch the fitted parabola as a check on the calculations.

(b) Do the data present sufficient evidence to indicate a lack of linearity in the relation between y and x? (Test the hypothesis that the quadratic coefficient, β_2, equals zero using $\alpha = .10$.)

(c) Find a 90 percent confidence interval for β_2.

11.4. A response, y, is a function of three independent variables, x_1, x_2, and x_3, that are related as follows:

$$Y = \beta_0 + \beta_1 x_1 + \beta_2 x_2 + \beta_3 x_3 + \varepsilon.$$

(a) Fit this model to the following $n = 7$ data points:

y	1	0	0	1	2	3	3
x_1	-3	-2	-1	0	1	2	3
x_2	5	0	-3	-4	-3	0	5
x_3	-1	1	1	0	-1	-1	1

(b) Predict y when $x_1 = 1$, $x_2 = -3$, $x_3 = -1$. Compare with the observed response in the original data. Why are these two not equal?

(c) Do the data present sufficient evidence to indicate that x_3 contributes information for the prediction of Y? (Test the hypothesis $\beta_3 = 0$, using $\alpha = .05$.)

(d) Find a 95 percent confidence interval for the expected value of Y, given $x_1 = 1$, $x_2 = -3$, $x_3 = -1$.

(e) Find a 95 percent prediction interval for Y, given $x_1 = 1$, $x_2 = -3$, $x_3 = -1$.

11.5. An experiment was conducted to investigate the effect of four factors: temperature, T_1; pressure, P; catalyst, C; and temperature, T_2, on the yield of a chemical.

(a) The levels of the four factors are as follows:

T_1	x_1	P	x_2	C	x_3	T_2	x_4
50	-1	10	-1	1	-1	100	-1
70	1	20	1	2	1	200	1

If each of the four factors is coded to produce the four variables, x_1, x_2, x_3, and x_4, respectively, give the linear equation relating each coded variable to its corresponding original.

(b) Fit the linear model

$$Y = \beta_0 + \beta_1 x_1 + \beta_2 x_2 + \beta_3 x_3 + \beta_4 x_4 + \varepsilon$$

to the following data:

				x_4			
				+1		−1	
				x_3		x_3	
				−1	1	−1	1
x_1	−1	x_2	−1	22.2	24.5	24.4	25.9
			1	19.4	24.1	25.2	18.4
	+1	x_2	−1	22.1	19.6	23.5	16.5
			1	14.2	12.7	19.3	16.0

(c) Do the data present sufficient evidence to indicate that T_1 contributes information for the prediction of Y? P? C? T_2? (Test the hypotheses, respectively, that $\beta_1 = 0$, $\beta_2 = 0$, $\beta_3 = 0$, and $\beta_4 = 0$, using $\alpha = .05$.)

(d) Find a 90 percent confidence interval for the expected value of Y, given $T_1 = 50$, $P = 20$, $C = 1$, and $T_2 = 200$.

(e) Find a 90 percent prediction interval for Y, given $T_1 = 50$, $P = 20$, $C = 1$, and $T_2 = 200$.

11.6. The following data come from the comparison of the growth rates for bacteria types A and B. The growth, Y, recorded at five equally spaced (and coded) points of time, is recorded below.

	Time				
Bacteria Type	−2	−1	0	1	2
A	8.0	9.0	9.1	10.2	10.4
B	10.0	10.3	12.2	12.6	13.9

(a) Fit the linear model

$$Y = \beta_0 + \beta_1 x_1 + \beta_2 x_2 + \beta_3 x_1 x_2 + \varepsilon$$

to the $n = 10$ data points. Let $x_1 = 1$ if the point refers to bacteria type B and $x_1 = 0$ if the point refers to type A. Let x_2 = coded time.

(b) Plot the data points and graph the two growth lines. Note that β_3 is the difference between the slopes of the two lines and represents slope–bacteria interaction.

(c) Predict the growth of bacteria type A at time $x_2 = 0$ and compare with the graph. Repeat the process for bacteria type B.

(d) Do the data present sufficient evidence to indicate a difference in the rates of growth for the two types of bacteria?

(e) Find a 90 percent confidence interval for the expected growth for bacteria type B at time $x_2 = 1$.

(f) Find a 90 percent prediction interval for the growth, Y, of bacteria type B at time $x_2 = 1$.

11.7. Suppose that an independent variable, T, takes values 20, 35, 50, 65, and 80 during an experiment. Find a coded variable, x, that is linearly related to T so that the values of x will be equally spaced about their origin, and assume values such that $|x| \leq 2$. (Give the relationship between x and T.)

11.8. If independent variables are equally spaced, what is the advantage of coding to new variables that represent symmetric spacing about the origin?

11.9. Suppose that you wish to fit a straight line to a set of n data points, where n is an even integer, and that you can select the n values of x in the intervals $-9 \leq x \leq 9$. How should you select the values of x so as to minimize $V(\hat{\beta}_1)$?

11.10. Refer to Exercise 11.9. It is common to employ equal spacing in the selection of the values of x. Suppose that $n = 10$. Find the relative efficiency of the estimator, $\hat{\beta}_1$, based on equal spacing versus the same estimator based on the spacing of Exercise 11.9. Assume that $-9 \leq x \leq 9$.

11.11. Using the assumptions on $\varepsilon_i, i = 1, 2, \ldots, n$, of Section 11.5, show that $\hat{\beta}_0$ and $\hat{\beta}_1$ are independent if $\sum_{i=1}^{n} x_i = 0$.

11.12. Use the properties of the least-squares estimators, Section 11.6, to derive the confidence interval for β_i (presented in Section 11.6). (*Hint*: Use Student's t to construct a pivotal statistic.)

11.13. Let $U = \sum_{i=1}^{k} a_i\hat{\beta}_i$ be a linear function of the parameter estimators as given in Section 11.7 and assume that $V(U) = \mathbf{a}'(\mathbf{X}'\mathbf{X})^{-1}\mathbf{a}\sigma^2$. Show that

$$\frac{U - E(U)}{S\sqrt{\mathbf{a}'(\mathbf{X}'\mathbf{X})^{-1}\mathbf{a}}}$$

has a Student's t distribution.

11.14. Suppose that $Y_1, \ldots, Y_n$ are independent normal random variables with $E(Y_i) = \beta_0 + \beta_1 x_i$ and $V(Y_i) = \sigma^2, i = 1, \ldots, n$. Show that the maximum likelihood estimators of β_0 and β_1 are the same as the least-squares estimators of Section 11.3.

11.15. Under the assumptions of Exercise 11.14 show that the likelihood ratio test of $H_0: \beta_1 = 0$ against $H_a: \beta_1 \neq 0$ is equivalent to the t test given in Section 11.7.

11.16. Let $Y_1, \ldots, Y_n$ be as in Exercise 11.14. Suppose that we have an additional set of independent random variables $W_1, \ldots, W_m$, where W_i is normally distributed with $E(W_i) = \gamma_0 + \gamma_1 c_i$ and $V(W_i) = \sigma^2$, $i = 1, \ldots, m$. Construct a test of $H_0: \beta_1 = \gamma_1$ against $H_a: \beta_1 \neq \gamma_1$.

11.17. For the linear model $Y = \beta_0 + \beta_1 x + \varepsilon$ (Section 11.3), show that

$$\text{SSE} = (1 - r^2) \sum_{i=1}^{n} (y_i - \bar{y})^2,$$

where r is as given in Section 11.12. Also, show that $\text{SSE} \leq \sum_{i=1}^{n} (y_i - \bar{y})^2$ and conclude that $0 \leq r^2 \leq 1$, or $-1 \leq r \leq 1$.

11.18. Let $(X_1, Y_1), \ldots, (X_n, Y_n)$ denote a random sample from the bivariate normal distribution. A test of $H_0: \rho = 0$ against $H_a: \rho \neq 0$ can be derived as follows.

(a) Let $S_{yy} = \sum_{i=1}^{n} (y_i - \bar{y})^2$ and $S_{xx} = \sum_{i=1}^{n} (x_i - \bar{x})^2$.

Show that

$$\hat{\beta}_1 = r\sqrt{\frac{S_{yy}}{S_{xx}}}.$$

(b) Conditional on $X_i = x_i$, $i = 1, \ldots, n$, show that, under $H_0: \rho = 0$,

$$\frac{\hat{\beta}_1 \sqrt{(n - 2)S_{xx}}}{\sqrt{S_{yy}(1 - r^2)}}$$

has a t distribution with $(n - 2)$ degrees of freedom.

(c) Conditional on $X_i = x_i, i = 1, \ldots, n$, conclude that

$$T = \frac{r\sqrt{n-2}}{\sqrt{1-r^2}}$$

has a t distribution with $(n-2)$ degrees of freedom, under $H_0: \rho = 0$. Hence conclude that T has the same distribution unconditionally.

11.19. Show that the least-squares prediction equation

$$\hat{Y} = \hat{\beta}_0 + \hat{\beta}_1 x_1 + \cdots + \hat{\beta}_k x_k$$

passes through the point $(\bar{x}_1, \bar{x}_2, \ldots, \bar{x}_k, \bar{Y})$.

11.20. Find the expression for $V(\hat{Y})$ (see Section 11.8) for the model

$$E(Y) = \beta_0 + \beta_1 x.$$

What is the functional form of $V(\hat{Y})$ as a function of x? At what x value will $V(\hat{Y})$ possess a minimum? [Assume that β_0 and β_1 are estimated from n points of the form (x_i, y_i).]

12

Considerations in Designing Experiments

12.1 *The Elements Affecting the Information in a Sample*

The information in a sample that is available to make an inference about a population parameter can be measured by the width (or half-width) of the confidence interval that could be constructed from the sample data. Thus a large-sample confidence interval for a population mean is

$$\bar{y} \pm 2\frac{\sigma}{\sqrt{n}}.$$

The widths of almost all the commonly employed confidence intervals are, like the confidence interval for a population mean, dependent on the population variance, σ^2, and the sample size, n. The less variation in the population, measured by σ^2, the smaller will be the confidence interval. Similarly, the width of the confidence interval will decrease as n increases. This interesting phenomenon would lead us to believe that two factors affect the quantity of information in a sample pertinent to a parameter, namely the variation of the data and the sample size, n. We will find this deduction to be slightly oversimplified but essentially true.

A strong similarity exists between the audio theory of communication and the theory of statistics. Both are concerned with the transmission of a message (signal) from one point to another and, consequently, both are theories of information. For example, the telephone engineer is responsible for transmitting a verbal message that might originate in New York City and be received in New Orleans. Or, equivalently, a speaker may wish to communicate with a large and noisy audience. If static or background noise is sizable for either example, the receiver may acquire only a sample of the complete signal, and from this partial information he must infer the nature of the complete message. Similarly, scientific experimentation is conducted to verify certain theories about natural phenomena, or simply to explore some aspect of nature and hopefully to deduce—either exactly or with a good approximation—the relationships among certain natural variables. Thus one might think of experimentation as the communication between nature and a scientist. The message about the natural phenomenon is contained, in garbled form, in the experimenter's sample data. Imperfections in his measuring instruments, nonhomogeneity of experimental material, and many other factors contribute background noise (or static) that tends to obscure nature's signal and cause the observed response to vary in a random manner. For both the communications engineer and the statistician, two elements affect the quantity of information in an experiment, the magnitude of the background noise (or variation) and the volume of the signal. The greater the noise or, equivalently, the variation, the less information will be contained in the sample. Likewise, the louder the signal, the greater the amplification will be, and hence it is more likely that the message will penetrate the noise and be received.

The design of experiments is a very broad subject concerned with methods of sampling to reduce the variation in an experiment, to amplify nature's signal, and thereby to acquire a *specified quantity* of information at minimum cost. Despite the complexity of the subject, some of the important considerations in the design of good experiments can be easily understood and should be presented to the beginner. We take these considerations as our objective in the succeeding discussion.

12.2 The Physical Process of Designing an Experiment

We commence our discussion of experimental design by clarifying terminology and then, through examples, by identifying the steps that one must take in designing an experiment.

> ***Definition 12.1:*** *The objects upon which measurements are taken are called* experimental units.

If an experimenter subjects a set of $n = 10$ rats to a stimulus and measures the response of each, a rat is the experimental unit. The collection of $n = 10$ measurements is a sample. Similarly, if a set of $n = 10$ items is selected from a list of hospital supplies in an inventory audit, each item is an experimental unit. The observation made on each experimental unit is the dollar value of the item actually in stock, and the set of ten measurements constitutes a sample.

What one does to the experimental units that makes them differ from one population to another is called a treatment. One might wish to study the density of a specific kind of cake when baked at $x = 350°F$, $x = 400°F$, and $x = 450°F$ in a given oven. An experimental unit would be a single mix of batter in the oven at a given point in time. The three temperatures, $x = 350, 400$, and 450°F, would represent three treatments. The millions and millions of cakes that conceptually *could* be baked at 350°F would generate a population of densities, and one could similarly generate populations corresponding to 400 and 450°F. The objective of the experiment would be to compare the cake density, Y, for the three populations. Or, we might wish to study the effect of temperature of baking, x, on cake density by fitting a linear or curvilinear model to the data points as described in Chapter 11.

In another experiment we might wish to compare tire wear for two manufacturers, A and B. Each tire–wheel combination tested at a particular time would represent an experimental unit, and each of the two manufacturers would represent a treatment. Note that one does not physically treat the tires to make them different. They receive two different treatments by the very fact that they are manufactured by two different companies in different locations and in different factories.

As a third example, consider an experiment conducted to investigate the effect of various amounts of nitrogen and phosphate on the yield of a variety of corn. An experimental unit would be a specified acreage, say 1 acre, of corn. A treatment would be a fixed number of pounds of nitrogen, x_1, and phosphate, x_2, applied to a given acre of corn. For example, one treatment might be to use $x_1 = 100$ pounds of nitrogen per acre and $x_2 = 200$ pounds of phosphate. A second treatment could be $x_1 = 150$ and $x_2 = 100$. Note that the experimenter could experiment with different amounts (x_1, x_2) of nitrogen and phosphate and that each combination would represent a treatment.

Most experiments involve a study of the effect of one or more independent variables on a response.

> ***Definition 12.2:*** *Independent experimental variables are called* factors.

Factors can be *quantitative* or *qualitative.*

> ***Definition 12.3:*** *A* quantitative factor *is one that can take values corresponding to points on a real line. Factors that are not quantitative are said to be* qualitative.

Oven temperature, pounds of nitrogen, and pounds of phosphate are examples of quantitative factors. In contrast, manufacturers, types of drugs, or physical locations are factors that cannot be quantified and are called qualitative.

> ***Definition 12.4:*** *The intensity setting of a factor is called a* level.

The three temperatures, 350, 400, and 450°F, represent three levels of the quantitative factor "oven temperature." Similarly, the two treatments, manufacturer *A* and manufacturer *B*, represent two levels of the qualitative factor "manufacturers." Note that a third or fourth tire manufacturer could have been included in the tire wear experiment and resulted in either three or four levels of the factor "manufacturers."

We noted previously that what one does to experimental units that makes them differ from one population to another is called a treatment. Since every treatment implies a combination of one or more factor levels, we have a more precise definition for the term.

> ***Definition 12.5:*** *A* treatment *is a specific combination of factor levels.*

The experiment may involve only a single factor such as temperature in the baking experiment. Or, it could be composed of combinations of levels of two (or more) factors as for the corn-fertilizing experiment. Each combination would represent a treatment. One of the early steps in the design of an experiment is the selection of factors to be studied and a decision regarding the combinations of levels (treatments) to be employed in the experiment.

The design of an experiment implies one final problem. After selecting the factor combinations (treatments) to be employed in an experiment, one must decide how the treatments should be assigned to the experimental units. Should the treatments be randomly assigned to the experimental units or should a semirandom pattern be employed? For example, should the tires corresponding to manufacturers A and B be randomly assigned to all the automobile wheels, or should one each of tire types A and B be assigned to the rear wheels of each car?

The foregoing discussion suggests that the design of an experiment involves four steps:

Steps Employed in Designing an Experiment

1. *Selecting the factors to be included in the experiment and a specification of the population parameter(s) of interest.*
2. *Deciding how much information is desired pertinent to the parameter(s) of interest. (For example, how accurately do you wish to estimate the parameters?)*
3. *Selecting the treatments to be employed in the experiment (combination of factor levels) and deciding on the number of experimental units to be assigned to each.*
4. *Deciding on the manner in which the treatments should be applied to the experimental units.*

Steps 3 and 4 correspond to the two elements that affect the quantity of information in an experiment. First, how one selects the treatments (combinations of factor levels) and the number of experimental units assigned to each affects the intensity of nature's signal pertinent to the population parameter(s) of interest to the experimenter. Second, the method of assigning the treatments to the experimental units affects the background noise or, equivalently, the variation of the experimental units. We will examine each of these assertions in detail in Sections 12.4 and 12.5 after digressing briefly in Section 12.3 to consider the implications of random sampling and how to draw a random sample.

12.3 Random Sampling and the Completely Randomized Design

Random sampling—that is, giving every possible sample in a population an equal probability of selection—has two purposes. First, it avoids the possibility of bias introduced by a nonrandom selection of sample elements. For example, a sample selection of voters from telephone directories in 1936 indicated a clear win for Landon. However, the sample did not represent a random selection from the whole population of eligible voters because the majority of telephone users in 1936 were Republicans. As another example, suppose that we sample to determine the percentage of homeowners favoring the construction of a new city park and modify our original random sample by ignoring owners who are not at home. The result may yield a biased response because those at home will likely have children and may be more inclined to favor the new park. The second purpose of random sampling is to provide a probabilistic basis for the selection of a sample. That is, it treats the selection of a sample as an experiment (Chapter 2), enabling the statistician to calculate the probability of an observed sample and to use this probability in making inferences. We learned that under fairly general conditions, the mean, $\bar{y}$, of a random sample of n elements will possess a probability distribution that is approximately normal when n is large (the central limit theorem). Fundamental to the proof of the central limit theorem is the assumption of random sampling. Similarly, the confidence intervals and tests of hypotheses based on Student's t (Chapters 8, 9, and 10) required the assumption of random sampling.

The random selection of independent samples to compare two or more populations is the simplest type of experimental design. The populations differ because we have applied different treatments (factor combinations) and we now wish to consider step 4 in the design of an experiment—deciding how to apply the treatments to the experimental units or, equivalently, how to select the samples.

> ***Definition 12.6:*** *The selection of independent random samples from p populations is called a* completely randomized design.

The comparison of five brands of aspirin, A, B, C, D, and E, by randomly selecting 100 pills from the production of each manufacturer would be a completely randomized design with $n_A = n_B = \cdots = n_E = 100$. Similarly, we might wish to compare five teaching techniques, A, B, C, D, and E, using 25 students in each class ($n_A = n_B = \cdots = n_E = 25$). The populations corresponding to A, B, C, D, and E are nonexistent (that is, they are conceptual) because students taught by the five techniques either do not exist or they are unavailable. Consequently, random samples are obtained for the five conceptual populations by randomly selecting 125 students of the type envisioned for the study and then *randomly* assigning 25 students to each of the five teaching techniques. This scheme will yield independent random samples of 25 students subjected to each technique and will result in a completely randomized design.

As you might surmise, not all designs are completely randomized, but all good experiments utilize randomization to some extent. Examples of designs that restrict randomization will be given in Section 12.5.

Before concluding this section on randomization, let us consider a method for selecting a random sample or randomly assigning experimental units to a set of treatments.

The simplest and most reliable way to select a random sample of n elements from a large population is to employ a table of random numbers such as that shown in Table 13, Appendix III. Random-number tables are constructed so that integers occur randomly and with equal frequency. For example, suppose that the population contains $N = 1000$ elements. Number the elements in sequence, 1 to 1000. Then turn to a table of random numbers such as the excerpt shown in Table 12.1.

Table 12.1 Portion of a Table of Random Numbers

15574	35026	98924
45045	36933	28630
03225	78812	50856
88292	26053	21121

Select n of the random numbers in order. The population elements to be included in the random sample will be given by the first three digits of the random numbers (unless the first four digits are 1000). If $n = 5$, we would include elements numbered 155, 450, 32, 882, and 350. So as not to use the same sequence of random numbers over and over again, the experiment should select different starting points in Table 13 to begin the selection of random numbers for different samples.

The random assignment of 40 experimental units, ten to each of four treatments, A, B, C, and D, is equally easy using the random-number table. Number the experimental units 1 to 40. Select a sequence of random numbers and refer only to the first two digits since $n = 40$. Discard random numbers greater than 40. Then assign the experimental units associated with the first ten numbers that appear to A, those associated with the second ten to B, and so on. To illustrate refer to Table 12.1. Starting in the first column and moving top to bottom, we acquire the numbers 15 and 3 (discard 45 and 88 because no experimental units possess those numbers). This procedure would be continued until all integers 1 to 40 were selected. The resulting numbers would occur in random order.

The analysis of data for a completely randomized design is treated in Section 13.3.

12.4 *Volume-Increasing Experimental Designs*

You will recall that designing to increase the volume of a signal, that is, the information pertinent to one or more population parameters, depends on the selection of treatments and the number of experimental units assigned to each (step 3 in designing an experiment). The treatments, or combinations of factor levels, identify points at which one or more response measurements will be made and indicate the general location in which the experimenter is focusing his attention. As we shall see, some designs contain more information concerning specific population parameters than others for the same number of observations. Other very costly experiments contain no information concerning certain population parameters. And no single design is best in acquiring information concerning all types of population parameters. Indeed, the problem of finding the best design for focusing information on a specific population parameter has been solved in only a few specific cases.

The purpose of this section is not to present a general theory or find the best selection of factor level combinations for a given experiment but rather to present a few examples to illustrate the principles involved. The optimal design providing the maximum amount of information pertinent to the parameter(s) of interest (for a fixed sample size, n) will be given for the following two examples.

The simplest example of an information-focusing experiment is the problem of estimating the difference between a pair of population means, $\mu_1 - \mu_2$, based on independent random samples. In this instance, the two treatments have already been selected and the question concerns the allocation of experimental units to the two samples. If the experimenter plans to invest money sufficient to sample a total of n experimental units, how many units should he select from populations 1 and 2, say n_1 and n_2 ($n_1 + n_2 = n$), respectively, so as to maximize the information in the data pertinent to $(\mu_1 - \mu_2)$? If $n = 10$, should he select $n_1 = n_2 = 5$ observations from each population or would an allocation of $n_1 = 4$ and $n_2 = 6$ be better?

Recall that the estimator of $(\mu_1 - \mu_2)$, $(\bar{Y}_1 - \bar{Y}_2)$, has a standard deviation

$$\sigma_{(\bar{Y}_1 - \bar{Y}_2)} = \sqrt{\frac{\sigma_1^2}{n_1} + \frac{\sigma_2^2}{n_2}}.$$

The smaller $\sigma_{(\bar{Y}_1 - \bar{Y}_2)}$, the smaller will be the corresponding error of estimation and the greater will be the quantity of information in the sample pertinent to $(\mu_1 - \mu_2)$. If, as we frequently assume, $\sigma_1^2 = \sigma_2^2 = \sigma^2$, then

$$\sigma_{(\bar{Y}_1 - \bar{Y}_2)} = \sigma\sqrt{\frac{1}{n_1} + \frac{1}{n_2}}.$$

You can verify that this quantity is a minimum when $n_1 = n_2$ and consequently that the sample contains a maximum of information on $(\mu_1 - \mu_2)$ when the n experimental units are equally divided between the two treatments. A more general case is considered in Example 12.1.

Example 12.1: *If n observations are to be used to estimate $(\mu_1 - \mu_2)$, find n_1 and n_2 so that $V(\bar{Y}_1 - \bar{Y}_2)$ is minimized (assume that $n_1 + n_2 = n$).*

Solution: *Let b denote the fraction of the n observations assigned to the sample from population 1, that is, $n_1 = bn$ and $n_2 = (1 - b)n$. Then*

$$V(\bar{Y}_1 - \bar{Y}_2) = \frac{\sigma_1^2}{bn} + \frac{\sigma_2^2}{(1 - b)n}.$$

To find the fraction, b, which minimizes this variance, we set the first derivative, with respect to b, equal to zero. This process yields

$$-\frac{\sigma_1^2}{n}\frac{1}{b^2} + \frac{\sigma_2^2}{n}\left(\frac{1}{1-b}\right)^2 = 0,$$

$$\frac{\sigma_1^2}{n}\frac{1}{b^2} = \frac{\sigma_2^2}{n}\left(\frac{1}{1-b}\right)^2,$$

$$\frac{b^2}{(1-b)^2} = \frac{\sigma_1^2}{\sigma_2^2},$$

or

$$\frac{b}{1-b} = \frac{\sigma_1}{\sigma_2}.$$

Solving for b, we obtain

$$b = \frac{\sigma_1}{\sigma_1 + \sigma_2}$$

and

$$1 - b = \frac{\sigma_2}{\sigma_1 + \sigma_2}.$$

Thus $V(\bar{Y}_1 - \bar{Y}_2)$ *is minimized when*

$$n_1 = \frac{\sigma_1}{\sigma_1 + \sigma_2} n.$$

That is, sample sizes are allocated proportional to the standard deviations. Note that $n_1 = n/2$ *if* $\sigma_1 = \sigma_2$.

As a second example, consider the problem of fitting a straight line through a set of n points using the least-squares method of Chapter 11 (see Figure 12.1). Further, suppose that we are primarily interested in the slope of the line, β_1, in the linear model

$$Y = \beta_0 + \beta_1 x_1 + \varepsilon.$$

If we have the option of selecting the n values of x for which y will be observed, which values of x will maximize the quantity of information on β_1? We have

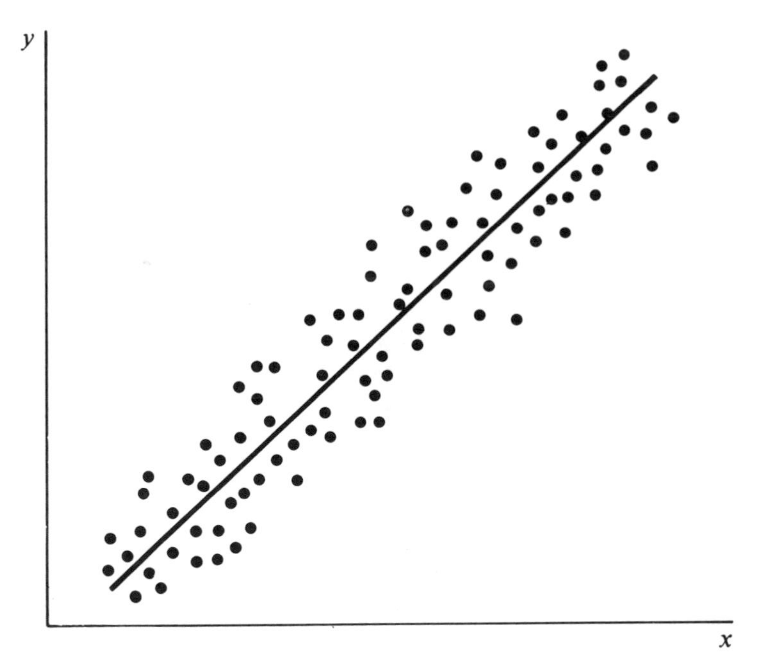

Figure 12.1 Fitting a Straight Line by the Method of Least Squares

one single quantitative factor, x, and have the problem of deciding on the levels to employ, $x_1, x_2, \ldots, x_n$, as well as the number of observations to be taken at each.

A strong lead to the best design for fitting a straight line can be achieved by viewing Figure 12.2(a) and (b). Suppose that y was linearly related to x and generated data similar to that shown in Figure 12.2(a) for the interval $x_1 < x < x_2$. Note the approximate range of variation for a given value of x. Now suppose that instead of the wide range for x employed in Figure 12.2(a), the experimenter selected data from the same population but over the very narrow range $x_3 < x < x_4$, as shown in Figure 12.2(b). The variation in Y, given x, is the same as for Figure 12.2(a). Which distribution of data points would provide the greater amount of information concerning the slope of the line β_1? You might guess (correctly) that the best estimate of slope will occur when the levels of x are selected farther apart, as shown in Figure 12.2(a). The data for Figure 12.2(b) could yield a very inaccurate estimate of the slope and, as a matter of fact, might leave a question as to whether the slope is positive or negative.

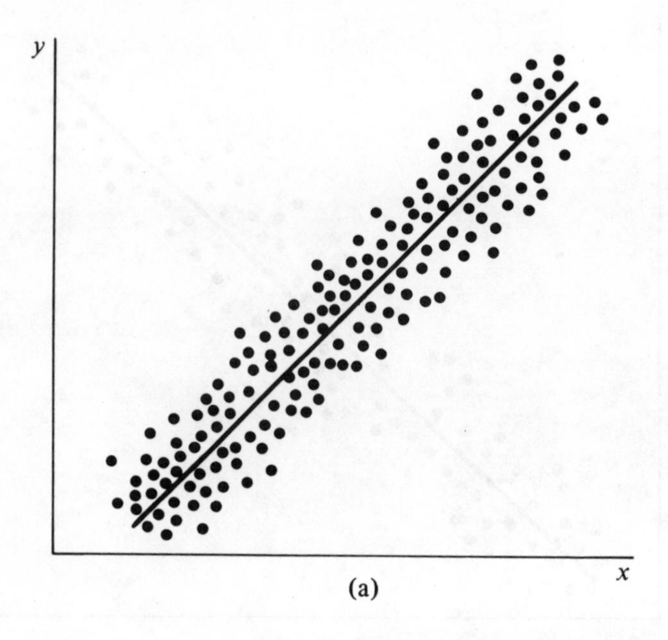

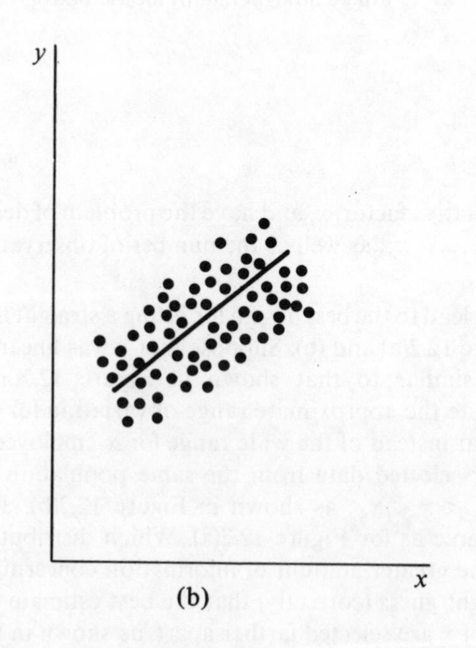

Figure 12.2 Two Different Level Selections for Fitting a Straight Line

The best design for estimating the slope, β_1, can be determined by considering the standard deviation of $\hat{\beta}_1$,

$$\sigma_{\hat{\beta}_1} = \frac{\sigma}{\sqrt{\sum_{i=1}^{n} (x_1 - \bar{x})^2}}.$$

The larger the sum of squares of deviations of $x_1, x_2, \ldots, x_n$ about their mean, the smaller will be the standard deviation of $\hat{\beta}_1$. The experimenter will usually have some experimental region, say $x_1 < x < x_2$, over which he wishes to observe Y, and this range will frequently be selected prior to experimentation. Then the smallest value for $\sigma_{\hat{\beta}_1}$ will occur when the n data points are equally divided with half located at the lower boundary of the region, x_1, and half at the upper boundary, x_2. (The proof is omitted.) If an experimenter wishes to fit a line using $n = 10$ data points in the interval $2 < x < 6$, he should select five data points at $x = 2$ and five at $x = 6$. Before concluding discussion of this example, you should note that observing all values of y at only two values of x will not provide information on curvature of the response curve in case the assumption of linearity in the relation of y and x is incorrect. It is frequently safer to select a few points (as few as one or two) somewhere near the middle of the experimental region to detect curvature if it should be present (see Figure 12.3).

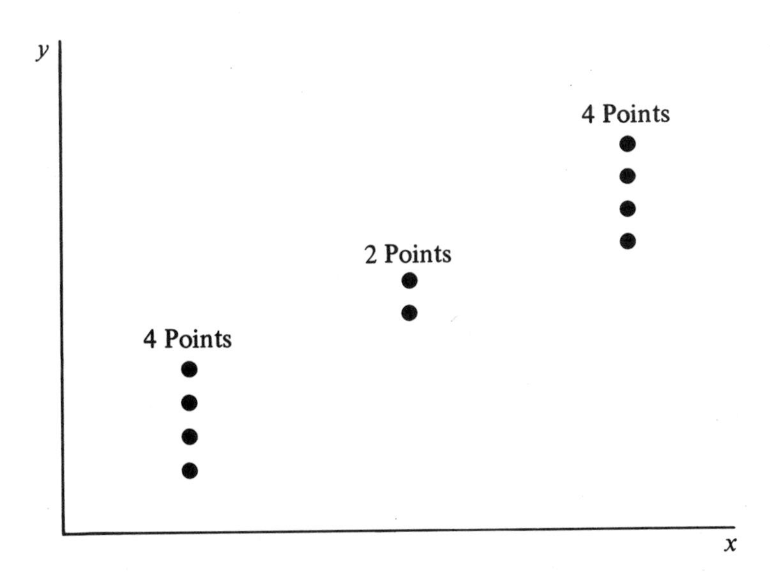

Figure 12.3 A Good Design for Fitting a Straight Line ($n = 10$)

To summarize, we have given optimal designs (factor-level combinations and the allocation of experimental units per combination) for comparing a pair of means and fitting a straight line. These two simple designs illustrate the manner in which information in an experiment can be increased or decreased by the selection of the factor-level combinations that represent treatments and by changing the allocation of observations to treatments. We have demonstrated that factor-level selection and the allocation of experimental units to treatments can greatly affect the information in an experiment pertinent to a particular population parameter and thereby can amplify nature's signal. Thus step 3, page 435 is an important consideration in the design of good experiments.

12.5 *Noise-Reducing Experimental Designs*

Noise-reducing experimental designs increase the information in an experiment by decreasing the background noise (variation) caused by uncontrolled nuisance variables. By serving as filters to screen out undesirable noise, they permit nature's signal to be received more clearly. Reduction of noise can be accomplished in step 4 of the design of an experiment, in the method for assigning treatments to the experimental units.

Designing for noise reduction is based on the single principle of making all comparisons of treatments within relatively homogeneous groups of experimental units. The more homogeneous the experimental units, the easier it is to detect a difference in treatments. Because noise-reducing, or filtering, designs work with blocks of relatively homogeneous experimental units, they are called *block* designs. We will illustrate with an example.

A manufacturer would like to compare two procedures, A and B, for measuring enzyme growth in a fermentation process. The precision of the two methods of measurement appears to be the same, but there is some question as to whether one method produces higher readings than the other.

The experiment could be performed as follows. Samples could be selected at random from the well-mixed vat of fermenting substance. Half of the samples could be measured by method A and half by method B. One could then estimate the difference between the means, $(\mu_A - \mu_B)$, or could test a hypothesis concerning their equality. Assuming near normality for the two populations of measurements, one would use estimation and test procedures based on the t distribution.

A second design for the experiment would choose only half as many samples. Each sample would be divided into two parts, one randomly assigned for analysis by method A and the other by method B.

Although both of the designs described above result in the same number of analyses and approximately the same cost, the second design would be preferable because it might yield a greater amount of information. The variability or noise for the first procedure would be composed of variability between samples of the fermenting substance in the vat and measurement error introduced by methods A and B. Admittedly, we expect the within-vat variability of the measurements to be small because the substance is mixed, but lack of homogeneity between samples, however small, will exist and may be considerably larger than the variability within samples and the measurement error. This within-vat variability would then contribute to the variability of the measurements.

We would like to protect against this contingency. If nonhomogeneity exists, we would expect samples of vat contents near each other to be more nearly homogeneous than those far apart. Hence we would divide each sample and compare A and B within this relatively homogeneous block of experimental material (that is, the substance in the sample). Comparisons between A and B would be made in the presence of the reduced noise of the more homogeneous sample.

The experiment, conducted according to the second design, produced the data listed in Table 12.2.

Table 12.2 Data for a Paired-Difference Experiment

Sample	*Method A*	*Method B*
1	327.6	327.6
2	327.7	327.7
3	327.7	327.6
4	327.9	327.8
5	327.4	327.4
6	327.7	327.6
7	327.8	327.8
8	327.8	327.7
9	327.4	327.3
	$\bar{y}_A = 327.667$	$\bar{y}_B = 327.611$

This is a paired-difference experiment in which the ith paired difference is

$$d_i = Y_{Ai} - Y_{Bi}, \qquad i = 1, 2, 3, \ldots, 9.$$

The expected value of the *i*th paired difference may be found using Theorem 5.8:

$$\mu_d = E(d_i) = E(Y_{Ai} - Y_{Bi}) = E(Y_{Ai}) - E(Y_{Bi})$$

or

$$\mu_d = \mu_A - \mu_B.$$

In other words, we may make inferences regarding the difference between the means of the methods of measurement, $(\mu_A - \mu_B)$, by making inferences regarding the single mean of the differences, μ_d.

The analysis of the paired-difference experiment utilizes the nine paired differences, $d_i, i = 1, 2, \ldots, 9$, shown in Table 12.3.

Table 12.3 The Paired Differences for the Data, Table 12.2

Sample	d_i
1	0
2	0
3	.1
4	.1
5	0
6	.1
7	0
8	.1
9	.1
$\bar{d}$	.056

The sample variance of the $n = 9$ differences is

$$s_d^2 = \sum_{i=1}^{n} \frac{(d_i - \bar{d})^2}{n - 1} = .002778.$$

The null hypothesis,

$$H_0: \mu_A - \mu_B = \mu_d = 0,$$

may be tested using Student's t statistic,

$$t = \frac{\bar{d} - \mu_d}{s_d/\sqrt{n}}.$$

For our example,

$$t = \frac{.056}{.053/\sqrt{9}} = 3.17.$$

Since we do not know whether $\mu_A > \mu_B$ or vice versa, we will employ a two-tailed test. The critical value of t (Table 4, Appendix III) based upon $n - 1 = 8$ degrees of freedom and $\alpha = .05$ is $t = 2.306$. Since the calculated t exceeds this value, we reject the hypothesis that $(\mu_A - \mu_B) = 0$ and conclude that the two methods have different mean responses. Furthermore, we would be reasonably confident that our decision was correct because the probability of a type I error for our test procedure is only $\alpha = .05$.

An indication of the difference in the quantity of information in the unpaired design versus the paired design can be obtained by comparing the relative variability involved in the two resulting methods of analysis. This can most easily be seen by comparing confidence intervals for the two procedures.

The $(1 - \alpha)$ confidence interval for $(\mu_A - \mu_B)$ using the paired differences is

$$\bar{d} \pm t_{\alpha/2}\frac{s_d}{\sqrt{n}}.$$

Substituting, we obtain a 95 percent confidence interval,

$$.056 \pm \frac{(2.306)(.053)}{\sqrt{9}}$$

or

$$.056 \pm .041.$$

Unfortunately, we cannot calculate the $(1 - \alpha)$ confidence interval for $(\mu_A - \mu_B)$ using the unpaired design because the experiment was not conducted in an unpaired manner. Although it is difficult to state exactly what might have happened if this design had been employed, one can approximate the confidence interval by using the variability of the paired-difference data to obtain a pooled estimate of the variance of the experimental error. Specifically, we would assume the population variances to be approximately equal,

$$\sigma_A^2 = \sigma_B^2 = \sigma^2,$$

and estimate σ^2 by a pooled estimate of variance,

$$s^2 = \frac{\sum_{i=1}^{n_1} (y_{Ai} - \bar{y}_A)^2 + \sum_{i=1}^{n_2} (y_{Bi} - \bar{y}_B)^2}{n_1 + n_2 - 2}.$$

Substituting into this formula, we obtain

$$s^2 = \frac{.2400 + .2289}{16} = .02931$$

and

$$s = .171.$$

The 95 percent confidence interval for $(\mu_A - \mu_B)$ from this unpaired analysis of the paired data is

$$(\bar{y}_A - \bar{y}_B) \pm t_{\alpha/2} s \sqrt{\frac{1}{n_1} + \frac{1}{n_2}}$$

$$(327.667 - 327.611) \pm (2.12)(.171)\sqrt{\frac{1}{9} + \frac{1}{9}}$$

or

$$.056 \pm .171.$$

We may now compare the 95 percent confidence intervals for $(\mu_A - \mu_B)$ for the paired and the unpaired analyses of the data for the paired design.

Paired analysis: $.056 \pm .041$,

Unpaired analysis: $.056 \pm .171$.

We reiterate that a paired design must utilize the paired-difference analysis and that an unpaired analysis of the same data is simply a procedure for calculating a confidence interval that would be a good approximation to what might have been obtained had the experiment been conducted in an unpaired manner.

A comparison of the two intervals reveals the substantial difference in the quantity of information concerning $(\mu_A - \mu_B)$ contained in the two designs. Specifically, the paired analysis produces an interval, and hence a bound on the error of estimation, of less than one-fourth that obtained by the unpaired analysis. Since the standard deviation of $(\bar{Y}_A - \bar{Y}_B)$ is inversely proportional to the square root of the sample size, it would require approximately $(4)^2 = 16$ times as many observations in the unpaired analysis to reduce the width of the unpaired confidence interval to that obtained from the paired design. In other words, the reduction in noise obtained by the paired design obtains the necessary information at approximately one-sixteenth the cost required for the unpaired design.

The paired design described above, a simple example of a randomized block design, illustrates the principle of noise reduction in increasing the information in an experiment.

> ***Definition 12.7:*** *A* randomized block design *containing b blocks and p treatments consists of b blocks of p experimental units each. The treatments are randomly assigned to the units in each block with each treatment appearing exactly once in every block.*

The difference between a randomized block design and the completely randomized design can be demonstrated by considering an experiment designed to compare subject reaction to a set of four stimuli (treatments) in a stimulus-response psychological experiment. We will denote the treatments as T_1, T_2, T_3, and T_4.

Suppose that eight subjects are to be randomly assigned to each of the four treatments. Random assignment of subjects to treatments (or vice versa) randomly distributes errors due to person-to-person variability to the four treatments and yields four samples that are, for all practical purposes, random and independent. This would be a completely randomized experimental design.

The experimental error associated with a completely randomized design is composed of a number of components. Some of these are due to the difference between subjects, to the failure of repeated measurements within a subject to be identical (due to the variations in physical and psychological conditions), to the failure of the experimenter to administer a given stimulus with exactly the same intensity in repeated measurements, and, finally, to errors of measurement. Reduction of any of these causes of error will increase the information in the experiment.

The subject-to-subject variation in the above experiment can be eliminated by using subjects as blocks. Each subject would receive each of the four treatments assigned in a random sequence. The resulting randomized block design would appear as in Figure 12.4. Now only eight subjects are required to obtain eight response measurements per treatment. Note that each treatment occurs exactly once in each block.

The word "randomization" in the name of the design implies that the treatments are randomly assigned within a block. For our experiment, position in the block would pertain to the position in the sequence when assigning the stimuli to a given subject over time. The purpose of the randomization (that is, position in the block) is to eliminate bias caused by fatigue or learning.

Blocks may represent time, location, or experimental material. If three treatments are to be compared and there is a suspected trend in the mean response over time, a substantial part of the time-trend variation may be removed by blocking. All three treatments would be randomly applied to experimental

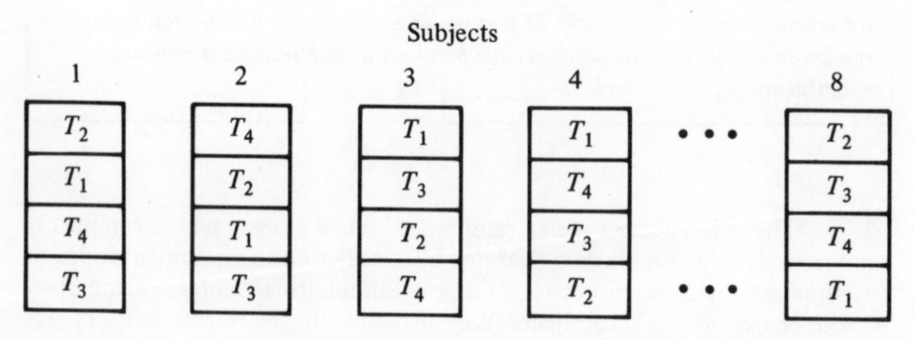

Figure 12.4 A Randomized Block Design

units in one small block of time. This procedure would be repeated in succeeding blocks of time until the required amount of data is collected. A comparison of the sale of competitive products in supermarkets should be made within supermarkets, thus using the supermarkets as blocks and removing market-to-market variability. Animal experiments in agriculture and medicine often utilize animal litters as blocks, applying all of the treatments, one each, to animals within a litter. Because of heredity, animals within a litter are more homogeneous than those between litters. This type of blocking removes the litter-to-litter variation. The analysis of data generated by a randomized block design is discussed in Section 13.7.

The randomized block design is only one of many types of block designs. Blocking in two directions can be accomplished using a Latin-square design. Suppose that the subjects of the preceding example became fatigued as the stimuli were applied so that the last stimulus always produced a lower response than the first. If this trend (and consequent lack of homogeneity of the experimental units in a block) were true for all subjects, a Latin-square design would be appropriate. The design would be constructed as shown in Figure 12.5. Each stimulus is applied once to each subject and occurs exactly once in each position of the order of presentation. All four stimuli occur in each row and in each column of the 4×4 configuration. The resulting design is a 4×4 Latin square. A Latin-square design for three treatments will require a 3×3 configuration and, in general, p treatments will require a $p \times p$ array of experimental units. If more observations are desired per treatment, the experimenter would utilize several Latin-square configurations in one experiment. In the example

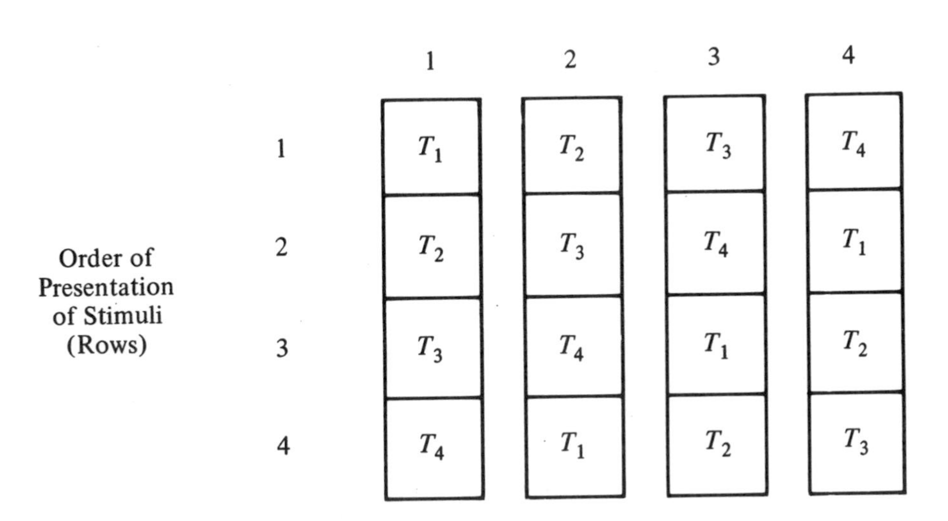

Figure 12.5 A Latin-Square Design

above, it would be necessary to run two Latin squares to obtain eight observations per treatment. The experiment would then contain the same number of observations per treatment as the randomized block design, Figure 12.4.

A comparison of means for any pair of stimuli would eliminate the effect of subject-to-subject variation but would, in addition, eliminate the effect of the fatigue trend within each stimulus because each treatment was applied in each position of the stimuli-time administering sequence. Consequently, the effect of the trend would be canceled in comparing the means.

A more extensive discussion of block designs and their analyses is contained in the references at the end of the chapter. The objective of this section is to make the reader aware of the existence of block designs, how they work, and how they can produce substantial increases in the quantity of information in an experiment by reducing nuisance variation.

12.6 Summary

The object of this chapter has been to identify the factors that affect the quantity of information in an experiment and to use this knowledge to design better experiments. The subject, design of experiments, is very broad and is certainly not susceptible to condensation into a single chapter in an introductory text. In contrast, the philosophy underlying design, the methods for varying information in an experiment, and desirable strategies for design are easily explained and constitute the objective of this chapter.

Two factors affect the quantity of information in an experiment—the volume of nature's signal and the magnitude of variation caused by uncontrolled variables. The volume of information pertinent to a parameter of interest depends on the selection of factor-level combinations (treatments) to be included in the experiment and on the allocation of the total number of experimental units to the treatments. This choice determines the focus of attention of the experimenter.

The second method for increasing the information in an experiment concerns the method for assigning treatments to the experimental units. Blocking, comparing treatments within relatively homogeneous blocks of experimental material, can be used to eliminate block-to-block variation when comparing treatments. As such, it serves as a filter to reduce the unwanted variation that tends to obscure nature's signal.

The selection of factors and the selection of factor levels are important considerations in shifting information in an experiment to amplify the information on a population parameter. The use of blocking in assigning treatments to experimental units reduces the noise created by uncontrolled variables, and, consequently, increases the information in an experiment.

The analysis of some elementary experimental designs is given in Chapter 13. A more extensive treatment of the design and analysis of experiments is a course in itself. The reader interested in exploring this subject is directed to the references at the end of the chapter.

References

1. Cochran, W. G., and G. Cox, *Experimental Designs*, 2nd ed. New York: John Wiley & Sons, Inc., 1957.

2. Hicks, C. R., *Fundamental Concepts in the Design of Experiments*. New York: Holt, Rinehart and Winston, Inc., 1964.

3. Mendenhall, W., *An Introduction to Linear Models and the Design and Analysis of Experiments.* North Scituate, Mass.: Duxbury Press, 1967.

4. Mendenhall, W., L. Ott, and R. L. Scheaffer, *Elementary Survey Sampling.* Belmont, Calif.: Wadsworth Publishing Company, Inc., 1971.

Exercises

12.1. How can one measure the information in a sample pertinent to a specific population parameter?

12.2. Give the two factors that affect the quantity of information in an experiment.

12.3. What is a random sample?

12.4. Give two reasons for the use of random samples.

12.5. A political analyst wishes to select a sample of $n = 20$ people from a population of 2000. Use the random-number table to identify the people to be included in the sample.

12.6. Two drugs, A and B, are to be applied to five rats each. Suppose that the rats are numbered from 1 to 10. Use the random-number table to randomly assign the rats to the two treatments.

12.7. Refer to Exercise 12.6 and suppose that the experiment involved three drugs, A, B, and C, with five rats assigned to each. Use the random-number table to randomly assign the 15 rats to the three treatments.

12.8. A population contains 50,000 voters. Use the random-number table to identify the voters to be included in a random sample of $n = 15$.

12.9. What is a factor?

12.10. If one were to design an experiment, what part of the design procedure would result in signal amplification?

12.11. Refer to Exercise 12.10. What part of the design procedure would result in noise reduction?

12.12. State the steps involved in designing an experiment.

12.13. Could an independent variable be a factor in one experiment and a nuisance variable (a noise contributor) in another?

12.14. Complete the assignment of treatments for the following 3×3 Latin-square design.

	A	
C		
		B

12.15. Suppose that one wishes to compare the mean for two populations and that $\sigma_1^2 = 9$, $\sigma_2^2 = 25$, and $n = 90$. What allocation of $n = 90$ to the two samples will result in the maximum amount of information on $(\mu_1 - \mu_2)$? Does this operation employ the principle of noise reduction or signal amplification?

12.16. Refer to Exercise 12.15. Suppose that we allocate $n_1 = n_2$ observations to each sample. How large must n_1 and n_2 be in order to obtain the same amount of information as that implied by the solution to Exercise 12.15?

12.17. Suppose that one wishes to study the effect of the stimulant digitalis on the blood pressure of rats over a dosage range of $x = 2$ to $x = 5$ units. The response is expected to be linear over the region. Six rats are available for the experiment and each rat can receive only one dose. What dosages of digitalis should be employed in the experiment and how many rats should be run at each dosage to maximize the quantity of information in the experiment? Which aspect of design—noise reduction or signal amplification—is implied in this experiment?

12.18. Refer to Exercise 12.17. Consider two methods for selecting the dosages. Method 1 assigns three rats to the dosage $x = 2$ and three rats to $x = 5$. Method 2 equally spaces the dosages between $x = 2$ and $x = 5$ ($x = 2$, 2.6, 3.2, 3.8, 4.4, 5.0). Suppose that σ is known and that the relationship between $E(Y)$ and x is truly linear (see Chapter 11). How much larger will be the confidence interval for the slope (β_1) for method 2 in comparison with method 1? Approximately how many observations would be required to obtain the same size of confidence interval as obtained by the optimal assignment of method 1?

12.19. Refer to Exercise 12.17. Why might it be advisable to assign one or two points at $x = 3.5$?

12.20. An experiment is to be conducted to compare the effect of digitalis on the contraction of the heart muscle of a rat. The experiment is conducted by removing the heart from a live rat, slicing the heart into thin layers, and treating the layers with a dosage of digitalis. The muscle contraction is then measured. If four dosages (A, B, C, and D) are to be employed, what advantage might be derived by applying A, B, C, and D to a slice of tissue from each rat heart? What principle of design is illustrated by this example?

12.21. Describe the factors that affect the quantity of information in an experiment and the design procedures that control these factors.

12.22. Refer to the paired-difference experiment of Section 12.5 and assume that the measurement receiving treatment i, $i = 1, 2$, in the jth pair, $j = 1, 2, \ldots, n$, is

$$Y_{ij} = \mu_i + P_j + \varepsilon_{ij},$$

where

μ_i = expected response for treatment i, $i = 1, 2, \ldots$

P_j = additive random effect (positive or negative) contributed by the jth pair of experimental units, $j = 1, 2, \ldots, n$

ε_{ij} = random error associated with the experimental unit in the jth pair that receives treatment i

Assume that ε_{ij} are independent normal random variables with $E(\varepsilon_{ij}) = 0$, $V(\varepsilon_{ij}) = \sigma^2$, and P_j are independent normal random variables with $E(P_j) = 0$, $V(P_j) = \sigma_p^2$. Also, assume that P_j and ε_{ij} are independent.

(a) Find $E(Y_{ij})$.

(b) Find $E(\bar{Y}_i)$ and $V(\bar{Y}_i)$ where $\bar{Y}_i$ is the mean of the n observations receiving treatment i, $i = 1, 2, \ldots$.

(c) Let $\bar{d} = \bar{Y}_1 - \bar{Y}_2$. Find $E(\bar{d})$, $V(\bar{d})$, and the probability distribution for $\bar{d}$.

12.23. Refer to Exercise 12.22. Prove that

$$\frac{\bar{d}\sqrt{n}}{S_d}$$

is a Student's t statistic, under $H_0: (\mu_1 - \mu_2) = 0$.

12.24. Refer to Exercise 12.22. Suppose that a completely randomized design is employed for the comparison of the two treatment means. Then a response could still be modeled by the expression

$$Y_{ij} = \mu_i + P_j + \varepsilon_{ij},$$

but the "pair effect" (which will still affect an experimental unit), P_j, will be randomly selected and will likely differ from one of the $2n$ observations to another. Further note that, in contrast to the paired-difference experiment, the "pair effects" will not cancel when you calculate $(\bar{Y}_1 - \bar{Y}_2)$. Compare $V(\bar{Y}_1 - \bar{Y}_2) = V(\bar{d})$ for this design with the paired-difference design of Exercise 12.22. Why is the variance for the completely randomized design usually larger?

12.25. Suppose that you wish to fit a model,

$$Y = \beta_0 + \beta_1 x + \beta_2 x^2 + \varepsilon,$$

to a set of n data points. If the n points are to be allocated at the design points, $x = -1, 0, 1$, what fraction should be assigned to each value of x so as to minimize $V(\hat{\beta}_2)$? (Assume that n is large and that k_1, k_2, and k_3, $k_1 + k_2 + k_3 = 1$, are the fractions of the total number of observations to be assigned at $x = -1, 0$, and 1, respectively.)

13

The Analysis of Variance

13.1 *Introduction*

Most experiments involve a study of the effect of one or more independent variables on a response. In Chapter 12 we learned that a response, Y, can be affected by two types of independent variables, quantitative and qualitative. Those independent variables that can be controlled in an experiment are called factors.

The analysis of data generated by a multivariable experiment requires identification of the independent variables in the experiment. These not only will be factors (controlled independent variables) but also will be directions of blocking. If one studies the wear for three types of tires, A, B, and C, on each of four automobiles, "tire types" is a factor representing a single qualitative variable at three levels. Automobiles are blocks and represent a single qualitative variable at four levels. The response for a Latin-square design depends upon the factors that represent treatments, but it is also affected by two qualitative independent block variables, "rows" and "columns."

It is not possible to present a comprehensive treatment of the analysis of multivariable experiments in a single chapter of an introductory text. However, it is possible to introduce the reasoning upon which one method of analysis, the analysis of variance, is based and to show how the technique is applied to a few common experimental designs.

The application of noise reduction and signal amplification to the design of experiments was illustrated in Chapter 12. In particular, the completely randomized and the randomized block designs were shown to be generalizations of simple designs for the unpaired and paired comparisons of means discussed in Chapters 8, 10, and 12. Treatments correspond to combinations of factor levels and identify the different populations of interest to the experimenter. Chapter 13 presents an introduction to the analysis of variance and gives methods for the analysis of the completely randomized, the randomized block, and the Latin-square designs.

13.2 The Analysis of Variance

The methodology for the analysis of experiments involving several independent variables can best be explained in terms of the linear probabilistic model of Chapter 11. Although elementary and unified, this approach is not susceptible to the condensation necessary for inclusion in an elementary text. Instead, we shall attempt an intuitive discussion using a procedure known as the *analysis of variance*. Actually, the two approaches are connected and the analysis of variance can easily be explained in a general way in terms of the linear model. The interested reader can consult the references at the end of the chapter.

As the name implies, the analysis-of-variance procedure attempts to analyze the variation of a response and to assign portions of this variation to each of a set of independent variables. The reasoning is that response variables vary only because of variation in a set of unknown independent variables. Since the experimenter will rarely, if ever, include all the variables affecting the response in his experiment, random variation in the response is observed even though all independent variables considered are held constant. The objective of the analysis of variance is to locate important independent variables in a study and to determine how they interact and affect the response.

The rationale underlying the analysis of variance can be indicated best with a symbolic discussion. The actual analysis of variance—that is, "how to do it"—can be illustrated with an example.

You will recall that the variability of a set of n measurements is proportional to the sum of squares of deviations, $\sum_{i=1}^{n} (y_i - \bar{y})^2$, and that this quantity is used to calculate the sample variance. The analysis of variance partitions the sum of squares of deviations, called the *total sum of squares of deviations*, into parts, each of which is attributed to one of the independent variables in the experiment, plus a remainder that is associated with random error. This can be shown diagrammatically as indicated in Figure 13.1 for three independent

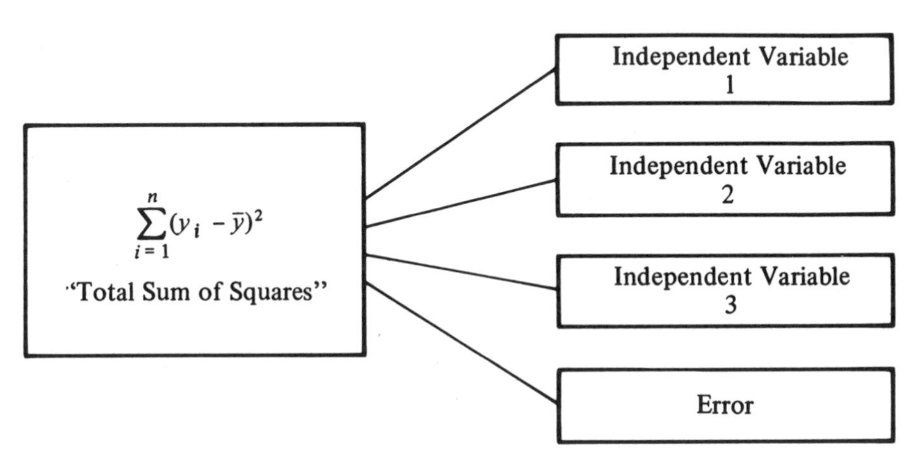

Figure 13.1 The Partitioning of the Total Sum of Squares of Deviations

variables. If a multivariate linear model were written for the response, as suggested in Chapter 11, the portion of the total sum of squares of deviations assigned to error would be labeled SSE.

For the cases that we consider, and when the independent variables are unrelated to the response, it can be shown that each of the pieces of the total sum of squares of deviations, divided by an appropriate constant, provides an independent and unbiased estimator of σ^2, the variance of the experimental error. When a variable is highly related to the response, its portion (called the "sum of squares" for that variable) will be inflated. This condition can be detected by comparing the estimate of σ^2 for a particular independent variable with that obtained from SSE using an F test (see Section 10.6). If the estimate for the independent variable is significantly larger, the F test will reject a hypothesis of "no effect for the independent variable" and produce evidence to indicate a relation to the response.

The mechanism involved in an analysis of variance can best be illustrated by considering a familiar example, say the comparison of means for the unpaired experiment for the special case where $n_1 = n_2$. This experiment, formerly analyzed by the use of Student's t, will now be approached from another point of view. The total variation of the response measurements about their mean for the two samples is

$$\text{Total SS} = \sum_{i=1}^{n} \sum_{j=1}^{n} (y_{ij} - \bar{y})^2,$$

where y_{ij} denotes the jth observation in the ith sample and $\bar{y}$ is the mean of all $2n_1 = n$ observations. This quantity can be partitioned into two parts. That is,

$$\text{Total SS} = \sum_{i=1}^{2} \sum_{j=1}^{n_1} (y_{ij} - \bar{y})^2$$

$$= n_1 \sum_{i=1}^{2} (\bar{y}_i - \bar{y})^2 + \sum_{i=1}^{2} \sum_{j=1}^{n_1} (y_{ij} - \bar{y}_i)^2$$

(proof deferred to Section 13.4), where $\bar{y}_i$ is the average of the observations in the ith sample, $i = 1, 2$, and the second quantity on the right-hand side of the equality is simply the pooled sum of squares of deviations used to calculate s^2. (Recall that we assume equal population variances for this statistical method.) We will denote this quantity as SSE.

Note that we have partitioned the "total sum of squares" of deviations into two parts. One part, SSE, can be used to obtain a pooled estimator of σ^2. The other part,

$$n_1 \sum_{i=1}^{2} (\bar{y}_i - \bar{y})^2 = \frac{n_1}{2}(\bar{y}_1 - \bar{y}_2)^2 = \text{SST},$$

which we will call the *sum of squares for treatments*, will increase as $(\bar{y}_1 - \bar{y}_2)$ increases. Hence, the larger SST, the greater the weight of evidence to indicate a difference in $(\mu_1 - \mu_2)$.

How large is "large?" When will SST be large enough to indicate a real difference between μ_1 and μ_2?

As indicated in Chapter 8,

$$s^2 = \text{MSE} = \frac{\text{SSE}}{n_1 + n_2 - 2} = \frac{\text{SSE}}{2n_1 - 2} \qquad (\text{since } n_1 = n_2)$$

provides an unbiased estimator of σ^2. (MSE denotes mean square for error.) Also, when the null hypothesis is true (that is, $\mu_1 = \mu_2$), SST divided by an appropriate number of degrees of freedom yields a second unbiased estimator of σ^2, which we will denote as MST. For this example, the number of degrees of freedom for MST is equal to 1. When the null hypothesis is true, MSE and MST (the mean square for treatments) will estimate the same quantity and should be "roughly" of the same magnitude. When the null hypothesis is false and $\mu_1 \neq \mu_2$, MST will tend to be larger than MSE. We will show in Section 13.4 that $E(\text{MST}) = \sigma^2$ when $\mu_1 = \mu_2$.

The preceding discussion, together with a review of the variance ratio (Section 10.6), suggests the use of

$$\frac{\text{MST}}{\text{MSE}}$$

as a test statistic to test the hypothesis, $\mu_1 = \mu_2$, against the alternative, $\mu_1 \neq \mu_2$. Indeed, when both populations are normally distributed, it can be shown that MST and MSE are independent and that

$$F = \frac{\text{MST}}{\text{MSE}}$$

follows the F probability distribution of Section 10.6. Disagreement with the null hypothesis is indicated by a large value of F, and hence the rejection region for a given α will be

$$F \geq F_{\alpha}.$$

The analysis-of-variance test results in a one-tailed F test. The degrees of freedom for F will be those associated with MST and MSE, which we will denote as ν_1 and ν_2, respectively. Although we have not indicated, in general, how one determines ν_1 and ν_2, $\nu_1 = 1$ and $\nu_2 = (2n_1 - 2)$ for the two-sample experiment described.

Proof of the independence of MST and MSE will not be considered here because it is a special case of a more general and more complicated theorem that applies to many different analyses of variance. The interested reader is referred to Graybill's text [1].

Example 13.1: *The coded values for the measure of elasticity in plastic, prepared by two different processes, for samples of six drawn randomly from each of the two processes, are as follows:*

A	*B*
6.1	9.1
7.1	8.2
7.8	8.6
6.9	6.9
7.6	7.5
8.2	7.9

Do the data present sufficient evidence to indicate a difference in mean elasticity for the two processes?

Solution: *Although the Student's t could be used as the test statistic for this example, we shall use our analysis-of-variance F test since it is more general and can be used to compare more than two means.*

The three desired sums of squares of deviations are

$$\text{Total SS} = \sum_{i=1}^{2}\sum_{j=1}^{6}(y_{ij} - \bar{y})^2 = \sum_{i=1}^{2}\sum_{j=1}^{6} y_{ij}^2 - \frac{\left(\sum_{i=1}^{2}\sum_{j=1}^{6} y_{ij}\right)^2}{12}$$

$$= 711.35 - \frac{(91.9)^2}{12} = 7.5492,$$

$$\text{SST} = n_1 \sum_{i=1}^{2}(\bar{y}_i - \bar{y})^2 = 6\sum_{i=1}^{2}(\bar{y}_i - \bar{y})^2 = 1.6875,$$

$$\text{SSE} = \sum_{i=1}^{2}\sum_{j=1}^{6}(y_{ij} - \bar{y}_i)^2 = 5.8617.$$

(You may verify that SSE *is the pooled sum of squares of the deviations for the two samples. Also, note that* Total SS = SST + SSE.*) The mean squares for treatment and error are, respectively,*

$$\text{MST} = \frac{\text{SST}}{1} = 1.6875,$$

$$\text{MSE} = \frac{\text{SSE}}{2n_1 - 2} = \frac{5.8617}{10} = .58617.$$

To test the null hypothesis, $\mu_1 = \mu_2$, we compute the test statistic

$$F = \frac{\text{MST}}{\text{MSE}} = \frac{1.6875}{.58617} = 2.88.$$

The critical value of the F statistic for $\alpha = .05$ is 4.96. Although the mean square for treatments is almost three times as large as the mean square for error, it is not large enough to reject the null hypothesis. Consequently, there is not sufficient evidence to indicate a difference between μ_1 and μ_2.

As noted, the purpose of the preceding example was to illustrate the computations involved in a simple analysis of variance. The F test for comparing *two means is equivalent to a Student's t test because an F statistic with one degree of freedom in the numerator is equal to t^2. You can easily verify that the square of $t_{.025} = 2.228$ (used for the two-tailed test with $\alpha = .05$ and $v = 10$ degrees of freedom) is equal to $F_{.05} = 4.96$. Similarly, the value of the t statistic for Example 13.1 would equal the square root of the computed $F = 2.88$.*

13.3 A Comparison of More Than Two Means

An analysis of variance to detect a difference in a set of more than two population means is a simple generalization of the analysis of variance of Section 13.2. The random selection of independent samples from p populations is known as a *completely randomized experimental design*.

Assume that independent random samples have been drawn from p normal populations with means $\mu_1, \mu_2, \ldots, \mu_p$, respectively, and variance σ^2. All populations are assumed to possess equal variances. To be completely general, we shall allow the sample sizes to be unequal and let n_i, $i = 1, 2, \ldots, p$, be the number in the sample drawn from the ith population. The total number of observations in the experiment will be $n = n_1 + n_2 + \cdots + n_p$.

Let y_{ij} denote the measured response on the jth experimental unit in the ith sample and let T_i and $\overline{T}_i$ represent the total and mean, respectively, for the observations in the ith sample. (The modification in the symbols for sample totals and averages will simplify the computing formulae for the sums of squares.) Then, as in the analysis of variance involving two means,

$$\text{Total SS} = \text{SST} + \text{SSE}$$

(proof deferred to Section 13.4), where

$$\text{Total SS} = \sum_{i=1}^{p} \sum_{j=1}^{n_i} (y_{ij} - \bar{y})^2 = \sum_{i=1}^{p} \sum_{j=1}^{n_i} y_{ij}^2 - \text{CM},$$

$$\text{CM} = \frac{(\text{total of all observations})^2}{n}$$

$$= \frac{\left(\sum_{i=1}^{p}\sum_{j=1}^{n_i} y_{ij}\right)^2}{n} = n\bar{y}^2$$

(the symbol CM denotes "correction for the mean"),

$$\text{SST} = \sum_{i=1}^{p} n_i(\bar{T}_i - \bar{y})^2 = \sum_{i=1}^{p} \frac{T_i^2}{n_i} - \text{CM},$$

$$\text{SSE} = \text{Total SS} - \text{SST}.$$

Although the easy way to compute SSE is by subtraction as shown above, it is interesting to note that SSE is the pooled sum of squares for all p samples and is equal to

$$\text{SSE} = \sum_{i=1}^{p}\sum_{j=1}^{n_i} (y_{ij} - \bar{T}_i)^2.$$

The unbiased estimator of σ^2 based on $(n_1 + n_2 + \cdots + n_p - p)$ degrees of freedom is

$$s^2 = \text{MSE} = \frac{\text{SSE}}{n_1 + n_2 + \cdots + n_p - p}.$$

The mean square for treatments will possess $(p - 1)$ degrees of freedom, that is, one less than the number of means, and

$$\text{MST} = \frac{\text{SST}}{p - 1}.$$

To test the null hypothesis,

$$H_0: \mu_1 = \mu_2 = \cdots = \mu_p,$$

against the alternative that at least one of the equalities does not hold, MST is compared with MSE using the F statistic based upon $\nu_1 = (p - 1)$ and $\nu_2 = \left(\sum_{i=1}^{p} n_i - p\right) = (n - p)$ degrees of freedom. The null hypothesis will be rejected if

$$F = \frac{\text{MST}}{\text{MSE}} > F_\alpha,$$

where F_α is the critical value of F for probability of a type I error, α.

Intuitively, the greater the difference between the observed treatment means, $\bar{T}_1, \bar{T}_2, \ldots, \bar{T}_p$, the greater will be the evidence to indicate a difference between their corresponding population means. It can be seen from the above expression that SST = 0 when all the observed treatment means are identical, because then $\bar{T}_1 = \bar{T}_2 = \cdots = \bar{T}_p = \bar{y}$ and the deviations appearing in SST, $(\bar{T}_i - \bar{y}), i = 1, 2, \ldots, p$, will equal zero. As the treatment means get farther apart, the deviations, $(\bar{T}_i - \bar{y})$, will increase in absolute value and SST will increase in magnitude. Consequently, the larger the value of SST, the greater will be the weight of evidence favoring a rejection of the null hypothesis. This same line of reasoning will apply to the F tests employed in the analyses of variances for all designed experiments.

The assumptions underlying the analysis-of-variance F tests should receive particular attention. The samples are assumed to have been randomly selected from the p populations in an independent manner. The populations are assumed to be normally distributed with equal variances, σ^2, and means, $\mu_1, \mu_2, \ldots, \mu_p$. Moderate departures from these assumptions will not seriously affect the properties of the test. This is particularly true of the normality assumption.

Example 13.2: *Four groups of students were subjected to different teaching techniques and tested at the end of a specified period of time. As a result of dropouts from the experimental groups (due to sickness, transfer, and so on), the number of students varied from group to group. Do the data shown present*

sufficient evidence to indicate a difference in the mean achievement for the four teaching techniques?

	1	*2*	*3*	*4*
	65	75	59	94
	87	69	78	89
	73	83	67	80
	79	81	62	88
	81	72	83	
	69	79	76	
		90		
T_i	454	549	425	351
n_i	6	7	6	4
$\bar{T}_i$	75.67	78.43	70.83	87.75

Solution:

$$\text{CM} = \frac{\left(\sum_{i=1}^{4}\sum_{j=1}^{n_i} y_{ij}\right)^2}{n} = \frac{(1779)^2}{23} = 137{,}601.8$$

$$\text{Total SS} = \sum_{i=1}^{4}\sum_{j=1}^{n_i} y_{ij}^2 - \text{CM}$$

$$= 139{,}511 - 137{,}601.8 = 1909.2,$$

$$\text{SST} = \sum_{i=1}^{4} \frac{T_i^2}{n_i} - \text{CM}$$

$$= 138{,}214.4 - 137{,}601.8 = 712.6,$$

$$\text{SSE} = \text{Total SS} - \text{SST} = 1196.6.$$

The mean squares for treatment and error are

$$\text{MST} = \frac{\text{SST}}{p-1} = \frac{712.6}{3} = 237.5,$$

$$\text{MSE} = \frac{\text{SSE}}{n_1 + n_2 + \cdots + n_p - p} = \frac{\text{SSE}}{n-p}$$

$$= \frac{1196.6}{19} = 63.0.$$

The test statistic for testing the hypothesis, $\mu_1 = \mu_2 = \mu_3 = \mu_4$, *is*

$$F = \frac{\text{MST}}{\text{MSE}} = \frac{237.5}{63.0} = 3.77,$$

where

$$\nu_1 = p - 1 = 3,$$

$$\nu_2 = \sum_{i=1}^{p} n_i - 4 = 19.$$

The critical value of F for $\alpha = .05$ *is* $F_{.05} = 3.13$. *Since the computed value of F exceeds* $F_{.05}$, *we reject the null hypothesis and conclude that the evidence is sufficient to indicate a difference in mean achievement for the four teaching procedures.*

You may feel that the above conclusion could have been made on the basis of visual observation of the treatment means. It is not difficult to construct a set of data that will lead the "visual" decision maker to erroneous results.

13.4 Proof of Additivity of the Sums of Squares and E(MST) for a Completely Randomized Design

The proof that

$$\text{Total SS} = \text{SST} + \text{SSE}$$

for the completely randomized design is presented in this section for the benefit of the interested reader. It may be omitted without loss of continuity.

The proof utilizes elementary results on summations that appear in the exercises for Chapter 1, and the device of adding and subtracting $\bar{T}_i$ within the expression for the Total SS. Thus

$$\begin{aligned}
\text{Total SS} &= \sum_{i=1}^{p}\sum_{j=1}^{n_i}(y_{ij} - \bar{y})^2 \\
&= \sum_{i=1}^{p}\sum_{j=1}^{n_i}(y_{ij} - \bar{T}_i + \bar{T}_i - \bar{y})^2 \\
&= \sum_{i=1}^{p}\sum_{j=1}^{n_i}[(y_{ij} - \bar{T}_i) + (\bar{T}_i - \bar{y})]^2 \\
&= \sum_{i=1}^{p}\sum_{j=1}^{n_i}[(y_{ij} - \bar{T}_i)^2 \\
&\qquad + 2(y_{ij} - \bar{T}_i)(\bar{T}_i - \bar{y}) + (\bar{T}_i - \bar{y})^2].
\end{aligned}$$

Summing first over j, we obtain

$$\begin{aligned}
\text{Total SS} = \sum_{i=1}^{p}\Bigg[&\sum_{j=1}^{n_i}(y_{ij} - \bar{T}_i)^2 \\
&+ 2(\bar{T}_i - \bar{y})\sum_{j=1}^{n_i}(y_{ij} - \bar{T}_i) + n_i(\bar{T}_i - \bar{y})^2\Bigg],
\end{aligned}$$

where

$$\sum_{j=1}^{n_i}(y_{ij} - \bar{T}_i) = T_i - n_i\bar{T}_i = T_i - T_i = 0.$$

Consequently, the middle term in the expression for the Total SS is equal to zero.

Then, summing over i, we obtain

$$\begin{aligned}
\text{Total SS} &= \sum_{i=1}^{p}\sum_{j=1}^{n_i}(y_{ij} - \bar{T}_i)^2 + \sum_{i=1}^{p} n_i(\bar{T}_i - \bar{y})^2, \\
&= \text{SSE} + \text{SST}.
\end{aligned}$$

The first expression is SSE, the pooled sum of squares of deviations of the sample measurements about their respective means. The second is the formula for SST.

Proof of the additivity of the analysis-of-variance sums of squares for other experimental designs can be obtained in a similar manner. The procedure is tedious.

We will now derive the expected value of MST for a completely randomized design. Assume that Y_{ij}, the jth sampled value from the ith population, has $E(Y_{ij}) = \mu_i$ and $V(Y_{ij}) = \sigma^2$, $i = 1, \ldots, p$, $j = 1, \ldots, n_i$. Since $\overline{T}_i$ is the average of n_i independent random variables, Y_{ij}, $j = 1, \ldots, n_i$, it follows that $E(\overline{T}_i) = \mu_i$ and $V(\overline{T}_i) = \sigma^2/n_i$. Similarly, $\overline{Y}$ is given by

$$\overline{Y} = \frac{1}{n}\sum_{i=1}^{p}\sum_{j=1}^{n_i} Y_{ij},$$

and hence

$$E(\overline{Y}) = \frac{1}{n}\sum_{i=1}^{p}\sum_{j=1}^{n_i} \mu_i = \frac{1}{n}\sum_{i=1}^{p} n_i\mu_i.$$

We will denote $E(\overline{Y})$ by $\bar{\mu}$.

From Theorem 5.8,

$$V(\overline{Y}) = \frac{1}{n^2}\sum_{i=1}^{p}\sum_{j=1}^{n_i} V(Y_{ij}) = \frac{\sigma^2}{n}.$$

Then

$$\begin{aligned}
E(\text{MST}) &= \frac{1}{p-1} E\left[\sum_{i=1}^{p} n_i(\overline{T}_i - \overline{Y})^2\right] \\
&= \frac{1}{p-1} E\left[\sum_{i=1}^{p} n_i(\overline{T}_i^2 - 2\overline{T}_i\overline{Y} + \overline{Y}^2)\right] \\
&= \frac{1}{p-1} E\left[\sum_{i=1}^{p} n_i\overline{T}_i^2 - n\overline{Y}^2\right] \\
&= \frac{1}{p-1}\left[\sum_{i=1}^{p} n_i E(\overline{T}_i^2) - nE(\overline{Y}^2)\right] \\
&= \frac{1}{p-1}\left[\sum_{i=1}^{p} n_i\left(\frac{\sigma^2}{n_i} + \mu_i^2\right) - n\left(\frac{\sigma^2}{n} + \bar{\mu}^2\right)\right],
\end{aligned}$$

on noting that, for any random variable U, $E(U^2) = V(U) + [E(U)]^2$. It then follows that

$$E(\text{MST}) = \frac{1}{p-1}\left(\sum_{i=1}^{p} \sigma^2 + \sum_{i=1}^{p} n_i\mu_i^2 - \sigma^2 - n\bar{\mu}^2\right)$$
$$= \frac{1}{p-1}\left[\sigma^2(p-1) + \sum_{i=1}^{p} n_i\mu_i^2 - n\bar{\mu}^2\right]$$
$$= \sigma^2 + \frac{1}{p-1}\sum_{i=1} n_i(\mu_i - \bar{\mu})^2.$$

Under $H_0: \mu_1 = \mu_2 = \cdots = \mu_p$, $\mu_i = \bar{\mu}$, $i = 1, \ldots, p$, and hence $E(\text{MST}) = \sigma^2$. Thus MST/MSE is a ratio of unbiased estimators of σ^2 when H_0 is true.

13.5 An Analysis-of-Variance Table for a Completely Randomized Design

The calculations of the analysis of variance are usually displayed in an analysis-of-variance (ANOVA or AOV) table. The table for the design of Section 13.3 involving p treatment means is shown in Table 13.1. The first

Table 13.1 ANOVA Table for a Completely Randomized Design

Source	*d.f.*	SS	MS	F
Treatments	$p-1$	SST	$\text{MST} = \frac{\text{SST}}{p-1}$	$\frac{\text{MST}}{\text{MSE}}$
Error	$n-p$	SSE	$\text{MSE} = \frac{\text{SSE}}{n-p}$	
Total	$n-1$	$\sum_{i=1}^{p}\sum_{j=1}^{n}(y_{ij}-\bar{y})^2$		

column shows the source of each sum of squares of deviations; the second column gives the respective degrees of freedom; the third and fourth columns give the corresponding sums of squares and mean squares, respectively. A calculated value of F, comparing MST and MSE, is usually shown in the fifth column. Note that the degrees of freedom and sums of squares add to their respective totals.

The ANOVA table for Example 13.2, shown in Table 13.2, gives a compact presentation of the appropriate computed quantities for the analysis of variance.

Table 13.2 ANOVA Table for Example 13.2

Source	*d.f.*	SS	MS	*F*
Treatments	3	712.6	237.5	3.77
Error	19	1196.6	63.0	
Total	22	1909.2		

13.6 Estimation for the Completely Randomized Design

Confidence intervals for a single treatment mean and the difference between a pair of treatment means, Section 13.3, are identical to those given in Chapter 8. The confidence interval for the mean of treatment i or the difference between treatments i and j are, respectively,

$$\bar{T}_i \pm \frac{t_{\alpha/2}s}{\sqrt{n_i}}$$

and

$$(\bar{T}_i - \bar{T}_j) \pm t_{\alpha/2}s\sqrt{\frac{1}{n_i} + \frac{1}{n_j}},$$

where

$$s = \sqrt{s^2} = \sqrt{\text{MSE}} = \sqrt{\frac{\text{SSE}}{n_1 + n_2 + \cdots + n_p - p}}$$

and $t_{\alpha/2}$ *is based upon* $(n - p)$ *degrees of freedom.*

Note that the confidence intervals stated above are appropriate for single treatment means or a comparison of a pair of means selected prior to observation of the data. The stated confidence coefficients are based on random sampling. If one were to look at the data and always compare the largest and smallest sample means, the assumption of randomness would be disturbed. Certainly the difference between the largest and smallest sample means is expected to be larger than for a pair selected at random.

Example 13.3: *Find a 95 percent confidence interval for the mean score for teaching technique 1, Example 13.2.*

Solution: *The 95 percent confidence interval for the mean score is*

$$\bar{T}_1 \pm \frac{t_{.025}s}{\sqrt{6}}$$

or

$$75.67 \pm \frac{(2.093)(7.94)}{\sqrt{6}}$$

or

$$75.67 \pm 6.78.$$

Example 13.4: *Find a 95 percent confidence interval for the difference in mean score for teaching techniques 1 and 4, Example 13.2.*

Solution: *The 95 percent confidence interval is*

$$(\bar{T}_1 - \bar{T}_4) \pm (2.093)(7.94)\sqrt{1/6 + 1/4}$$

or

$$-12.08 \pm 10.74.$$

Hence the 95 percent confidence interval for $(\mu_1 - \mu_4)$ *is* -22.82 *to* -1.34. *This suggests that* $\mu_4 > \mu_1$.

13.7 The Analysis of Variance for a Randomized Block Design

The method for constructing a randomized block design was presented in Section 12.5.

The randomized block design implies the presence of two qualitative independent variables, "blocks" and "treatments." Consequently, the total sum of squares of deviations of the response measurements about their mean may be partitioned into three parts, the sums of squares for blocks, treatments, and error.

Denote the total and average of all observations in block i as B_i and $\bar{B}_i$, respectively. Similarly, let T_j and $\bar{T}_j$ represent the total and the average for all observations receiving treatment j. Then, for a randomized block design involving b blocks and p treatments,

$$\text{Total SS} = \text{SSB} + \text{SST} + \text{SSE},$$

$$= \sum_{i=1}^{b} \sum_{j=1}^{p} (y_{ij} - \bar{y})^2 = \sum_{i=1}^{b} \sum_{j=1}^{p} y_{ij}^2 - \text{CM},$$

$$\text{SSB} = p \sum_{i=1}^{b} (\bar{B}_i - \bar{y})^2 = \frac{\sum_{i=1}^{b} B_i^2}{p} - \text{CM},$$

$$\text{SST} = b \sum_{j=1}^{p} (\bar{T}_j - \bar{y})^2 = \frac{\sum_{j=1}^{p} T_j^2}{b} - \text{CM},$$

$$\text{SSE} = \text{Total SS} - \text{SSB} - \text{SST}.$$

In the formulas,

$$\bar{y} = (\text{average of all } n = bp \text{ observations})$$

$$= \frac{\sum_{i=1}^{b} \sum_{j=1}^{p} y_{ij}}{n}$$

and

$$\text{CM} = \frac{(\text{total of all observations})^2}{n}$$

$$= \frac{\left(\sum_{i=1}^{b} \sum_{j=1}^{p} y_{ij}\right)^2}{n}.$$

The analysis of variance for the randomized block design is presented in Table 13.3. The degrees of freedom associated with each sum of squares is

Table 13.3 ANOVA Table for a Randomized Block Design

Source	*d.f.*	SS	MS
Blocks	$b - 1$	SSB	$\frac{\text{SSB}}{b-1}$
Treatments	$p - 1$	SST	$\frac{\text{SST}}{p-1}$
Error	$n - b - p + 1$	SSE	MSE
Total	$n - 1$	Total SS	

shown in the second column. Mean squares are calculated by dividing the sums of squares by their respective degrees of freedom.

To test the null hypothesis "there is no difference in treatment means," we use the F statistic,

$$F = \frac{\text{MST}}{\text{MSE}},$$

and reject if $F > F_\alpha$ based on $\nu_1 = (p - 1)$ and $\nu_2 = (n - b - p + 1)$ degrees of freedom.

Blocking not only reduces the experimental error, it also provides an opportunity to see whether evidence exists to indicate a difference in the mean response for blocks. Under the null hypothesis that there is no difference in mean response for blocks, MSB provides an unbiased estimator for σ^2 based on $(b - 1)$ degrees of freedom. Where real differences exist among block means, MSB will tend to be inflated in comparison with MSE and

$$F = \frac{\text{MSB}}{\text{MSE}}$$

provides a test statistic. As in the test for treatments, the rejection region for the test will be

$$F > F_\alpha,$$

based on $\nu_1 = (b - 1)$ and $\nu_2 = (n - b - p + 1)$ degrees of freedom.

Example 13.5: *A stimulus–response experiment involving three treatments was laid out in a randomized block design using four subjects. The response was the*

Table 13.4

Subjects			
1	*2*	*3*	*4*
① 1.7	③ 2.1	① .1	② 2.2
③ 2.3	① 1.5	② 2.3	① .6
② 3.4	② 2.6	③ .8	③ 1.6

length of time to reaction measured in seconds. The data, arranged in blocks, are shown in Table 13.4. The treatment number is circled and shown above each observation. Do the data present sufficient evidence to indicate a difference in the mean response for stimuli (treatments)? Subjects?

Solution: *The sums of squares for the analysis of variance are shown individually below and jointly in Table 13.5. Thus*

$$\text{CM} = \frac{(\text{total})^2}{n} = \frac{(21.2)^2}{12} = 37.45,$$

$$\text{Total SS} = \sum_{i=1}^{4} \sum_{j=1}^{3} (y_{ij} - \bar{y})^2 = \sum_{i=1}^{4} \sum_{j=1}^{3} y_{ij}^2 - \text{CM}$$

$$= 46.86 - 37.45 = 9.41,$$

$$\text{SSB} = \frac{\sum_{i=1}^{4} B_i^2}{3} - \text{CM} = 40.93 - 37.45 = 3.48,$$

$$\text{SST} = \frac{\sum_{j=1}^{3} T_j^2}{4} - \text{CM} = 42.93 - 37.45 = 5.48,$$

$$\text{SSE} = \text{Total SS} - \text{SSB} - \text{SST}$$

$$= 9.41 - 3.48 - 5.48 = .45.$$

Table 13.5 ANOVA Table

Source	*d.f.*	SS	MS	*F*
Blocks	3	3.48	1.160	15.47
Treatments	2	5.48	2.740	36.53
Error	6	.45	.075	
Total	11	9.41		

We use the ratio of mean-square treatments to mean-square error to test a hypothesis of no difference in the expected response for treatments. Thus

$$F = \frac{\text{MST}}{\text{MSE}} = \frac{2.74}{.075} = 36.53.$$

The critical value of the F statistic ($\alpha = .05$) *for* $v_1 = 2$ *and* $v_2 = 6$ *degrees of freedom is* $F_{.05} = 5.14$. *Since the computed value of F exceeds the critical value, there is sufficient evidence to reject the null hypothesis and conclude that real differences do exist among the expected responses for the three stimuli.*

A similar test may be conducted for the null hypothesis that no difference exists in the mean response for subjects. Rejection of this hypothesis would imply that subject-to-subject variability does exist, and that blocking is desirable. The computed value of F based on $v_1 = 3$ *and* $v_2 = 6$ *degrees of freedom is*

$$F = \frac{\text{MSB}}{\text{MSE}} = \frac{1.16}{.075} = 15.47.$$

Since this value of F exceeds the corresponding tabulated critical value, $F_{.05} = 4.76$, we reject the null hypothesis and conclude that a real difference exists in the expected response for the group of subjects.

13.8 Estimation for the Randomized Block Design

The confidence interval for the difference between a pair of means is exactly the same as for the completely randomized design, Section 13.6. It is

$$(\bar{T}_i - \bar{T}_j) \pm t_{\alpha/2} s \sqrt{\frac{2}{b}},$$

where $n_i = n_j = b$, the number of observations contained in a treatment mean, and $s = \sqrt{\text{MSE}}$. The difference between the confidence intervals for the completely randomized and the randomized block designs is that s, appearing in the expression above, will tend to be smaller than for the completely randomized design.

Similarly, one may construct a $(1 - \alpha)$ confidence interval for the difference between a pair of block means. Each block contains p observations corresponding to the p treatments. Therefore, the confidence interval is

$$(\bar{B}_i - \bar{B}_j) \pm t_{\alpha/2} s \sqrt{\frac{2}{p}}.$$

Example 13.6: *Construct a 95 percent confidence interval for the difference between treatments 1 and 2, Example 13.5.*

Solution: *The confidence interval for the difference in mean response for a pair of treatments is*

$$(\bar{T}_i - \bar{T}_j) \pm t_{\alpha/2} s \sqrt{\frac{2}{b}},$$

where for our example $t_{.025}$ is based upon six degrees of freedom. For treatments 1 and 2 we have

$$(.98 - 2.63) \pm (2.447)(.27)\sqrt{\frac{2}{4}}$$

or

$$-1.65 \pm .47.$$

13.9 The Analysis of Variance for a Latin-Square Design

The method for constructing a Latin-square design for comparing p treatments is presented in Section 12.5. The purpose of the design is to remove unwanted variation as might occur in the mechanized application of icing to cakes on a conveyor belt. Variation in the thickness of icing could occur across the belt due to the variation in pressure at the applicator nozzles. Similarly, the thickness of icing could vary somewhat along the length of the belt due to variations in the consistency of the icing supplied to the machine. Now suppose that we wish to compare three different types of cake mixes, A, B, and C, that result in different porosities which affect absorption of the icing into the cakes. Then the thickness of the resulting icing, y, could be compared for the three treatments (mixes) by employing a 3 × 3 Latin-square design. Each mix would appear in each column (across the conveyor belt) and in each row as one proceeds down the belt. The design configuration is shown in Figure 13.2.

Columns (Positions Across the Belt)

Rows (Positions Down the Belt)	1	2	3
	B	*A*	*C*
	C	*B*	*A*
	A	*C*	*B*

Conveyor Belt

Figure 13.2 A 3 × 3 Latin-Square Design

The three independent variables in a Latin-square design are "rows," "columns," and treatments. All are qualitative variables, although the treatments could be levels of a single quantitative factor or combinations of levels for two or more factors. Thus the total variation in an analysis of variance can be partitioned into four parts, one each corresponding to the variation in rows, columns, treatments, and experimental error. The analysis-of-variance table for a $p \times p$ Latin-square design is shown in Table 13.6. As for previous designs, the four sums of squares add to the total sum of squares of deviations.

Table 13.6 ANOVA Table for a Latin-Square Design

Source	*d.f.*	SS	MS
Rows	$p - 1$	SSR	$\frac{\text{SSR}}{p-1}$
Columns	$p - 1$	SSC	$\frac{\text{SSC}}{p-1}$
Treatments	$p - 1$	SST	$\frac{\text{SST}}{p-1}$
Error	$n - 3p + 2$	SSE	MSE
Total	$n - 1$	Total SS	

The formulas for computing the total sums of squares, SSR, SSC, and SST, are identical to the corresponding formulas given for the randomized block design. Let y_{ij} denote an observation in row i and column j. Then

$$\text{Total SS} = \sum_{j=1}^{p}\sum_{i=1}^{p}(y_{ij} - \bar{y})^2 = \sum_{j=1}^{p}\sum_{i=1}^{p} y_{ij}^2 - \text{CM},$$

where

$$\text{CM} = \frac{(\text{total of all observations})^2}{n}$$

$$= \frac{\left(\sum_{j=1}^{p}\sum_{i=1}^{p} y_{ij}\right)^2}{n}.$$

Similarly, let R_i and $\bar{R}_i$, C_j and $\bar{C}_j$, and T_k and $\bar{T}_k$ represent the total and average for all observations in row i, column j, and treatment k, respectively. Then the sums of squares for rows, columns, and treatments are

$$\text{SSR} = p \sum_{i=1}^{p} (\bar{R}_i - \bar{y})^2 = \frac{\sum_{i=1}^{p} R_i^2}{p} - \text{CM},$$

$$\text{SSC} = p \sum_{i=1}^{p} (\bar{C}_j - \bar{y})^2 = \frac{\sum_{j=1}^{p} C_j^2}{p} - \text{CM},$$

and

$$\text{SST} = p \sum_{i=1}^{p} (\bar{T}_k - \bar{y})^2 = \frac{\sum_{k=1}^{p} T_k^2}{p} - \text{CM}.$$

The sums of squares for error, SSE, can be obtained by subtraction. Thus

$$\text{Total SS} = \text{SSR} + \text{SSC} + \text{SST} + \text{SSE}.$$

Hence

$$\text{SSE} = \text{Total SS} - \text{SSR} - \text{SSC} - \text{SST}.$$

The mean squares corresponding to rows, columns, and treatments can be obtained by dividing the respective sum of squares by $(p - 1)$. Thus

$$\text{MST} = \frac{\text{SST}}{p - 1}.$$

The hypothesis "no difference in mean response for treatments" is tested using the F statistic,

$$F = \frac{\text{MST}}{\text{MSE}}.$$

Similarly, the F statistic can be used to test a hypothesis of no difference between rows (or columns) by using the ratio of MSR (or MSC) to MSE. We will illustrate with an example.

Example 13.7: *An experiment was conducted to investigate the difference in mean time to assemble four different electronic devices, 1, 2, 3, and 4. Two sources of unwanted variation affect the response—the variation between people and the effect of fatigue if a person assembles a series of the devices over time. Consequently, four assemblers were selected and each assembled all four of the devices in the Latin-square design of Table 13.7. (The observed responses, in*

Table 13.7

Rows (Position in Assembly Sequence)	*Columns (Assemblers)*				*Total*
	1	*2*	*3*	*4*	
1	③ 44	① 41	② 30	④ 40	155
2	② 41	③ 42	④ 49	① 49	181
3	① 59	④ 41	③ 59	② 34	193
4	④ 58	② 37	① 53	③ 59	207
Total	202	161	191	182	736

minutes, are shown in the cells. Circled numbers above the observations indicate the treatments employed.) Do the data provide sufficient evidence to indicate a difference in mean time to assemble the four devices? A difference in the mean time to assemble for people? Is there evidence of a fatigue factor (a difference in mean response for positions in the assembly sequence)?

Solution: *The totals for rows, columns, and treatments are as shown in Table 13.8*

Table 13.8

	i			
	1	*2*	*3*	*4*
R_i	155	181	193	207
C_i	202	161	191	182
T_i	202	142	204	188

and

$$\text{CM} = \frac{(736)^2}{16} = \frac{541{,}696}{16} = 33{,}856.0.$$

Then

$$\text{Total SS} = \sum_{j=1}^{4} \sum_{i=1}^{4} y_{ij}^2 - \text{CM}$$

$$= 35{,}186.0 - 33{,}856.0 = 1330.0,$$

$$\text{SSR} = \frac{\sum_{i=1}^{4} R_i^2}{4} - \text{CM}$$

$$= \frac{(155)^2 + (181)^2 + (193)^2 + (207)^2}{4} - \text{CM}$$

$$= 34{,}221.0 - 33{,}856.0 = 365.0,$$

$$\text{SSC} = \frac{\sum_{j=1}^{4} C_j^2}{4} - \text{CM}$$

$$= \frac{(202)^2 + (161)^2 + (191)^2 + (182)^2}{4} - \text{CM}$$

$$= 34{,}082.5 - 33{,}856.0 = 226.5,$$

$$\text{SST} = \frac{\sum_{k=1}^{4} T_k^2}{4} - \text{CM}$$

$$= \frac{(202)^2 + (142)^2 + (204)^2 + (188)^2}{4} - \text{CM}$$

$$= 34{,}482.0 - 33{,}856.0 = 626.0.$$

Finally,

$$\text{SSE} = \text{Total SS} - \text{SSR} - \text{SSC} - \text{SST}$$

$$= 1330.0 - 365.0 - 226.5 - 626.0$$

$$= 112.5.$$

The analysis-of-variance table for this example is shown as Table 13.9. Note that the mean squares were obtained by dividing the sums of squares by their respective degrees of freedom.

Table 13.9

Source	d.f.	SS	MS	F
Rows	3	365.0	121.67	6.49
Columns	3	226.5	75.50	4.03
Treatments	3	626.0	208.67	11.13
Errors	6	112.5	18.75	
Total	15	1330.0		

All the computed F statistics are based on $v_1 = 3$ and $v_2 = 6$ degrees of freedom. The corresponding tabulated critical value is $F_{3,6} = 4.76$ ($\alpha = .05$). A comparison of the computed F statistics with the tabulated value indicates that the computed F's for both rows and treatments exceed the critical value. Thus the data provide sufficient evidence to indicate a difference in the mean time to assemble the four devices. The data also show a difference in mean time to assemble for rows or, equivalently, positions in the sequences of assembly. It appears that fatigue, boredom, or some other factor increases the mean time to assemble as the length of employment increases.

13.10 Estimation for the Latin-Square Design

The $(1 - \alpha)$ confidence interval for the difference between a pair of treatment means is obtained in the same manner as for the completely randomized and the randomized block designs. Since each treatment mean will contain p observations, $n_1 = n_2 = p$ and the standard deviation of the difference between a pair of means is

$$\sigma_{(\bar{T}_i - \bar{T}_j)} = \sigma\sqrt{\frac{1}{p} + \frac{1}{p}}.$$

The $(1 - \alpha)$ confidence interval is

$$(\bar{T}_i - \bar{T}_j) \pm t_{\alpha/2} s\sqrt{\frac{2}{p}}.$$

Corresponding confidence intervals for the difference between a pair of row or column means are, respectively,

$$(\bar{R}_i - \bar{R}_j) \pm t_{\alpha/2} s\sqrt{\frac{2}{p}}$$

and

$$(\bar{C}_i - \bar{C}_j) \pm t_{\alpha/2} s\sqrt{\frac{2}{p}}.$$

Example 13.8: *Refer to Example 13.7. Estimate the difference in mean response between treatments 1 and 2 using a 95 percent confidence interval.*

Solution: *From Table 13.10 observe that* $s^2 = \text{MSE} = 18.75$ *is based on six degrees of freedom. Consequently, the tabulated value,* $t_{\alpha/2}$, *with six degrees of freedom is* $t_{.025} = 2.447$. *Then the 95 percent confidence interval for the* *difference between the treatment means,* $(\mu_1 - \mu_2)$, *is*

$$(\bar{T}_i - \bar{T}_j) \pm t_{\alpha/2} s \sqrt{\frac{2}{p}}$$

or

$$(50.5 - 35.5) \pm (2.447)(4.33)\sqrt{\frac{2}{4}}$$

or

$$15 \pm 7.49.$$

Thus we estimate the difference in mean time to assemble the two devices to be between 7.51 *and* 22.49 *minutes.*

13.11 Selecting the Sample Size

Selecting the sample size for the completely randomized or the randomized block design and the Latin-square design is an extension of the procedures of Section 8.7. We confine our attention to the case of equal sample sizes, $n_1 = n_2 = \cdots = n_p$, for the treatments of the completely randomized design. The number of observations per treatment is equal to b for the randomized block design and for a $b \times b$ Latin-square design. Thus the problem is to select n_1 or b for these three designs so as to purchase a specified quantity of information.

The selection of sample size follows a similar procedure for all three designs; we will outline a general method. First the experimenter must decide on the parameter (or parameters) of major interest. Usually, he will wish to compare a pair of treatment means. Second, he must specify a bound on the error of estimation that he is willing to tolerate. Once determined, he need only select n_i (the number of observations in a treatment mean) or, correspondingly, b (the number of observations in a treatment mean for a randomized block or Latin-square design) that will reduce the half-width of the confidence interval for the parameter so that it is less than or equal to the specified bound on the error of estimation. It should be emphasized that the sample-size solution will *always* be an approximation, since σ is unknown and s is unknown until the sample is acquired. The best available value will be used for s in order to produce an approximate solution. We will illustrate the procedure with an example.

Example 13.9: *A completely randomized design is to be conducted to compare five teaching techniques in classes of equal size. Estimation of the difference in mean response on an achievement test is desired correct to within* 30 *test-score* *points, with probability equal to .95. It is expected that the test scores for a given teaching technique will possess a range approximately equal to 240. Find the approximate number of observations required for each sample in order to acquire the specified information.*

Solution: *The confidence interval for the difference between a pair of treatment means is*

$$(\bar{T}_i - \bar{T}_j) \pm t_{\alpha/2} s \sqrt{\frac{1}{n_i} + \frac{1}{n_j}}.$$

Therefore, we will wish to select n_i and n_j so that

$$t_{\alpha/2} s \sqrt{\frac{1}{n_i} + \frac{1}{n_j}} \leq 30.$$

The value of σ is unknown and s is a random variable. However, an approximate solution for $n_i = n_j$ can be obtained by guessing s to be roughly equal to one-fourth of the range. Thus $s \approx 240/4 = 60$. The value of $t_{\alpha/2}$ will be based upon $(n_1 + n_2 + \cdots + n_5 - 5)$ degrees of freedom, and for even moderate values of n_i, $t_{.025}$ will approximately equal 2. Then

$$t_{.025} s \sqrt{\frac{1}{n_i} + \frac{1}{n_j}} = (2)(60)\sqrt{\frac{2}{n_i}} \leq 30$$

or

$$n_i = 32, \qquad i = 1, 2, \ldots, 5.$$

Example 13.10: *An experiment is to be conducted to compare the toxic effect of three chemicals on the skin of rats. The resistance to the chemicals was expected to vary substantially from rat to rat. Therefore, all three chemicals were to be tested on each rat, thereby blocking out rat-to-rat variability.*

The standard deviation of the experimental error was unknown, but prior experimentation involving several applications of a given chemical on the same rat suggested a range of response measurements equal to 5 units.

Find a value for b such that the error of estimating the difference between a pair of treatment means is less than one unit, with probability equal to .95.

Solution: *A very approximate value for s would be one-fourth the range, or $s \approx 1.25$. Then we wish to select b so that*

$$t_{.025} s \sqrt{\frac{1}{b} + \frac{1}{b}} = t_{.025} s \sqrt{\frac{2}{b}} \leq 1.$$

Since $t_{.025}$ will depend upon the degrees of freedom associated with s^2, which will be $(n - b - p + 1)$, we will guess $t_{.025} \approx 2$. Then

$$(2)(1.25)\sqrt{\frac{2}{b}} \leq 1$$

or

$$b \approx 13, \qquad i = 1, 2, 3.$$

Approximately 13 rats will be required to obtain the desired information.

The degrees of freedom associated with s^2 will be 24, based on this solution. Therefore, the guessed value of t would seem to be adequate for this approximate solution.

The sample-size solutions for Examples 13.9 and 13.10 are very approximate, and are intended to provide only a rough estimate of approximate size and consequent cost of the experiment. The experimenter will obtain information on σ as the data are being collected and can recalculate a better approximation to n as he proceeds.

Selecting the sample size for Latin-square designs follows essentially the same procedure as for the completely randomized and the randomized block designs. The only difference is that one must decide on the number of $p \times p$ Latin squares that will be needed to acquire the desired information. We omit discussion of this topic because we have not shown how to conduct an analysis of variance for more than one $p \times p$ Latin square. The reader interested in this topic should consult the references at the end of the chapter.

13.12 Summary and Comments

The completely randomized, the randomized block, and the Latin-square designs are illustrations of experiments involving one, two, and three qualitative independent variables, respectively. The analysis of variance partitions the total sum of squares of deviations of the response measurements about their mean into portions associated with each independent variable and the experimental error. The former may be compared with the sum of squares for error, using mean squares and the F statistics, to see whether the mean squares for the independent variables are unusually large and thereby indicative of an effect on the response.

In this chapter we have presented a very brief introduction to the analysis of variance and its associated subject, the design of experiments. Experiments can be designed to investigate the effect of many quantitative and qualitative variables on a response. These may be variables of primary interest to the experimenter as well as nuisance variables, such as blocks, which we attempt to separate from the experimental error. These experiments are subject to an analysis of variance when properly designed. A more extensive coverage of the basic concepts of experimental design and the analysis of experiments will be found in the references.

References

1. Graybill, F., *An Introduction to Linear Statistical Models*, Vol. 1. New York: McGraw-Hill Book Company, 1961.
2. Guenther, W. C., *Analysis of Variance*. Englewood Cliffs, N.J.: Prentice-Hall, Inc., 1964.
3. Hicks, C. R., *Fundamental Concepts in the Design of Experiments*. New York: Holt, Rinehart and Winston, Inc., 1964.
4. Li, J. C. R., *Introduction to Statistical Inference*. Ann Arbor, Mich.: J. W. Edwards, Publisher, Inc., 1961.
5. Mendenhall, W., *An Introduction to Linear Models and the Design and Analysis of Experiments*. North Scituate, Mass.: Duxbury Press, 1967.

Exercises

13.1. State the assumptions underlying the analysis of variance of a completely randomized design.

13.2. Refer to Example 13.2. Calculate SSE by pooling the sums of squares of deviations within each of the four samples, and compare with the value obtained by subtraction. Note that this is an extension of the pooling procedure used in the two-sample case discussed in Section 13.2.

13.3. To compare the strengths of concrete produced by four experimental mixes, three specimens were prepared from each type of mix. Each of the

12 specimens was subjected to increasing compressive loads until breakdown. The following compressive loads in tons per square inch were attained at breakdown. Specimen numbers 1–12 are indicated in parentheses for identification purposes.

Mix A	*Mix B*	*Mix C*	*Mix D*
(1) 2.30	(2) 2.20	(3) 2.15	(4) 2.25
(5) 2.20	(6) 2.10	(7) 2.15	(8) 2.15
(9) 2.25	(10) 2.20	(11) 2.20	(12) 2.25

Assuming that the requirements for a completely randomized design are met, analyze the data. State whether there is statistical support at the $\alpha = .05$ level of significance for the conclusion that the four types of concrete differ in average strength.

13.4. Refer to Exercise 13.3. Let μ_A and μ_B denote the mean strengths of concrete specimens prepared for mix *A* and mix *B*, respectively.

(a) Find a 90 percent confidence interval for μ_A.
(b) Find a 95 percent confidence interval for $(\mu_A - \mu_B)$.

13.5. A clinical psychologist wished to compare three methods for reducing hostility levels in university students. A certain psychological test (HLT) was used to measure the degree of hostility. High scores on this test were taken to indicate great hostility. Eleven students obtaining high and nearly equal scores were used in the experiment. Five were selected at random from among the 11 problem cases and treated by method *A*. Three were taken at random from the remaining 6 students and treated by method *B*. The other 3 students were treated by method *C*. All treatments continued throughout a semester. Each student was given the HLT test again at the end of the semester, with the following results:

Method A	*Method B*	*Method C*
73	54	79
83	74	95
76	71	87
68		
80		

(a) Perform an analysis of variance for this experiment.
(b) Do the data provide sufficient evidence to indicate a difference in mean student response for the three methods after treatment?

13.6. Refer to Exercise 13.5. Let μ_A and μ_B, respectively, denote the mean scores at the end of the semester for the populations of extremely hostile students who were treated throughout that semester by method *A* and method *B*.
(a) Find a 95 percent confidence interval for μ_A.
(b) Find a 95 percent confidence interval for μ_B.
(c) Find a 95 percent confidence interval for $(\mu_A - \mu_B)$.
(d) Is it correct to claim that the confidence intervals found in parts (a), (b), and (c) are jointly valid?

13.7. Refer to Exercise 13.3. Suppose that the sand used in the mixes for samples 1, 2, 3, and 4 came from pit *A*, that the sand used for samples 5, 6, 7, and 8 came from pit *B*, and that the sand for the other samples came from pit *C*. Analyze the data, assuming that the requirements for a randomized block are met with three blocks consisting respectively of samples 1, 2, 3, and 4; samples 5, 6, 7, and 8; and samples 9, 10, 11, and 12.
(a) At the 5 percent level, is there evidence of differences in concrete strength due to the sand used?
(b) Is there evidence at the 5 percent level that the four types of concrete differ in average strength?
(c) Does the conclusion of part (b) contradict the conclusion obtained in Exercise 13.3?

13.8. Refer to Exercise 13.7. Let μ_A and μ_B, respectively, denote the mean strengths of concrete specimens prepared from mix *A* and mix *B*.
(a) Find a 95 percent confidence interval for $(\mu_A - \mu_B)$.
(b) Is the interval found in part (a) the same interval found in Exercise 13.4(b)? Explain.

13.9. A study was initiated to investigate the effect of two drugs, administered simultaneously, in reducing human blood pressure. It was decided to utilize three levels of each drug and to include all nine combinations in the experiment. Nine high-blood-pressure patients were selected for the experiment, and one was randomly assigned to each of the nine drug combinations. The response observed was a drop in blood pressure over a fixed interval of time.
(a) Is this a randomized block design?
(b) Suppose that two patients were assigned to each of the nine drug combinations. What type of experimental design is this?

13.10. Refer to Exercise 13.9. Suppose that prior experimentation suggests that $\sigma = 20$.

(a) How many replications would be required to estimate any treatment (drug combination) mean correct to within ± 10 with probability .95?
(b) How many degrees of freedom will be available for estimating σ^2 when using the number of replications determined in part (a)?
(c) Give the approximate half-widths of a confidence interval for the difference in mean response for two treatments when using the number of replications determined in part (a).

13.11. A dealer has in stock three cars (car A, car B, and car C) of the same make and model. Wishing to compare these cars in gas consumption, a customer arranged to test each car with each of three brands of gasoline (brand A, brand B, and brand C). In each trial, a gallon of gasoline was added to an empty tank and the car was driven without stopping until it ran out of gasoline. The following table shows the number of miles covered in each of the nine trials.

Brand of Gasoline	*Distance* (miles)		
	Car A	*Car B*	*Car C*
A	22.4	17.0	19.2
B	20.8	19.4	20.2
C	21.5	18.7	21.2

(a) Should the customer conclude that the three cars differ in gas mileage? Test at the $\alpha = .05$ level.
(b) Do the data indicate that the brand of gasoline affects gas mileage?

13.12. Refer to Exercise 13.11. Suppose that the gas mileage is unrelated to brand of gasoline; carry out an analysis of the data appropriate for a completely randomized design with three treatments.
(a) Should the customer conclude that the three cars differ in gas mileage? Test at the $\alpha = .05$ level.
(b) Comparing your answer for part (a) in Exercise 13.11 with your answer for part (a) above, can you suggest a reason why blocking may be unwise in certain cases?

13.13. A portion of a questionnaire was constructed to enable judges to evaluate a certain aspect of observed classroom teaching. Four films portraying teaching performances, and differing markedly in the teaching characteristic under study, were viewed by each of eight judges. The order of viewing the four films was assigned in a random manner to each judge.

The data obtained are as follows:

	Judges							
Films	*1*	*2*	*3*	*4*	*5*	*6*	*7*	*8*
1	9	10	7	5	12	7	8	6
2	4	9	3	0	6	8	2	4
3	12	16	10	9	11	10	10	14
4	9	11	7	8	12	7	7	8

(a) Give the type of design employed for this experiment and justify your diagnosis.
(b) How many degrees of freedom are available for estimating σ^2? Perform an analysis of variance on the data.
(c) Do the data provide sufficient evidence to indicate that the mean questionnaire score varies from film to film? Test using $\alpha = .05$.
(d) Suppose that the data did provide sufficient evidence to indicate differences among the mean questionnaire scores for the four films. Would this imply that the questionnaire was able to detect a difference in the teaching characteristic exhibited in the four films?

13.14. Refer to Exercise 13.3. About how many specimens per mix should be prepared to allow estimation of the difference in mean strengths for a preselected pair of specimen types to within .02 tons per square inch? Assume knowledge of the data given in Exercise 13.3.

13.15. Refer to Exercise 13.13. About how many judges should be used to allow estimation of the difference in mean questionnaire scores for a preselected pair of films to within one unit? Assume knowledge of the data given in Exercise 13.13.

13.16. A completely randomized design was conducted to compare the effect of five stimuli on reaction time. Twenty-seven people were employed in the experiment, which was conducted using a completely randomized design. Regardless of the results of the analysis of variance, it is desired to compare stimuli *A* and *D*. The reaction times (in seconds) were as follows:

	Stimulus				
	A	*B*	*C*	*D*	*E*
	.8	.7	1.2	1.0	.6
	.6	.8	1.0	.9	.4
	.6	.5	.9	.9	.4
	.5	.5	1.2	1.1	.7
		.6	1.3	.7	.3
		.9	.8		
		.7			
Total	2.5	4.7	6.4	4.6	2.4
Mean	.625	.643	1.067	.920	.48

(a) Conduct an analysis of variance and test for a difference in mean reaction time due to the five stimuli.

(b) Compare stimuli *A* and *D* to see if there is a difference in mean reaction time.

13.17. The experiment in Exercise 13.16 might have been more effectively conducted using a randomized block design with people as blocks since we would expect mean reaction time to vary from one person to another. Hence four people were used in a new experiment and each person was subjected to each of the five stimuli in a random order. The reaction times (in seconds) were as follows:

	Stimulus				
Subject	*A*	*B*	*C*	*D*	*E*
1	.7	.8	1.0	1.0	.5
2	.6	.6	1.1	1.0	.6
3	.9	1.0	1.2	1.1	.6
4	.6	.8	.9	1.0	.4

Conduct an analysis of variance and test for differences in treatments (stimuli).

13.18. An experiment was conducted to determine the effect of three methods of soil preparation on the first-year growth of slash pine seedlings. Four locations (state forest lands) were selected and each location was divided into three plots. Since it was felt that soil fertility within a location was more homogeneous than between locations, a randomized block design was employed using locations as blocks. The methods of soil preparation were A (no preparation), B (light fertilization), and C (burning). Each soil preparation was randomly applied to a plot within each location. On each plot the same number of seedlings were planted, and the observation recorded was the average first-year growth (in centimeters) of the seedlings on each plot.

(a) Conduct an analysis of variance. Do the data provide sufficient evidence to indicate a difference in the mean growth for the three soil preparations?
(b) Is there evidence to indicate a difference in mean growth for the four locations?
(c) Use a 90 percent confidence interval to estimate the difference in mean growth for methods A and B.

Soil Preparation	*Location*			
	1	*2*	*3*	*4*
A	11	13	16	10
B	15	17	20	12
C	10	15	13	10

13.19. Give the analysis of variance for the following 3×3 Latin-square design.

Rows	*Columns*		
	1	*2*	*3*
1	B 12	A 7	C 17
2	C 10	B 7	A 4
3	A 2	C 8	B 12

Find a 95 percent confidence interval for the difference in mean response for treatments A and C assuming that this is a preplanned comparison.

13.20. The following measurements are of the thickness of cake icing for the Latin-square design discussed in Section 13.9. The only difference between this exercise and the text discussion is that five (not three) mixes, *A*, *B*, *C*, *D*, and *E*, were employed in a 5 × 5 Latin-square design. Thickness measurements are given in hundredths of an inch.

	Columns				
Rows	*1*	*2*	*3*	*4*	*5*
1	*D* 9	*E* 18	*C* 6	*A* 8	*B* 11
2	*B* 12	*D* 17	*E* 10	*C* 4	*A* 5
3	*A* 6	*B* 16	*D* 10	*E* 9	*C* 4
4	*C* 4	*A* 13	*B* 11	*D* 8	*E* 13
5	*E* 14	*C* 11	*A* 7	*B* 10	*D* 15

Do the data provide sufficient evidence to indicate a difference in mean thickness of icing for the five cake mixtures? Estimate the difference in mean thickness for mixtures *A* and *B*.

13.21. An experiment was conducted to investigate the toxic effect of three chemicals, *A*, *B*, and *C*, on the skin of rats. One-inch squares of skin were treated with the chemicals and then scored from 0 to 10, depending on the degree of irritation. Three adjacent 1-inch squares were marked on the backs of eight rats, and each of the three chemicals was applied to each rat. The experiment was blocked on rats to eliminate the variation in skin sensitivity from rat to rat. The data are as follows:

1	*2*	*3*	*4*	*5*	*6*	*7*	*8*
B 5	*A* 9	*A* 6	*C* 6	*B* 8	*C* 5	*C* 5	*B* 7
A 6	*C* 4	*B* 9	*B* 8	*C* 8	*A* 5	*B* 7	*A* 6
C 3	*B* 9	*C* 3	*A* 5	*A* 7	*B* 7	*A* 6	*C* 7

(a) Do the data provide sufficient evidence to indicate a difference in the toxic effect of the three chemicals?
(b) Estimate the difference in mean score for chemicals A and B using a 95 percent confidence interval.

13.22. Refer to Exercise 13.21. Approximately how many rats would be required to estimate the difference in mean scores for two chemicals, correct to within one unit?

13.23. Consider the following one-way classification consisting of three treatments, A, B, and C, where the number of observations per treatment varies from treatment to treatment.

A	B	C
24.2	24.5	26.0
27.5	22.7	
25.9		
24.7		

Do the data present sufficient evidence to indicate a difference between treatments?

13.24. Show that

$$\text{total SS} = \text{SST} + \text{SSB} + \text{SSE}$$

for a randomized block design, where $\text{SSE} = \sum_{i=1}^{b} \sum_{j=1}^{p} (Y_{ij} - \bar{B}_i - \bar{T}_j + \bar{Y})^2$."

13.25. Consider the following model for the response measured for a randomized block design containing b blocks and p treatments:

$$Y_{ij} = \mu + \beta_i + \tau_j + \varepsilon_{ij},$$

where

Y_{ij} = response taken on treatment j in block i,

β_i = nonrandom additive effect due to the ith block,

$$\sum_{i=1}^{b} \beta_i = 0,$$

τ_j = nonrandom additive effect due to the jth treatment,

$$\sum_{j=1}^{p} \tau_j = 0,$$

and ε_{ij}, $i = 1, 2, \ldots, b$ and $j = 1, 2, \ldots, p$, are independent normal random variables with $E(\varepsilon_{ij}) = 0$, $V(\varepsilon_{ij}) = \sigma^2$.

(a) Give the expected value and variance of Y_{ij}.

(b) Let T_j and $\overline{T}_j$ be the total and mean of all observations receiving treatment j. Find the expected value and variance of $\overline{T}_j$. Is $\overline{T}_j$ an unbiased estimator of the mean response for treatment j? Explain.

(c) Let $(\overline{T}_j - \overline{T}_k)$ be the difference in means for the observations receiving treatments j and k. Find the expected value and variance of $(\overline{T}_j - \overline{T}_k)$. Is $(\overline{T}_j - \overline{T}_k)$ an unbiased estimator of the differences in the effects of treatments j and k?

13.26. Refer to Exercise 13.25.

(a) Find E(MST).

(b) Find E(MSB).

(c) Find E(MSE).

(Note that these quantities appear in the F statistic used to test for differences in mean responses among blocks and among treatments.)

13.27. Suppose that $Y_1, \ldots, Y_n$ is a random sample from a normal distribution with mean μ and variance σ^2. The independence of $\sum_{i=1}^{n} (Y_i - \overline{Y})^2$ and $\overline{Y}$ can be shown as follows. Define an $n \times n$ matrix $\mathbf{A}$ by

$$\mathbf{A} = \begin{bmatrix} \frac{1}{\sqrt{n}} & \frac{1}{\sqrt{n}} & & \cdots & & \frac{1}{\sqrt{n}} \\ \frac{1}{\sqrt{2}} & \frac{-1}{\sqrt{2}} & 0 & \cdots & 0 & 0 \\ \frac{1}{\sqrt{2 \cdot 3}} & \frac{1}{\sqrt{2 \cdot 3}} & \frac{-2}{\sqrt{2 \cdot 3}} & 0 \cdots & 0 & 0 \\ \vdots & \vdots & \vdots & & & \vdots \\ \frac{1}{\sqrt{(n-1)n}} & \frac{1}{\sqrt{(n-1)n}} & & \cdots & & \frac{-(n-1)}{\sqrt{(n-1)n}} \end{bmatrix}$$

and note that $\mathbf{A}'\mathbf{A} = \mathbf{I}$, the identity matrix. Then

$$\sum_{i=1}^{n} Y_i^2 = \mathbf{Y}'\mathbf{Y} = \mathbf{Y}'\mathbf{A}'\mathbf{A}\mathbf{Y},$$

where $\mathbf{Y}$ is the vector of Y_i's.

(a) Show that

$$\mathbf{AY} = \begin{bmatrix} \sqrt{n}\,\overline{Y} \\ u_1 \\ u_2 \\ \vdots \\ u_{n-1} \end{bmatrix},$$

where $u_1, u_2, \ldots, u_{n-1}$ are linear functions of $Y_1, \ldots, Y_n$. Thus

$$\sum_{i=1}^{n} Y_i^2 = n\overline{Y}^2 + \sum_{i=1}^{n-1} u_i^2.$$

(b) Show that the linear functions $\sqrt{n}\,\overline{Y}, u_1, \ldots, u_{n-1}$, are pairwise orthogonal, and hence independent under the normality assumption.

(c) Show that

$$\sum_{i=1}^{n} (Y_i - \overline{Y})^2 = \sum_{i=1}^{n-1} u_i^2$$

and conclude that this quantity is independent of $\overline{Y}$.

(d) Using the results of (c), show that

$$\frac{\sum_{i=1}^{n} (Y_i - \overline{Y})^2}{\sigma^2} = \frac{(n-1)S^2}{\sigma^2}$$

has a χ^2 distribution with $(n - 1)$ degrees of freedom.

13.28. Consider a completely randomized design with p treatments. Assume that Y_{ij} is the jth response receiving treatment i and that Y_{ij} is normal with mean μ_i and variance σ^2, $j = 1, \ldots, n_i$, $i = 1, \ldots, p$.

(a) Use Exercise 13.27 to justify that $\overline{Y}_1, \overline{Y}_2, \ldots, \overline{Y}_p$ are independent of SSE.

(b) Show that MST/MSE has an F distribution with $\nu_1 = p - 1$ and $\nu_2 = n_1 + n_2 + \cdots + n_p - p$ degrees of freedom under H_0: $\mu_1 = \mu_2 = \cdots = \mu_p$. (You may assume, for simplicity, that $n_1 = n_2 = \cdots = n_p$.)

13.29. Let T have a Student's t distribution with ν degrees of freedom. Show that T^2 has an F distribution with 1 and ν degrees of freedom.

14

Analysis of Enumerative Data

14.1 *A Description of the Experiment*

Many experiments, particularly in the social sciences, result in enumerative (or count) data. For instance, the classification of people into five income brackets would result in an enumeration or count corresponding to each of the five income classes. Or, we might be interested in studying the reaction of a mouse to a particular stimulus in a psychological experiment. If a mouse will react in one of three ways when the stimulus is applied and if a large number of mice were subjected to the stimulus, the experiment would yield three counts, indicating the number of mice falling in each of the reaction classes. Similarly, a traffic study might require a count and classification of the type of motor vehicles using a section of highway. An industrial process manufactures items that fall into one of three quality classes: acceptable, seconds, and rejects. A student of the arts might classify paintings in one of k categories according to style and period in order to study trends in style over time. We might wish to classify ideas in a philosophical study or style in the field of literature. The results of an advertising campaign would yield count data indicating a classification of consumer reaction. Indeed, many observations in the physical sciences are not amenable to measurement on a continuous scale and hence result in enumerative or classificatory data.

The illustrations in the preceding paragraph exhibit, to a reasonable degree of approximation, the characteristics of a *multinomial* experiment (Section 5.9). To refresh your memory, the multinomial experiment is analogous to tossing n balls at k boxes, where each ball must fall in one of the boxes. The boxes are arranged such that the probability, p_i, that a ball will fall in box i, $i = 1, \ldots, k$, remains the same in repeated tosses. Finally, the balls are tossed in such a way that the trials are independent. At the conclusion of the experiment, we observe n_1 balls in the first box, n_2 in the second, and n_k in the kth. The total number of balls is equal to

$$\sum_{i=1}^{k} n_i = n.$$

You will note the similarity between the binomial and multinomial experiments and, in particular, that the binomial experiment represents the special case for the multinomial experiment when $k = 2$. The single parameter of the binomial experiment, p, is replaced by the k parameters, $p_1, p_2, \ldots, p_k$, of the multinomial, where

$$\sum_{i=1}^{k} p_i = 1.$$

In this chapter, inferences concerning $p_1, p_2, \ldots, p_k$ will be expressed in terms of a statistical test of a hypothesis concerning their specific numerical values or their relationship, one to another.

If we were to proceed as in Chapter 10, we would derive the probability of the observed sample $(n_1, n_2, \ldots, n_k)$ for use in calculating the probability of type I and type II errors associated with a statistical test. Fortunately, we have been relieved of this chore by the British statistician Karl Pearson, who proposed a very useful test statistic for testing hypotheses concerning $p_1, p_2, \ldots, p_k$, and gave its approximate probability distribution in repeated sampling.

14.2 The Chi-Square Test

Suppose that $n = 100$ balls were tossed at the cells and that we knew that p_1 was equal to .1. How many balls would be expected to fall in the first cell? Referring to Chapter 5 and utilizing knowledge of the multinomial experiment, we would calculate

$$E(n_1) = np_1 = (100)(.1) = 10.$$

In like manner, the expected number falling in the remaining cells may be calculated using the formula

$$E(n_i) = np_i, \qquad i = 1, 2, \ldots, k.$$

Now suppose that we hypothesize values for $p_1, p_2, \ldots, p_k$ and calculate the expected value for each cell. Certainly, if our hypothesis is true, the cell counts, n_i, should not deviate greatly from their expected values, np_i ($i = 1, 2, \ldots, k$). Hence it would seem intuitively reasonable to use a test statistic involving the k deviations,

$$n_i - np_i, \qquad i = 1, 2, \ldots, k.$$

In 1900, Karl Pearson proposed the following test statistic, which is a function of the squares of the deviations of the observed counts from their expected values, weighted by the reciprocals of their expected values:

$$X^2 = \sum_{i=1}^{k} \frac{[n_i - E(n_i)]^2}{E(n_i)}$$

$$= \sum_{i=1}^{k} \frac{[n_i - np_i]^2}{np_i}$$

Although the mathematical proof is beyond the scope of this text, it can be shown that when n is large, X^2 will possess, approximately, a chi-square probability distribution in repeated sampling. We can easily show this result for the case $k = 2$ as follows. If $k = 2$, then $n_2 = n - n_1$ and $p_1 + p_2 = 1$. Thus

$$X^2 = \sum_{i=1}^{2} \frac{[n_i - E(n_i)]^2}{E(n_i)}$$

$$= \frac{(n_1 - np_1)^2}{np_1} + \frac{(n_2 - np_2)^2}{np_2}$$

$$= \frac{(n_1 - np_1)^2}{np_1} + \frac{[(n - n_1) - n(1 - p_1)]^2}{np_2}$$

$$= \frac{(n_1 - np_1)^2}{np_1} + \frac{(-n_1 + np_1)^2}{np_2}$$

$$= \frac{p_2(n_1 - np_1)^2 + p_1(-n_1 + np_1)^2}{np_1(1 - p_1)}$$

$$= \frac{(n_1 - np_1)^2}{np_1(1 - p_1)}.$$

We have seen (Chapter 7) that

$$\frac{n_1 - np_1}{\sqrt{np_1(1 - p_1)}}$$

is approximately standard normal in distribution for large n. Thus, for large n, X^2 as given above is approximately a χ^2 random variable with one degree of freedom. Recall that the square of a standard normal random variable has a χ^2 distribution (see Example 6.4).

Experience has shown that the cell counts, n_i, should not be too small in order that the chi-square distribution provide an adequate approximation to the distribution of σ^2. As a rule of thumb, we will require that all expected cell counts equal or exceed five, although Cochran [2] has noted that this value can be as low as 1 for some situations.

You will recall the use of the chi-square probability distribution for testing a hypothesis concerning a population variance, σ^2, in Section 10.6. Particularly, we stated that the shape of the chi-square distribution would vary depending upon the number of degrees of freedom associated with s^2, and we discussed the use of Table 5, Appendix III, which presents the critical values of χ^2 corresponding to various right-hand-tail areas of the distribution. Therefore, we must know which χ^2 distribution to use—that is, the number of degrees of freedom—in approximating the distribution of X^2, and we must know whether to use a one-tailed or two-tailed test in locating the rejection region for the test. The latter problem may be solved directly. Since large deviations of the observed cell counts from those expected would tend to contradict the null hypothesis concerning the cell probabilities, $p_1, p_2, \ldots, p_k$, we would reject the null hypothesis when X^2 is large and employ a one-tailed statistical test using the upper tail values of χ^2 to locate the rejection region.

The determination of the appropriate number of degrees of freedom to be employed for the test can be rather difficult and therefore will be specified for the physical applications described in the following sections. In addition, we will state the principle involved (which is fundamental to the mathematical proof of the approximation) so that the reader may understand why the number of degrees of freedom changes with various applications. This states that the appropriate number of degrees of freedom will equal the number of cells, k, less one degree of freedom for each independent linear restriction placed upon the observed cell counts. For example, one linear restriction is present because the sum of the cell counts must equal n; that is,

$$n_1 + n_2 + n_3 + \cdots + n_k = n.$$

Other restrictions will be introduced for some applications because of the necessity for estimating unknown parameters required in the calculation of the expected cell frequencies or because of the method in which the sample is collected. When unknown parameters must be estimated in order to compute X^2, a maximum likelihood estimator should be employed. The degrees of freedom for the approximating chi-square distribution will be reduced by one for each parameter estimated. These cases will become apparent as we consider various practical examples.

14.3 *A Test of a Hypothesis Concerning Specified Cell Probabilities: A Goodness-of-Fit Test*

The simplest hypothesis concerning the cell probabilities would be one that specifies numerical values for each. For example, consider the following experiment.

Example 14.1: *A group of rats, one by one, proceed down a ramp to one of three doors. We wish to test the hypothesis that the rats have no preference concerning the choice of a door and therefore that*

$$H_0: p_1 = p_2 = p_3 = 1/3,$$

where p_i *is the probability that a rat will choose door* i, $i = 1, 2$, *or* 3.

Suppose that the rats were sent down the ramp $n = 90$ *times and that the three observed cell frequencies were* $n_1 = 23$, $n_2 = 36$, *and* $n_3 = 31$. *The expected cell frequency would be the same for each cell:* $E(n_i) = np_i = (90)(1/3) = 30$. *The observed and expected cell frequencies are presented in Table 14.1. Note the discrepancy between the observed and expected cell frequency. Do the data present sufficient evidence to warrant rejection of the hypothesis of no preference?*

Table 14.1 Observed and Expected Cell Counts

	Door		
	1	*2*	*3*
Observed cell frequency	$n_1 = 23$	$n_2 = 36$	$n_3 = 31$
Expected cell frequency	(30)	(30)	(30)

Solution: *The chi-square test statistic for our example will possess* $(k - 1) = 2$ *degrees of freedom since the only linear restriction on the cell frequencies is that*

$$n_1 + n_2 + \cdots + n_k = n,$$

or, for our example,

$$n_1 + n_2 + n_3 = 90.$$

Therefore, if we choose $\alpha = .05$, *we would reject the null hypothesis when* $X^2 > 5.991$ *(see Table 5, Appendix III).*

Substituting into the formula for X^2, we obtain

$$X^2 = \sum_{i=1}^{k} \frac{[n_i - E(n_i)]^2}{E(n_i)} = \sum_{i=1}^{k} \frac{(n_i - np_i)^2}{np_i}$$

$$= \frac{(23 - 30)^2}{30} + \frac{(36 - 30)^2}{30} + \frac{(31 - 30)^2}{30}$$

$$= 2.87.$$

Since X^2 is less than the tabulated critical value of χ^2, the null hypothesis is not rejected, and we conclude that the data do not present sufficient evidence to indicate that the rats have a preference for a particular door.

Example 14.2: *The number of accidents per week, Y, at a certain intersection was checked for n = 50 weeks with the following results:*

y	*Frequency*
0	32
1	12
2	6
3 or more	0

Test the hypothesis that the random variable, Y, has a Poisson distribution assuming the observations to be independent. Use $\alpha = .05$.

Solution: *The null hypothesis, H_0, states that Y has the Poisson distribution, given by*

$$p(y) = \frac{\lambda^y e^{-\lambda}}{y!}, \qquad y = 0, 1, 2, \ldots.$$

Since λ is unknown, we must find its maximum likelihood estimator, which turns out to be $\hat{\lambda} = \bar{Y}$ (see Exercise 9.4). For the given data, $\hat{\lambda}$ has the value $\bar{y} = 24/50 = .48$.

We have, for the given data, three cells with five or more observations—the cells defined by $Y = 0$, $Y = 1$, and $Y \geq 2$. Under H_0, the probabilities for *these cells are*

$$p_1 = P(Y = 0) = e^{-\lambda},$$

$$p_2 = P(Y = 1) = \lambda e^{-\lambda},$$

and

$$p_3 = P(Y \geq 2) = 1 - e^{-\lambda} - \lambda e^{-\lambda}.$$

These probabilities are estimated by replacing λ with $\hat{\lambda}$, which gives

$$\hat{p}_1 = e^{-.48} = .619,$$

$$\hat{p}_2 = .48e^{-.48} = .297,$$

and

$$\hat{p}_3 = 1 - \hat{p}_1 - \hat{p}_2 = .084.$$

If the observations are independent, the cell frequencies, n_1, n_2, and n_3, have a multinomial distribution with parameters p_1, p_2, and p_3. Thus $E(n_i) = np_i$, and the estimated expected cell frequencies are given by

$$\hat{E}(n_1) = n\hat{p}_1 = 30.95,$$

$$\hat{E}(n_2) = n\hat{p}_2 = 14.85,$$

and

$$\hat{E}(n_3) = n\hat{p}_3 = 4.20.$$

The test statistic is now given by

$$X^2 = \sum_{i=1}^{3} \frac{[n_i - \hat{E}(n_i)]^2}{\hat{E}(n_i)},$$

which has approximately a χ^2 distribution with $(k - 2) = 1$ degree of freedom. (One degree of freedom is lost because λ had to be estimated, the other because $\sum_{i=1}^{3} n_i = n$.)

On computing X^2 we find

$$X^2 = \frac{(32 - 30.95)^2}{30.95} + \frac{(12 - 14.85)^2}{14.85} + \frac{(6 - 4.20)^2}{4.20}$$

$$= 1.354.$$

Since $\chi^2_{.05} = 3.841$, with one degree of freedom, we do not reject H_0. The data do not present sufficient evidence to contradict our hypothesis that Y possesses a Poisson distribution.

14.4 Contingency Tables

A problem frequently encountered in the analysis of count data concerns the independence of two methods of classification of observed events. For example, we might wish to classify defects found on furniture produced in a manufacturing plant according to (1) the type of defect and (2) the production shift. Ostensibly, we wish to investigate a contingency—a dependence between the two classifications. Do the proportions of various types of defects vary from shift to shift?

A total of $n = 309$ furniture defects was recorded and the defects were classified according to one of four types: A, B, C, or D. At the same time, each piece of furniture was identified according to the production shift in which it was manufactured. These counts are presented in Table 14.2, which is known as a *contingency table*. (*Note*: Numbers in parentheses are the estimated expected cell frequencies.)

Table 14.2 A Contingency Table

	Type of Defect				
Shift	*A*	*B*	*C*	*D*	*Total*
1	15 (22.51)	21 (20.99)	45 (38.94)	13 (11.56)	94
2	26 (22.99)	31 (21.44)	34 (39.77)	5 (11.81)	96
3	33 (28.50)	17 (26.57)	49 (49.29)	20 (14.63)	119
Total	74	69	128	38	309

Let p_A equal the unconditional probability that a defect will be of type A. Similarly, define p_B, p_C, and p_D as the probabilities of observing the three other types of defects. Then these probabilities, which we will call the column probabilities of Table 14.2, will satisfy the requirement

$$p_A + p_B + p_C + p_D = 1.$$

In like manner, let p_i ($i = 1, 2,$ or 3) equal the row probability that a defect will have occurred on shift i, where

$$p_1 + p_2 + p_3 = 1.$$

If the two classifications are independent of each other, a cell probability will equal the product of its respective row and column probabilities in accordance with the multiplicative law of probability. For example, the probability that a particular defect will occur on shift 1 and be of type A is $(p_1)(p_A)$. We observe that the numerical values of the cell probabilities are unspecified in the problem under consideration. The null hypothesis specifies only that each cell probability will equal the product of its respective row and column probabilities and therefore imply independence of the two classifications.

The analysis of the data obtained from a contingency table differs from Example 14.1 because we must *estimate* the row and column probabilities in order to estimate the expected cell frequencies.

As we have noted, the estimated expected cell frequencies may be substituted for the $E(n_i)$ in X^2, and X^2 will continue to possess a distribution in repeated sampling that is approximated by the chi-square probability distribution.

The maximum likelihood estimator for any row or column probability is found as follows. Let n_{ij} denote the observed frequency in row i and column j of the contingency table, and let p_{ij} denote the probability of an observation falling into this cell. If observations are independently selected, then the cell frequencies have a multinomial distribution and the maximum likelihood estimator of p_{ij} is simply the observed relative frequency for that cell. That is,

$$\hat{p}_{ij} = \frac{n_{ij}}{n}, \qquad i = 1, \ldots, r; j = 1, \ldots, c$$

(see Exercise 9.7).

Likewise, viewing row i as a single cell, the probability for row i is given by p_i, and hence

$$\hat{p}_i = \frac{r_i}{n}$$

(where r_i denotes the number of observations in row i) is the maximum likelihood estimator of p_i.

By analogous arguments, the maximum likelihood estimator of the jth-column probability is c_j/n, where c_j denotes the number of observations in column j.

Now let us compute the maximum likelihood estimator of the row and column probabilities:

$$\hat{p}_A = \frac{c_1}{n} = \frac{74}{309},$$

$$\hat{p}_B = \frac{c_2}{n} = \frac{69}{309},$$

$$\hat{p}_C = \frac{c_3}{n} = \frac{128}{309},$$

$$\hat{p}_D = \frac{c_4}{n} = \frac{38}{309}.$$

The row probabilities, p_1, p_2, and p_3, can be estimated using the row totals, r_1, r_2, and r_3:

$$\hat{p}_1 = \frac{r_1}{n} = \frac{94}{309},$$

$$\hat{p}_2 = \frac{r_2}{n} = \frac{96}{309},$$

$$\hat{p}_3 = \frac{r_3}{n} = \frac{119}{309}.$$

Under the null hypothesis, the estimated expected value of n_{11} is

$$\hat{E}(n_{11}) = n(\hat{p}_1 \cdot \hat{p}_A) = n\frac{r_1}{n}\frac{c_1}{n}$$

$$= \frac{r_1 \cdot c_1}{n}.$$

In other words, we observe that the estimated expected value of the observed cell frequency, n_{ij}, for a contingency table is equal to the product of its respective row and column totals divided by the total frequency; that is,

$$\hat{E}(n_{ij}) = \frac{r_i c_j}{n}.$$

The estimated expected cell frequencies for our example are shown in parentheses in Table 14.2.

We may now use the expected and observed cell frequencies shown in Table 14.2 to calculate the value of the test statistic:

$$X^2 = \sum_{j=1}^{4} \sum_{i=1}^{3} \frac{[n_{ij} - \hat{E}(n_{ij})]^2}{\hat{E}(n_{ij})}$$

$$= \frac{(15 - 22.51)^2}{22.51} + \frac{(26 - 22.99)^2}{22.99}$$

$$+ \cdots + \frac{(20 - 14.63)^2}{14.63}$$

$$= 19.17.$$

The only remaining obstacle involves the determination of the appropriate number of degrees of freedom associated with the test statistic. We will give this as a rule which we will attempt to justify. *The degrees of freedom associated with a contingency table possessing r rows and c columns will always equal* $(r - 1)(c - 1)$. *For our example, we will compare* X^2 *with the critical value of* χ^2 *with* $(r - 1)(c - 1) = (3 - 1)(4 - 1) = 6$ *degrees of freedom.*

You will recall that the number of degrees of freedom associated with the χ^2 statistic will equal the number of cells (in this case, $k = rc$) less one degree of freedom for each independent linear restriction placed upon the observed cell frequencies. The total number of cells for the data of Table 14.2 is $k = 12$. From this we subtract one degree of freedom because the sum of the observed cell frequencies must equal n; that is,

$$n_{11} + n_{12} + \cdots + n_{34} = 309.$$

In addition, we used the cell frequencies to estimate three of the four column probabilities. Note that the estimate of the fourth column probability will be determined once we have estimated p_A, p_B, and p_C, because

$$p_A + p_B + p_C + p_D = 1.$$

Thus we lose $c - 1 = 3$ degrees of freedom for estimating the column probabilities.

Finally, we used the cell frequencies to estimate $(r - 1) = 2$ row probabilities, and therefore we lose $(r - 1) = 2$ additional degrees of freedom. The total number of degrees of freedom remaining will be

$$\text{d.f.} = 12 - 1 - 3 - 2 = 6.$$

And, in general, we see that the total number of degrees of freedom associated with an $r \times c$ contingency table will be

$$\begin{aligned} \text{d.f.} &= rc - 1 - (c - 1) - (r - 1) \\ &= (r - 1)(c - 1). \end{aligned}$$

Therefore, if we use $\alpha = .05$, we will reject the null hypothesis that the two classifications are independent if $X^2 > 12.592$. Since the value of the test statistic, $X^2 = 19.17$, exceeds the critical value of χ^2, we will reject the null hypothesis. The data present sufficient evidence to indicate that the proportion of the various types of defects varies from shift to shift. A study of the production operations for the three shifts would probably reveal the cause.

Example 14.3: *A survey was conducted to evaluate the effectiveness of a new flu vaccine that had been administered in a small community. The vaccine was provided free of charge in a two-shot sequence over a period of two weeks to those wishing to avail themselves of it. Some people received the two-shot sequence, some appeared only for the first shot, and the others received neither.*

A survey of 1000 *local inhabitants in the following spring provided the information shown in Table 14.3. Do the data present sufficient evidence to indicate that the vaccine was successful in reducing the number of flu cases in the community?*

Table 14.3 Data Tabulation for Example 14.3

	No Vaccine	One Shot	Two Shots	Total
Flu	24 (14.4)	9 (5.0)	13 (26.6)	46
No flu	289 (298.6)	100 (104.0)	565 (551.4)	954
Total	313	109	578	1000

Solution: *The question stated above asks whether the data provide sufficient evidence to indicate a dependence between the vaccine classification and the occurrence or nonoccurrence of flu. We therefore analyze the data as a contingency table.*

The estimated expected cell frequencies may be calculated using the appropriate row and column totals,

$$\hat{E}(n_{ij}) = \frac{r_i c_j}{n}.$$

Thus

$$\hat{E}(n_{11}) = \frac{r_1 c_1}{n} = \frac{(46)(313)}{1000} = 14.4,$$

$$\hat{E}(n_{12}) = \frac{r_1 c_2}{n} = \frac{(46)(109)}{1000} = 5.0.$$

These values are shown in parentheses in Table 14.3.

The value of the test statistic, X^2, will now be computed and compared with the critical value of χ^2 possessing $(r - 1)(c - 1) = (1)(2) = 2$ degrees of freedom. Then, for $\alpha = .05$, we will reject the null hypothesis when $X^2 > 5.991$. Substituting into the formula for X^2, we obtain

$$X^2 = \frac{(24 - 14.4)^2}{14.4} + \frac{(289 - 298.6)^2}{298.6} + \cdots + \frac{(565 - 551.4)^2}{551.4}$$

$$= 17.35.$$

Observing that X^2 falls in the rejection region, we reject the null hypothesis of independence of the two classifications. A comparison of the percentage incidence of flu for each of the three categories would suggest that those receiving the two-shot sequence were less susceptible to the disease. Further analysis of the data could be obtained by deleting one of three categories, the second column, for example, to compare the effect of the vaccine with that of no vaccine. This could be done by using either a 2×2 contingency table or treating the two categories as two binomial populations and using the methods of Section 10.3. Or, we might wish to analyze the data by comparing the results of the two-shot vaccine sequence with those of the combined no vaccine–one shot group. That is, we would combine the first two columns of the 2×3 table into one.

14.5 $r \times c$ Tables with Fixed Row or Column Totals

In the previous section we have described the analysis of an $r \times c$ contingency table using examples which, for all practical purposes, fit the multinomial experiment described in Section 14.1. Although the methods of collecting data in many surveys may obviously adhere to the requirements of a multinomial experiment, other methods do not. For example, we might not wish to sample randomly the population described in Example 14.3 because we might find that, due to chance, one category is completely missing. People who have received no flu shots might fail to appear in the sample. We might decide beforehand to interview a specified number of people in each column category, thereby fixing the column totals in advance. Although these restrictions tend to disturb somewhat our visualization of the experiment in the multinomial context, they have no effect on the analysis of the data. As long as we wish to test the hypothesis of independence of two classifications and none of the row or column probabilities are specified in advance, we can analyze the data as an $r \times c$ contingency table. It can be shown that the resulting X^2 will possess a probability distribution in repeated sampling that is approximated by a chi-square distribution with $(r - 1)(c - 1)$ degrees of freedom.

To illustrate, suppose that we wish to test a hypothesis concerning the equivalence of four binomial populations as indicated in the following example.

Example 14.4: *A survey of voter sentiment was conducted in four midcity political wards to compare the fraction of voters favoring candidate A. Random samples of 200 voters were polled in each of the four wards, with results as shown in Table 14.4. Do the data present sufficient evidence to indicate that the fractions of voters favoring candidate A differ in the four wards?*

Table 14.4 Data Tabulation for Example 14.4

	Ward				
	1	*2*	*3*	*4*	*Total*
Favor A	76 (59)	53 (59)	59 (59)	48 (59)	236
Do not favor A	124 (141)	147 (141)	141 (141)	152 (141)	564
Total	200	200	200	200	800

Solution: *You will observe that the test of a hypothesis concerning the equivalence of the parameters of the four binomial populations corresponding to the four wards is identical to a hypothesis implying independence of the row and column classifications. If we denote the fraction of voters favoring A as p and hypothesize that p is the same for all four wards, we imply that the first- and second-row probabilities are equal to p and* $(1 - p)$*, respectively. The probability that a member of the sample of* $n = 800$ *voters falls in a particular ward will equal one-fourth since this was fixed in advance. Then the cell probabilities for the table would be obtained by multiplication of the appropriate row and column probabilities under the null hypothesis and be equivalent to a test of independence of the two classifications.*

The estimated expected cell frequencies, calculated using the row and column totals, appear in parentheses in Table 14.4. We see that

$$X^2 = \frac{\sum_{j=1}^{4} \sum_{i=1}^{2} [n_{ij} - \hat{E}(n_{ij})]^2}{\hat{E}(n_{ij})}$$

$$= \frac{(76 - 59)^2}{59} + \frac{(124 - 141)^2}{141} + \cdots + \frac{(152 - 141)^2}{141}$$

$$= 10.72.$$

The critical value of χ^2 *for* $\alpha = .05$ *and* $(r - 1)(c - 1) = (1)(3) = 3$ *degrees of freedom is 7.815. Since* X^2 *exceeds this critical value, we reject the null hypothesis and conclude that the fraction of voters favoring candidate A is not the same for all four wards.*

This example was worked in Exercise 10.29 by the likelihood ratio method. Note that the conclusions are the same.

14.6 *Other Applications*

The applications of the chi-square test in analyzing enumerative data described in Sections 14.3, 14.4, and 14.5 represent only a few of the interesting classificatory problems that may be approximated by the multinomial experiment and for which our method of analysis is appropriate. By and large these applications are complicated to a greater or lesser degree because the numerical values of the cell probabilities are unspecified and hence require the estimation of one or more population parameters. Then, as in Sections 14.4 and 14.5, we can estimate the cell probabilities. Although we omit the mechanics of the statistical tests, several additional applications of the chi-square test are worth mention as a matter of interest.

For example, suppose that we wish to test a hypothesis stating that a population possesses a normal probability distribution. The cells of a sample frequency histogram would correspond to the k cells of the multinomial experiment and the observed cell frequencies would be the number of measurements falling in each cell of the histogram. Given the hypothesized normal probability distribution for the population, we could use the areas under the normal curve to calculate the theoretical cell probabilities and hence the expected cell frequencies. Maximum likelihood estimators must be employed when μ and σ are unspecified for the normal population, and these parameters must be estimated to obtain the estimated cell probabilities.

The construction of a two-way table to investigate dependency between two classifications can be extended to three or more classifications. For example, if we wish to test the mutual independence of three classifications, we would employ a three-dimensional "table" or rectangular parallelepiped. The reasoning and methodology associated with the analysis of both the two- and three-way tables are identical, although the analysis of the three-way table is a bit more complex.

A third and interesting application of our methodology would be its use in the investigation of the rate of change of a multinomial (or binomial) population as a function of time. For example, we might study the decision-making ability of a human (or any animal) as he is subjected to an educational program and tested over time. If, for instance, he is tested at prescribed intervals of time and the test is of the yes or no type, yielding a number of correct answers, y, that would follow a binomial probability distribution, we would be interested in the behavior of the probability of a correct response, p, as a function of time. If the number of correct responses was recorded for c time periods, the data would fall in a $2 \times c$ table similar to that in Example 14.4 (Section 14.5). We would then be interested in testing the hypothesis that p is equal to a constant, that is, that no learning has occurred, and we would then proceed to more interesting hypotheses to determine whether the data present sufficient evidence to indicate a gradual (say, linear) change over time as opposed to an abrupt change at some point in time. The procedures we have described could be extended to decisions involving more than two alternatives.

You will observe that our learning example is common to business, to industry, and to many other fields, including the social sciences. For example, we might wish to study the rate of consumer acceptance of a new product for various types of advertising campaigns as a function of the length of time that the campaign has been in effect. Or, we might wish to study the trend in the lot fraction defective in a manufacturing process as a function of time. Both of these examples, as well as many others, require a study of the behavior of a binomial (or multinomial) process as a function of time.

The examples we have just described are intended to suggest the relatively broad .application of the chi-square analysis of enumerative data, a fact that should be borne in mind by the experimenter concerned with this type of data. The statistical test employing X^2 as a test statistic is often called a "goodness-of-fit" test. Its application for some of these examples requires care in the determination of the appropriate estimates and the number of degrees of freedom for X^2, which, for some of these problems, may be rather complex.

14.7 Summary

The material in this chapter has been concerned with a test of a hypothesis regarding the cell probabilities associated with a multinomial experiment. When the number of observations, n, is large, the test statistic, X^2, can be shown to possess, approximately, a chi-square probability distribution in repeated sampling, the number of degrees of freedom being dependent upon the particular application. In general we assume that n is large and that the minimum expected cell frequency is equal to or is greater than 5.

Several words of caution concerning the use of the X^2 statistic as a method of analyzing enumerative type data are appropriate. The determination of the correct number of degrees of freedom associated with the X^2 statistic is very important in locating the rejection region. If the number is incorrectly specified, erroneous conclusions might result. Also, note that nonrejection of the null hypothesis does not imply that it should be accepted. We would have difficulty in stating a meaningful alternative hypothesis for many practical applications, and therefore we would lack knowledge of the probability of making a type II error. For example, we hypothesize that the two classifications of a contingency table are independent. A specific alternative would have to specify a measure of dependence that may or may not possess practical significance to the experimenter. Finally, if parameters are missing and the expected cell frequencies must be estimated, missing parameters should be estimated by the method of maximum likelihood in order that the test be valid. In other words, the application of the chi-square test for other than the simple applications outlined in Sections 14.3, 14.4, and 14.5 will require experience beyond the scope of this introductory presentation of the subject.

References

1. Anderson, R. L., and T. A. Bancroft, *Statistical Theory in Research.* New York: McGraw-Hill Book Company, 1952, Chap. 12.
2. Cochran, W. G., "The χ^2 Test of Goodness of Fit." *Annals of Mathematical Statistics*, Vol. 23 (1952), pp. 315–345.
3. Dixon, W. J., and F. J. Massey, Jr., *Introduction to Statistical Analysis.* New York: McGraw-Hill Book Company, 1957, Chap. 13.
4. Kendall, M. G., and A. Stuart, *The Advanced Theory of Statistics*, Vol. 2. New York: Hafner Publishing Company, Inc., 1961, Chap. 30.

Exercises

14.1. List the characteristics of a multinomial experiment.

14.2. A city expressway utilizing four lanes in each direction was studied to see whether drivers preferred to drive on the inside lanes. A total of 1000 automobiles were observed during the heavy early morning traffic and their respective lanes recorded. The results were as follows:

Lane	1	2	3	4
Observed Count	294	276	238	192

Do the data present sufficient evidence to indicate that some lanes are preferred over others? (Test the hypothesis that $p_1 = p_2 = p_3 = p_4 = 1/4$ using $\alpha = .05$.)

14.3. A die was rolled 600 times with the following results:

Observed Number	1	2	3	4	5	6
Frequency	89	113	98	104	117	79

Do these data present sufficient evidence to indicate that the die is unbalanced? Test using $\alpha = .05$.

14.4. After inspecting the data in Exercise 14.3, one might wish to test the hypothesis that the probability of a 6 is 1/6 against the alternative that this probability is less than 1/6.
(a) Carry out the above test using $\alpha = .05$.
(b) What tenet of good statistical practice is violated in the test of part (a)?

14.5. A study to determine the effectiveness of a drug (serum) for arthritis resulted in the comparison of two groups each consisting of 200 arthritic patients. One group was inoculated with the serum while the other received a placebo (an inoculation that appears to contain serum but actually is nonactive). After a period of time, each person in the study was asked to state whether his arthritic condition was improved. The following results were observed:

	Treated	*Untreated*
Improved	117	74
Not improved	83	126

Do these data present sufficient evidence to indicate that the serum was effective in improving the condition of arthritic patients?
(a) Test using the X^2 statistic. Use $\alpha = .05$.
(b) Test using the z test of Section 10.3 and $\alpha = .05$.

14.6. A radio station conducted a survey to study the relationship between the number of radios per household and family income. The survey, based upon $n = 1000$ interviews, produced the following results:

	Family Income (in dollars)			
No. Radios per Household	*Less Than 4000*	*4000–7000*	*7000–10,000*	*More Than 10,000*
1	126	362	129	78
2 or more	29	138	82	56

Do the data present sufficient evidence to indicate that the number of radios per household is dependent upon family income? Test at the $\alpha = .10$ level of significance.

14.7. A group of 306 people were interviewed to determine their opinion concerning a particular current American foreign-policy issue. At the same time their political affiliation was recorded. The data are as follows:

	Approve of Policy	*Do Not Approve of Policy*	*No Opinion*
Republicans	114	53	17
Democrats	87	27	8

Do the data present sufficient evidence to indicate a dependence between party affiliation and the opinion expressed for the sampled population?

14.8. A survey of student opinion concerning a resolution presented to the student council was studied to determine whether the resulting opinion was independent of fraternity and sorority affiliation. Two hundred students were interviewed, with the following results:

	In Favor	*Opposed*	*Undecided*
Fraternity	37	16	5
Sorority	30	22	8
No affiliation	32	44	6

Do these data present sufficient evidence to indicate that student opinion concerning the resolution was dependent upon affiliation with fraternities or sororities? Test using $\alpha = .05$.

14.9. The responses for the data in Exercise 14.8 were reclassified according to whether the student was male or female.

	In Favor	*Opposed*	*Undecided*
Female	39	46	9
Male	60	36	10

Do the data present sufficient evidence to indicate that student reaction to the resolution varied for the various opinion categories depending upon whether the student was male or female?

14.10. A manufacturer of buttons wished to determine whether the fraction of defective buttons produced by three machines varied from machine to machine. Samples of 400 buttons were selected from each of the three machines and the number of defectives counted for each sample. The results were as follows:

Machine Number	1	2	3
No. Defectives	16	24	9

Do these data present sufficient evidence to indicate that the fraction of defective buttons varies from machine to machine? Test, using $\alpha = .05$, with
(a) A χ^2 test.
(b) A likelihood ratio test.

14.11. A survey was conducted by an auto repairman to determine whether various auto ills were dependent upon the make of the auto. His survey, restricted to this year's model, produced the following results:

	Type of Repair		
Make	*Electrical*	*Fuel Supply*	*Other*
A	17	19	7
B	14	7	9
C	6	21	12
D	33	44	19
E	7	9	6

Do these data present sufficient evidence to indicate a dependency between auto makes and type of repair for these new-model cars? Note that the repairman was not utilizing all the information available when he conducted his survey. In conducting a study of this type, what other factors should be recorded?

14.12. A manufacturer of floor polish conducted a consumer-preference experiment to see whether a new floor polish, *A*, was superior to those produced by four of his competitors. A sample of 100 housewives viewed five patches of flooring that had received the five polishes and each indicated the patch that she considered superior in appearance. The lighting, background, and so on, were approximately the same for all five patches. The results of the survey were as follows:

Polish	*A*	*B*	*C*	*D*	*E*
Frequency	27	17	15	22	19

Do these data present sufficient evidence to indicate a preference for one or more of the polished patches of floor over the others? If one were to reject the hypothesis of "no preference" for this experiment, would this imply that polish A is superior to the others? Can you suggest a better method of conducting the experiment?

14.13. A sociologist conducted a survey to determine whether the incidence of various types of crime varied from one part of a particular city to another. The city was partitioned into three regions and the crimes classified as homicide, car theft, grand larceny, and others. An analysis of 1599 cases produced the following results:

	Type of Crime				
City Region	*Homicide*	*Auto Theft*	*Grand Larceny (Omitting Auto Theft)*	*Petty Larceny*	*Other*
1	12	239	191	122	47
2	17	163	278	201	54
3	7	98	109	44	17

Do these data present sufficient evidence to indicate that the occurrence of various types of crime is dependent upon city region?

14.14. A survey was conducted to investigate interest of middle-aged adults in physical-fitness programs in Rhode Island, Colorado, California, and Florida. The objective of the investigation was to determine whether adult participation in physical-fitness programs varies from one region of the United States to another. A random sample of people were interviewed in each state and the following data were recorded:

	Rhode Island	*Colorado*	*California*	*Florida*
Participate	46	63	108	121
Do not participate	149	178	192	179

Do the data indicate a difference in adult participation in physical-fitness programs from one state to another?

14.15. A carpet company was interested in comparing the fraction of new-home builders favoring carpet over other floor coverings for homes in three different areas of a city. The objective was to decide how to allocate sales effort to the areas. A survey was conducted and the data were as follows:

	Areas		
	1	*2*	*3*
Carpet	69	126	16
Other material	78	99	27

Do the data indicate a difference in the percentage favoring carpet from one region of the city to another?

14.16. Refer to Exercise 14.15. Estimate the difference in the fractions of new-home builders favoring carpet between 1 and 2. Use a 95 percent confidence interval.

14.17. A survey was conducted to determine student, faculty, and administration attitudes on a new university parking policy. The distribution of those favoring or opposed to the policy was as follows:

	Student	*Faculty*	*Administration*
Favor	252	107	43
Oppose	139	81	40

Do the data provide sufficient evidence to indicate that attitudes regarding the parking policy are independent of student, faculty, or administration status?

14.18. An analysis of accident data was made to determine the distribution of numbers of fatal accidents for automobiles of three sizes. The data for 346 accidents were as follows:

	Size of Auto		
	Small	*Medium*	*Large*
Fatal	67	26	16
Not fatal	128	63	46

Do the data indicate that the frequency of fatal accidents is dependent on the size of automobiles?

14.19. The chi-square test used in Exercise 14.5 is equivalent to the two-tailed z test of Section 10.3 provided α is the same for the two tests. Show algebraically that the chi-square test statistic, X^2, is the square of the test statistic, z, for the equivalent test.

14.20. It is often not clear whether all properties of a binomial experiment are actually met in a given application. A goodness-of-fit test is desirable for such cases. Suppose that an experiment consisting of four trials was repeated 100 times. The number of repetitions on which a given number of successes was obtained is recorded in the following table:

Possible Results (No. Successes)	0	1	2	3	4
No. Times Obtained	11	17	42	21	9

Estimate p (assuming that the experiment was binomial), obtain estimates of the expected cell frequencies, and test for goodness of fit. To determine the appropriate number of degrees of freedom for X^2, note that p was estimated by a linear combination of the observed frequencies.

14.21. Refer to the $r \times c$ contingency table of Section 14.4. Show that the maximum likelihood estimator of the probability for row i, p_i, is $\hat{p}_i = r_i/n$, $i = 1, 2, \ldots, r$.

14.22. A genetic model states that the proportions of offspring in three classes should be p^2, $2p(1 - p)$, and $(1 - p)^2$, for a parameter p, $0 \leq p \leq 1$. An experiment yielded frequencies of 30, 40, and 30 for the respective classes.
(a) Does the model fit the data?
(b) Suppose the hypothesis states that the model holds with $p = .5$. Do the data contradict this hypothesis?

14.23. The Mendelian theory states that the number of a certain type of peas falling into the classifications round and yellow, wrinkled and yellow, round and green, and wrinkled and green should be in the ratio 9:3:3:1. Suppose that 100 such peas revealed 56, 19, 17, and 8 in the respective classes. Are these data consistent with the model? Use $\alpha = .05$.

14.24. According to the genetic model for the relationship between sex and colorblindness, the four categories, male and normal, female and normal, male and colorblind, female and colorblind, should have probabilities given by $p/2$, $(p^2/2) + pq$, $q/2$, and $q^2/2$, respectively, where $q = 1 - p$. A sample of 2000 people revealed 880, 1032, 80, and 8 in the respective categories. Do these data agree with the model? Use $\alpha = .05$.

14.25. Suppose that $(Y_1, \ldots, Y_k)$ has a multinomial distribution with parameters $n, p_1, p_2, \ldots, p_k$, and $(X_1, \ldots, X_k)$ has a multinomial distribution with parameters $m, p_1^*, p_2^*, \ldots, p_k^*$. Construct a test of the null hypothesis that the two multinomial distributions are identical; that is, test H_0: $p_1 = p_1^*, \ldots, p_k = p_k^*$.

14.26. In an insecticide experiment, the probability of insect survival was expected to be linearly related to the dosage, D, over the region of experimentation. That is, $p = 1 + \beta D$. An experiment was conducted using four levels of dosage, 1, 2, 3, and 4, and 1000 insects in each group. The resulting data were as follows:

Dosage	1	2	3	4
No. Survivors	820	650	310	50

Do these data contradict the hypothesis that $p = 1 + \beta D$?

15

Nonparametric Statistics

15.1 Introduction

Some experiments yield response measurements that defy quantification. That is, they generate response measurements that can be ordered (ranked), but the location of the response on a scale of measurement is arbitrary. Although experiments of this type occur in almost all fields of study, they are particularly evident in social-science research and in studies of consumer preference. For example, suppose that a judge is employed to evaluate and rank the instructional abilities of four teachers, the edibility and taste characteristics of five brands of cornflakes, or the relative merits of 53 Miss America contestants. Since it is clearly impossible to give an exact measure of teacher competence, "tastability" of food, or the merit of beauty, the response measurements differ completely in character from those presented in preceding chapters. *Nonparametric* statistical methods are useful for analyzing this type of data.

The word "nonparametric" evolves from the type of hypothesis usually tested when dealing with ranked data of the type described above. *Parametric hypotheses* are those concerned with the population parameters. *Nonparametric hypotheses* do not involve the population parameters but are concerned with the form of the population frequency distribution. Tests of hypotheses concerning the binomial parameter, p, the test concerning μ and σ^2, and the analysis-of-variance tests were parametric. On the other hand, a hypothesis that a particular population possesses a normal distribution (without specification of parameter values) would be nonparametric. Similarly, a hypothesis that the distributions for two populations are identical would be nonparametric.

The latter hypothesis would be pertinent to the three ranking problems previously described. Even though we do not have an exact measure of teacher competence, we can imagine that one exists and that in repeated performances a given teacher would generate a population of such measurements. A hypothesis that the four probability distributions for the populations (associated with the four teachers) are identical would imply no difference in the instructional ability of the teachers. Similarly, we would imagine that a scale of tastability for cornflakes does exist (even if unknown to us) and that a population of responses representing the reactions of a very large set of prospective consumers corresponds to each brand. A hypothesis that the distributions of "tastability" for the five brands are identical implies no difference in consumer preference for these products.

The preceding illustrations suggest that the hypotheses not only are nonparametric but indicate that the underlying population distributions are unknown. Although Kendall and Stuart [2] suggest that statistical procedures appropriate for this latter condition be called *distribution-free*, it has become common to classify statistical methods for either nonparametric hypotheses or populations of unknown distributional form as *nonparametric methods*—in spite of the fact that these two conditions are in many respects unrelated (one deals with the null hypothesis and the other with the dependence of a test procedure on knowledge of the population distributional forms). However, we will conform with common usage of the term "nonparametric statistical methods" and will think of procedures applicable to situations where the form of the population distributions is unknown or where the case involves a nonparametric hypothesis.

Nonparametric statistical procedures apply not only to data that are difficult to quantify. They are particularly useful in making inferences in situations where serious doubt exists about the assumptions underlying standard methodology. For example, the t test for comparing a pair of means, Section 10.5, is based on the assumption that both populations are normally distributed with equal variances. Now, admittedly, the experimenter will never know whether these assumptions hold in a practical situation, but he will often be reasonably certain that departures from the assumptions will be small enough so that the properties of his statistical procedure will be undisturbed. That is, α and β will be approximately what he thinks they are. On the other hand, it is not uncommon for the experimenter seriously to question his assumptions and wonder whether he is using a valid statistical procedure. This difficulty may be circumvented by using a nonparametric statistical test and thereby avoiding reliance on a very uncertain set of assumptions.

It is a hard fact of life that one rarely gets "something for nothing." Circumventing the assumptions usually requires a rephrasing of one's hypothesis, a price that often is not difficult to pay. Instead of hyothesizing that $\mu_1 = \mu_2$, we hypothesize that "the population distributions are identical." Note that the practical implications of the two hypotheses are not equivalent because the latter hypothesis is less clearly defined. Two distributions could differ and still possess the same mean. A second and less obvious price exacted by the nonparametric procedures is that they usually use a smaller amount of information in the sample than do corresponding parametric methods. Consequently, they may be less powerful. This means that in testing with a given α, the probability of a type II error, β, for a specified alternative could be larger for the nonparametric test. We will have more to say on this point after giving an example of a nonparametric procedure.

15.2 *The Sign Test for Comparing Two Populations*

Suppose we wish to compare two populations, A and B, based on random samples of equal size, and wish to employ a nonparametric statistical test. A simple and rapid test of the hypothesis "the two population probability distributions are identical" can be accomplished by pairing the observations, one from each sample, and computing the difference as was done for the paired-difference experiment of Section 12.5. For the test statistic, we will employ the number of positive (or negative) signs (Y) associated with the paired differences. Thus, regardless of the probability distributions for the two sampled populations, the probability that an A observation exceeds its paired mate from B will be $p = .5$ when the null hypothesis is true (that is, when the distributions for A and B are identical). Then Y will possess a binomial probability distribution and a rejection region for Y can be obtained using the binomial probability distribution of Chapter 3. This test, based on Y, is known as a *sign test.*

The sign test is appropriate for the comparison of two populations using either paired or unpaired data. We will illustrate with an example.

Example 15.1: *The number of defective electrical fuses proceeding from each of two production lines, A and B, was recorded daily for a period of* 10 *days with the results shown in Table 15.1. Assume that both production lines produced the same daily output. Compare the number of defectives produced by A and B each day and let Y equal the number of days when A exceeded B. Do the data present sufficient evidence to indicate that production line B produces more defectives, on the average, than A? State the null hypothesis to be tested and use Y as a test statistic.*

Table 15.1

Day	*A*	*B*
1	172	201
2	165	179
3	206	159
4	184	192
5	174	177
6	142	170
7	190	182
8	169	179
9	161	169
10	200	210

Solution: *Pair the observations as they appear in the data tabulation and let Y be the number of times that the observation for production line A exceeds that for B in a given day. Under the null hypothesis that the two distributions of defectives are identical, the probability, p, that A exceeds B for a given pair is $p = .5$. Consequently, the null hypothesis is equivalent to a hypothesis that the binomial parameter, p, equals .5.*

Very large or very small values of Y are most contradictory to the null hypothesis. Therefore, the rejection region for the test will be located by including the most extreme values of Y that at the same time provide an α that is feasible for the test.

Suppose that we would like α to be somewhere on the order of .05 or .10. We would commence the selection of the rejection region by including $Y = 0$ and $Y = 10$ and calculate the α associated with this region using $p(y)$ (the probability distribution for the binomial random variable, Chapter 3). With $n = 10$, $p = .5$,

$$\alpha = p(0) + p(10) = \binom{10}{0}(.5)^{10} + \binom{10}{10}(.5)^{10} = .002.$$

Since this value of α is too small, the region will be expanded by including the next pair of Y values most contradictory to the null hypothesis, $Y = 1$ and $Y = 9$. The value of α for this region ($Y = 0, 1, 9, 10$) can be obtained from Table 1, Appendix III.

$$\alpha = p(0) + p(1) + p(9) + p(10) = .022.$$

This also is too small, so we will again expand the region to include $Y = 0, 1, 2, 8, 9, 10$. You can verify that the corresponding value of α is .11. We will suppose *that this value of* α *is acceptable to the experimenter and will employ* $Y = 0, 1, 2, 8, 9, 10$ *as the rejection region for the test.*

From the data we observe that $y = 2$, *and therefore we reject the null hypothesis. We conclude that sufficient evidence exists to indicate that the population distributions of number of defects are not identical and that the number for production line B tends to exceed that for A. The probability of rejecting the null hypothesis when true is only* $\alpha = .11$, *and therefore we are reasonably confident of our conclusion.*

The experimenter in this example is using the test procedure as a rough tool for detecting faulty production lines. The rather large value of α *is not likely to disturb him because he can easily collect additional data if he is concerned about making a type I error in reaching his conclusion.*

Example 15.1 was a paired or blocked experimental design because the number of defectives for the two production lines were paired by collecting them for a given day. Although the nonparametric procedure is exactly the same for either the completely randomized or the blocked experimental designs, the latter will contain more information (if, indeed, a substantial difference does exist between blocks) and this may increase the probability of rejecting the null hypothesis if it is false.

The values of α associated with the sign test can be obtained by using the normal approximation to the binomial probability distribution discussed in Section 7.4. You can verify (by comparison of exact probabilities with their approximations) that these approximations will be quite adequate for n as small as 25. This is due to the symmetry of the binomial probability distribution for $p = .5$. For $n \geq 25$, the z test of Chapter 10 will be quite adequate where

$$Z = \frac{Y - np}{\sqrt{npq}} = \frac{Y - n/2}{(1/2)\sqrt{n}}.$$

This would be testing the null hypothesis $p = .5$ against the alternative, $p \neq .5$, and would utilize the familiar rejection regions of Chapter 10.

What should be done in case of ties between the paired observations? This difficulty is circumvented by omitting tied pairs and thereby reducing n, the number of pairs.

Suppose that the two populations were normally distributed with equal variances. Will the sign test detect a difference in the populations as effectively as Student's t test? Intuitively, we would suspect that the answer is "no," and this is correct, because Student's t test utilizes comparatively more information. In addition to giving the sign of the difference, the t test uses the magnitudes of the observations to obtain a more accurate value of the sample means and variances. One might say that the sign test is not as powerful as Student's t test, but this statement is meaningful only if the populations conform to the assumptions stated above; that is, they are normally distributed with equal variances. The sign test might be more efficient for some other population distributions.

Summarizing, the sign test is a very easily applied nonparametric procedure for comparing two populations. No assumptions are made concerning the underlying population distributions. The value of the test statistic can be quickly obtained by a visual count and the rejection region can be easily located by using a table of binomial probabilities. Furthermore, the data need not be ordinal. That is, we need not know the exact values of pairs of responses.

15.3 The Rank-Sum Test for Comparing Two Population Distributions

The sign test for comparing two population distributions ignores the actual magnitudes of the paired observations and thereby discards information that would be useful in detecting a departure from the null hypothesis. A statistical test that partially circumvents this loss by utilizing the relative magnitudes of the observations was proposed by F. Wilcoxon as a rank-sum test. The procedure will be illustrated with an example.

Example 15.2: *The bacteria counts per unit volume are shown in Table 15.2 for two types of cultures, A and B. Four observations were made for each culture.*

Table 15.2

A	*B*
27	32
31	29
26	35
25	28

Let n_1 and n_2 represent the number of observations in samples A and B, respectively. Then the rank sum procedure ranks each of the $(n_1 + n_2) = n$

measurements in order of magnitude and uses W, the sum of the ranks for A or B as the test statistic. For the data given in Table 15.2, the corresponding ranks are

Table 15.3

	A	B
	3	7
	6	5
	2	8
	1	4
Rank sum	12	24

as shown in Table 15.3. Do these data present sufficient evidence to indicate a difference in the population distributions for A and B?

Solution: *Let W equal the rank sum for sample A (for this sample, $W = 12$). Certainly very small or very large values of W will provide evidence to indicate a difference between the two population distributions, and hence, W, the* rank sum, *will be employed as a test statistic.*

The rejection region for a given test will be obtained in the same manner as for the sign test. We will commence by selecting the most contradictory values of W as the rejection region and will add to these until α is of sufficient size.

The minimum rank sum would include the ranks 1, 2, 3, 4, or $W = 10$. Similarly, the maximum would include 5, 6, 7, 8, with $W = 26$. We will commence by including these two values of W in the rejection region. What is the corresponding value of α?

Finding the value of α is a probability problem that can be solved by using the methods of Chapter 2. If the populations are identical, every permutation of the eight ranks will represent a sample point and will be equally likely. Then α will represent the sum of the probabilities of the sample points (arrangements) which imply $W = 10$ or $W = 26$ in sample A. The total number of permutations of the eight ranks is $8!$ The number of different arrangements of the ranks, 1, 2, 3, 4 in sample A with the 5, 6, 7, 8 of sample B is $(4!4!)$. Similarly, the number of arrangements that place the maximum value of W in sample A (ranks 5, 6, 7, 8) is $(4!4!)$. Then the probability that $W = 10$ or $W = 26$ is

$$p(10) + p(26) = \frac{(2)(4!)(4!)}{8!} = \frac{2}{\binom{8}{4}} = \frac{1}{35} = .029.$$

If this value of α is too small, the rejection region can be enlarged to include the next smallest and largest rank sums, $W = 11$ and $W = 25$. The rank sum $W = 11$ will include the ranks 1, 2, 3, 5, and

$$p(11) = \frac{4!4!}{8!} = \frac{1}{70}.$$

Similarly,

$$p(25) = \frac{1}{70}.$$

Then

$$\alpha = p(10) + p(11) + p(25) + p(26) = \frac{2}{35} = .057.$$

Expansion of the rejection region to include 12 and 24 will substantially increase the value of α. The set of sample points giving a rank of 12 will be all sample points associated with rankings of (1, 2, 3, 6) and (1, 2, 4, 5). Thus

$$p(12) = \frac{(2)(4!)(4!)}{8!} = \frac{1}{35}$$

and

$$\alpha = p(10) + p(11) + p(12) + p(24) + p(25) + p(26)$$

$$= \frac{1}{70} + \frac{1}{70} + \frac{1}{35} + \frac{1}{35} + \frac{1}{70} + \frac{1}{70} = \frac{4}{35} = .114.$$

This value of α might be considered too large for practical purposes. In this case, we would be better satisfied with the rejection region $W = 10, 11, 25$, and 26.

The rank sum for the sample, $W = 12$, falls in the nonrejection region, and hence we do not have sufficient evidence to reject the hypothesis that the population distribution of bacteria counts for the two cultures are identical.

Mann and Whitney in 1947 proposed a nonparametric test statistic for the comparison of two population distributions that at first glance appears to be different from Wilcoxon's statistic but actually is a function of W. Consequently, the two tests are equivalent. As a matter of interest, we will present the

Mann–Whitney U test in Section 15.4, together with tables of critical values of U for various sample sizes. This will eliminate the need for a similar table for Wilcoxon's W.

15.4 The Mann–Whitney U Test

The Mann–Whitney statistic, U, is obtained by ordering all $(n_1 + n_2)$ observations according to their magnitude and counting the number of observations in sample A that precede each observation in sample B. The U statistic is the sum of these counts.

For example, the eight observations of Example 15.2 are

$$\begin{array}{cccccccc} 25, & 26, & 27, & 28, & 29, & 31, & 32, & 35. \\ A & A & A & B & B & A & B & B \end{array}$$

The smallest B observation is 28, and $u_1 = 3$ observations from sample A precede it. Similarly, $u_2 = 3$ A observations precede the second B observation and $u_3 = 4$ A observations precede the third and fourth B observations, respectively (32 and 35). Then

$$U = u_1 + u_2 + u_3 + u_4 = 14.$$

Very large or small values of U will imply a separation of the ordered A and B observations and will provide evidence to indicate a difference between the population distributions for A and B.

The probabilities required for calculating α for small samples (n_1 and $n_2 < 10$) are included in Table 8, Appendix III. A large-sample approximation is given in the latter part of this section. Entries in Table 8 are $P(U \leq U_0)$. For example, $P(U \leq 1) = .024$ for $n_1 = 3$ and $n_2 = 6$. The rejection region for a two-tailed test such as that in Example 15.2 would require doubling the tabulated one-tailed probability to find α. The test procedure would require calculating the smaller value of U and using the lower tail of the U distribution as the rejection region.

If $n_1 = 6$ and $n_2 = 8$, one could use the smaller value of U and reject when $U \leq 9$. The tabulated value gives $P(U \leq 9) = .030$. This probability must be doubled in calculating α to take into account the corresponding large values of U that would be included in the rejection region for a two-tailed test. Then $\alpha = .060$.

A shortcut procedure for calculating U can be obtained by ranking the observations as for the Wilcoxon rank sum test, letting the smallest observation have a rank of 1, and using the formula

$$U = n_1 n_2 + \frac{n_1(n_1 + 1)}{2} - W_A$$

or

$$U = n_1 n_2 + \frac{n_2(n_2 + 1)}{2} - W_B,$$

where W_A and W_B are the rank sums of samples A and B, respectively.

The first value of U will be the number of A observations preceding the B observations, and the second will be the total count of the B observations preceding the A's. For our example,

$$U = n_1 n_2 + \frac{n_1(n_1 + 1)}{2} - W_A$$

$$= (4)(4) + \frac{(4)(5)}{2} - 12 = 14$$

and

$$U = n_1 n_2 + \frac{n_2(n_2 + 1)}{2} - W_B$$

$$= (4)(4) + \frac{(4)(5)}{2} - 24 = 2.$$

We have agreed to use the smaller sum and hence would use $U = 2$ as the observed value of the test statistic. If we decide to reject H_0 when $U \leq 1$, then for a two-tailed test, $\alpha = 2P(U \leq 1) = (2)(.0286) = .0572$. Since $U = 2$ is not in the rejection region, we cannot reject H_0.

The expressions given above not only give an easy way of calculating U but also indicate the very simple relationship between the U statistic and the rank sum, W, of the Wilcoxon test. For all practical purposes, the tests are equivalent.

Ties in the observations can be handled by averaging the ranks that would have been assigned to the tied observations and assigning this average to each. If three observations are tied and are due to receive ranks 3, 4, 5, we would assign the rank of 4 to all three. The next observation in the sequence would receive the rank of 6, and ranks 3 and 5 would not appear. Similarly, if two observations are tied for ranks 3 and 4, each would receive a rank of 3.5, and ranks 3 and 4 would not appear.

Example 15.3: *Test the hypothesis of no difference in the distribution of strengths for two types of packaging materials. The data, based on two independent random samples, are shown in Table 15.4.*

Table 15.4

		A	B	
	(15)	1.25	.89	(1)
	(11)	1.16	1.01	(7)
	(18)	1.33	.97	(4)
	(10)	1.15	.95	(3)
	(14)	1.23	.94	(2)
	(12)	1.20	1.02	(8)
	(17)	1.32	.98	(5.5)
	(16)	1.28	1.06	(9)
	(13)	1.21	.98	(5.5)
Rank sum	126			45

Solution: *See Table 15.4. The rank associated with each observation is given alongside, in parentheses. Although this is not a blocked (paired) experiment, you will observe that each A observation in a pair exceeds its corresponding B measurement. Knowledge of the nonparametric sign test would quickly indicate that A exceeds B in nine out of nine pairs and would imply rejection of the hypothesis of no difference in the population distribution of strengths for the two types of packaging material. Certainly this procedure, which required less than 30 seconds to discuss, was much more rapid than the calculations involved in Student's t test, Chapter 10.*

Although we have tested our hypothesis using the sign test, we will show that the same result can be quickly obtained using the more powerful Mann–

Whitney U test. Thus

$$n_1 = n_2 = 9$$

and

$$U = n_1 n_2 + \frac{n_1(n_1 + 1)}{2} - W_A$$

$$= (9)(9) + \frac{(9)(10)}{2} - 126 = 0.$$

From Table 8, $P(U = 0)$ for $n_1 = n_2 = 9$ is .000. The value of α for a two-tailed test and a rejection region of $U = 0$ and $U = 81$ is $\alpha = .000$.

A simplified large sample test can be obtained by using the familiar Z statistic of Chapter 10. When the population distributions are identical, it can be shown that the U statistic has expected value and variance,

$$E(U) = \frac{n_1 n_2}{2},$$

$$V(U) = \frac{n_1 n_2(n_1 + n_2 + 1)}{12}.$$

The distribution of

$$Z = \frac{U - E(U)}{\sqrt{V(U)}}$$

tends to normality with mean zero and variance equal to 1 as n_1 and n_2 become large. This approximation will be adequate when n_1 and n_2 are both larger than, say, 10. For a two-tailed test with $\alpha = .05$, we would reject the null hypothesis if $|z| \geq 1.96$.

What constitutes an "adequate" approximation is a matter of opinion. Observe that the Z statistic will reach the same conclusion as the exact U test for

Example 15.3. Thus

$$z = \frac{0 - \frac{(9)(9)}{2}}{\sqrt{\frac{(9)(9)(9 + 9 + 1)}{12}}} = \frac{-45}{\sqrt{128 \cdot 25}} = -3.97.$$

This value of Z falls in the rejection region ($|z|) \geq 1.96$) and hence agrees with the sign test and the U test using the exact tabulated values for the rejection region.

15.5 The Wilcoxon Rank-Sum Test for a Paired Experiment

The Wilcoxon test (or, equivalently, the Mann–Whitney U test) can be adapted to the paired-difference experiment of Section 12.5 by considering the paired differences of the two treatments, A and B. Under the null hypothesis of no difference in the distributions for A and B, the expected number of negative differences between pairs would be $n/2$ (where n is the number of pairs), and positive and negative differences of equal absolute magnitude should occur with equal probability. Thus, if one were to order the differences according to their absolute values and rank them from smallest to largest, the expected rank sums for the negative and positive differences would be equal. Sizable departures of the rank sum of the positive (or negative) differences from its expected value would provide evidence to indicate a difference between the distributions of responses for the two treatments, A and B.

The rank sum, T, of the positive (or negative) differences contains the information provided by the sign test statistic and, in addition, gives information on the relative magnitude of the differences. For example, a low rank sum for negative differences indicates that both the number and absolute values of negative differences are small.

The probabilities associated with rank sums for the paired-difference experiment can be calculated using the probability laws of Chapter 2. If we let

$$n = \text{number of pairs},$$

$$Y = \text{number of positive differences},$$

$$T = \text{rank sum for the positive differences},$$

and observe that a particular value of T is the union of the mutually exclusive events ($Y = 0$ and the rank sum equals T), ($Y = 1$ and the rank sum equals T), $\ldots$, ($Y = n$ and the rank sum equals T), then

$$P(T) = P(Y = 0, T) + P(Y = 1, T) + \cdots + P(Y = n, T).$$

Each of the events $(Y = j, T), j = 1, \ldots, n$, is an intersection of the events $Y = j$ and "the rank sum equals T." Then

$$P(Y = j, T) = P(Y = j)P(T/Y = j),$$

where $p(y) = \binom{n}{y}(1/2)^n$ is the binomial probability distribution with $p = 1/2$ when the null hypothesis is true and the two population distributions are identical. Since T and U are related by a one-to-one correspondence, $P(T/Y = j)$ can be obtained (with a bit of difficulty) from Table 8, Appendix III.

Or, rather than use T, we could use the Mann–Whitney U and calculate

$$P(U) = P(Y = 0, U) + P(Y = 1, U) + \cdots + P(Y = n, U).$$

The probabilities would be equivalent for corresponding values of T and U.

The critical values of T for the Wilcoxon paired signed-ranks test are given in Table 9, where T is the smaller sum of ranks taking the same sign. For example, with $n = 17$ pairs, we would reject the hypothesis that the two population distributions are identical when $T \leq 35$ with $\alpha = .05$ using a two-tailed test. In calculating T, differences equal to zero are eliminated and the number of pairs, n, is reduced accordingly. Ties are treated in the same manner as for the unpaired comparison of two distributions.

Example 15.4: *Test a hypothesis of no difference in population distribution of cake density for a paired-difference experiment involving six baked cakes, one using mix A and one using mix B in each pair.*

Solution: *The original data and differences in density (in ounces/cubic inch) for the six pairs of cakes are shown in Table 15.5.*

Table 15.5

A	B	Difference, $A - B$	Rank
.135	.129	.006	3
.102	.120	−.018	5
.098	.112	−.004	1.5
.141	.152	−.011	4
.131	.135	−.004	1.5
.144	.163	−.019	6

The rank sum is smaller for positive differences, and hence $T = 3$. From Table 9, the critical value of T for a two-tailed test, $\alpha = .10$, is $T = 2$. Since the observed value of T exceeds the critical value of T, there is not sufficient evidence to indicate a difference in the population distribution of density for the two types of cake mix. (Values of T less than or equal to the critical value imply rejection.) The lower power of the nonparametric test in comparison with Student's t test is illustrated by this example because a t test will lead to rejection of the null hypothesis for these data.

When n is large (say 25), T will be approximately normally distributed with mean and variance:

$$E(T) = \frac{n(n+1)}{4},$$

$$V(T) = \frac{n(n+1)(2n+1)}{24}.$$

Then the Z statistic,

$$Z = \frac{T - E(T)}{\sqrt{V(T)}} = \frac{T - \dfrac{n(n+1)}{4}}{\sqrt{\dfrac{n(n+1)(2n+1)}{24}}},$$

can be used as a test statistic. For a two-tailed test and $\alpha = .05$, we would reject the hypothesis of "identical population distributions" when $|z| \geq 1.96$.

15.6 The Runs Test: A Test for Randomness

Consider a production process in which manufactured items emerge in sequence and each is classified as either defective (D) or nondefective (N). We have studied how one might compare the fraction defective over two equal time intervals using the normal deviate test (Chapter 10) and extended this to a test of a hypothesis of constant p over two or more time intervals using the chi-square test of Chapter 14. The purpose of these tests was to detect a change or trend in the fraction defective, p. Evidence to indicate increasing fraction defective might indicate the need for a process study to locate the source of difficulty. A decreasing value might suggest that a process quality-control program was having a beneficial effect in reducing the fraction defective.

Trends in fraction defective (or other quality measures) are not the only indication of lack of process control. A process may be causing periodic runs of defectives with the average fraction defective remaining constant, for all practical purposes, over long periods of time. For example, photoflash lamps are manufactured on a rotating machine with a fixed number of positions for bulbs. A bulb is placed on the machine at a given position, the air is removed, oxygen is pumped into the bulb, and the glass base is flame sealed. If a machine contains 20 positions, and several adjacent positions are faulty (too much heat in the sealing process), surges of defective lamps will emerge from the process in a periodic manner. Tests to compare the process fraction defective over equal intervals of time will not detect this periodic difficulty in the process. The periodicity, indicated by runs of defectives, is indicative of nonrandomness in the occurrence of defectives over time and can be detected by a *test for randomness.* The statistical test that we present, known as the runs test, is discussed in detail by Wald and Wolfowitz [5]. Other practical applications of the runs test will follow.

As the name implies, the runs test studies a sequence of events where each element in the sequence may assume one of two outcomes, say success (S) or failure (F). If we think of the sequences of items emerging from a manufacturing process as defective (F) or nondefective (S), the observation of 20 items might yield

$$S\ S\ S\ S\ S\ F\ F\ S\ S\ S$$

$$F\ F\ F\ S\ S\ S\ S\ S\ S\ S.$$

We notice the groupings of defectives and nondefectives and wonder whether this implies nonrandomness and, consequently, lack of process control.

> ***Definition 15.1:*** *A* run *is defined to be a maximal subsequence of like elements.*

For example, the first five successes is a subsequence of five like elements and it is maximal in the sense that it includes the maximum number of like elements before encountering an F. (The first four elements form a subsequence of like elements, but it is not maximal because the fifth element could also be included.) Consequently, the 20 elements shown above are arranged in five runs, the first containing five S's, the second containing two F's, and so on.

A very small or very large number of runs in a sequence would indicate nonrandomness. Therefore, let R (the number of runs in a sequence) be the test statistic, and let the rejection region be $R \leq k_1$ and $R \geq k_2$, as indicated in Figure 15.1. Let m denote the maximum possible number of runs. We must then find the probability distribution for R, $P(R = r)$, in order to calculate α and to locate a suitable rejection region for the test.

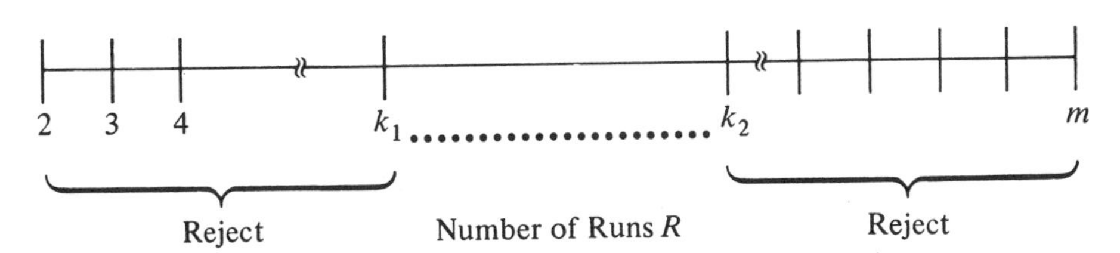

Figure 15.1 The Rejection Region for the Runs Test

Suppose that the complete sequence contains n_1 S elements, and n_2 F elements, resulting in Y_1 S runs and Y_2 F runs, where $(Y_1 + Y_2) = R$. Then, for a given Y_1, Y_2 can equal Y_1, $(Y_1 - 1)$, or $(Y_1 + 1)$, and $m = 2n_1$ if $n_1 = n_2$ and $m = (2n_1 + 1)$ if $n_1 < n_2$. We will suppose that every distinguishable arrangement of the $(n_1 + n_2)$ elements in the sequence constitutes a simple event for the experiment and that the sample points are equiprobable. It then remains for us to count the number of sample points that imply R runs.

The total number of distinguishable arrangements of n_1 S elements and n_2 F elements is $\binom{n_1 + n_2}{n_1}$, and therefore the probability per sample point is $1\Big/\binom{n_1 + n_2}{n_1}$. The number of ways of achieving y_1 S runs is equal to the number of distinguishable arrangements of n_1 indistinguishable elements in y_1 cells, none of which is empty. The situation is represented in Figure 15.2. You will note that this is equal to the number of ways of distributing the $(y_1 - 1)$ identical inner bars in the $(n_1 - 1)$ spaces between the S elements (the outer two bars remain fixed). Consequently, it is equal to the number of ways of selecting $(y_1 - 1)$ spaces (for the bars) out of the $(n_1 - 1)$ spaces available. This will equal

$$\binom{n_1 - 1}{y_1 - 1}.$$

| S | S S S S | S S...| S S | S S S | S |

Figure 15.2 The Distribution of n_1 S Elements in y_1 Cells (None Empty)

The number of ways of observing y_1 S runs and y_2 F runs, obtained by applying the *mn* rule, is

$$\binom{n_1 - 1}{y_1 - 1}\binom{n_2 - 1}{y_2 - 1}.$$

This gives the number of sample points in the event "y_1 S runs and y_2 F runs." Then multiplying this number by the probability per sample point, we obtain the probability of exactly y_1 S runs and y_2 F runs to be

$$p(y_1, y_2) = \frac{\binom{n_1 - 1}{y_1 - 1}\binom{n_2 - 1}{y_2 - 1}}{\binom{n_1 + n_2}{n_1}}.$$

Then $P(R = r)$ equals the sum of $P(y_1, y_2)$ over all values of y_1 and y_2 such that $(y_1 + y_2) = r$.

To illustrate the use of the formula, the event $R = 4$ could occur when $y_1 = 2$ and $y_2 = 2$ with either the S or F elements commencing the sequences. Consequently,

$$p(4) = 2P(Y_1 = 2, Y_2 = 2).$$

On the other hand, $R = 5$ could occur when $y_1 = 2$ and $y_2 = 3$, or $y_1 = 3$ and $y_2 = 2$, and these points are mutually exclusive. Then

$$p(5) = P(Y_1 = 3, Y_2 = 2) + P(Y_1 = 2, Y_2 = 3).$$

Example 15.5: *Suppose that a sequence consists of $n_1 = 5$ S elements, and $n_2 = 3$ F elements. Calculate the probability of observing $R = 3$ runs. Also, calculate $P(R \leq 3)$.*

Solution: *Three runs could occur when $y_1 = 2$ and $y_2 = 1$, or $y_1 = 1$ and $y_2 = 2$. Then*

$$p(3) = P(Y_1 = 2, Y_2 = 1) + P(Y_1 = 1, Y_2 = 2)$$

$$= \frac{\binom{4}{1}\binom{2}{0}}{\binom{8}{5}} + \frac{\binom{4}{0}\binom{2}{1}}{\binom{8}{5}}$$

$$= \frac{4}{56} + \frac{2}{56} = .107.$$

Next we require that $P(R \leq 3) = p(2) + p(3)$. Accordingly, note that

$$p(2) = 2P(Y_1 = 1, Y_2 = 1)$$

$$= 2\frac{\binom{4}{0}\binom{2}{0}}{\binom{8}{5}} = \frac{2}{56} = .036.$$

Thus the probability of 3 or less runs is .143.

The values of $P(R \leq a)$ are given in Table 10, Appendix III, for all combinations of n_1 and n_2 where n_1 and n_2 are less than or equal to 10. These can be used to locate the rejection regions for a one- or two-tailed test. We will illustrate with an example.

Example 15.6: *A true–false examination was constructed with the answers running in the following sequence:*

T F F T F T F T T F T F F T F T F T T F.

Does this sequence indicate a departure from randomness in the arrangement of T *and* F *answers?*

Solution: *The sequence contains* $n_1 = 10$ T *and* $n_2 = 10$ F *answers with* $y = 16$ *runs. Nonrandomness can be indicated by either an unusually small or an unusually large number of runs, and consequently we will be concerned with a two-tailed test.*

Suppose that we wish to use α *approximately equal to .05 with .025 or less in each tail of the rejection region. Then from Table 10 with* $n_1 = n_2 = 10$, *we note that* $P(R \leq 6) = .019$ *and* $P(R \leq 15) = .981$. *Then* $P(R \geq 16) = 1 - P(R \leq 15) = .019$, *and we would reject the hypothesis of randomness if* $R \leq 6$ *or* $R \geq 16$. *Since* $R = 16$ *for the observed data, we conclude that evidence exists to indicate nonrandomness in the professor's arrangement of answers. His attempt to mix the answers was overdone.*

A second application of the runs test is in detecting nonrandomness of a sequence of quantitative measurements over time. These sequences, known as *time series*, occur in many fields. For example, the measurement of a quality characteristic of an industrial product, blood pressure of a human, and the price of a stock on the stock market all vary over time. Departures in randomness in a series, caused either by trends or periodicities, can be detected by examining the deviations of the time-series measurements from their average. Negative and positive deviations could be denoted by S and F, respectively, and we could then test this time sequence of deviations for nonrandomness. We will illustrate with an example.

Example 15.7: *Paper is produced in a continuous process. Suppose that a brightness measurement, Y, is made on the paper once every hour and that the results appear as shown in Figure 15.3.*

The average for the 15 sample measurements, $\bar{y}$, *appears as shown. Note the deviations about* $\bar{y}$. *Do these indicate a lack of randomness and thereby suggest periodicity in the process and lack of control?*

Solution: *The sequence of negative (S) and positive (F) deviations as indicated in Figure 15.3 is*

$$S\ S\ S\ S\ F\ F\ S\ F\ F\ S\ F\ S\ S\ S\ S.$$

Then $n_1 = 10$, $n_2 = 5$, and $R = 7$. Consulting Table 10, $P(R \leq 7) = .455$. This value of R is not improbable, assuming the hypothesis of randomness to be true. Consequently, there is not sufficient evidence to indicate nonrandomness in the sequence of brightness measurements.

The runs test can also be used to compare two population frequency distributions for a two-sample unpaired experiment. Thus it provides an alternative to the sign test (Section 15.2) and the Mann–Whitney U test (Section 15.4). If the measurements for the two samples are arranged in order of magnitude, they will form a sequence. The measurements for samples 1 and 2 can be denoted as S and F, respectively, and we are once again concerned with a test for randomness. If all measurements for sample 1 are smaller than those for sample 2, the sequence will result in $SSSS \cdots SFFF \cdots F$, or $R = 2$ runs. A small value of R will provide evidence of a difference in population frequency distributions, and the rejection region chosen would be $R \leq a$. This rejection region would imply a one-tailed statistical test. An illustration of the application of the runs test to compare two population frequency distributions will be left as an exercise for the reader.

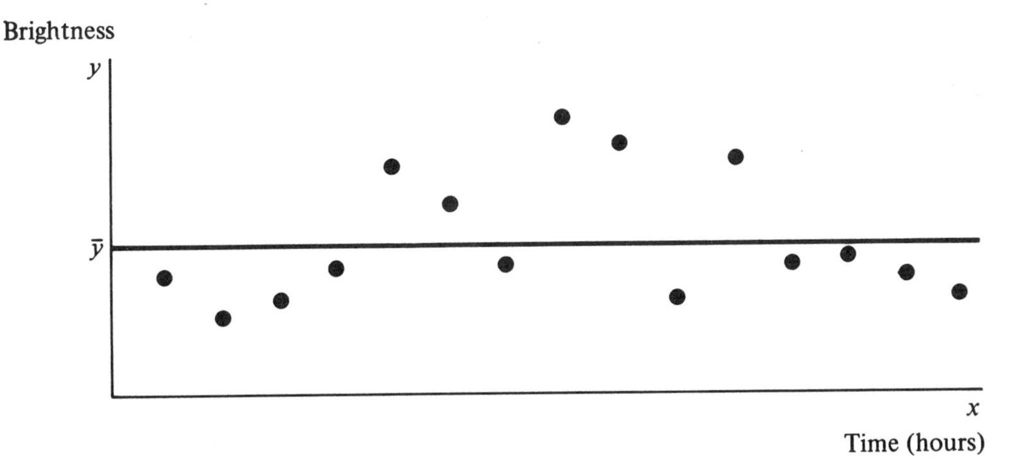

Figure 15.3 Paper Brightness Versus Time

As in the case of the other nonparametric test statistics studied in earlier sections of this chapter, the probability distribution for R tends to normality as n_1 and n_2 become large. The approximation is good when n_1 and n_2 are both greater than 10. Consequently, we may use the Z statistic as a large-sample test statistic where

$$Z = \frac{R - E(R)}{\sqrt{V(R)}},$$

and

$$E(R) = \frac{2n_1n_2}{n_1 + n_2} + 1,$$

$$V(R) = \frac{2n_1n_2(2n_1n_2 - n_1 - n_2)}{(n_1 + n_2)^2(n_1 + n_2 - 1)}$$

are the expected value and variance of R, respectively. The rejection region for a two-tailed test, $\alpha = .05$, is $|z| \geq 1.96$.

15.7 Rank Correlation Coefficient

In the preceding sections we have used ranks to indicate the relative magnitude of observations in nonparametric tests for comparison of treatments. We will now employ the same technique in testing for a relation between two ranked variables. Two common rank correlation coefficients are the Spear-

Table 15.6

Teacher	Judge's Rank	Examination Score
1	7	44
2	4	72
3	2	67
4	6	70
5	1	93
6	3	82
7	8	67
8	5	80

man r_s and the Kendall τ. We will present the Spearman r_s because its computation is identical to that for the sample correlation coefficient, r, of Chapter 11. Kendall's rank correlation coefficient is discussed in detail in Kendall and Stuart [2].

Suppose that eight elementary science teachers have been ranked by a judge according to their teaching ability, and all have taken a "national teachers' examination." The data are given in Table 15.6. Do the data suggest an agreement between the judge's ranking and the examination score? Or one might express this question by asking whether a correlation exists between ranks and test scores.

The two variables of interest are rank and test score. The former is already in rank form, and the test scores may be ranked similarly, as shown in Table 15.7. The ranks for tied observations are obtained by averaging the ranks that the tied observations would occupy as for the Mann–Whitney U statistic.

Table 15.7

Teacher	*Judge's Rank* (x_i)	*Test Rank* (y_i)
1	7	1
2	4	5
3	2	2.5
4	6	4
5	1	8
6	3	7
7	8	2.5
8	5	6

The Spearman rank correlation coefficient, r_s, is calculated using the ranks as the paired measurements on the two variables, x and y, in the formula for r, Chapter 11. Thus

$$r_s = \frac{n \sum_{i=1}^{n} x_i y_i - \left(\sum_{i=1}^{n} x_i\right)\left(\sum_{i=1}^{n} y_i\right)}{\sqrt{\left[n \sum_{i=1}^{n} x_i^2 - \left(\sum_{i=1}^{n} x_i\right)^2\right]\left[n \sum_{i=1}^{n} y_i^2 - \left(\sum_{i=1}^{n} y_i\right)^2\right]}}.$$

The above expression for r_s algebraically reduces to the simpler expression

$$r_s = 1 - \frac{6 \sum_{i=1}^{n} d_i^2}{n(n^2 - 1)}, \quad \textit{where } d_i = x_i - y_i.$$

We will leave proof of this simplification as an exercise for the reader and will illustrate the use of the formula by an example.

Example 15.8: *Calculate r_s for the teacher–judge test-score data.*

Solution: *The differences and squares of differences between the two rankings are shown in Table 15.8.*

Table 15.8

Teacher	x_i	y_i	d_i	d_i^2
1	7	1	6	36
2	4	5	−1	1
3	2	2.5	−.5	.25
4	6	4	2	4
5	1	8	−7	49
6	3	7	−4	16
7	8	2.5	5.5	30.25
8	5	6	−1	1
				137.5

Substituting into the formula for r_s,

$$r_s = 1 - \frac{6 \sum_{i=1}^{n} d_i^2}{n(n^2 - 1)} = 1 - \frac{(6)(137.5)}{(8)(64 - 1)}$$

$$= -.637.$$

The Spearman rank correlation coefficient may be employed as a test statistic to test a hypothesis of "no association" between two populations. We assume that the n pairs of observations, (x_i, y_i), have been randomly selected and therefore "no association between the populations" would imply a random assignment of the n ranks within each sample. Each random assignment (for the two samples) would represent a sample point associated with the experiment and a value of r_s could be calculated for each. It is possible to calculate the probability that r_s assumes a large absolute value due solely to chance and thereby suggests an association between populations when none exists.

We will spare you the details involved in calculating the probability that r_s exceeds some critical value, say r_0, and will present the critical values for $\alpha = .05$ and $\alpha = .01$ in Table 11. The tabulated values give r_0 such that $P(r_s > r_0) \approx .05$ or .01, as indicated. Therefore, they represent the critical values for a one-tailed test. The α for a two-tailed test would require doubling the tabulated probabilities.

Example: 15.9: *Test a hypothesis of "no association" between the populations for Example 15.8.*

Solution: *The critical value of r_s for a one-tailed test with $\alpha = .05$ and $n = 8$ is .643. Let us assume that a correlation between judge's rank and the teachers' test scores could not possibly be positive. (Low rank means good teaching and should be associated with a high test score if the judge and test measure teaching ability.) The alternative hypothesis would be that the population correlation coefficient, ρ, is less than zero, and we would be concerned with a one-tailed statistical test. Thus α for the test would be the value .05, and we would reject the null hypothesis if $r_s \leq -.643$.*

The calculated value of the test statistic, $r_s = -.637$, is slightly larger than the critical value for $\alpha = .05$. Hence the null hypothesis would not be rejected at the $\alpha = .05$ level of significance but would fall well within the rejection region for $\alpha = .10$. It appears that some agreement does exist between the judge's rankings and the test scores. However, it should be noted that this agreement could exist when *neither* provides an adequate yardstick for measuring teaching ability. For example, the association could exist if both the judge and those who constructed the teacher's examination possessed a completely erroneous, but identical, concept of the characteristics of good teaching.

15.8 Some General Comments on Nonparametric Statistical Tests

The nonparametric statistical tests presented in the preceding pages represent only a few of the many nonparametric statistical methods of inference available. A much larger collection of nonparametric test procedures, along with worked examples, is given by Siegel [4] and by Hajek [1].

We have indicated that nonparametric statistical procedures are particularly useful when the experimental observations are susceptible to ordering but cannot be measured on a quantitative scale. Parametric statistical procedures usually cannot be applied to this type of data. Hence all inferential procedures must be based on nonparametric methods.

A second application of nonparametric statistical methods is in testing hypotheses associated with populations of quantitative data when uncertainty exists concerning the satisfaction of assumptions about the form of the population distributions. Just how useful are nonparametric methods for this situation?

It is well known that many statistical test and estimation procedures are only slightly affected by moderate departures from the assumed form of the population frequency distribution and that parametric test procedures are more powerful than their nonparametric equivalents when the assumptions underlying parametric tests are true. Furthermore, it is sometimes possible to transform statistical data and thereby make the population distributions of the transformed data satisfy the assumptions basic to a parametric procedure. In spite of these positive arguments in favor of parametric procedures, nonparametric statistical methods possess definite advantages in some experimental situations.

Nonparametric statistical methods are rapid and often lead to an immediate decision in testing hypotheses, as indicated in Example 15.3. When experimental conditions depart substantially from the basic assumptions underlying parametric tests, the response measurements can often be transformed to alleviate the condition, but an unfortunate consequence often develops. That is, the transformed response is no longer meaningful from a practical point of view, and analysis of the transformed data no longer answers the objectives of the experimenter. The use of nonparametric methods will often circumvent this difficulty. Finally, one should note that many nonparametric methods are nearly as efficient as their parametric counterparts when the assumptions underlying the parametric procedures are true, and, as noted earlier, they could be more efficient when the assumptions are unsatisfied. These reasons suggest that nonparametric techniques play a very useful role in statistical methodology.

References

1. Hajek, J., *Nonparametric Statistics*. San Francisco: Holden-Day, Inc., 1969.
2. Kendall, M. G., and A. Stuart, *The Advanced Theory of Statistics*, Vol. 2. New York: Hafner Publishing Company, Inc., 1961.
3. Savage, I. R., "Bibliography of Nonparametric Statistics and Related Topics." *Journal of the American Statistical Association*, Vol. 48 (1953), pp. 844–906.
4. Siegel, S., *Nonparametric Statistics for the Behavioral Sciences*. New York: McGraw-Hill Book Company, Inc., 1956.
5. Wald, A., and J. Wolfowitz, "On a Test Whether Two Samples Are from the Same Population." *Annals of Mathematical Statistics*, Vol. 2 (1940), pp. 147–162.

Exercises

15.1. What significance levels between $\alpha = .01$ and $\alpha = .15$ are available for a two-tailed sign test utilizing 25 paired observations? (Make use of tabulated values in Table 1, $n = 25$.) What are the corresponding rejection regions?

15.2. Two plastics, each produced by a different process, were tested for ultimate strength. The measurements represent breaking load in units of 1000 pounds per square inch:

Plastic 1	*Plastic 2*
15.3	21.2
18.7	22.4
22.3	18.3
17.6	19.3
19.1	17.1
14.8	27.7

Do the data present evidence of a difference between the mean ultimate strengths for the two plastics? Solve by using a sign test with a level of significance as near as possible to $\alpha = .10$.

15.3. The coded values for a measure of brightness in paper (light reflectivity), prepared by two different processes, are as follows for samples of size 9 drawn randomly from each of the two processes:

A	*B*
6.1	9.1
9.2	8.2
8.7	8.6
8.9	6.9
7.6	7.5
7.1	7.9
9.5	8.3
8.3	7.8
9.0	8.9

Do the data present sufficient evidence ($\alpha = .10$) to indicate a difference in the populations of brightness measurements for the two processes?
(a) Use the sign test.
(b) Use Student's t test.

15.4. To compare two junior high schools, A and B, in academic effectiveness, an experiment was designed requiring the use of 10 sets of identical twins, each twin having just completed the sixth grade. In each case, the twins in the same set had obtained their schooling in the same classrooms at each grade level. One child was selected at random from each set and assigned to school A. The other was sent to school B. Near the end of the ninth grade, a certain achievement test was given to each child in the experiment. The results are shown in the following table:

Twin Pair	*A*	*B*	*Twin Pair*	*A*	*B*
1	67	39	6	50	52
2	80	75	7	63	56
3	65	69	8	81	72
4	70	55	9	86	89
5	86	74	10	60	47

(a) Test (using the sign test) the hypothesis that the two schools are the same in academic effectiveness, as measured by scores on the achievement test, against the alternative that the schools are not equally effective. Use a level of significance as near as possible to $\alpha = .05$.

(b) Suppose it were known that junior high school A had a superior faculty and better learning facilities. Test the hypothesis of equal academic effectiveness against the alternative that school A is superior. Use a level of significance as near as possible to $\alpha = .05$.

15.5. Refer to Exercise 15.3. What answer is obtained if the Mann–Whitney U test is used in analyzing the data? Compare with the answers to Exercise 15.3.

15.6. Refer to Exercise 15.3. What answer is obtained if the runs test is used in analyzing the data? Compare with the answer to Exercise 15.5.

15.7. Refer to Exercise 15.4. What answers are obtained if Wilcoxon's T test is used in analyzing the data? Compare with the answers to Exercise 15.4.

15.8. A psychological experiment was conducted to compare the lengths of response time (in seconds) for two different stimuli. In order to remove natural person-to-person variability in the responses, both stimuli were applied to each of nine subjects, thus permitting an analysis of the difference between stimuli *within* each person.

Subject	*Stimulus 1*	*Stimulus 2*
1	9.4	10.3
2	7.8	8.9
3	5.6	4.1
4	12.1	14.7
5	6.9	8.7
6	4.2	7.1
7	8.8	11.3
8	7.7	5.2
9	6.4	7.8

(a) Use the sign test to determine whether sufficient evidence exists to indicate a difference in mean response for the two stimuli. Use a rejection region for which $\alpha \leq .05$.

(b) Test the hypothesis of no difference in mean response using Student's t test.

15.9. Refer to Exercise 15.8. Test the hypothesis that no difference exists in the distributions of responses for the two stimuli, using the Wilcoxon T test. Use a rejection region for which α is as near as possible to the α achieved in Exercise 15.8, part (a).

15.10. Four Republican and four Democratic politicians attended a civic banquet and selected their seats in the following order (left to right): *R, D, D, D, D, R, R, R*. Does this seating sequence provide sufficient evidence to imply nonrandomness in the seating selection and a consequent grouping by party affiliation?

15.11. Fifteen experimental batteries were selected at random from a lot at pilot plant *A*, and 15 standard batteries were selected at random from production at plant *B*. All 30 batteries were simultaneously placed under an electrical load of the same magnitude. The first battery to fail was an *A*, the second a *B*, the third a *B*, and so on. The following sequence shows the order of failure for the 30 batteries:

A B B B A B A A B B B B A B A

B B B B A A B A A A B A A A A.

(a) Using the large sample theory for the *U* test, determine (use $\alpha = .05$) if there is sufficient evidence to conclude that the mean life for the experimental batteries is greater than the mean life for the standard batteries.

(b) If, indeed, the experimental batteries have the greater mean life, what would be the effect on the expected number of runs? Using the large-sample theory for the runs test, test (using $\alpha = .05$) whether there is a difference in the distribution of battery life for the two populations.

15.12. The conditions (*D* for diseased, *S* for sound) of the individual trees in a row of ten poplars were found to be from left to right: *S, S, D, D, S, D, D, D, S, S*. Is there sufficient evidence to indicate nonrandomness in the sequence and therefore the possibility of contagion?

15.13. Items emerging from a continuous production process were classified as defective or nondefective. A sequence of items, observed over time, was as follows: *D, N, N, N, N, N, N, D, D, N, N, N, N, N, N, D, D, D, N, N, N, N, N, D, N, N, N, D, D, N, N, N, D, D*.

(a) Give the appropriate probability that $R \leq 1$ where $n_1 = 11$ and $n_2 = 23$.

(b) Do these data suggest lack of randomness in the occurrence of defectives (*D*) and nondefectives (*N*)? Use the large sample approximation for the runs test.

15.14. A quality-control chart has been maintained for a certain measurable characteristic of items taken from a conveyor belt at a certain point in a production line. The measurements obtained today in order of time are: 68.2, 71.6, 69.3, 71.6, 70.4, 65.0, 63.6, 64.7, 65.3, 64.2, 67.6, 68.6, 66.8, 68.9, 66.8, 70.1.

(a) Classify the measurements in this time series as above or below the sample mean and determine (use the runs test) whether consecutive observations suggest lack of stability in the production process.

(b) Divide the time period into two equal parts and compare the means, using Student's t test. Do the data provide evidence of a shift in the mean level of the quality characteristics?

15.15. If (as in the case of measurements produced by two well-calibrated measuring instruments) the means of two populations are equal, it is possible to use the Mann–Whitney U statistic for testing hypotheses concerning the population variances as follows:

(1) Rank the combined sample.

(2) Number the ranked observations "from the outside in"; that is, number the smallest observation 1; the largest 2; the next to smallest, 3; the next to largest, 4; and so on. This final sequence of numbers induces an ordering on the symbols A (population A items) and B (population B items). If $\sigma_A^2 > \sigma_B^2$, one would expect to find a preponderance of A's near the first of the sequences, and thus a relatively small "sum of ranks" for the A observations.

(a) Given the following measurements produced by well-calibrated precision instruments A and B, test at near the $\alpha = .05$ level to determine whether the more expensive instrument, B, is more precise than A. (Note that this would imply a one-tailed test.) Use the Mann–Whitney U test.

A	B
1060.21	1060.24
1060.34	1060.28
1060.27	1060.32
1060.36	1060.30
1060.40	

(b) Test, using the F statistic of Section 10.6.

15.16. A large corporation selects graduates for employment, using both interviews and a psychological-achievement test. Interviews conducted at the home office of the company were far more expensive than the test which could be conducted on campus. Consequently, the personnel office was interested in determining whether the test scores were correlated with interview ratings and whether the tests could be substituted for interviews. The idea was not to eliminate interviews but to reduce their number. To determine whether correlation was present, ten prospects were ranked during inverviews, and tested. The paired scores were as follows:

Subject	*Interview Rank*	*Test Score*
1	8	74
2	5	81
3	10	66
4	3	83
5	6	66
6	1	94
7	4	96
8	7	70
9	9	61
10	2	86

Calculate the Spearman rank correlation coefficient, r_s. Rank 1 is assigned to the candidate judged to be the best.

15.17. Refer to Exercise 15.16. Do the data present sufficient evidence to indicate that the correlation between interview rankings and test scores is less than zero? If this evidence does exist, can we say that tests could be used to reduce the number of interviews?

15.18. A political scientist wished to examine the relationship of the voter image of a conservative political candidate and the distance between the residences of the voter and the candidate. Each of 12 voters rated the candidate on a scale of 1 to 20. The data were as follows:

Voter	*Rating*	*Distance*
1	12	75
2	7	165
3	5	300
4	19	15
5	17	180
6	12	240
7	9	120
8	18	60
9	3	230
10	8	200
11	15	130
12	4	130

Calculate the Spearman rank correlation coefficient, r_s.

15.19. Refer to Exercise 15.18. Do these data provide sufficient evidence to indicate a negative correlation between rating and distance?

15.20. A comparison of reaction times (in seconds) for two different stimuli in a psychological word-association experiment produced the following results when applied to a random sample of 16 people:

Stimulus 1	*Stimulus 2*
1	4
3	2
2	3
1	3
2	1
1	2
3	3
2	3

Do the data present sufficient evidence to indicate a difference in mean reaction time for the two stimuli? Use the Mann–Whitney U statistic and test using $\alpha = .05$. (*Note*: This test was conducted using Student's t in the exercises for Chapter 10. Compare your results.)

15.21. The following table gives the scores of a group of 15 students in mathematics and art:

Student	*Math*	*Art*
1	22	53
2	37	68
3	36	42
4	38	49
5	42	51
6	58	65
7	58	51
8	60	71
9	62	55
10	65	74
11	66	68
12	56	64
13	66	67
14	67	73
15	62	65

Use Wilcoxon's signed rank test to determine if the distributions of scores for these students differ significantly for the two subjects.

15.22. Refer to Exercise 15.21. Compute Spearman's rank correlation coefficient for these data and test $H_0: \rho_s = 0$ at the 10 percent level of significance.

15.23. Calculate the probability that $U \leq 2$ for $n_1 = n_2 = 5$. Assume that no ties will be present and that H_0 is true.

15.24. Calculate the probability that the Wilcoxon T (Section 15.5) is less than or equal to 2 for $n = 3$ pairs. Assume that no ties will be present and that H_0 is true.

15.25. Calculate $P(R \leq 6)$ for the runs test where $n_1 = n_2 = 8$ and H_0 is true.

15.26. Consider a Wilcoxon rank sum test for the comparison of two probability distributions based on independent random samples of $n_1 = n_2 = 5$. Find $P(W \leq 17)$ assuming that H_0 is true.

15.27. Consider a runs test based on $n_1 = n_2 = 5$ elements. Assuming H_0 to be true, find
(a) $P(R = 2)$.
(b) $P(R \leq 3)$.
(c) $P(R \leq 4)$.

15.28. Let U denote the Mann–Whitney statistic and W_A the Wilcoxon rank-sum statistic for the sample from population A. Show that

$$U = n_1 n_2 + (1/2)n_1(n_1 + 1) - W_A.$$

15.29. Refer to Exercise 15.28. Show that
(a) $E(U) = (1/2)n_1 n_2$ when H_0 is true.
(b) $V(U) = (1/12)[n_1 n_2(n_1 + n_2 + 1)]$ when H_0 is true, where H_0 states that the two populations are identical.

15.30. Let T denote the Wilcoxon signed-rank test for n pairs of observations. Show that $E(T) = (1/4)n(n + 1)$ and $V(T) = (1/24)[n(n + 1)(2n + 1)]$ when the two populations are identical. Observe that these properties do not depend on whether T is constructed from negative or positive differences.

15.31. Suppose $Y_1, \ldots, Y_n$ is a random sample from a continuous distribution function $F(y)$. It is desired to test a hypothesis concerning the median, ξ, of $F(y)$. Construct a test of $H_0: \xi = \xi_0$ against $H_a: \xi \neq \xi_0$, where ξ_0 is a specified constant, by making use of
(a) The sign test.
(b) The Wilcoxon signed-rank test.

15.32. Refer to the Spearman rank correlation coefficient of Section 15.7. Show that

$$r_s = \frac{n \sum_{i=1}^{n} x_i y_i - \left(\sum_{i=1}^{n} x_i\right)\left(\sum_{i=1}^{n} y_i\right)}{\sqrt{\left[n \sum_{i=1}^{n} x_i^2 - \left(\sum_{i=1}^{n} x_i\right)^2\right]\left[n \sum_{i=1}^{n} y_i^2 - \left(\sum_{i=1}^{n} y_i\right)^2\right]}}$$

$$= 1 - \frac{6 \sum_{i=1}^{n} d_i^2}{n(n^2 - 1)}$$

where $d_i = x_i - y_i$.

Appendix I

Matrices

I.1 *Matrices and Matrix Algebra*

The following represents a very elementary and condensed discussion of matrices and matrix operations. The reader seeking a more comprehensive introduction to the subject should consult the references indicated at the end of Chapter 11.

We will define a *matrix* as a rectangular array (arrangement) of real numbers and will indicate specific matrices symbolically with capital letters. The numbers in the matrix, *elements*, appear in specific row–column positions, all of which are filled. The number of rows and columns may vary from one matrix to another, so we conveniently describe the size of a matrix by giving its *dimensions*—that is, the number of its rows and columns. Thus matrix **A**

$$\underset{2\times 3}{\mathbf{A}} = \begin{bmatrix} 6 & 0 & -1 \\ 4 & 2 & 7 \end{bmatrix}$$

possesses dimensions 2×3 because it contains two rows and three columns.

Similarly, for

$$\underset{4\times 1}{\mathbf{B}} = \begin{bmatrix} 1 \\ -3 \\ 0 \\ 7 \end{bmatrix}, \qquad \underset{2\times 2}{\mathbf{C}} = \begin{bmatrix} 2 & 0 \\ -1 & 4 \end{bmatrix},$$

the dimensions of **B** and **C** are 4×1 and 2×2, respectively. Note that the row dimension always appears first and that the dimensions may be written below the identifying symbol of the matrix as indicated for matrices **A**, **B**, and **C**.

As in ordinary algebra, an element of a matrix may be indicated by a symbol, $a, b, \ldots$, and its row–column position identified by means of a double subscript. Thus a_{21} would be the element in the second row, first column. Rows are numbered in order from top to bottom and columns from left to right. In matrix **A**, $a_{21} = 4$, $a_{13} = -1$, and so on.

Elements in a particular row are identified by their column subscript and hence are numbered from left to right. The first element in a row is on the left. Likewise, elements in a particular column are identified by their row subscript and therefore are identified from the top element in the column to the bottom. For example, the first element in column 2 of matrix **A** is 0, the second is 2. The first, second, and third elements of row 1 are 6, 0, and -1, respectively.

The term "matrix algebra" involves, as the name implies, an algebra dealing with matrices much as the ordinary algebra deals with real numbers or symbols representing real numbers. Hence we will wish to state rules for the addition and multiplication of matrices as well as to define other elements of an algebra. In so doing, we will point out the similarities as well as the dissimilarities between matrix and ordinary algebra. Finally, we will use our matrix operations to state and solve a very simple *matrix equation*. This, as the reader may suspect, will be the solution for the least-squares equations that we desire.

I.2 *Addition of Matrices*

Two matrices, say **A** and **B**, can be added *only* if they are of the same dimensions. The sum of the two matrices will be a matrix obtained by adding *corresponding* elements of matrices **A** and **B**—that is, elements in corresponding positions. This being the case, the resulting sum will be a matrix of the same dimensions as **A** and **B**.

Example I.1: *Find the indicated sum of matrices* **A** *and* **B**:

$$\underset{2\times 3}{\mathbf{A}} = \begin{bmatrix} 2 & 1 & 4 \\ -1 & 6 & 0 \end{bmatrix}, \qquad \underset{2\times 3}{\mathbf{B}} = \begin{bmatrix} 0 & -1 & 1 \\ 6 & -3 & 2 \end{bmatrix}.$$

Solution:

$$\begin{aligned} \mathbf{A} + \mathbf{B} &= \begin{bmatrix} 2 & 1 & 4 \\ -1 & 6 & 0 \end{bmatrix} + \begin{bmatrix} 0 & -1 & 1 \\ 6 & -3 & 2 \end{bmatrix} \\ &= \begin{bmatrix} (2+0) & (1-1) & (4+1) \\ (-1+6) & (6-3) & (0+2) \end{bmatrix} = \begin{bmatrix} 2 & 0 & 5 \\ 5 & 3 & 2 \end{bmatrix}. \end{aligned}$$

Example I.2: *Find the sum of the matrices*

$$\underset{3\times 3}{\mathbf{A}} = \begin{bmatrix} 1 & 0 & 3 \\ 1 & -1 & 4 \\ 2 & -1 & 0 \end{bmatrix}, \qquad \underset{3\times 3}{\mathbf{B}} = \begin{bmatrix} 4 & 2 & -1 \\ 1 & 0 & 6 \\ 3 & 1 & 4 \end{bmatrix}.$$

Solution:

$$\mathbf{A} + \mathbf{B} = \begin{bmatrix} 5 & 2 & 2 \\ 2 & -1 & 10 \\ 5 & 0 & 4 \end{bmatrix}.$$

Note that $(\mathbf{A} + \mathbf{B}) = (\mathbf{B} + \mathbf{A})$, as in ordinary algebra, and remember that we never add matrices of unlike dimensions.

I.3 Multiplication of a Matrix by a Real Number

We desire a rule for multiplying a matrix by a real number, for example 3**A**, where

$$\mathbf{A} = \begin{bmatrix} 2 & 1 \\ 4 & 6 \\ -1 & 0 \end{bmatrix}.$$

Certainly, we would want 3**A** to equal (**A** + **A** + **A**) to conform with the addition rule. Hence 3**A** would mean that each element in the **A** matrix must be multiplied by the multiplier, 3, and

$$3\mathbf{A} = \begin{bmatrix} 3(2) & 3(1) \\ 3(4) & 3(6) \\ 3(-1) & 3(0) \end{bmatrix} = \begin{bmatrix} 6 & 3 \\ 12 & 18 \\ -3 & 0 \end{bmatrix}.$$

In general, given a real number c and a matrix **A** with elements a_{ij}, the product $c\mathbf{A}$ will be a matrix whose elements are equal to ca_{ij}.

I.4 Matrix Multiplication

The rule for matrix multiplication requires "row–column multiplication," which we will define subsequently. The procedure may seem a bit complicated to the novice but should not prove too difficult after practice. We will illustrate with an example. Let **A** and **B** be

$$\mathbf{A} = \begin{bmatrix} 2 & 0 \\ 1 & 4 \end{bmatrix}, \quad \mathbf{B} = \begin{bmatrix} 5 & 2 \\ -1 & 3 \end{bmatrix}.$$

An element in the *ith row* and *jth column* of the product, **AB**, is obtained by multiplying the *ith row of* **A** by the *jth column of* **B**. Thus the element in the first row, first column of **AB** is obtained by multiplying the first row of **A** by the first column of **B**. Likewise, the element in the first row, second column would be the product of the first row of **A** and the second column of **B**. Notice that we always use the rows of **A** and the columns of **B**, where **A** is the matrix to the left of **B** in the product, **AB**.

Row–column multiplication is relatively easy. Obtain the products, first row element by first column element, second row element by second column element, third by third, and so on, and then sum. Remember that row and column elements are numbered from left to right and top to bottom, respectively.

Applying these rules to our example,

$$\underset{2\times 2}{\mathbf{A}}\;\underset{2\times 2}{\mathbf{B}} = \begin{bmatrix} 2 & 0 \\ 1 & 4 \end{bmatrix}\begin{bmatrix} 5 & 2 \\ -1 & 3 \end{bmatrix} = \begin{bmatrix} \textcircled{10} & 4 \\ 1 & 14 \end{bmatrix}.$$

The first row–first column product would be $(2)(5) + (0)(-1) = 10$, which is located (and circled) in the first row, first column of **AB**. Likewise, the element in the first row, second column is equal to the product of the first row of **A** and the second column of **B**, or $(2)(2) + (0)(3) = 4$. The second row–first column product is $(1)(5) + (4)(-1) = 1$ and is located in the second row, first column of **AB**. Finally, the second row–second column product is $(1)(2) + (4)(3) = 14$.

Example I.3: *Find the products* **AB** *and* **BA**, *where*

$$\mathbf{A} = \begin{bmatrix} 2 & 1 \\ 1 & -1 \\ 0 & 4 \end{bmatrix} \quad \text{and} \quad \mathbf{B} = \begin{bmatrix} 4 & -1 & -1 \\ 2 & 0 & 2 \end{bmatrix}.$$

Solution:

$$\underset{3\times 2}{\mathbf{A}}\;\underset{2\times 3}{\mathbf{B}} = \begin{bmatrix} 2 & 1 \\ 1 & -1 \\ 0 & 4 \end{bmatrix}\begin{bmatrix} 4 & -1 & -1 \\ 2 & 0 & 2 \end{bmatrix} = \begin{bmatrix} 10 & -2 & 0 \\ 2 & -1 & -3 \\ 8 & 0 & 8 \end{bmatrix},$$

and

$$\underset{2\times 3}{\mathbf{B}}\;\underset{3\times 2}{\mathbf{A}} = \begin{bmatrix} 4 & -1 & -1 \\ 2 & 0 & 2 \end{bmatrix}\begin{bmatrix} 2 & 1 \\ 1 & -1 \\ 0 & 4 \end{bmatrix} = \begin{bmatrix} 7 & 1 \\ 4 & 10 \end{bmatrix}.$$

Note that in matrix algebra, unlike ordinary algebra, **AB** does not equal **BA**. Since **A** contains three rows and **B** contains three columns, we can form (3)(3) = 9 row–column combinations and hence nine elements for **AB**. In contrast, **B** contains only two rows, **A** two columns, and hence the product **BA** will possess only (2)(2) = 4 elements, corresponding to the four different row–column combinations.

Furthermore, we observe that row–column multiplication is predicated on the assumption that the rows of the matrix on the left contain the same number of elements as the columns of the matrix on the right so that "corresponding elements" will exist for the row–column multiplication. What do we do when this condition is not satisfied? We agree never to multiply two matrices, say **AB**, where the rows of **A** and the columns of **B** contain an unequal number of elements.

An examination of the dimensions of the matrices will tell whether they can be multiplied as well as give the dimensions of the product. Writing the dimensions underneath the two matrices,

$$\underset{m \times p}{\mathbf{A}}\ \underset{p \times q}{\mathbf{B}} = \underset{m \times q}{\mathbf{AB}}$$

we observe that the inner two numbers, giving the number of elements in a row of **A** and column of **B**, respectively, must be equal. The outer two numbers, indicating the number of rows of **A** and columns of **B**, give the dimensions of the product matrix. The reader may verify the operation of this rule for Example I.3.

Example I.4: *Obtain the product* **AB**:

$$\underset{1 \times 3}{\mathbf{A}}\ \underset{3 \times 2}{\mathbf{B}} = [2 \quad 1 \quad 0]\begin{bmatrix} 2 & 0 \\ 0 & 3 \\ -1 & 0 \end{bmatrix} = [4 \quad 3].$$

Note that product **AB** is (1 × 2) and that **BA** is undefined because of the respective dimensions of **A** and **B**.

Example I.5: *Find the product* **AB**, *where*

$$\mathbf{A} = [1 \quad 2 \quad 3 \quad 4], \qquad \mathbf{B} = \begin{bmatrix} 1 \\ 2 \\ 3 \\ 4 \end{bmatrix}.$$

Solution:

$$\underset{1\times 4}{\mathbf{A}}\;\underset{4\times 1}{\mathbf{B}} = [1 \quad 2 \quad 3 \quad 4]\begin{bmatrix} 1 \\ 2 \\ 3 \\ 4 \end{bmatrix} = [30].$$

Note that this example produces a different method for writing a sum of squares.

I.5 *Identity Elements*

The identity elements for addition and multiplication in the ordinary algebra are 0 and 1, respectively. In addition, 0 plus any other element, say a, is identically equal to a; that is,

$$0 + 2 = 2, \qquad 0 + (-9) = -9.$$

Similarly, the multiplication of the identity element, 1, by any other element, say a, is equal to a; that is,

$$(1)(5) = 5, \qquad (1)(-4) = -4.$$

In matrix algebra, two matrices are said to be equal when all corresponding elements are equal. With this in mind we will define the identity matrices in a manner similar to that employed in the ordinary algebra. Hence, if **A** is any matrix, a matrix **B** will be an identity matrix for addition if

$$\mathbf{A} + \mathbf{B} = \mathbf{A} \qquad \text{and} \qquad \mathbf{B} + \mathbf{A} = \mathbf{A}.$$

It can easily be seen that the identity matrix for addition is one in which every element is equal to zero. This matrix is of interest but of no practical importance in our work.

Similarly, if **A** is any matrix, the identity matrix for multiplication is a matrix **I** which satisfies the relation

$$\mathbf{AI} = \mathbf{A} \qquad \text{and} \qquad \mathbf{IA} = \mathbf{A}.$$

This matrix, called the *identity matrix*, is the *square matrix*

$$\underset{n \times n}{\mathbf{I}} = \begin{bmatrix} 1 & 0 & 0 & \cdot & \cdots & \cdot \\ 0 & 1 & 0 & \cdot & \cdots & \cdot \\ 0 & 0 & 1 & \cdot & \cdots & \cdot \\ 0 & 0 & 0 & 1 & \cdots & \cdot \\ \cdot & \cdot & \cdot & & & \\ \cdot & \cdot & \cdot & & & \\ 0 & 0 & 0 & \cdot & \cdots & 1 \end{bmatrix}.$$

That is, all elements in the *main diagonal* of the matrix, running from top left to bottom right, are equal to 1; all other elements equal zero. Note that the identity matrix is always indicated by the symbol **I**.

Unlike the ordinary algebra, which contains only one identity element for multiplication, matrix algebra must contain an infinitely large number of identity matrices. Thus we must have matrices with dimensions 1×1, 2×2, 3×3, 4×4, and so on, so as to provide an identity of the correct dimensions to permit multiplication. All will be of the pattern indicated above.

That the **I** matrix satisfies the relation

$$\mathbf{IA} = \mathbf{AI} = \mathbf{A}$$

can be easily shown by an example.

Example I.6: *Let*

$$\mathbf{A} = \begin{bmatrix} 2 & 1 & 0 \\ -1 & 6 & 3 \end{bmatrix}.$$

Show that **IA** = **A** *and* **AI** = **A**.

Solution:

$$\underset{2\times 2}{\mathbf{I}}\ \underset{2\times 3}{\mathbf{A}} = \begin{bmatrix} 1 & 0 \\ 0 & 1 \end{bmatrix} \begin{bmatrix} 2 & 1 & 0 \\ -1 & 6 & 3 \end{bmatrix} = \begin{bmatrix} 2 & 1 & 0 \\ -1 & 6 & 3 \end{bmatrix} = \mathbf{A},$$

and

$$\underset{2\times 3}{\mathbf{A}}\ \underset{3\times 3}{\mathbf{I}} = \begin{bmatrix} 2 & 1 & 0 \\ -1 & 6 & 3 \end{bmatrix} \begin{bmatrix} 1 & 0 & 0 \\ 0 & 1 & 0 \\ 0 & 0 & 1 \end{bmatrix} = \begin{bmatrix} 2 & 1 & 0 \\ -1 & 6 & 3 \end{bmatrix} = \mathbf{A}.$$

I.6 ***The Inverse of a Matrix***

In order that matrix algebra be useful, we must be able to construct and solve matrix equations for a matrix of unknowns in a manner similar to that employed in ordinary algebra. This, in turn, requires a method of performing "division."

For example, we would solve the simple equation in ordinary algebra,

$$2x = 6,$$

by "dividing" both sides of the equation by 2 and obtaining $x = 3$. Another way to view this operation is to define the reciprocal of each element in an algebraic system and to think of "division" as multiplication by the reciprocal of an element. We could solve the equation

$$2x = 6$$

by multiplying both sides of the equation by the reciprocal of 2. Since every element in the real system possesses a reciprocal with the exception of 0, the multiplication operation eliminates the need for division.

The reciprocal of a number, c, in ordinary algebra is a number, b, that satisfies the relation

$$cb = 1;$$

that is, the product of a number by its reciprocal must equal the identity element for multiplication. For example, the reciprocal of 2 is $\frac{1}{2}$ and $(2)(\frac{1}{2}) = 1$.

A reciprocal in matrix algebra is called the inverse of a matrix and is defined as follows:

Definition I.1: *Let* $\mathbf{A}_{n \times n}$ *be a square matrix. If a matrix* $\mathbf{A}^{-1}$ *can be found such that*

$$\mathbf{A}\mathbf{A}^{-1} = \mathbf{I} \quad \text{and} \quad \mathbf{A}^{-1}\mathbf{A} = \mathbf{I},$$

then $\mathbf{A}^{-1}$ *is called the* inverse *of* $\mathbf{A}$.

Note that the requirement for an inverse in matrix algebra is the same as in ordinary algebra—that is, the product of $\mathbf{A}$ by its inverse must equal the identity matrix for multiplication. Furthermore, the inverse is undefined for nonsquare matrices, and hence many matrices in matrix algebra do not have inverses (recall that 0 was the only element in the real number system without an inverse). Finally, we state without proof that many square matrices do not possess inverses. Those which do will be identified in Section I.9, and a method will be given for finding the inverse of a matrix.

I.7 The Transpose of a Matrix

We have just discussed a relationship between a matrix and its inverse. A second useful matrix relationship defines the *transpose* of a matrix.

> ***Definition I.2:*** *Let* $\mathbf{A}_{p \times q}$ *be a matrix of dimensions* $p \times q$. *Then* $\mathbf{A}'$, *called the* transpose *of* $\mathbf{A}$, *is defined to be a matrix obtained by interchanging corresponding rows and columns of* $\mathbf{A}$, *that is, first with first, second with second, and so on.*

For example, let

$$\underset{3 \times 2}{\mathbf{A}} = \begin{bmatrix} 2 & 0 \\ 1 & 1 \\ 4 & 3 \end{bmatrix}.$$

Then

$$\underset{2 \times 3}{\mathbf{A}'} = \begin{bmatrix} 2 & 1 & 4 \\ 0 & 1 & 3 \end{bmatrix}.$$

Note that the first and second rows of $\mathbf{A}'$ are identical with the first and second columns, respectively, of $\mathbf{A}$.

As a second example, let

$$\mathbf{Y} = \begin{bmatrix} y_1 \\ y_2 \\ y_3 \end{bmatrix}.$$

Then $\mathbf{Y}' = [y_1 \quad y_2 \quad y_3]$. As a point of interest, we observe that $\mathbf{Y}'\mathbf{Y} = \sum_{i=1}^{3} y_i^2$.

Finally, if

$$\mathbf{A} = \begin{bmatrix} 2 & 1 & 4 \\ 0 & 2 & 3 \\ 1 & 6 & 9 \end{bmatrix},$$

then

$$\mathbf{A}' = \begin{bmatrix} 2 & 0 & 1 \\ 1 & 2 & 6 \\ 4 & 3 & 9 \end{bmatrix}.$$

I.8 A Matrix Expression for a System of Simultaneous Linear Equations

We will now introduce the reader to one of the very simple and important applications of matrix algebra. Let

$$2v_1 + v_2 = 5,$$

$$v_1 - v_2 = 1$$

be a pair of simultaneous linear equations in the two variables, v_1 and v_2. We will then define three matrices:

$$\underset{2\times 2}{\mathbf{A}} = \begin{bmatrix} 2 & 1 \\ 1 & -1 \end{bmatrix}, \quad \underset{2\times 1}{\mathbf{V}} = \begin{bmatrix} v_1 \\ v_2 \end{bmatrix}, \quad \underset{2\times 1}{\mathbf{G}} = \begin{bmatrix} 5 \\ 1 \end{bmatrix}.$$

Note that **A** is the matrix of coefficients of the unknowns when the equations are each written with the variables appearing in the same order, reading left to right, and with the constants on the right-hand side of the equality sign. The **V** matrix gives the unknowns in a column and in the same order as they appear in the equations. Finally, the **G** matrix contains the constants in a column exactly as they occur in the set of equations.

The simultaneous system of two linear equations may now be written in matrix algebra as

$$\mathbf{AV} = \mathbf{G},$$

a statement that can easily be verified by multiplying **A** and **V** and then comparing with **G**.

$$\mathbf{AV} = \begin{bmatrix} 2 & 1 \\ 1 & -1 \end{bmatrix} \begin{bmatrix} v_1 \\ v_2 \end{bmatrix} = \begin{bmatrix} 2v_1 + v_2 \\ v_1 - v_2 \end{bmatrix} = \begin{bmatrix} 5 \\ 1 \end{bmatrix} = \mathbf{G}.$$

The reader will observe that corresponding elements in **AV** and **G** are equal—that is, $2v_1 + v_2 = 5$ and $v_1 - v_2 = 1$. Therefore, $\mathbf{AV} = \mathbf{G}$.

The method for writing a pair of linear equations in two unknowns as a matrix equation can easily be extended to a system of r equations in r unknowns.

For example, if the equations are:

$$\begin{aligned}
a_{11}v_1 + a_{12}v_2 + a_{13}v_3 + \cdots + a_{1r}v_r &= g_1, \\
a_{21}v_1 + a_{22}v_2 + a_{23}v_3 + \cdots + a_{2r}v_r &= g_2, \\
a_{31}v_1 + a_{32}v_2 + a_{33}v_3 + \cdots + a_{3r}v_r &= g_3, \\
\vdots \qquad \vdots \qquad \vdots \qquad\quad \vdots \quad & \quad \vdots \\
a_{r1}v_1 + a_{r2}v_2 + a_{r3}v_3 + \cdots + a_{rr}v_r &= g_r,
\end{aligned}$$

define

$$\mathbf{A} = \begin{bmatrix} a_{11} & a_{12} & a_{13} & \cdots & a_{1r} \\ a_{21} & a_{22} & a_{23} & \cdots & a_{2r} \\ a_{31} & a_{32} & a_{33} & \cdots & a_{3r} \\ \vdots & \vdots & \vdots & & \vdots \\ a_{r1} & a_{r2} & a_{r3} & & a_{rr} \end{bmatrix},$$

$$\mathbf{V} = \begin{bmatrix} v_1 \\ v_2 \\ v_3 \\ \vdots \\ v_r \end{bmatrix}, \qquad \mathbf{G} = \begin{bmatrix} g_1 \\ g_2 \\ g_3 \\ \vdots \\ g_r \end{bmatrix}.$$

Observe that $\mathbf{A}$ is, once again, a square matrix of variable coefficients, while $\mathbf{V}$ and $\mathbf{G}$ are column matrices containing the variables and constants, respectively. Then $\mathbf{AV} = \mathbf{G}$.

Regardless of how large the system of equations, if we possess n linear equations, in n unknowns, the system may be written as the simple matrix equation, $\mathbf{AV} = \mathbf{G}$.

The reader will observe that the matrix $\mathbf{V}$ contains all the unknowns, while $\mathbf{A}$ and $\mathbf{G}$ are "constant" matrices.

Our objective, of course, is to solve for the matrix of unknowns, $\mathbf{V}$, where the equation $\mathbf{AV} = \mathbf{G}$ is similar to the equation

$$2v = 6$$

in ordinary algebra. This being true, we would not be too surprised to find that the methods of solution are the same. In ordinary algebra both sides of the equation are multiplied by the reciprocal of 2; in matrix algebra both sides of the equation are multiplied by $\mathbf{A}^{-1}$.

Then

$$\mathbf{A}^{-1}(\mathbf{AV}) = \mathbf{A}^{-1}\mathbf{G}$$

or

$$\mathbf{A}^{-1}\mathbf{AV} = \mathbf{A}^{-1}\mathbf{G}.$$

But, $\mathbf{A}^{-1}\mathbf{A} = \mathbf{I}$ and $\mathbf{IV} = \mathbf{V}$. Therefore, $\mathbf{V} = \mathbf{A}^{-1}\mathbf{G}$. In other words, the solution to the system of simultaneous linear equations can be obtained by finding $\mathbf{A}^{-1}$ and then obtaining the product, $\mathbf{A}^{-1}\mathbf{G}$. The solution values of $v_1, v_2, v_3, \ldots, v_r$ will appear in sequence in the column matrix, $\mathbf{V} = \mathbf{A}^{-1}\mathbf{G}$.

I.9 *Inverting a Matrix*

We have indicated in Section I.8 that the key to the solution of a system of simultaneous linear equations by the method of matrix algebra rests on the acquisition of the inverse of the **A** matrix. Many methods exist for inverting matrices. The method that we present is not the best from a computational point of view, but it works very well for the matrices associated with most experimental designs and it is one of the easiest to present to the novice. It will depend upon a theorem in matrix algebra and the use of *row operations.*

Before defining row operations on matrices, one must state what is meant by the "addition" of two rows of a matrix and the multiplication of a row by a constant. We will illustrate with the **A** matrix for the system of two simultaneous linear equations,

$$\mathbf{A} = \begin{bmatrix} 2 & 1 \\ 1 & -1 \end{bmatrix}.$$

Two rows of a matrix may be added by adding corresponding elements. Thus if the two rows of the **A** matrix are added, one obtains a new row with elements $[(2 + 1)\ (1 - 1)] = [3 \quad 0]$. Multiplication of a row by a constant means that each element in the row is multiplied by the constant. Twice the first row of the **A** matrix would generate the row $[4 \quad 2]$. With these ideas in mind, we will define three ways to operate on a row in a matrix.

1. *A row may be multiplied by a constant.*
2. *A row may be multiplied by a constant and added to or subtracted from another row (which is identified as the one upon which the operation is performed).*
3. *Two rows may be interchanged.*

Given matrix **A**, it is quite easy to see that one might perform a series of row operations that would yield some new matrix **B**. In this connection, we state without proof a surprising and interesting theorem from matrix algebra; namely, there exists some matrix **C** such that

$$\mathbf{CA} = \mathbf{B}.$$

In other words, a series of row operations on a matrix **A** is equivalent to multiplying **A** by a matrix **C**. We will use this principle to invert a matrix.

Place the matrix **A**, which is to be inverted, alongside an identity matrix of the same dimensions:

$$\mathbf{A} = \begin{bmatrix} 2 & 1 \\ 1 & -1 \end{bmatrix}, \quad \mathbf{I} = \begin{bmatrix} 1 & 0 \\ 0 & 1 \end{bmatrix}.$$

Then perform identically the same row operations on **A** *and* **I** *in such a way that* **A** *changes to an identity matrix.* In doing so, we must have multiplied **A** by a matrix **C** so that **CA** = **I**. Therefore, **C** must be the inverse of **A**! The problem, of course, is to find the unknown matrix **C** and, fortunately, this proves to be of little difficulty. Since we performed identically the same row operations on **A** and **I**, the identity matrix must have changed to $\mathbf{CI} = \mathbf{C} = \mathbf{A}^{-1}$.

$$\mathbf{A} = \begin{bmatrix} 2 & 1 \\ 1 & -1 \end{bmatrix}, \qquad \mathbf{I} = \begin{bmatrix} 1 & 0 \\ 0 & 1 \end{bmatrix}.$$

(same row operations)

$$\mathbf{CA} = \mathbf{I} \qquad \mathbf{CI} = \mathbf{C} = \mathbf{A}^{-1}$$

We will illustrate with the following example.

Example I.7: *Invert the matrix*

$$\mathbf{A} = \begin{bmatrix} 2 & 1 \\ 1 & -1 \end{bmatrix}.$$

Solution:

$$\mathbf{A} = \begin{bmatrix} 2 & 1 \\ 1 & -1 \end{bmatrix}, \quad \mathbf{I} = \begin{bmatrix} 1 & 0 \\ 0 & 1 \end{bmatrix}.$$

Step 1: Operate on row 1 by multiplying row 1 by 1/2. (Note: *It is helpful to the beginner to identify the row upon which he is operating* since all other rows will remain unchanged, even though they may be used in the operation. *We will star the row upon which the operation is being performed.*)

$$\begin{matrix} * \\ \end{matrix}\begin{bmatrix} 1 & 1/2 \\ 1 & -1 \end{bmatrix} \quad \begin{bmatrix} 1/2 & 0 \\ 0 & 1 \end{bmatrix}.$$

Step 2: Operate on row 2 *by subtracting row* 1 *from row* 2.

$$\begin{matrix} \\ * \end{matrix}\begin{bmatrix} 1 & 1/2 \\ 0 & -3/2 \end{bmatrix} \quad \begin{bmatrix} 1/2 & 0 \\ -1/2 & 1 \end{bmatrix}.$$

(*Note that row* 2 *is simply used to operate on row* 1 *and hence remains unchanged.*)

Step 3: Multiply row 2 *by* (−2/3).

$$\begin{matrix} \\ * \end{matrix}\begin{bmatrix} 1 & 1/2 \\ 0 & 1 \end{bmatrix} \quad \begin{bmatrix} 1/2 & 0 \\ 1/3 & -2/3 \end{bmatrix}.$$

Step 4: Operate on row 1 *by multiplying row* 2 *by* 1/2 *and subtracting from row* 1.

$$\begin{matrix} * \\ \end{matrix}\begin{bmatrix} 1 & 0 \\ 0 & 1 \end{bmatrix} \quad \begin{bmatrix} 1/3 & 1/3 \\ 1/3 & -2/3 \end{bmatrix}.$$

(*Note that row* 2 *is simply used to operate on row* 1 *and hence remains unchanged.*) *Hence the inverse of* **A** *must be*

$$\mathbf{A}^{-1} = \begin{bmatrix} 1/3 & 1/3 \\ 1/3 & -2/3 \end{bmatrix}.$$

A ready check on the calculations for the inversion procedure is available because $\mathbf{A}^{-1}\mathbf{A}$ *must equal the identity matrix,* **I**. *Thus*

$$\mathbf{A}^{-1}\mathbf{A} = \begin{bmatrix} 1/3 & 1/3 \\ 1/3 & -2/3 \end{bmatrix}\begin{bmatrix} 2 & 1 \\ 1 & -1 \end{bmatrix} = \begin{bmatrix} 1 & 0 \\ 0 & 1 \end{bmatrix}.$$

Example I.8: *Invert the matrix*

$$\mathbf{A} = \begin{bmatrix} 2 & 0 & 1 \\ 1 & -1 & 2 \\ 1 & 0 & 0 \end{bmatrix}$$

and check the results.

Solution:

$$\mathbf{A} = \begin{bmatrix} 2 & 0 & 1 \\ 1 & -1 & 2 \\ 1 & 0 & 0 \end{bmatrix}, \qquad \mathbf{I} = \begin{bmatrix} 1 & 0 & 0 \\ 0 & 1 & 0 \\ 0 & 0 & 1 \end{bmatrix}.$$

Step 1: *Multiply row 1 by 1/2.*

$$\begin{matrix} * \\ \\ \\ \end{matrix}\begin{bmatrix} 1 & 0 & 1/2 \\ 1 & -1 & 2 \\ 1 & 0 & 0 \end{bmatrix} \qquad \begin{bmatrix} 1/2 & 0 & 0 \\ 0 & 1 & 0 \\ 0 & 0 & 1 \end{bmatrix}$$

Step 2: *Operate on row 2 by subtracting row 1 from row 2.*

$$\begin{matrix} \\ * \\ \\ \end{matrix}\begin{bmatrix} 1 & 0 & 1/2 \\ 0 & -1 & 3/2 \\ 1 & 0 & 0 \end{bmatrix} \qquad \begin{bmatrix} 1/2 & 0 & 0 \\ -1/2 & 1 & 0 \\ 0 & 0 & 1 \end{bmatrix}$$

Step 3: *Operate on row 3 by subtracting row 1 from row 3.*

$$\begin{matrix} \\ \\ * \\ \end{matrix}\begin{bmatrix} 1 & 0 & 1/2 \\ 0 & -1 & 3/2 \\ 0 & 0 & -1/2 \end{bmatrix} \qquad \begin{bmatrix} 1/2 & 0 & 0 \\ -1/2 & 1 & 0 \\ -1/2 & 0 & 1 \end{bmatrix}$$

Step 4: *Operate on row 2 by multiplying row 3 by 3 and adding to row 2.*

$$\begin{matrix} \\ * \\ \\ \end{matrix}\begin{bmatrix} 1 & 0 & 1/2 \\ 0 & -1 & 0 \\ 0 & 0 & -1/2 \end{bmatrix} \qquad \begin{bmatrix} 1/2 & 0 & 0 \\ -2 & 1 & 3 \\ -1/2 & 0 & 1 \end{bmatrix}$$

Step 5: Multiply row 2 by (−1).

$$\begin{bmatrix} 1 & 0 & 1/2 \\ 0 & 1 & 0 \\ 0 & 0 & -1/2 \end{bmatrix} \begin{bmatrix} 1/2 & 0 & 0 \\ 2 & -1 & -3 \\ -1/2 & 0 & 1 \end{bmatrix}$$

Step 6: Operate on row 1 by adding row 3 to row 1.

$$\begin{bmatrix} 1 & 0 & 0 \\ 0 & 1 & 0 \\ 0 & 0 & -1/2 \end{bmatrix} \begin{bmatrix} 0 & 0 & 1 \\ 2 & -1 & -3 \\ -1/2 & 0 & 1 \end{bmatrix}$$

Step 7: Multiply row 3 by (−2).

$$\begin{bmatrix} 1 & 0 & 0 \\ 0 & 1 & 0 \\ 0 & 0 & 1 \end{bmatrix} \begin{bmatrix} 0 & 0 & 1 \\ 2 & -1 & -3 \\ 1 & 0 & -2 \end{bmatrix} = \mathbf{A}^{-1}.$$

The seven row operations have changed the **A** *matrix to the identity matrix and, barring errors of calculation, have changed the identity to* $\mathbf{A}^{-1}$*. Checking,*

$$\mathbf{A}^{-1}\mathbf{A} = \begin{bmatrix} 0 & 0 & 1 \\ 2 & -1 & -3 \\ 1 & 0 & -2 \end{bmatrix} \begin{bmatrix} 2 & 0 & 1 \\ 1 & -1 & 2 \\ 1 & 0 & 0 \end{bmatrix} = \begin{bmatrix} 1 & 0 & 0 \\ 0 & 1 & 0 \\ 0 & 0 & 1 \end{bmatrix}.$$

We see that $\mathbf{A}^{-1}\mathbf{A} = \mathbf{I}$ *and hence that the calculations are correct.*

Note that the sequence of row operations required to convert **A** to **I** is not unique. One person might achieve the inverse by using five row operations while another might require ten, and the end result will be the same. However, in the interests of efficiency it is desirable to employ a system.

The reader will observe that the inversion process utilizes row operations to change off-diagonal elements in the **A** matrix to zeros and the main diagonal elements to 1's. One systematic procedure is as follows. Change the top left element into a 1 and then perform row operations to change all other elements in the *first* column to 0. Then move to the diagonal element in the second row, second column, change it into a 1, and eliminate all elements in the *second* column *below* the main diagonal. This process is repeated, moving down the main diagonal from top left to bottom right until all elements below the main diagonal have been changed to 0's. To eliminate nonzero elements above the main diagonal, operate on all elements in the last column, changing each to zero; then move to the next to last column and repeat the process. Continue this procedure until you arrive at the first element in the first column, which was the starting point. This procedure is indicated diagrammatically in Figure I.1.

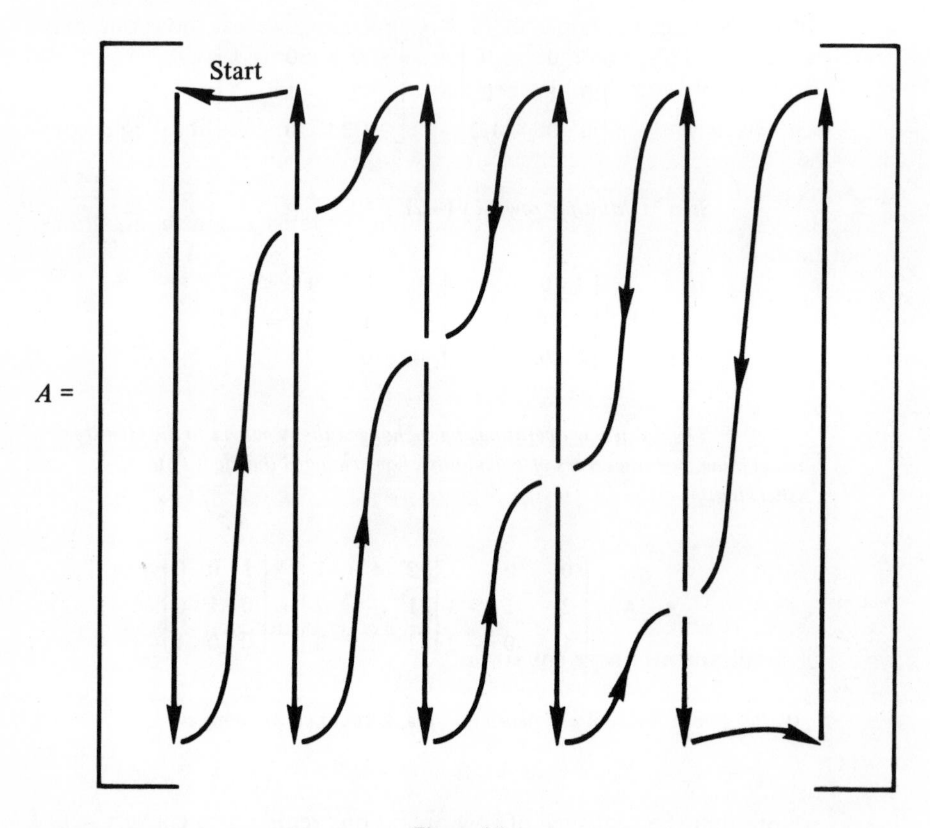

Figure I.1

Matrix inversion is a tedious process, at best, and requires every bit as much labor as the solution of a system of simultaneous equations by elimination or substitution. The reader will be pleased to learn that we do not expect him to develop a facility for matrix inversion. Fortunately, most matrices associated with designed experiments follow patterns and are easily inverted.

It will be beneficial to the reader to invert a few 2×2 and 3×3 matrices. Matrices lacking pattern, particularly large matrices, are inverted most efficiently and economically using an electronic computer. (Programs for matrix inversion have been developed for most electronic computers.)

We emphasize that obtaining the solution for the least-squares equations (Chapter 11) by matrix inversion has distinct advantages that may or may not be apparent. Not the least of these is the fact that the inversion procedure is systematic and hence is particularly suitable for electronic computation. However, the major advantage is that the inversion procedure will automatically produce the variances of the estimators of all parameters in the linear model.

Before leaving the topic of matrix inversion, we ask how one may identify a matrix that has an inverse. Reference to a discussion of linear equations in ordinary algebra should reveal the answer.

Clearly, a unique solution for a system of simultaneous linear equations cannot be obtained unless the equations are independent. Thus, if one of the equations is a linear combination of the others, the equations are dependent. Coefficient matrices associated with dependent systems of linear equations do not possess an inverse.

I.10 Solving a System of Simultaneous Linear Equations

We have finally obtained all the ingredients necessary for solving a system of simultaneous linear equations,

$$2v_1 + v_2 = 5,$$

$$v_1 - v_2 = 1.$$

Recalling that the matrix solution to the system of equations $\mathbf{AV} = \mathbf{G}$ is $\mathbf{V} = \mathbf{A}^{-1}\mathbf{G}$, we obtain

$$\mathbf{V} = \mathbf{A}^{-1}\mathbf{G} = \begin{bmatrix} 1/3 & 1/3 \\ 1/3 & -2/3 \end{bmatrix} \begin{bmatrix} 5 \\ 1 \end{bmatrix} = \begin{bmatrix} 2 \\ 1 \end{bmatrix}.$$

Hence the solution is

$$\mathbf{V} = \begin{bmatrix} v_1 \\ v_2 \end{bmatrix} = \begin{bmatrix} 2 \\ 1 \end{bmatrix}$$

—that is, $v_1 = 2$ and $v_2 = 1$, a fact that may be verified by substitution of these values in the original linear equations.

Example I.9: *Solve the system of simultaneous linear equations.*

$$2v_1 + v_3 = 4,$$
$$v_1 - v_2 + 2v_3 = 2,$$
$$v_1 = 1.$$

Solution: *The coefficient matrix for these equations,*

$$\mathbf{A} = \begin{bmatrix} 2 & 0 & 1 \\ 1 & -1 & 2 \\ 1 & 0 & 0 \end{bmatrix},$$

appeared in Example I.8. In that example we found that

$$\mathbf{A}^{-1} = \begin{bmatrix} 0 & 0 & 1 \\ 2 & -1 & -3 \\ 1 & 0 & 0 \end{bmatrix}.$$

Solving,

$$\mathbf{V} = \mathbf{A}^{-1}\mathbf{G} = \begin{bmatrix} 0 & 0 & 1 \\ 2 & -1 & -3 \\ 1 & 0 & -2 \end{bmatrix} \begin{bmatrix} 4 \\ 2 \\ 1 \end{bmatrix} = \begin{bmatrix} 1 \\ 3 \\ 2 \end{bmatrix}.$$

Thus $v_1 = 1$, $v_2 = 3$, and $v_3 = 2$ give the solution to the set of three simultaneous linear equations.

Appendix II

Common Probability Distributions, Means, Variances, and Moment-Generating Functions

II.1 Discrete Distributions

Distribution	*Probability Function*	*Mean*	*Variance*	*Moment-Generating Function*
Binomial	$p(y) = \binom{n}{y} p^y (1-p)^{n-y}$; $y = 0, 1, \ldots, n$	np	$np(1-p)$	$[pe^t + (1-p)]^n$
Geometric	$p(y) = p(1-p)^{y-1}$; $y = 1, 2, \ldots$	$\frac{1}{p}$	$\frac{1-p}{p^2}$	$\frac{pe^t}{1-(1-p)e^t}$
Hypergeometric	$p(y) = \frac{\binom{r}{y}\binom{N-r}{n-y}}{\binom{N}{n}}$; $y = 0, 1, \ldots, n$ if $n \le r$, $y = 0, 1, \ldots, r$ if $n > r$.	$\frac{nr}{N}$	$n\left(\frac{r}{N}\right)\left(\frac{N-r}{N}\right)\left(\frac{N-n}{N-1}\right)$	

Distribution	Probability Function	Mean	Variance	Moment-Generating Function
Poisson	$p(y) = \dfrac{\lambda^y e^{-\lambda}}{y!}$; $y = 0, 1, 2, \ldots$	λ	λ	$\exp[\lambda(e^t - 1)]$
Negative binomial	$p(y) = \binom{y-1}{r-1} p^r (1-p)^{y-r}$; $y = r, r+1, \ldots$	$\dfrac{r}{p}$	$\dfrac{r(1-p)}{p^2}$	$\left[\dfrac{pe^t}{1-(1-p)e^t}\right]^r$

II.2 Continuous Distributions

Distribution	Probability Function	Mean	Variance	Moment-Generating Function
Uniform	$f(y) = \dfrac{1}{\theta_2 - \theta_1}$; $\theta_1 \le y \le \theta_2$	$\dfrac{\theta_1 + \theta_2}{2}$	$\dfrac{(\theta_2 - \theta_1)^2}{12}$	$\dfrac{e^{t\theta_2} - e^{t\theta_1}}{t(\theta_2 - \theta_1)}$
Normal	$f(y) = \dfrac{1}{\sigma\sqrt{2\pi}} \exp\left[-\left(\dfrac{1}{2\sigma^2}\right)(y-\mu)^2\right]$; $-\infty < y < +\infty$	μ	σ^2	$\exp\left(\mu t + \dfrac{t^2\sigma^2}{2}\right)$
Gamma	$f(y) = \dfrac{1}{\Gamma(\alpha)\beta^\alpha} y^{\alpha-1} e^{-y/\beta}$; $0 < y < \infty$	$\alpha\beta$	$\alpha\beta^2$	$(1 - \beta t)^{-\alpha}$
Chi square	$f(\chi^2) = \dfrac{(\chi^2)^{(v/2)-1} e^{-\chi^2/2}}{2^{v/2}\Gamma(v/2)}$; $\chi^2 > 0$	v	$2v$	$(1 - 2t)^{-v/2}$
Beta	$f(y) = \dfrac{\Gamma(\alpha+\beta)}{\Gamma(\alpha)\Gamma(\beta)} y^{\alpha-1}(1-y)^{\beta-1}$; $0 < y < 1$	$\dfrac{\alpha}{\alpha+\beta}$	$\dfrac{\alpha\beta}{(\alpha+\beta)^2(\alpha+\beta+1)}$	Does not exist in closed form

Appendix III

Tables

Table 1 Binomial Probability Tables

Tabulated values are $\sum_{y=0}^{a} p(Y)$.

(Computations are rounded at third decimal place.)

(a) $n = 5$

a	0.01	0.05	0.10	0.20	0.30	0.40	0.50	0.60	0.70	0.80	0.90	0.95	0.99	a
0	.951	.774	.590	.328	.168	.078	.031	.010	.002	.000	.000	.000	.000	0
1	.999	.977	.919	.737	.528	.337	.188	.087	.031	.007	.000	.000	.000	1
2	1.000	.999	.991	.942	.837	.683	.500	.317	.163	.058	.009	.001	.000	2
3	1.000	1.000	1.000	.993	.969	.913	.812	.663	.472	.263	.081	.023	.001	3
4	1.000	1.000	1.000	1.000	.998	.990	.969	.922	.832	.672	.410	.226	.049	4

(b) $n = 10$

a	0.01	0.05	0.10	0.20	0.30	0.40	0.50	0.60	0.70	0.80	0.90	0.95	0.99	a
0	.904	.599	.349	.107	.028	.006	.001	.000	.000	.000	.000	.000	.000	0
1	.996	.914	.736	.376	.149	.046	.011	.002	.000	.000	.000	.000	.000	1
2	1.000	.988	.930	.678	.383	.167	.055	.012	.002	.000	.000	.000	.000	2
3	1.000	.999	.987	.879	.650	.382	.172	.055	.011	.001	.000	.000	.000	3
4	1.000	1.000	.998	.967	.850	.633	.377	.166	.047	.006	.000	.000	.000	4
5	1.000	1.000	1.000	.994	.953	.834	.623	.367	.150	.033	.002	.000	.000	5
6	1.000	1.000	1.000	.999	.989	.945	.828	.618	.350	.121	.013	.001	.000	6
7	1.000	1.000	1.000	1.000	.998	.988	.945	.833	.617	.322	.070	.012	.000	7
8	1.000	1.000	1.000	1.000	1.000	.998	.989	.954	.851	.624	.264	.086	.004	8
9	1.000	1.000	1.000	1.000	1.000	1.000	.999	.994	.972	.893	.651	.401	.096	9

(c) $n = 15$

a	P 0.01	0.05	0.10	0.20	0.30	0.40	0.50	0.60	0.70	0.80	0.90	0.95	0.99	a
0	.860	.463	.206	.035	.005	.000	.000	.000	.000	.000	.000	.000	.000	0
1	.990	.829	.549	.167	.035	.005	.000	.000	.000	.000	.000	.000	.000	1
2	1.000	.964	.816	.398	.127	.027	.004	.000	.000	.000	.000	.000	.000	2
3	1.000	.995	.944	.648	.297	.091	.018	.002	.000	.000	.000	.000	.000	3
4	1.000	.999	.987	.836	.515	.217	.059	.009	.001	.000	.000	.000	.000	4
5	1.000	1.000	.998	.939	.722	.403	.151	.034	.004	.000	.000	.000	.000	5
6	1.000	1.000	1.000	.982	.869	.610	.304	.095	.015	.001	.000	.000	.000	6
7	1.000	1.000	1.000	.996	.950	.787	.500	.213	.050	.004	.000	.000	.000	7
8	1.000	1.000	1.000	.999	.985	.905	.696	.390	.131	.018	.000	.000	.000	8
9	1.000	1.000	1.000	1.000	.996	.966	.849	.597	.278	.061	.002	.000	.000	9
10	1.000	1.000	1.000	1.000	.999	.991	.941	.783	.485	.164	.013	.001	.000	10
11	1.000	1.000	1.000	1.000	1.000	.998	.982	.909	.703	.352	.056	.005	.000	11
12	1.000	1.000	1.000	1.000	1.000	1.000	.996	.973	.873	.602	.184	.036	.000	12
13	1.000	1.000	1.000	1.000	1.000	1.000	1.000	.995	.965	.833	.451	.171	.010	13
14	1.000	1.000	1.000	1.000	1.000	1.000	1.000	1.000	.995	.965	.794	.537	.140	14

(d) $n = 20$

a	P = 0.01	0.05	0.10	0.20	0.30	0.40	0.50	0.60	0.70	0.80	0.90	0.95	0.99	a
0	.818	.358	.122	.002	.001	.000	.000	.000	.000	.000	.000	.000	.000	0
1	.983	.736	.392	.069	.008	.001	.000	.000	.000	.000	.000	.000	.000	1
2	.999	.925	.677	.206	.035	.004	.000	.000	.000	.000	.000	.000	.000	2
3	1.000	.984	.867	.411	.107	.016	.001	.000	.000	.000	.000	.000	.000	3
4	1.000	.997	.957	.630	.238	.051	.006	.000	.000	.000	.000	.000	.000	4
5	1.000	1.000	.989	.804	.416	.126	.021	.002	.000	.000	.000	.000	.000	5
6	1.000	1.000	.998	.913	.608	.250	.058	.006	.000	.000	.000	.000	.000	6
7	1.000	1.000	1.000	.968	.772	.416	.132	.021	.001	.000	.000	.000	.000	7
8	1.000	1.000	1.000	.990	.887	.596	.252	.057	.005	.000	.000	.000	.000	8
9	1.000	1.000	1.000	.997	.952	.755	.412	.128	.017	.001	.000	.000	.000	9
10	1.000	1.000	1.000	.999	.983	.872	.588	.245	.048	.003	.000	.000	.000	10
11	1.000	1.000	1.000	1.000	.995	.943	.748	.404	.113	.010	.000	.000	.000	11
12	1.000	1.000	1.000	1.000	.999	.979	.868	.584	.228	.032	.000	.000	.000	12
13	1.000	1.000	1.000	1.000	1.000	.994	.942	.750	.392	.087	.002	.000	.000	13
14	1.000	1.000	1.000	1.000	1.000	.998	.979	.874	.584	.196	.011	.000	.000	14
15	1.000	1.000	1.000	1.000	1.000	1.000	.994	.949	.762	.370	.043	.003	.000	15
16	1.000	1.000	1.000	1.000	1.000	1.000	.999	.984	.893	.589	.133	.016	.000	16
17	1.000	1.000	1.000	1.000	1.000	1.000	1.000	.996	.965	.794	.323	.075	.001	17
18	1.000	1.000	1.000	1.000	1.000	1.000	1.000	.999	.992	.931	.608	.264	.017	18
19	1.000	1.000	1.000	1.000	1.000	1.000	1.000	1.000	.999	.988	.878	.642	.182	19

(e) $n = 25$

	P													
a	0.01	0.05	0.10	0.20	0.30	0.40	0.50	0.60	0.70	0.80	0.90	0.95	0.99	a
0	.778	.277	.072	.004	.000	.000	.000	.000	.000	.000	.000	.000	.000	0
1	.974	.642	.271	.027	.002	.000	.000	.000	.000	.000	.000	.000	.000	1
2	.998	.873	.537	.098	.009	.000	.000	.000	.000	.000	.000	.000	.000	2
3	1.000	.966	.764	.234	.033	.002	.000	.000	.000	.000	.000	.000	.000	3
4	1.000	.993	.902	.421	.090	.009	.000	.000	.000	.000	.000	.000	.000	4
5	1.000	.999	.967	.617	.193	.029	.002	.000	.000	.000	.000	.000	.000	5
6	1.000	1.000	.991	.780	.341	.074	.007	.000	.000	.000	.000	.000	.000	6
7	1.000	1.000	.998	.891	.512	.154	.022	.001	.000	.000	.000	.000	.000	7
8	1.000	1.000	1.000	.953	.677	.274	.054	.004	.000	.000	.000	.000	.000	8
9	1.000	1.000	1.000	.983	.811	.425	.115	.013	.000	.000	.000	.000	.000	9
10	1.000	1.000	1.000	.994	.902	.586	.212	.034	.002	.000	.000	.000	.000	10
11	1.000	1.000	1.000	.998	.956	.732	.345	.078	.006	.000	.000	.000	.000	11
12	1.000	1.000	1.000	1.000	.983	.846	.500	.154	.017	.000	.000	.000	.000	12
13	1.000	1.000	1.000	1.000	.994	.922	.655	.268	.044	.002	.000	.000	.000	13
14	1.000	1.000	1.000	1.000	.998	.966	.788	.414	.098	.006	.000	.000	.000	14
15	1.000	1.000	1.000	1.000	1.000	.987	.885	.575	.189	.017	.000	.000	.000	15
16	1.000	1.000	1.000	1.000	1.000	.996	.946	.726	.323	.047	.000	.000	.000	16
17	1.000	1.000	1.000	1.000	1.000	.999	.978	.846	.488	.109	.002	.000	.000	17
18	1.000	1.000	1.000	1.000	1.000	1.000	.993	.926	.659	.220	.009	.000	.000	18
19	1.000	1.000	1.000	1.000	1.000	1.000	.998	.971	.807	.383	.033	.001	.000	19
20	1.000	1.000	1.000	1.000	1.000	1.000	1.000	.991	.910	.579	.098	.007	.000	20
21	1.000	1.000	1.000	1.000	1.000	1.000	1.000	.998	.967	.766	.236	.034	.000	21
22	1.000	1.000	1.000	1.000	1.000	1.000	1.000	1.000	.991	.902	.463	.127	.002	22
23	1.000	1.000	1.000	1.000	1.000	1.000	1.000	1.000	.998	.973	.729	.358	.026	23
24	1.000	1.000	1.000	1.000	1.000	1.000	1.000	1.000	1.000	.996	.928	.723	.222	24

Table 2 Table of e^{-x}

x	e^{-x}	x	e^{-x}	x	e^{-x}	x	e^{-x}
0.00	1.000000	2.60	.074274	5.10	.006097	7.60	.000501
0.10	.904837	2.70	.067206	5.20	.005517	7.70	.000453
0.20	.818731	2.80	.060810	5.30	.004992	7.80	.000410
0.30	.740818	2.90	.055023	5.40	.004517	7.90	.000371
0.40	.670320	3.00	.049787	5.50	.004087	8.00	.000336
0.50	.606531	3.10	.045049	5.60	.003698	8.10	.000304
0.60	.548812	3.20	.040762	5.70	.003346	8.20	.000275
0.70	.496585	3.30	.036883	5.80	.003028	8.30	.000249
0.80	.449329	3.40	.033373	5.90	.002739	8.40	.000225
0.90	.406570	3.50	.030197	6.00	.002479	8.50	.000204
1.00	.357879	3.60	.027324	6.10	.002243	8.60	.000184
1.10	.332871	3.70	.024724	6.20	.002029	8.70	.000167
1.20	.301194	3.80	.022371	6.30	.001836	8.80	.000151
1.30	.272532	3.90	.020242	6.40	.001661	8.90	.000136
1.40	.246597	4.00	.018316	6.50	.001503	9.00	.000123
1.50	.223130	4.10	.016573	6.60	.001360	9.10	.000112
1.60	.201897	4.20	.014996	6.70	.001231	9.20	.000101
1.70	.182684	4.30	.013569	6.80	.001114	9.30	.000091
1.80	.165299	4.40	.012277	6.90	.001008	9.40	.000083
1.90	.149569	4.50	.011109	7.00	.000912	9.50	.000075
2.00	.135335	4.60	.010052	7.10	.000825	9.60	.000068
2.10	.122456	4.70	.009095	7.20	.000747	9.70	.000061
2.20	.110803	4.80	.008230	7.30	.000676	9.80	.000056
2.30	.100259	4.90	.007447	7.40	.000611	9.90	.000050
2.40	.090718	5.00	.006738	7.50	.000553	10.00	.000045
2.50	.082085						

Table 3 Normal Curve Areas

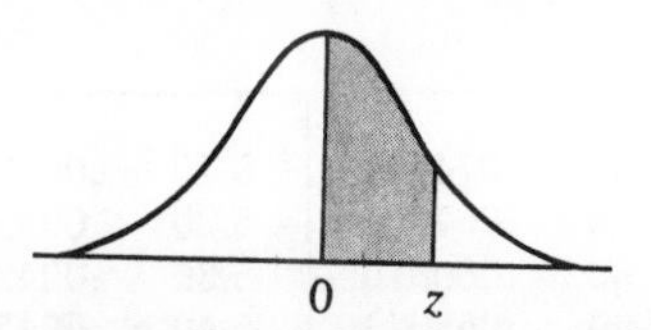

z	.00	.01	.02	.03	.04	.05	.06	.07	.08	.09
0.0	.0000	.0040	.0080	.0120	.0160	.0199	.0239	.0279	.0319	.0359
0.1	.0398	.0438	.0478	.0517	.0557	.0596	.0636	.0675	.0714	.0753
0.2	.0793	.0832	.0871	.0910	.0948	.0987	.1026	.1064	.1103	.1141
0.3	.1179	.1217	.1255	.1293	.1331	.1368	.1406	.1443	.1480	.1517
0.4	.1554	.1591	.1628	.1664	.1700	.1736	.1772	.1808	.1844	.1879
0.5	.1915	.1950	.1985	.2019	.2054	.2088	.2123	.2157	.2190	.2224
0.6	.2257	.2291	.2324	.2357	.2389	.2422	.2454	.2486	.2517	.2549
0.7	.2580	.2611	.2642	.2673	.2704	.2734	.2764	.2794	.2823	.2852
0.8	.2881	.2910	.2939	.2967	.2995	.3023	.3051	.3078	.3106	.3133
0.9	.3159	.3186	.3212	.3238	.3264	.3289	.3315	.3340	.3365	.3389
1.0	.3413	.3438	.3461	.3485	.3508	.3531	.3554	.3577	.3599	.3621
1.1	.3643	.3665	.3686	.3708	.3729	.3749	.3770	.3790	.3810	.3830
1.2	.3849	.3869	.3888	.3907	.3925	.3944	.3962	.3980	.3997	.4015
1.3	.4032	.4049	.4066	.4082	.4099	.4115	.4131	.4147	.4162	.4177
1.4	.4192	.4207	.4222	.4236	.4251	.4265	.4279	.4292	.4306	.4319
1.5	.4332	.4345	.4357	.4370	.4382	.4394	.4406	.4418	.4429	.4441
1.6	.4452	.4463	.4474	.4484	.4495	.4505	.4515	.4525	.4535	.4545
1.7	.4554	.4564	.4573	.4582	.4591	.4599	.4608	.4616	.4625	.4633
1.8	.4641	.4649	.4656	.4664	.4671	.4678	.4686	.4693	.4699	.4706
1.9	.4713	.4719	.4726	.4732	.4738	.4744	.4750	.4756	.4761	.4767
2.0	.4772	.4778	.4783	.4788	.4793	.4798	.4803	.4808	.4812	.4817
2.1	.4821	.4826	.4830	.4834	.4838	.4842	.4846	.4850	.4854	.4857
2.2	.4861	.4864	.4868	.4871	.4875	.4878	.4881	.4884	.4887	.4890
2.3	.4893	.4896	.4898	.4901	.4904	.4906	.4909	.4911	.4913	.4916
2.4	.4918	.4920	.4922	.4925	.4927	.4929	.4931	.4932	.4934	.4936
2.5	.4938	.4940	.4941	.4943	.4945	.4946	.4948	.4949	.4951	.4952
2.6	.4953	.4955	.4956	.4957	.4959	.4960	.4961	.4962	.4963	.4964
2.7	.4965	.4966	.4967	.4968	.4969	.4970	.4971	.4972	.4973	.4974
2.8	.4974	.4975	.4976	.4977	.4977	.4978	.4979	.4979	.4980	.4981
2.9	.4981	.4982	.4982	.4983	.4984	.4984	.4985	.4985	.4986	.4986
3.0	.4987	.4987	.4987	.4988	.4988	.4989	.4989	.4989	.4990	.4990

This table is abridged from Table I of *Statistical Tables and Formulas*, by A. Hald (New York: John Wiley & Sons, Inc., 1952). Reproduced by permission of A. Hald and the publishers, John Wiley & Sons, Inc.

Table 4 Critical Values of t

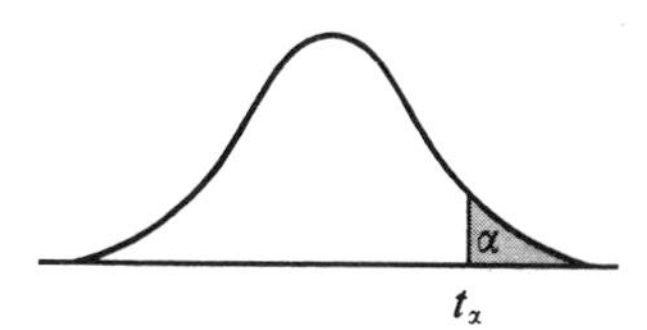

n	$t_{.100}$	$t_{.050}$	$t_{.025}$	$t_{.010}$	$t_{.005}$	*d.f.*
2	3.078	6.314	12.706	31.821	63.657	1
3	1.886	2.920	4.303	6.965	9.925	2
4	1.638	2.353	3.182	4.541	5.841	3
5	1.533	2.132	2.776	3.747	4.604	4
6	1.476	2.015	2.571	3.365	4.032	5
7	1.440	1.943	2.447	3.143	3.707	6
8	1.415	1.895	2.365	2.998	3.499	7
9	1.397	1.860	2.306	2.896	3.355	8
10	1.383	1.833	2.262	2.821	3.250	9
11	1.372	1.812	2.228	2.764	3.169	10
12	1.363	1.796	2.201	2.718	3.106	11
13	1.356	1.782	2.179	2.681	3.055	12
14	1.350	1.771	2.160	2.650	3.012	13
15	1.345	1.761	2.145	2.624	2.977	14
16	1.341	1.753	2.131	2.602	2.947	15
17	1.337	1.746	2.120	2.583	2.921	16
18	1.333	1.740	2.110	2.567	2.898	17
19	1.330	1.734	2.101	2.552	2.878	18
20	1.328	1.729	2.093	2.539	2.861	19
21	1.325	1.725	2.086	2.528	2.845	20
22	1.323	1.721	2.080	2.518	2.831	21
23	1.321	1.717	2.074	2.508	2.819	22
24	1.319	1.714	2.069	2.500	2.807	23
25	1.318	1.711	2.064	2.492	2.797	24
26	1.316	1.708	2.060	2.485	2.787	25
27	1.315	1.706	2.056	2.479	2.779	26
28	1.314	1.703	2.052	2.473	2.771	27
29	1.313	1.701	2.048	2.467	2.763	28
30	1.311	1.699	2.045	2.462	2.756	29
inf.	1.282	1.645	1.960	2.326	2.576	inf.

From "Table of Percentage Points of the t-Distribution." Computed by Maxine Merrington, *Biometrika*, Vol. 32 (1941), p. 300. Reproduced by permission of Professor E. S. Pearson.

Table 5 Critical Values of Chi Square

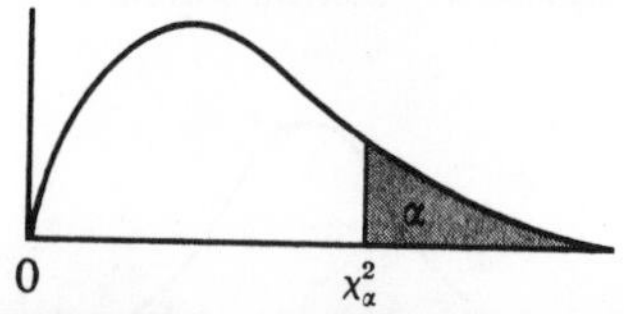

d.f.	$\chi^2$0.995	$\chi^2$0.990	$\chi^2$0.975	$\chi^2$0.950	$\chi^2$0.900
1	0.0000393	0.0001571	0.0009821	0.0039321	0.0157908
2	0.0100251	0.0201007	0.0506356	0.102587	0.210720
3	0.0717212	0.114832	0.215795	0.351846	0.584375
4	0.206990	0.297110	0.484419	0.710721	1.063623
5	0.411740	0.554300	0.831211	1.145476	1.61031
6	0.675727	0.872085	1.237347	1.63539	2.20413
7	0.989265	1.239043	1.68987	2.16735	2.83311
8	1.344419	1.646482	2.17973	2.73264	3.48954
9	1.734926	2.087912	2.70039	3.32511	4.16816
10	2.15585	2.55821	3.24697	3.94030	4.86518
11	2.60321	3.05347	3.81575	4.57481	5.57779
12	3.07382	3.57056	4.40379	5.22603	6.30380
13	3.56503	4.10691	5.00874	5.89186	7.04150
14	4.07468	4.66043	5.62872	6.57063	7.78953
15	4.60094	5.22935	6.26214	7.26094	8.54675
16	5.14224	5.81221	6.90766	7.96164	9.31223
17	5.69724	6.40776	7.56418	8.67176	10.0852
18	6.26481	7.01491	8.23075	9.39046	10.8649
19	6.84398	7.63273	8.90655	10.1170	11.6509
20	7.43386	8.26040	9.59083	10.8508	12.4426
21	8.03366	8.89720	10.28293	11.5913	13.2396
22	8.64272	9.54249	10.9823	12.3380	14.0415
23	9.26042	10.19567	11.6885	13.0905	14.8479
24	9.88623	10.8564	12.4011	13.8484	15.6587
25	10.5197	11.5240	13.1197	14.6114	16.4734
26	11.1603	12.1981	13.8439	15.3791	17.2919
27	11.8076	12.8786	14.5733	16.1513	18.1138
28	12.4613	13.5648	15.3079	16.9279	18.9392
29	13.1211	14.2565	16.0471	17.7083	19.7677
30	13.7867	14.9535	16.7908	18.4926	20.5992
40	20.7065	22.1643	24.4331	26.5093	29.0505
50	27.9907	29.7067	32.3574	34.7642	37.6886
60	35.5346	37.4848	40.4817	43.1879	46.4589
70	43.2752	45.4418	48.7576	51.7393	55.3290
80	51.1720	53.5400	57.1532	60.3915	64.2778
90	59.1963	61.7541	65.6466	69.1260	73.2912
100	67.3276	70.0648	74.2219	77.9295	82.3581

$\chi^2_{0.100}$	$\chi^2_{0.050}$	$\chi^2_{0.025}$	$\chi^2_{0.010}$	$\chi^2_{0.005}$	*d.f.*
2.70554	3.84146	5.02389	6.63490	7.87944	1
4.60517	5.99147	7.37776	9.21034	10.5966	2
6.25139	7.81473	9.34840	11.3449	12.8381	3
7.77944	9.48773	11.1433	13.2767	14.8602	4
9.23635	11.0705	12.8325	15.0863	16.7496	5
10.6446	12.5916	14.4494	16.8119	18.5476	6
12.0170	14.0671	16.0128	18.4753	20.2777	7
13.3616	15.5073	17.5346	20.0902	21.9550	8
14.6837	16.9190	19.0228	21.6660	23.5893	9
15.9871	18.3070	20.4831	23.2093	25.1882	10
17.2750	19.6751	21.9200	24.7250	26.7569	11
18.5494	21.0261	23.3367	26.2170	28.2995	12
19.8119	22.3621	24.7356	27.6883	29.8194	13
21.0642	23.6848	26.1190	29.1413	31.3193	14
22.3072	24.9958	27.4884	30.5779	32.8013	15
23.5418	26.2962	28.8454	31.9999	34.2672	16
24.7690	27.5871	30.1910	33.4087	35.7185	17
25.9894	28.8693	31.5264	34.8053	37.1564	18
27.2036	30.1435	32.8523	36.1908	38.5822	19
28.4120	31.4104	34.1696	37.5662	39.9968	20
29.6151	32.6705	35.4789	38.9321	41.4010	21
30.8133	33.9244	36.7807	40.2894	42.7956	22
32.0069	35.1725	38.0757	41.6384	44.1813	23
33.1963	36.4151	39.3641	42.9798	45.5585	24
34.3816	37.6525	40.6465	44.3141	46.9278	25
35.5631	38.8852	41.9232	45.6417	48.2899	26
36.7412	40.1133	43.1944	46.9630	49.6449	27
37.9159	41.3372	44.4607	48.2782	50.9933	28
39.0875	42.5569	45.7222	49.5879	52.3356	29
40.2560	43.7729	46.9792	50.8922	53.6720	30
51.8050	55.7585	59.3417	63.6907	66.7659	40
63.1671	67.5048	71.4202	76.1539	79.4900	50
74.3970	79.0819	83.2976	88.3794	91.9517	60
85.5271	90.5312	95.0231	100.425	104.215	70
96.5782	101.879	106.629	112.329	116.321	80
107.565	113.145	118.136	124.116	128.299	90
118.498	124.342	129.561	135.807	140.169	100

From "Tables of the Percentage Points of the χ^2-Distribution." *Biometrika*, Vol. 32 (1941), pp. 188–189, by Catherine M. Thompson. Reproduced by permission of Professor E. S. Pearson.

Table 6 Percentage Points of the F Distribution

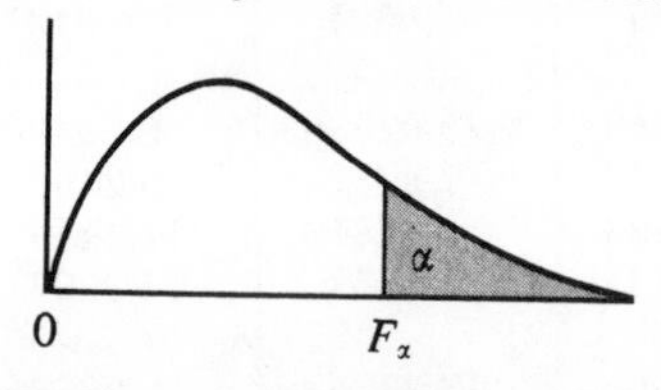

$\alpha = .05$

Degrees of Freedom

ν_2 \ ν_1	1	2	3	4	5	6	7	8	9
1	161.4	199.5	215.7	224.6	230.2	234.0	236.8	238.9	240.5
2	18.51	19.00	19.16	19.25	19.30	19.33	19.35	19.37	19.38
3	10.13	9.55	9.28	9.12	9.01	8.94	8.89	8.85	8.81
4	7.71	6.94	6.59	6.39	6.26	6.16	6.09	6.04	6.00
5	6.61	5.79	5.41	5.19	5.05	4.95	4.88	4.82	4.77
6	5.99	5.14	4.76	4.53	4.39	4.28	4.21	4.15	4.10
7	5.59	4.74	4.35	4.12	3.97	3.87	3.79	3.73	3.68
8	5.32	4.46	4.07	3.84	3.69	3.58	3.50	3.44	3.39
9	5.12	4.26	3.86	3.63	3.48	3.37	3.29	3.23	3.18
10	4.96	4.10	3.71	3.48	3.33	3.22	3.14	3.07	3.02
11	4.84	3.98	3.59	3.36	3.20	3.09	3.01	2.95	2.90
12	4.75	3.89	3.49	3.26	3.11	3.00	2.91	2.85	2.80
13	4.67	3.81	3.41	3.18	3.03	2.92	2.83	2.77	2.71
14	4.60	3.74	3.34	3.11	2.96	2.85	2.76	2.70	2.65
15	4.54	3.68	3.29	3.06	2.90	2.79	2.71	2.64	2.59
16	4.49	3.63	3.24	3.01	2.85	2.74	2.66	2.59	2.54
17	4.45	3.59	3.20	2.96	2.81	2.70	2.61	2.55	2.49
18	4.41	3.55	3.16	2.93	2.77	2.66	2.58	2.51	2.46
19	4.38	3.52	3.13	2.90	2.74	2.63	2.54	2.48	2.42
20	4.35	3.49	3.10	2.87	2.71	2.60	2.51	2.45	2.39
21	4.32	3.47	3.07	2.84	2.68	2.57	2.49	2.42	2.37
22	4.30	3.44	3.05	2.82	2.66	2.55	2.46	2.40	2.34
23	4.28	3.42	3.03	2.80	2.64	2.53	2.44	2.37	2.32
24	4.26	3.40	3.01	2.78	2.62	2.51	2.42	2.36	2.30
25	4.24	3.39	2.99	2.76	2.60	2.49	2.40	2.34	2.28
26	4.23	3.37	2.98	2.74	2.59	2.47	2.39	2.32	2.27
27	4.21	3.35	2.96	2.73	2.57	2.46	2.37	2.31	2.25
28	4.20	3.34	2.95	2.71	2.56	2.45	2.36	2.29	2.24
29	4.18	3.33	2.93	2.70	2.55	2.43	2.35	2.28	2.22
30	4.17	3.32	2.92	2.69	2.53	2.42	2.33	2.27	2.21
40	4.08	3.23	2.84	2.61	2.45	2.34	2.25	2.18	2.12
60	4.00	3.15	2.76	2.53	2.37	2.25	2.17	2.10	2.04
120	3.92	3.07	2.68	2.45	2.29	2.17	2.09	2.02	1.96
∞	3.84	3.00	2.60	2.37	2.21	2.10	2.01	1.94	1.88

ν_1 10	12	15	20	24	30	40	60	120	∞	ν_2
241.9	243.9	245.9	248.0	249.1	250.1	251.1	252.2	253.3	254.3	1
19.40	19.41	19.43	19.45	19.45	19.46	19.47	19.48	19.49	19.50	2
8.79	8.74	8.70	8.66	8.64	8.62	8.59	8.57	8.55	8.53	3
5.96	5.91	5.86	5.80	5.77	5.75	5.72	5.69	5.66	5.63	4
4.74	4.68	4.62	4.56	4.53	4.50	4.46	4.43	4.40	4.36	5
4.06	4.00	3.94	3.87	3.84	3.81	3.77	3.74	3.70	3.67	6
3.64	3.57	3.51	3.44	3.41	3.38	3.34	3.30	3.27	3.23	7
3.35	3.28	3.22	3.15	3.12	3.08	3.04	3.01	2.97	2.93	8
3.14	3.07	3.01	2.94	2.90	2.86	2.83	2.79	2.75	2.71	9
2.98	2.91	2.85	2.77	2.74	2.70	2.66	2.62	2.58	2.54	10
2.85	2.79	2.72	2.65	2.61	2.57	2.53	2.49	2.45	2.40	11
2.75	2.69	2.62	2.54	2.51	2.47	2.43	2.38	2.34	2.30	12
2.67	2.60	2.53	2.46	2.42	2.38	2.34	2.30	2.25	2.21	13
2.60	2.53	2.46	2.39	2.35	2.31	2.27	2.22	2.18	2.13	14
2.54	2.48	2.40	2.33	2.29	2.25	2.20	2.16	2.11	2.07	15
2.49	2.42	2.35	2.28	2.24	2.19	2.15	2.11	2.06	2.01	16
2.45	2.38	2.31	2.23	2.19	2.15	2.10	2.06	2.01	1.96	17
2.41	2.34	2.27	2.19	2.15	2.11	2.06	2.02	1.97	1.92	18
2.38	2.31	2.23	2.16	2.11	2.07	2.03	1.98	1.93	1.88	19
2.35	2.28	2.20	2.12	2.08	2.04	1.99	1.95	1.90	1.84	20
2.32	2.25	2.18	2.10	2.05	2.01	1.96	1.92	1.87	1.81	21
2.30	2.23	2.15	2.07	2.03	1.98	1.94	1.89	1.84	1.78	22
2.27	2.20	2.13	2.05	2.01	1.96	1.91	1.86	1.81	1.76	23
2.25	2.18	2.11	2.03	1.98	1.94	1.89	1.84	1.79	1.73	24
2.24	2.16	2.09	2.01	1.96	1.92	1.87	1.82	1.77	1.71	25
2.22	2.15	2.07	1.99	1.95	1.90	1.85	1.80	1.75	1.69	26
2.20	2.13	2.06	1.97	1.93	1.88	1.84	1.79	1.73	1.67	27
2.19	2.12	2.04	1.96	1.91	1.87	1.82	1.77	1.71	1.65	28
2.18	2.10	2.03	1.94	1.90	1.85	1.81	1.75	1.70	1.64	29
2.16	2.09	2.01	1.93	1.89	1.84	1.79	1.74	1.68	1.62	30
2.08	2.00	1.92	1.84	1.79	1.74	1.69	1.64	1.58	1.51	40
1.99	1.92	1.84	1.75	1.70	1.65	1.59	1.53	1.47	1.39	60
1.91	1.83	1.75	1.66	1.61	1.55	1.50	1.43	1.35	1.25	120
1.83	1.75	1.67	1.57	1.52	1.46	1.39	1.32	1.22	1.00	∞

From "Tables of Percentage Points of the Inverted Beta (F) Distribution," *Biometrika*, Vol. 33 (1943), pp. 73–88, by Maxine Merrington and Catherine M. Thompson. Reproduced by permission of Professor E. S. Pearson.

Table 7 Percentage Points of the F Distribution

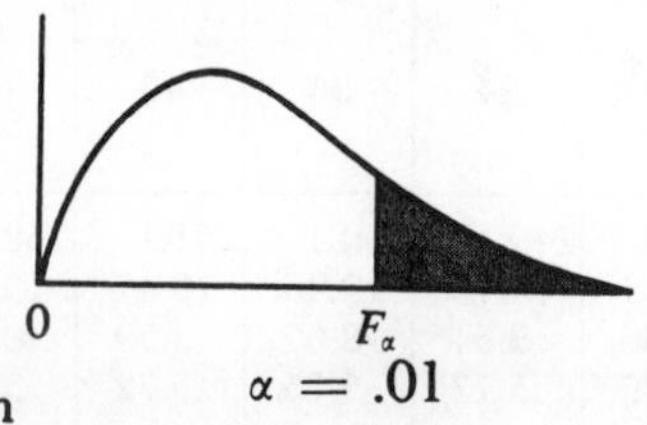

$\alpha = .01$

Degrees of Freedom

ν_2 \ ν_1	1	2	3	4	5	6	7	8	9
1	4052	4999.5	5403	5625	5764	5859	5928	5982	6022
2	98.50	99.00	99.17	99.25	99.30	99.33	99.36	99.37	99.39
3	34.12	30.82	29.46	28.71	28.24	27.91	27.67	27.49	27.35
4	21.20	18.00	16.69	15.98	15.52	15.21	14.98	14.80	14.66
5	16.26	13.27	12.06	11.39	10.97	10.67	10.46	10.29	10.16
6	13.75	10.92	9.78	9.15	8.75	8.47	8.26	8.10	7.98
7	12.25	9.55	8.45	7.85	7.46	7.19	6.99	6.84	6.72
8	11.26	8.65	7.59	7.01	6.63	6.37	6.18	6.03	5.91
9	10.56	8.02	6.99	6.42	6.06	5.80	5.61	5.47	5.35
10	10.04	7.56	6.55	5.99	5.64	5.39	5.20	5.06	4.94
11	9.65	7.21	6.22	5.67	5.32	5.07	4.89	4.74	4.63
12	9.33	6.93	5.95	5.41	5.06	4.82	4.64	4.50	4.39
13	9.07	6.70	5.74	5.21	4.86	4.62	4.44	4.30	4.19
14	8.86	6.51	5.56	5.04	4.69	4.46	4.28	4.14	4.03
15	8.68	6.36	5.42	4.89	4.56	4.32	4.14	4.00	3.89
16	8.53	6.23	5.29	4.77	4.44	4.20	4.03	3.89	3.78
17	8.40	6.11	5.18	4.67	4.34	4.10	3.93	3.79	3.68
18	8.29	6.01	5.09	4.58	4.25	4.01	3.84	3.71	3.60
19	8.18	5.93	5.01	4.50	4.17	3.94	3.77	3.63	3.52
20	8.10	5.85	4.94	4.43	4.10	3.87	3.70	3.56	3.46
21	8.02	5.78	4.87	4.37	4.04	3.81	3.64	3.51	3.40
22	7.95	5.72	4.82	4.31	3.99	3.76	3.59	3.45	3.35
23	7.88	5.66	4.76	4.26	3.94	3.71	3.54	3.41	3.30
24	7.82	5.61	4.72	4.22	3.90	3.67	3.50	3.36	3.26
25	7.77	5.57	4.68	4.18	3.85	3.63	3.46	3.32	3.22
26	7.72	5.53	4.64	4.14	3.82	3.59	3.42	3.29	3.18
27	7.68	5.49	4.60	4.11	3.78	3.56	3.39	3.26	3.15
28	7.64	5.45	4.57	4.07	3.75	3.53	3.36	3.23	3.12
29	7.60	5.42	4.54	4.04	3.73	3.50	3.33	3.20	3.09
30	7.56	5.39	4.51	4.02	3.70	3.47	3.30	3.17	3.07
40	7.31	5.18	4.31	3.83	3.51	3.29	3.12	2.99	2.89
60	7.08	4.98	4.13	3.65	3.34	3.12	2.95	2.82	2.72
120	6.85	4.79	3.95	3.48	3.17	2.96	2.79	2.66	2.56
∞	6.63	4.61	3.78	3.32	3.02	2.80	2.64	2.51	2.41

	ν_1									
10	12	15	20	24	30	40	60	120	∞	ν_2
6056	6106	6157	6209	6235	6261	6287	6313	6339	6366	1
99.40	99.42	99.43	99.45	99.46	99.47	99.47	99.48	99.49	99.50	2
27.23	27.05	26.87	26.69	26.60	26.50	26.41	26.32	26.22	26.13	3
14.55	14.37	14.20	14.02	13.93	13.84	13.75	13.65	13.56	13.46	4
10.05	9.89	9.72	9.55	9.47	9.38	9.29	9.20	9.11	9.02	5
7.87	7.72	7.56	7.40	7.31	7.23	7.14	7.06	6.97	6.88	6
6.62	6.47	6.31	6.16	6.07	5.99	5.91	5.82	5.74	5.65	7
5.81	5.67	5.52	5.36	5.28	5.20	5.12	5.03	4.95	4.86	8
5.26	5.11	4.96	4.81	4.73	4.65	4.57	4.48	4.40	4.31	9
4.85	4.71	4.56	4.41	4.33	4.25	4.17	4.08	4.00	3.91	10
4.54	4.40	4.25	4.10	4.02	3.94	3.86	3.78	3.69	3.60	11
4.30	4.16	4.01	3.86	3.78	3.70	3.62	3.54	3.45	3.36	12
4.10	3.96	3.82	3.66	3.59	3.51	3.43	3.34	3.25	3.17	13
3.94	3.80	3.66	3.51	3.43	3.35	3.27	3.18	3.09	3.00	14
3.80	3.67	3.52	3.37	3.29	3.21	3.13	3.05	2.96	2.87	15
3.69	3.55	3.41	3.26	3.18	3.10	3.02	2.93	2.84	2.75	16
3.59	3.46	3.31	3.16	3.08	3.00	2.92	2.83	2.75	2.65	17
3.51	3.37	3.23	3.08	3.00	2.92	2.84	2.75	2.66	2.57	18
3.43	3.30	3.15	3.00	2.92	2.84	2.76	2.67	2.58	2.49	19
3.37	3.23	3.09	2.94	2.86	2.78	2.69	2.61	2.52	2.42	20
3.31	3.17	3.03	2.88	2.80	2.72	2.64	2.55	2.46	2.36	21
3.26	3.12	2.98	2.83	2.75	2.67	2.58	2.50	2.40	2.31	22
3.21	3.07	2.93	2.78	2.70	2.62	2.54	2.45	2.35	2.26	23
3.17	3.03	2.89	2.74	2.66	2.58	2.49	2.40	2.31	2.21	24
3.13	2.99	2.85	2.70	2.62	2.54	2.45	2.36	2.27	2.17	25
3.09	2.96	2.81	2.66	2.58	2.50	2.42	2.33	2.23	2.13	26
3.06	2.93	2.78	2.63	2.55	2.47	2.38	2.29	2.20	2.10	27
3.03	2.90	2.75	2.60	2.52	2.44	2.35	2.26	2.17	2.06	28
3.00	2.87	2.73	2.57	2.49	2.41	2.33	2.23	2.14	2.03	29
2.98	2.84	2.70	2.55	2.47	2.39	2.30	2.21	2.11	2.01	30
2.80	2.66	2.52	2.37	2.29	2.20	2.11	2.02	1.92	1.80	40
2.63	2.50	2.35	2.20	2.12	2.03	1.94	1.84	1.73	1.60	60
2.47	2.34	2.19	2.03	1.95	1.86	1.76	1.66	1.53	1.38	120
2.32	2.18	2.04	1.88	1.79	1.70	1.59	1.47	1.32	1.00	∞

From "Tables of Percentage Points of the Inverted Beta (F) Distribution," *Biometrika*, Vol. 33 (1943), pp. 73–88, by Maxine Merrington and Catherine M. Thompson. Reproduced by permission of Professor E. S. Pearson.

Table 8 Distribution Function of U

$P(U \leq U_0)$; U_0 is the argument
$n_1 \leq n_2$; $3 \leq n_2 \leq 10$

$n_2 = 3$

U_0 \ n_1	1	2	3
0	.25	.10	.05
1	.50	.20	.10
2		.40	.20
3		.60	.35
4			.50

$n_2 = 4$

U_0 \ n_1	1	2	3	4
0	.2000	.0667	.0286	.0143
1	.4000	.1333	.0571	.0286
2	.6000	.2667	.1143	.0571
3		.4000	.2000	.1000
4		.6000	.3143	.1714
5			.4286	.2429
6			.5714	.3429
7				.4429
8				.5571

$n_2 = 5$

U_0 \ n_1	1	2	3	4	5
0	.1667	.0476	.0179	.0079	.0040
1	.3333	.0952	.0357	.0159	.0079
2	.5000	.1905	.0714	.0317	.0159
3		.2857	.1250	.0556	.0278
4		.4286	.1964	.0952	.0476
5		.5714	.2857	.1429	.0754
6			.3929	.2063	.1111
7			.5000	.2778	.1548
8				.3651	.2103
9				.4524	.2738
10				.5476	.3452
11					.4206
12					.5000

$n_2 = 6$

U_0 \ n_1	1	2	3	4	5	6
0	.1429	.0357	.0119	.0048	.0022	.0011
1	.2857	.0714	.0238	.0095	.0043	.0022
2	.4286	.1429	.0476	.0190	.0087	.0043
3	.5714	.2143	.0833	.0333	.0152	.0076
4		.3214	.1310	.0571	.0260	.0130
5		.4286	.1905	.0857	.0411	.0206
6		.5714	.2738	.1286	.0628	.0325
7			.3571	.1762	.0887	.0465
8			.4524	.2381	.1234	.0660
9			.5476	.3048	.1645	.0898
10				.3810	.2143	.1201
11				.4571	.2684	.1548
12				.5429	.3312	.1970
13					.3961	.2424
14					.4654	.2944
15					.5346	.3496
16						.4091
17						.4686
18						.5314

$n_2 = 7$

U_0 \ n_1	1	2	3	4	5	6	7
0	.1250	.0278	.0083	.0030	.0013	.0006	.0003
1	.2500	.0556	.0167	.0061	.0025	.0012	.0006
2	.3750	.1111	.0333	.0121	.0051	.0023	.0012
3	.5000	.1667	.0583	.0212	.0088	.0041	.0020
4		.2500	.0917	.0364	.0152	.0070	.0035
5		.3333	.1333	.0545	.0240	.0111	.0055
6		.4444	.1917	.0818	.0366	.0175	.0087
7		.5556	.2583	.1152	.0530	.0256	.0131
8			.3333	.1576	.0745	.0367	.0189
9			.4167	.2061	.1010	.0507	.0265
10			.5000	.2636	.1338	.0688	.0364
11				.3242	.1717	.0903	.0487
12				.3939	.2159	.1171	.0641
13				.4636	.2652	.1474	.0825
14				.5364	.3194	.1830	.1043
15					.3775	.2226	.1297
16					.4381	.2669	.1588
17					.5000	.3141	.1914
18						.3654	.2279
19						.4178	.2675
20						.4726	.3100
21						.5274	.3552
22							.4024
23							.4508
24							.5000

$n_2 = 8$

U_0 \ n_1	1	2	3	4	5	6	7	8
0	.1111	.0222	.0061	.0020	.0008	.0003	.0002	.0001
1	.2222	.0444	.0121	.0040	.0016	.0007	.0003	.0002
2	.3333	.0889	.0242	.0081	.0031	.0013	.0006	.0003
3	.4444	.1333	.0424	.0141	.0054	.0023	.0011	.0005
4	.5556	.2000	.0667	.0242	.0093	.0040	.0019	.0009
5		.2667	.0970	.0364	.0148	.0063	.0030	.0015
6		.3556	.1394	.0545	.0225	.0100	.0047	.0023
7		.4444	.1879	.0768	.0326	.0147	.0070	.0035
8		.5556	.2485	.1071	.0466	.0213	.0103	.0052
9			.3152	.1414	.0637	.0296	.0145	.0074
10			.3879	.1838	.0855	.0406	.0200	.0103
11			.4606	.2303	.1111	.0539	.0270	.0141
12			.5394	.2848	.1422	.0709	.0361	.0190
13				.3414	.1772	.0906	.0469	.0249
14				.4040	.2176	.1142	.0603	.0325
15				.4667	.2618	.1412	.0760	.0415
16				.5333	.3108	.1725	.0946	.0524
17					.3621	.2068	.1159	.0652
18					.4165	.2454	.1405	.0803
19					.4716	.2864	.1678	.0974
20					.5284	.3310	.1984	.1172
21						.3773	.2317	.1393
22						.4259	.2679	.1641
23						.4749	.3063	.1911
24						.5251	.3472	.2209
25							.3894	.2527
26							.4333	.2869
27							.4775	.3227
28							.5225	.3605
29								.3992
30								.4392
31								.4796
32								.5204

$n_2 = 9$

U_0 \ n_1	1	2	3	4	5	6	7	8	9
0	.1000	.0182	.0045	.0014	.0005	.0002	.0001	.0000	.0000
1	.2000	.0364	.0091	.0028	.0010	.0004	.0002	.0001	.0000
2	.3000	.0727	.0182	.0056	.0020	.0008	.0003	.0002	.0001
3	.4000	.1091	.0318	.0098	.0035	.0014	.0006	.0003	.0001
4	.5000	.1636	.0500	.0168	.0060	.0024	.0010	.0005	.0002
5		.2182	.0727	.0252	.0095	.0038	.0017	.0008	.0004
6		.2909	.1045	.0378	.0145	.0060	.0026	.0012	.0006
7		.3636	.1409	.0531	.0210	.0088	.0039	.0019	.0009
8		.4545	.1864	.0741	.0300	.0128	.0058	.0028	.0014
9		.5455	.2409	.0993	.0415	.0180	.0082	.0039	.0020
10			.3000	.1301	.0559	.0248	.0115	.0056	.0028
11			.3636	.1650	.0734	.0332	.0156	.0076	.0039
12			.4318	.2070	.0949	.0440	.0209	.0103	.0053
13			.5000	.2517	.1199	.0567	.0274	.0137	.0071
14				.3021	.1489	.0723	.0356	.0180	.0094
15				.3552	.1818	.0905	.0454	.0232	.0122
16				.4126	.2188	.1119	.0571	.0296	.0157
17				.4699	.2592	.1361	.0708	.0372	.0200
18				.5301	.3032	.1638	.0869	.0464	.0252
19					.3497	.1942	.1052	.0570	.0313
20					.3986	.2280	.1261	.0694	.0385
21					.4491	.2643	.1496	.0836	.0470
22					.5000	.3035	.1755	.0998	.0567
23						.3445	.2039	.1179	.0680
24						.3878	.2349	.1383	.0807
25						.4320	.2680	.1606	.0951
26						.4773	.3032	.1852	.1112
27						.5227	.3403	.2117	.1290
28							.3788	.2404	.1487
29							.4185	.2707	.1701
30							.4591	.3029	.1933
31							.5000	.3365	.2181
32								.3715	.2447
33								.4074	.2729
34								.4442	.3024
35								.4813	.3332
36								.5187	.3652
37									.3981
38									.4317
39									.4657
40									.5000

$n_2 = 10$

U_0 \ n_1	1	2	3	4	5	6	7	8	9	10
0	.0909	.0152	.0035	.0010	.0003	.0001	.0001	.0000	.0000	.0000
1	.1818	.0303	.0070	.0020	.0007	.0002	.0001	.0000	.0000	.0000
2	.2727	.0606	.0140	.0040	.0013	.0005	.0002	.0001	.0000	.0000
3	.3636	.0909	.0245	.0070	.0023	.0009	.0004	.0002	.0001	.0000
4	.4545	.1364	.0385	.0120	.0040	.0015	.0006	.0003	.0001	.0001
5	.5455	.1818	.0559	.0180	.0063	.0024	.0010	.0004	.0002	.0001
6		.2424	.0804	.0270	.0097	.0037	.0015	.0007	.0003	.0002
7		.3030	.1084	.0380	.0140	.0055	.0023	.0010	.0005	.0002
8		.3788	.1434	.0529	.0200	.0080	.0034	.0015	.0007	.0004
9		.4545	.1853	.0709	.0276	.0112	.0048	.0022	.0011	.0005
10		.5455	.2343	.0939	.0376	.0156	.0068	.0031	.0015	.0008
11			.2867	.1199	.0496	.0210	.0093	.0043	.0021	.0010
12			.3462	.1518	.0646	.0280	.0125	.0058	.0028	.0014
13			.4056	.1868	.0823	.0363	.0165	.0078	.0038	.0019
14			.4685	.2268	.1032	.0467	.0215	.0103	.0051	.0026
15			.5315	.2697	.1272	.0589	.0277	.0133	.0066	.0034
16				.3177	.1548	.0736	.0351	.0171	.0086	.0045
17				.3666	.1855	.0903	.0439	.0217	.0110	.0057
18				.4196	.2198	.1099	.0544	.0273	.0140	.0073
19				.4725	.2567	.1317	.0665	.0338	.0175	.0093
20				.5275	.2970	.1566	.0806	.0416	.0217	.0116
21					.3393	.1838	.0966	.0506	.0267	.0144
22					.3839	.2139	.1148	.0610	.0326	.0177
23					.4296	.2461	.1349	.0729	.0394	.0216
24					.4765	.2811	.1574	.0864	.0474	.0262
25					.5235	.3177	.1819	.1015	.0564	.0315
26						.3564	.2087	.1185	.0667	.0376
27						.3962	.2374	.1371	.0782	.0446
28						.4374	.2681	.1577	.0912	.0526
29						.4789	.3004	.1800	.1055	.0615
30						.5211	.3345	.2041	.1214	.0716
31							.3698	.2299	.1388	.0827
32							.4063	.2574	.1577	.0952
33							.4434	.2863	.1781	.1088
34							.4811	.3167	.2001	.1237
35							.5189	.3482	.2235	.1399
36								.3809	.2483	.1575
37								.4143	.2745	.1763
38								.4484	.3019	.1965
39								.4827	.3304	.2179
40								.5173	.3598	.2406
41									.3901	.2644
42									.4211	.2894
43									.4524	.3153
44									.4841	.3421
45									.5159	.3697
46										.3980
47										.4267
48										.4559
49										.4853
50										.5147

Computed by M. Pagano, Department of Statistics, University of Florida.

Table 9 Critical Values of T in the Wilcoxon Matched-Pairs Signed-Ranks Test

$n = 5(1)50$

One sided	Two-sided	$n = 5$	$n = 6$	$n = 7$	$n = 8$	$n = 9$	$n = 10$
$P = .05$	$P = .10$	1	2	4	6	8	11
$P = .025$	$P = .05$		1	2	4	6	8
$P = .01$	$P = .02$			0	2	3	5
$P = .005$	$P = .01$				0	2	3

One-sided	Two-sided	$n = 11$	$n = 12$	$n = 13$	$n = 14$	$n = 15$	$n = 16$
$P = .05$	$P = .10$	14	17	21	26	30	36
$P = .025$	$P = .05$	11	14	17	21	25	30
$P = .01$	$P = .02$	7	10	13	16	20	24
$P = .005$	$P = .01$	5	7	10	13	16	19

One-sided	Two-sided	$n = 17$	$n = 18$	$n = 19$	$n = 20$	$n = 21$	$n = 22$
$P = .05$	$P = .10$	41	47	54	60	68	75
$P = .025$	$P = .05$	35	40	46	52	59	66
$P = .01$	$P = .02$	28	33	38	43	49	56
$P = .005$	$P = .01$	23	28	32	37	43	49

One-sided	Two-sided	$n = 23$	$n = 24$	$n = 25$	$n = 26$	$n = 27$	$n = 28$
$P = .05$	$P = .10$	83	92	101	110	120	130
$P = .025$	$P = .05$	73	81	90	98	107	117
$P = .01$	$P = .02$	62	69	77	85	93	102
$P = .005$	$P = .01$	55	68	68	76	84	92

One-sided	Two-sided	$n = 29$	$n = 30$	$n = 31$	$n = 32$	$n = 33$	$n = 34$
$P = .05$	$P = .10$	141	152	163	175	188	201
$P = .025$	$P = .05$	127	137	148	159	171	183
$P = .01$	$P = .02$	111	120	130	141	151	162
$P = .005$	$P = .01$	100	109	118	128	138	149

One-sided	Two-sided	$n = 35$	$n = 36$	$n = 37$	$n = 38$	$n = 39$
$P = .05$	$P = .10$	214	228	242	256	271
$P = .025$	$P = .05$	195	208	222	235	250
$P = .01$	$P = .02$	174	186	198	211	224
$P = .005$	$P = .01$	160	171	183	195	208

One-sided	Two-sided	$n = 40$	$n = 41$	$n = 42$	$n = 43$	$n = 44$	$n = 45$
$P = .05$	$P = .10$	287	303	319	336	353	371
$P = .025$	$P = .05$	264	279	295	311	327	344
$P = .01$	$P = .02$	238	252	267	281	297	313
$P = .005$	$P = .01$	221	234	248	262	277	292

One-sided	Two-sided	$n = 46$	$n = 47$	$n = 48$	$n = 49$	$n = 50$
$P = .05$	$P = .10$	389	408	427	446	466
$P = .025$	$P = .05$	361	379	397	415	434
$P = .01$	$P = .02$	329	345	362	380	398
$P = .005$	$P = .01$	307	323	339	356	373

From "Some Rapid Approximate Statistical Procedures" (1964), 28, F. Wilcoxon and R. A. Wilcox. Reproduced with the kind permission of R. A. Wilcox and the Lederle Laboratories.

Table 10 Distribution of the Total Number of Runs R in Samples of Size (n_1, n_2); $P(R \leq a)$

	a								
(n_1, n_2)	2	3	4	5	6	7	8	9	10
(2,3)	.200	.500	.900	1.000					
(2,4)	.133	.400	.800	1.000					
(2,5)	.095	.333	.714	1.000					
(2,6)	.071	.286	.643	1.000					
(2,7)	.056	.250	.583	1.000					
(2,8)	.044	.222	.533	1.000					
(2,9)	.036	.200	.491	1.000					
(2,10)	.030	.182	.455	1.000					
(3,3)	.100	.300	.700	.900	1.000				
(3,4)	.057	.200	.543	.800	.971	1.000			
(3,5)	.036	.143	.429	.714	.929	1.000			
(3,6)	.024	.107	.345	.643	.881	1.000			
(3,7)	.017	.083	.283	.583	.833	1.000			
(3,8)	.012	.067	.236	.533	.788	1.000			
(3,9)	.009	.055	.200	.491	.745	1.000			
(3,10)	.007	.045	.171	.455	.706	1.000			
(4,4)	.029	.114	.371	.629	.886	.971	1.000		
(4,5)	.016	.071	.262	.500	.786	.929	.992	1.000	
(4,6)	.010	.048	.190	.405	.690	.881	.976	1.000	
(4,7)	.006	.033	.142	.333	.606	.833	.954	1.000	
(4,8)	.004	.024	.109	.279	.533	.788	.929	1.000	
(4,9)	.003	.018	.085	.236	.471	.745	.902	1.000	
(4,10)	.002	.014	.068	.203	.419	.706	.874	1.000	
(5,5)	.008	.040	.167	.357	.643	.833	.960	.992	1.000
(5,6)	.004	.024	.110	.262	.522	.738	.911	.976	.998
(5,7)	.003	.015	.076	.197	.424	.652	.854	.955	.992
(5,8)	.002	.010	.054	.152	.347	.576	.793	.929	.984
(5,9)	.001	.007	.039	.119	.287	.510	.734	.902	.972
(5,10)	.001	.005	.029	.095	.239	.455	.678	.874	.958
(6,6)	.002	.013	.067	.175	.392	.608	.825	.933	.987
(6,7)	.001	.008	.043	.121	.296	.500	.733	.879	.966
(6,8)	.001	.005	.028	.086	.226	.413	.646	.821	.937
(6,9)	.000	.003	.019	.063	.175	.343	.566	.762	.902
(6,10)	.000	.002	.013	.047	.137	.288	.497	.706	.864
(7,7)	.001	.004	.025	.078	.209	.383	.617	.791	.922
(7,8)	.000	.002	.015	.051	.149	.296	.514	.704	.867
(7,9)	.000	.001	.010	.035	.108	.231	.427	.622	.806
(7,10)	.000	.001	.006	.024	.080	.182	.355	.549	.743
(8,8)	.000	.001	.009	.032	.100	.214	.405	.595	.786
(8,9)	000	.001	.005	.020	.069	.157	.319	.500	.702
(8,10)	.000	.000	.003	.013	.048	.117	.251	.419	.621
(9,9)	.000	.000	.003	.012	.044	.109	.238	.399	.601
(9,10)	.000	.000	.002	.008	.029	.077	.179	.319	.510
(10,10)	.000	.000	.001	.004	.019	.051	.128	.242	.414

	a									
(n_1, n_2)	11	12	13	14	15	16	17	18	19	20
(2,3)										
(2,4)										
(2,5)										
(2,6)										
(2,7)										
(2,8)										
(2,9)										
(2,10)										
(3,3)										
(3,4)										
(3,5)										
(3,6)										
(3,7)										
(3,8)										
(3,9)										
(3,10)										
(4,4)										
(4,5)										
(4,6)										
(4,7)										
(4,8)										
(4,9)										
(4,10)										
(5,5)										
(5,6)	1.000									
(5,7)	1.000									
(5,8)	1.000									
(5,9)	1.000									
(5,10)	1.000									
(6,6)	.998	1.000								
(6,7)	.992	.999	1.000							
(6,8)	.984	.998	1.000							
(6,9)	.972	.994	1.000							
(6,10)	.958	.990	1.000							
(7,7)	.975	.996	.999	1.000						
(7,8)	.949	.988	.998	1.000	1.000					
(7,9)	.916	.975	.994	.999	1.000					
(7,10)	.879	.957	.990	.998	1.000					
(8,8)	.900	.968	.991	.999	1.000	1.000				
(8,9)	.843	.939	.980	.996	.999	1.000	1.000			
(8,10)	.782	.903	.964	.990	.998	1.000	1.000			
(9,9)	.762	.891	.956	.988	.997	1.000	1.000	1.000		
(9,10)	.681	.834	.923	.974	.992	.999	1.000	1.000	1.000	
(10,10)	.586	.758	.872	.949	.981	.996	.999	1.000	1.000	1.000

From "Tables for Testing Randomness of Grouping in a Sequence of Alternatives," C. Eisenhart and F. Swed, *Annals of Mathematical Statistics*, Volume 14 (1943). Reproduced with the kind permission of the Editor, *Annals of Mathematical Statistics.*

Table 11 Critical Values of Spearman's Rank Correlation Coefficient

n	$\alpha = .05$	$\alpha = .025$	$\alpha = .01$	$\alpha = .005$
5	0.900	—	—	—
6	0.829	0.886	0.943	—
7	0.714	0.786	0.893	—
8	0.643	0.738	0.833	0.881
9	0.600	0.683	0.783	0.833
10	0.564	0.648	0.745	0.794
11	0.523	0.623	0.736	0.818
12	0.497	0.591	0.703	0.780
13	0.475	0.566	0.673	0.745
14	0.457	0.545	0.646	0.716
15	0.441	0.525	0.623	0.689
16	0.425	0.507	0.601	0.666
17	0.412	0.490	0.582	0.645
18	0.399	0.476	0.564	0.625
19	0.388	0.462	0.549	0.608
20	0.377	0.450	0.534	0.591
21	0.368	0.438	0.521	0.576
22	0.359	0.428	0.508	0.562
23	0.351	0.418	0.496	0.549
24	0.343	0.409	0.485	0.537
25	0.336	0.400	0.475	0.526
26	0.329	0.392	0.465	0.515
27	0.323	0.385	0.456	0.505
28	0.317	0.377	0.448	0.496
29	0.311	0.370	0.440	0.487
30	0.305	0.364	0.432	0.478

From "Distribution of Sums of Squares of Rank Differences for Small Samples," E. G. Olds, *Annals of Mathematical Statistics*, Volume 9 (1938). Reproduced with the kind permission of the Editor, *Annals of Mathematical Statistics.*

Table 12 Squares, Cubes, and Roots

Roots of numbers other than those given directly may be found by the following relations:

$$\sqrt{100n} = 10\sqrt{n}; \quad \sqrt{1000n} = 10\sqrt{10n}; \quad \sqrt{\frac{1}{10}n} = \frac{1}{10}\sqrt{10n};$$

$$\sqrt{\frac{1}{100}n} = \frac{1}{10}\sqrt{n}; \quad \sqrt{\frac{1}{1000}n} = \frac{1}{100}\sqrt{10n}; \quad \sqrt[3]{1000n} = 10\sqrt[3]{n};$$

$$\sqrt[3]{10{,}000n} = 10\sqrt[3]{10n}; \quad \sqrt[3]{100{,}000n} = 10\sqrt[3]{100n};$$

$$\sqrt[3]{\frac{1}{10}n} = \frac{1}{10}\sqrt[3]{100n}; \quad \sqrt[3]{\frac{1}{100}n} = \frac{1}{10}\sqrt[3]{10n}; \quad \sqrt[3]{\frac{1}{1000}n} = \frac{1}{10}\sqrt[3]{n}.$$

n	n^2	$\sqrt{n}$	$\sqrt{10n}$	n	n^2	$\sqrt{n}$	$\sqrt{10n}$
				30	900	5.477 226	17.32051
1	1	1.000 000	3.162 278	31	961	5.567 764	17.60682
2	4	1.414 214	4.472 136	32	1 024	5.656 854	17.88854
3	9	1.732 051	5.477 226	33	1 089	5.744 563	18.16590
4	16	2.000 000	6.324 555	34	1 156	5.830 952	18.43909
5	25	2.236 068	7.071 068	35	1 225	5.916 080	18.70829
6	36	2.449 490	7.745 967	36	1 296	6.000 000	18.97367
7	49	2.645 751	8.366 600	37	1 369	6.082 763	19.23538
8	64	2.828 427	8.944 272	38	1 444	6.164 414	19.49359
9	81	3.000 000	9.486 833	39	1 521	6.244 998	19.74842
10	100	3.162 278	10.00000	**40**	1 600	6.324 555	20.00000
11	121	3.316 625	10.48809	41	1 681	6.403 124	20.24846
12	144	3.464 102	10.95445	42	1 764	6.480 741	20.49390
13	169	3.605 551	11.40175	43	1 849	6.557 439	20.73644
14	196	3.741 657	11.83216	44	1 936	6.633 250	20.97618
15	225	3.872 983	12.24745	45	2 025	6.708 204	21.21320
16	256	4.000 000	12.64911	46	2 116	6.782 330	21.44761
17	289	4.123 106	13.03840	47	2 209	6.855 655	21.67948
18	324	4.242 641	13.41641	48	2 304	6.928 203	21.90890
19	361	4.358 899	13.78405	49	2 401	7.000 000	22.13594
20	400	4.472 136	14.14214	**50**	2 500	7.071 068	22.36068
21	441	4.582 576	14.49138	51	2 601	7.141 428	22.58318
22	484	4.690 416	14.83240	52	2 704	7.211 103	22.80351
23	529	4.795 832	15.16575	53	2 809	7.280 110	23.02173
24	576	4.898 979	15.49193	54	2 916	7.348 469	23.23790
25	625	5.000 000	15.81139	55	3 025	7.416 198	23.45208
26	676	5.099 020	16.12452	56	3 136	7.483 315	23.66432
27	729	5.196 152	16.43168	57	3 249	7.549 834	23.87467
28	784	5.291 503	16.73320	58	3 364	7.615 773	24.08319
29	841	5.385 165	17.02939	59	3 418	7.618 146	24.28992

n	n^2	$\sqrt{n}$	$\sqrt{10n}$	n	n^2	$\sqrt{n}$	$\sqrt{10n}$
60	3 600	7.745 967	24.49490	**100**	10 000	10.00000	31.62278
61	3 721	7.810 250	24.69818	101	10 201	10.04998	31.78050
62	3 844	7.874 008	24.89980	102	10 404	10.09950	31.93744
63	3 969	7.937 254	25.09980	103	10 609	10.14889	32.09361
64	4 096	8.000 000	25.29822	104	10 816	10.19804	32.24903
65	4 225	8.062 258	25.49510	105	11 025	10.24695	32.40370
66	4 356	8.124 038	25.69047	106	11 236	10.29563	32.55764
67	4 489	8.185 353	25.88436	107	11 449	10.34408	32.71085
68	4 624	8.246 211	26.07681	108	11 664	10.39230	32.86335
69	4 761	8.306 624	26.26785	109	11 881	10.44031	33.01515
70	4 900	8.366 600	26.45751	**110**	12 100	10.48809	33.16625
71	5 041	8.426 150	26.64583	111	12 321	10.53565	33.31666
72	5 184	8.485 281	26.83282	112	12 544	10.58301	33.46640
73	5 329	8.544 004	27.01851	113	12 769	10.63015	33.61547
74	5 476	8.602 325	27.20294	114	12 996	10.67708	33.76389
75	5 625	8.660 254	27.38613	115	13 225	10.72381	33.91165
76	5 776	8.717 798	27.56810	116	13 456	10.77033	34.05877
77	5 929	8.774 964	27.74887	117	13 689	10.81665	34.20526
78	6 084	8.831 761	27.92848	118	13 924	10.86278	34.35113
79	6 241	8.888 194	28.10694	119	14 161	10.90871	34.49638
80	6 400	8.944 272	28.28427	**120**	14 400	10.95445	34.64102
81	6 561	9.000 000	28.46050	121	14 641	11.00000	34.78505
82	6 724	9.055 385	28.63564	122	14 884	11.04536	34.92850
83	6 889	9.110 434	28.80972	123	15 129	11.09054	35 07136
84	7 056	9.165 151	28.98275	124	15 376	11.13553	35 21363
85	7 225	9.219 544	29.15476	125	15 625	11.18034	35.35534
86	7 396	9.273 618	29.32576	126	15 876	11.22497	35.49648
87	7 569	9.327 379	29.49576	127	16 129	11.26943	35.63706
88	7 744	9.380 832	29.66479	128	16 384	11.31371	35.77709
89	7 921	9.433 981	29.83287	129	16 641	11.35782	35.91657
90	8 100	9.486 833	30.00000	**130**	16 900	11.40175	36.05551
91	8 281	9.539 392	30.16621	131	17 161	11.44552	36.19392
92	8 464	9.591 663	30.33150	132	17 424	11.48913	36.33180
93	8 649	9.643 651	30.49590	133	17 689	11.53256	36.46917
94	8 836	9.695 360	30.65942	134	17 956	11.57584	36.60601
95	9 025	9.746 794	30.82207	135	18 225	11.61895	36.74235
96	9 216	9.797 959	30.98387	136	18 496	11.66190	36.87818
97	9 409	9.848 858	31.14482	137	18 769	11.70470	37.01351
98	9 604	9.899 495	31.30495	138	19 044	11.74734	37.14835
99	9 801	9.949 874	31.46427	139	19 321	11.78983	37.28270

n	n^2	$\sqrt{n}$	$\sqrt{10n}$	n	n^2	$\sqrt{n}$	$\sqrt{10n}$
140	19 600	11.83216	37.41657	**180**	32 400	13.41641	42.42641
141	19 881	11.87434	37.54997	181	32 761	13.45362	42.54409
142	20 164	11.91638	37.68289	182	33 124	13.49074	42.66146
143	20 449	11.95826	37.81534	183	33 489	13.52775	42.77850
144	20 736	12.00000	37.94733	184	33 856	13.56466	42.89522
145	21 025	12.04159	38.07887	185	34 225	13.60147	43.01163
146	21 316	12.08305	38.20995	186	34 596	13.63818	43.12772
147	21 609	12.12436	38.34058	187	34 969	13.67479	43.24350
148	21 904	12.16553	38.47077	188	35 344	13.71131	43.35897
149	22 201	12.20656	38.60052	189	35 721	13.74773	43.47413
150	22 500	12.24745	38.72983	**190**	36 100	13.78405	43.58899
151	22 801	12.28821	38.85872	191	36 481	13.82027	43.70355
152	23 104	12.32883	38.98718	192	36 864	13.85641	43.81780
153	23 409	12.36932	39.11521	193	37 249	13.89244	43.93177
154	23 716	12.40967	39.24283	194	37 636	13.92839	44.04543
155	24 025	12.44990	39.37004	195	38 025	13.96424	44.15880
156	24 336	12.49000	39.49684	196	38 416	14.00000	44.27189
157	24 649	12.52996	39.62323	197	38 809	14.03567	44.38468
158	24 964	12.56981	39.74921	198	39 204	14.07125	44.49719
159	25 281	12.60952	39.87480	199	39 601	14.10674	44.60942
160	25 600	12.64911	40.00000	**200**	40 000	14.14214	44.72136
161	25 921	12.68858	40.12481	201	40 401	14.17745	44.83302
162	26 244	12.72792	40.24922	202	40 804	14.21267	44.94441
163	26 569	12.76715	40.37326	203	41 209	14.24781	45.05552
164	26 806	12.80625	40.49691	204	41 616	14.28286	45.16636
165	27 225	12.84523	40.62019	205	42 025	14.31782	45.27693
166	27 556	12.88410	40.74310	206	42 436	14.35270	45.38722
167	27 889	12.92285	40.86563	207	42 849	14.38749	45.49725
168	28 224	12.96148	40.98780	208	43 264	14.42221	45.60702
169	28 561	13.00000	41.10961	209	43 681	14.45683	45.71652
170	28 900	13.03840	41.23106	**210**	44 100	14.49138	45.82576
171	29 241	13.07670	41.35215	211	44 521	14.52584	45.93474
172	29 584	13.11488	41.47288	212	44 944	14.56022	46.04346
173	29 929	13.15295	41.59327	213	45 369	14.59452	46.15192
174	30 276	13.19091	41.71331	214	45 796	14.62874	46.26013
175	30 625	13.22876	41.83300	215	46 225	14.66288	46.36809
176	30 976	13.26650	41.95235	216	46 656	14.69694	46.47580
177	31 329	13.30413	42.07137	217	47 089	14.73092	46.58326
178	31 684	13.34166	42.19005	218	47 524	14.76482	46.69047
179	32 041	13.37909	42.30829	219	47 961	14.79865	46.79744

n	n^2	$\sqrt{n}$	$\sqrt{10n}$	n	n^2	$\sqrt{n}$	$\sqrt{10n}$
220	48 400	14.83240	46.90416	**260**	67 600	16.12452	50.99020
221	48 841	14.86607	47.01064	261	68 121	16.15549	51.08816
222	49 284	14.89966	47.11688	262	68 644	16.18641	51.18594
223	49 729	14.93318	47.22288	263	69 169	16.21727	51.28353
224	50 176	14.96663	47.32864	264	69 696	16.24808	51.38093
225	50 625	15.00000	47.43416	265	70 225	16.27882	51.47815
226	51 076	15.03330	47.53946	266	70 756	16.30951	51.57519
227	51 529	15.06652	47.64452	267	71 289	16.34013	51.67204
228	51 984	15.09967	47.74935	268	71 824	16.37071	51.76872
229	52 441	15.13275	47.85394	269	72 361	16.40122	51.86521
230	52 900	15.16575	47.95832	**270**	72 900	16.43168	51.96152
231	53 361	15.19868	48.06246	271	73 441	16.46208	52.05766
232	53 824	15.23155	48.16638	272	73 984	16.49242	52.15362
233	54 289	15.26434	48.27007	273	74 529	16.52271	52.24940
234	54 756	15.29706	48.37355	274	75 076	16.55295	52.34501
235	55 225	15.32971	48.47680	275	75 625	16.58312	52.44044
236	55 696	15.36229	48.57983	276	76 176	16.61235	52.53570
237	56 169	15.39480	48.68265	277	76 729	16.64332	52.63079
238	56 644	15.42725	48.78524	278	77 284	16.67333	52.72571
239	57 121	15.45962	48.88763	279	77 841	16.70329	52.82045
240	57 600	15.49193	48.98979	**280**	78 400	16.73320	52.91503
241	58 081	15.52417	49.09175	281	78 961	16.76305	53.00943
242	58 564	15.55635	49.19350	282	79 524	16.79286	53.10367
243	59 049	15.58846	49.29503	283	80 089	16.82260	53.19774
244	59 536	15.62050	49.39636	284	80 656	16.85230	53.29165
245	60 025	15.65248	49.49747	285	81 225	16.88194	53.38539
246	60 516	15.68439	49.59839	286	81 796	16.91153	53.47897
247	61 009	15.71623	49.69909	287	82 369	16.94107	53.57238
248	61 504	15.74902	49.79960	288	82 944	16.97056	53.66563
249	62 001	15.77973	49.89990	289	83 521	17.00000	53.75872
250	62 500	15.81139	50.00000	**290**	84 100	17.02939	53.85165
251	63 001	15.84298	50.09990	291	84 681	17.05872	53.94442
252	63 504	15.87451	50.19960	292	85 264	17.08801	54.03702
253	64 009	15.90597	50.29911	293	85 849	17.11724	54.12947
254	64 516	15.93738	50.39841	294	86 436	17.14643	54.22177
255	65 025	15.96872	50.49752	295	87 025	17.17556	54.31390
256	65 536	16.00000	50.59644	296	87 616	17.20465	54.40588
257	66 049	16.03122	50.69517	297	88 209	17.23369	54.49771
258	66 564	16.06238	50.79370	298	88 804	17.26268	54.58938
259	67 081	16.09348	50.89204	299	89 401	17.29162	54.68089

n	n^2	$\sqrt{n}$	$\sqrt{10n}$	n	n^2	$\sqrt{n}$	$\sqrt{10n}$
300	90 000	17.32051	54.77226	**340**	115 600	18.43909	58.30952
301	90 601	17.34935	54.86347	341	116 281	18.46619	58.39521
302	91 204	17.37815	54.95453	342	116 964	18.49324	58.48077
303	91 809	17.40690	55.04544	343	117 649	18.52026	58.56620
304	92 416	17.43560	55.13620	344	118 336	18.54724	58.65151
305	93 025	17.46425	55.22681	345	119 025	18.57418	58.73670
306	93 636	17.49286	55.31727	346	119 716	18.60108	58.82176
307	94 249	17.52142	55.40758	347	120 409	18.62794	58.90671
308	94 864	17.54993	55.49775	348	121 104	18.65476	58.99152
309	95 481	17.57840	55.58777	349	121 801	18.68154	59.07622
310	96 100	17.60682	55.67764	**350**	122 500	18.70829	59.16080
311	96 721	17.63519	55.76737	351	123 201	18.73499	59.24525
312	97 344	17.66352	55.85696	352	123 904	18.76166	59.32959
313	97 969	17.69181	55.94640	353	124 609	18.78829	59.41380
314	98 596	17.72005	56.03570	354	125 316	18.81489	59.49790
315	99 225	17.74824	56.12486	355	126 025	18.84144	59.58188
316	99 856	17.77639	56.21388	356	126 736	18.86796	59.66574
317	100 489	17.80449	56.30275	357	127 449	18.89444	59.74948
318	101 124	17.83255	56.39149	358	128 164	18.92089	59.83310
319	101 761	17.86057	56.48008	359	128 881	18.94730	59.91661
320	102 400	17.88854	56.56854	**360**	129 600	18.97367	60.00000
321	103 041	17.91647	56.65686	361	130 321	19.00000	60.08328
322	103 684	17.94436	56.74504	362	131 044	19.02630	60.16644
323	104 329	17.97220	56.83309	363	131 769	19.05256	60.24948
324	104 976	18.00000	56.92100	364	132 496	19.07878	60.33241
325	105 625	18.02776	57.00877	365	133 225	19.10497	60.41523
326	106 276	18.05547	57.09641	366	133 956	19.13113	60.49793
327	106 929	18.08314	57.18391	367	134 689	19.15724	60.58052
328	107 584	18.11077	57.27128	368	135 424	19.18333	60.66300
329	108 241	18.13836	57.35852	369	136 161	19.20937	60.74537
330	108 900	18.16590	57.44563	**370**	136 900	19.23538	60.82763
331	109 561	18.19341	57.53260	371	137 641	19.26136	60.90977
332	110 224	18.22087	57.61944	372	138 384	19.28730	60.99180
333	110 889	18.24829	57.70615	373	139 129	19.31321	61.07373
334	111 556	18.27567	57.79273	374	139 876	19.33908	61.15554
335	112 225	18.30301	57.87918	375	140 625	19.36492	61.23724
336	112 896	18.33030	57.96551	376	141 376	19.39072	61.31884
337	113 569	18.35756	58.05170	377	142 129	19.41649	61.40033
338	114 244	18.38478	58.13777	378	142 884	19.44222	61.48170
339	114 921	18.41195	58.22371	379	143 641	19.46792	61.56298

n	n^2	$\sqrt{n}$	$\sqrt{10n}$	n	n^2	$\sqrt{n}$	$\sqrt{10n}$
380	144 400	19.49359	61.64414	420	176 400	20.49390	64.80741
381	145 161	19.51922	61.72520	421	177 241	20.51828	64.88451
382	145 924	19.54482	61.80615	422	178 084	20.54264	64.96153
383	146 689	19.57039	61.88699	423	178 929	20.56696	65.03845
384	147 456	19.59592	61.96773	424	179 776	20.59126	65.11528
385	148 225	19.62142	62.04837	425	180 625	20.61553	65.19202
386	148 996	19.64688	62.12890	426	181 476	20.63977	65.26868
387	149 769	19.67232	62.20932	427	182 329	20.66398	65.34524
388	150 544	19.69772	62.28965	428	183 184	20.68816	65.42171
389	151 321	19.72308	62.36986	429	184 041	20.71232	65.49809
390	152 100	19.74842	62.44998	**430**	184 900	20.73644	65.57439
391	152 881	19.77372	62.52999	431	185 761	20.76054	65.65059
392	153 664	19.79899	62.60990	432	186 624	20.78461	65.72671
393	154 449	19.82423	62.68971	433	187 489	20.80865	65.80274
394	155 236	19.84943	62.76942	434	188 356	20.83267	65.87868
395	156 025	19.87461	62.84903	435	189 225	20.85665	65.95453
396	156 816	19.89975	62.92853	436	190 096	20.88061	66.03030
397	157 609	19.92486	63.00794	437	190 969	20.90454	66.10598
398	158 404	19.94994	63.08724	438	191 844	20.92845	66.18157
399	159 201	19.97498	63.16645	439	192 721	20.95233	66.25708
400	160 000	20.00000	63.24555	**440**	193 600	20.97618	66.33250
401	160 801	20.02498	63.32456	441	194 481	21.00000	66.40783
402	161 604	20.04994	63.40347	442	195 364	21.02380	66.48308
403	162 409	20.07486	63.48228	443	196 249	21.04757	66.55825
404	163 216	20.09975	63.56099	444	197 136	21.07131	66.63332
405	164 025	20.12461	63.63961	445	198 025	21.09502	66.70832
406	164 836	20.14944	63.71813	446	198 916	21.11871	66.78323
407	165 649	20.17424	63.79655	447	199 809	21.14237	66.85806
408	166 464	20.19901	63.87488	448	200 704	21.16601	66.93280
409	167 281	20.22375	63.95311	449	201 601	21.18962	67.00746
410	168 100	20.24864	64.03124	**450**	202 500	21.21320	67.08204
411	168 921	20.27313	64.10928	451	203 401	21.23676	67.15653
412	169 744	20.29778	64.18723	452	204 304	21.26029	67.23095
413	170 569	20.32240	54.26508	453	205 209	21.28380	67.30527
414	171 396	20.34699	64.34283	454	206 116	21.30728	67.37952
415	172 225	20.37155	64.42049	455	207 025	21.33073	67.45369
416	173 056	20.39608	64.49806	456	207 936	21.35416	67.52777
417	173 889	20.42058	64.57554	457	208 849	21.37756	67.60178
418	174 724	20.44505	64.65292	458	209 764	21.40093	67.67570
419	175 561	20.46949	64.73021	459	210 681	21.42429	67.74954

n	n^2	$\sqrt{n}$	$\sqrt{10n}$	n	n^2	$\sqrt{n}$	$\sqrt{10n}$
460	211 600	21.44761	67.82330	**500**	250 000	22.36068	70.71068
461	212 521	21.47091	67.89698	501	251 001	22.38303	70.78135
462	213 444	21.49419	67.97058	502	252 004	22.40536	70.85196
463	214 369	21.51743	68.04410	503	253 009	22.42766	70.92249
464	215 296	21.54066	68.11755	504	254 016	22.44994	70.99296
465	216 225	21.56386	68.19091	505	255 025	22.47221	71.06335
466	217 156	21.58703	68.26419	506	256 036	22.49444	71.13368
467	218 089	21.61018	68.33740	507	257 049	22.51666	71.20393
468	219 024	21.63331	68.41053	508	258 064	22.53886	71.27412
469	219 961	21.65641	68.48957	509	259 081	22.56103	71.34424
470	220 900	21.67948	68.55655	**510**	260 100	22.58318	71.41428
471	221 841	21.70253	68.62944	511	261 121	22.60531	71.48426
472	222 784	21.72556	68.70226	512	262 144	22.62742	71.55418
473	223 729	21.74856	68.77500	513	263 169	22.64950	71.62402
474	224 676	21.77154	68.84766	514	264 196	22.67157	71.69379
475	225 625	21.79449	68.92024	515	265 225	22.69361	71.76350
476	226 576	21.81742	68.99275	516	266 256	22.71563	71.83314
477	227 529	21.84033	69.06519	517	267 289	22.73763	71.90271
478	228 484	21.86321	69.13754	518	268 324	22.75961	71.97222
479	229 441	21.88607	69.20983	519	269 361	22.78157	72.04165
480	230 400	21.90890	69.28203	**520**	270 400	22.80351	72.11103
481	231 361	21.93171	69.35416	521	271 441	22.82542	72.18033
482	232 324	21.95450	69.42622	522	272 484	22.84732	72.24957
483	233 289	21.97726	69.49820	523	273 529	22.86919	72.31874
484	234 256	22.00000	69.57011	524	274 576	22.89105	72.38784
485	235 225	22.02272	69.64194	525	275 625	22.91288	72.45688
486	236 196	22.04541	69.71370	526	276 676	22.93469	72.52586
487	237 169	22.06808	69.78539	527	277 729	22.95648	72.59477
488	238 144	22.09072	69.85700	528	278 784	22.97825	72.66361
489	239 121	22.11334	69.92853	529	279 841	23.00000	72.73239
490	240 100	22.13594	70.00000	**530**	280 900	23.02173	72.80110
491	241 081	22.15852	70.07139	531	281 961	23.04344	72.86975
492	242 064	22.18107	70.14271	532	283 024	23.06513	72.93833
493	243 049	22.20360	70.21396	533	284 089	23.08679	73.00685
494	244 036	22.22611	70.28513	534	285 156	23.10844	73.07530
495	245 025	22.24860	70.35624	535	286 225	23.13007	73.14369
496	246 016	22.27106	70.42727	536	287 296	23.15167	73.21202
497	247 009	22.29350	70.49823	537	288 369	23.17326	73.28028
498	248 004	22.31591	70.56912	538	289 444	23.19483	73.34848
499	249 001	22.33831	70.63993	539	290 521	23.21637	73.41662

n	n^2	$\sqrt{n}$	$\sqrt{10n}$	n	n^2	$\sqrt{n}$	$\sqrt{10n}$
540	291 600	23.23790	73.48469	**580**	336 400	24.08319	76.15773
541	292 681	23.25941	73.55270	581	337 561	24.10394	76.22336
542	293 764	23.28089	73.62065	582	338 724	24.12468	76.28892
543	294 849	23.30236	73.68853	583	339 889	24.14539	76.35444
544	295 936	23.32381	73.75636	584	341 056	24.16609	76.41989
545	297 025	23.34524	73.82412	585	342 225	24.18677	76.48529
546	298 116	23.36664	73.89181	586	343 396	24.20744	76.55064
547	299 209	23.38803	73.95945	587	344 569	24.22808	76.61593
548	300 304	23.40940	74.02702	588	345 744	24.24871	76.68116
549	301 401	23.43075	74.09453	589	346 921	24.26932	76.74634
550	302 500	23.45208	74.16198	**590**	348 100	24.28992	76.81146
551	303 601	23.47339	74.22937	591	349 281	24.31049	76.87652
552	304 704	23.49468	74.29670	592	350 464	24.33105	76.94154
553	305 809	23.51595	74.36397	593	351 649	24.35159	77.00649
554	306 916	23.53720	74.43118	594	352 836	24.37212	77.07140
555	308 025	23.55844	74.49832	595	354 025	24.39262	77.13624
556	309 136	23.57965	74.56541	596	355 216	24.41311	77.20104
557	310 249	23.60085	74.63243	597	356 409	24.43358	77.26578
558	311 364	23.62202	74.69940	598	357 604	24.45404	77.33046
559	312 481	23.64318	74.76630	599	358 801	24.47448	77.39509
560	313 600	23.66432	74.83315	**600**	360 000	24.49490	77.45967
561	314 721	23.68544	74.89993	601	361 201	24.51530	77.52419
562	315 844	23.70654	74.96666	602	362 404	24.53569	77.58866
563	316 969	23.72762	75.03333	603	363 609	24.55606	77.65307
564	318 096	23.74868	75.09993	604	364 816	24.57641	77.71744
565	319 225	23.76973	75.16648	605	366 025	24.59675	77.78175
566	320 356	23.79075	75.23297	606	367 236	24.61707	77.84600
567	321 489	23.81176	75.29940	607	368 449	24.63737	77.91020
568	322 624	23.83275	75.36577	608	369 664	24.65766	77.97435
569	323 761	23.85372	75.43209	609	370 881	24.67793	78.03845
570	324 900	23.87467	75.49834	**610**	372 100	24.69818	78.10250
571	326 041	23.89561	75.56454	611	373 321	24.71841	78.16649
572	327 184	23.91652	75.63068	612	374 544	24.73863	78.23043
573	328 329	23.93742	75.69676	613	375 769	24.75884	78.29432
574	329 476	23.95830	75.76279	614	376 996	24.77902	78.35815
575	330 625	23.97916	75.82875	615	378 225	24.79919	78.42194
576	331 776	24.00000	75.89466	616	379 456	24.81935	78.48567
577	332 929	24.02082	75.96052	617	380 689	24.83948	78.54935
578	334 084	24.04163	76.02631	618	381 924	24.85961	78.61298
579	335 241	24.06242	76.09205	619	383 161	24.87971	78.67655

n	n^2	$\sqrt{n}$	$\sqrt{10n}$	n	n^2	$\sqrt{n}$	$\sqrt{10n}$
620	384 400	24.89980	78.74008	**660**	435 600	25.69047	81.24038
621	385 641	24.91987	78.80355	661	436 921	25.70992	81.30191
622	386 884	24.93993	78.86698	662	438 244	25.72936	81.36338
623	388 129	24.95997	78.93035	663	439 569	25.74879	81.42481
624	389 376	24.97999	78.99367	664	440 896	25.76820	81.48620
625	390 625	25.00000	79.05694	665	442 225	25.78759	81.54753
626	391 876	25.01999	79.12016	666	443 556	25.80698	81.60882
627	393 129	25.03997	79.18333	667	444 889	25.82634	81.67007
628	394 384	25.05993	79.24645	668	446 224	25.84570	81.73127
629	395 641	25.07987	79.30952	669	447 561	25.86503	81.79242
630	396 900	25.09980	79.37254	**670**	448 900	25.88436	81.85353
631	398 161	25.11971	79.43551	671	450 241	25.90367	81.91459
632	399 424	25.13961	79.49843	672	451 584	25.92296	81.97561
633	400 689	25.15949	79.56130	673	452 929	25.94224	82.03658
634	401 956	25.17936	79.62412	674	454 276	25.96151	82.09750
635	403 225	25.19921	79.68689	675	455 625	25.98076	82.15838
636	404 496	25.21904	79.74961	676	456 976	26.00000	82.21922
637	405 769	25.23886	79.81228	677	458 329	26.01922	82.28001
638	407 044	25.25866	79.87490	678	459 684	26.03843	82.34076
639	408 321	25.27845	79.93748	679	461 041	26.05763	82.40146
640	409 600	25.29822	80.00000	**680**	462 400	26.07681	82.46211
641	410 881	25.31798	80.06248	681	463 761	26.09598	82.52272
642	412 164	25.33772	80.12490	682	465 124	26.11513	82.58329
643	413 449	25.35744	80.18728	683	466 489	26.13427	82.64381
644	414 736	25.37716	80.24961	684	467 856	26.15339	82.70429
645	416 025	25.39685	80.31189	685	469 225	26.17250	82.76473
646	417 316	25.41653	80.37413	686	470 596	26.19160	82.82512
647	418 609	25.43619	80.43631	687	471 969	26.21068	82.88546
648	419 904	25.45584	80.49845	688	473 344	26.22975	82.94577
649	421 201	25.47548	80.56054	689	474 721	26.24881	83.00602
650	422 500	25.49510	80.62258	**690**	476 100	26.26785	83.06624
651	423 801	25.51470	80.68457	691	477 481	26.28688	83.12641
652	425 104	25.53429	80.74652	692	478 864	26.30589	83.18654
653	426 409	25.55386	80.80842	693	480 249	26.32489	83.24662
654	427 716	25.57342	80.87027	694	481 636	26.34388	83.30666
655	429 025	25.59297	80.93207	695	483 025	26.36285	83.36666
656	430 336	25.61250	80.99383	696	484 416	26.38181	83.42661
657	431 649	25.63201	81.05554	697	485 809	26.40076	83.48653
658	432 964	25.65151	81.11720	698	487 204	26.41969	83.54639
659	434 281	25.67100	81.17881	699	488 601	26.43861	83.60622

n	n^2	$\sqrt{n}$	$\sqrt{10n}$	n	n^2	$\sqrt{n}$	$\sqrt{10n}$
700	490 000	26.45751	83.66600	**740**	547 600	27.20294	86.02325
701	491 401	26.47640	83.72574	741	549 081	27.22132	86.08136
702	492 804	26.49528	83.78544	742	550 564	27.23968	86.13942
703	494 209	26.51415	83.84510	743	552 049	27.25803	86.19745
704	495 616	26.53300	83.90471	744	553 536	27.27636	86.25543
705	497 025	26.55184	83.96428	745	555 025	27.29469	86.31338
706	498 436	26.57066	84.02381	746	556 516	27.31300	86.37129
707	499 849	26.58947	84.08329	747	558 009	27.33130	86.42916
708	501 264	26.60827	84.14274	748	559 504	27.34959	86.48699
709	502 681	26.62705	84.20214	749	561 001	27.36786	86.54479
710	504 100	26.64583	84.26150	**750**	562 500	27.38613	86.60254
711	505 521	26.66458	84.32082	751	564 001	27.40438	86.66026
712	506 944	26.68333	84.38009	752	565 504	27.42262	86.71793
713	508 369	26.70206	84.43933	753	567 009	27.44085	86.77557
714	509 796	26.72078	84.49852	754	568 516	27.45906	86.83317
715	511 225	26.73948	84.55767	755	570 025	27.47726	86.89074
716	512 656	26.75818	84.61678	756	571 536	27.49545	86.94826
717	514 089	26.77686	84.67585	757	573 049	27.51363	87.00575
718	515 524	26.79552	84.73488	758	574 564	27.53180	87.06320
719	516 961	26.81418	84.79387	759	576 081	27.54995	87.12061
720	518 400	26.83282	84.85281	**760**	577 600	27.56810	87.17798
721	519 841	26.85144	84.91172	761	579 121	27.58623	87.23531
722	521 284	26.87006	84.97058	762	580 644	27.60435	87.29261
723	522 729	26.88866	85.02941	763	582 169	27.62245	87.34987
724	524 176	26.90725	85.08819	764	583 696	27.64055	87.40709
725	525 625	26.92582	85.14693	765	585 225	27.65863	87.46428
726	527 076	26.94439	85.20563	766	586 756	27.67671	87.52143
727	528 529	26.96294	85.26429	767	588 289	27.69476	87.57854
728	529 984	26.98148	85.32292	768	589 824	27.71281	87.63561
729	531 441	27.00000	85.38150	769	591 361	27.73085	87.69265
730	532 900	27.01851	85.44004	**770**	592 900	27.74887	87.74964
731	534 361	27.03701	85.49854	771	594 441	27.76689	87.80661
732	535 824	27.05550	85.55700	772	595 984	27.78489	87.86353
733	537 289	27.07397	85.61542	773	597 529	27.80288	87.92042
734	538 756	27.09243	85.67380	774	599 076	27.82086	87.97727
735	540 225	27.11088	85.73214	775	600 625	27.83882	88.03408
736	541 696	27.12932	85.79044	776	602 176	27.85678	88.09086
737	543 169	27.14774	85.84870	777	603 729	27.87472	88.14760
738	544 644	27.16616	85.90693	778	605 284	27.89265	88.20431
739	546 121	27.18455	85.96511	779	606 841	27.91057	88.26098

n	n^2	$\sqrt{n}$	$\sqrt{10n}$	n	n^2	$\sqrt{n}$	$\sqrt{10n}$
780	608 400	27.92848	88.31761	**820**	672 400	28.63564	90.55385
781	609 961	27.94638	88.37420	821	674 041	28.65310	90.60905
782	611 524	27.96426	88.43076	822	675 684	28.67054	90.66422
783	613 089	27.98214	88.48729	823	677 329	28.68798	90.71935
784	614 656	28.00000	88.54377	824	678 976	28.70540	90.77445
785	616 225	28.01785	88.60023	825	680 625	28.72281	90.82951
786	617 796	28.03569	88.65664	826	682 726	28.74022	90.88454
787	619 369	28.05352	88.71302	827	683 929	28.75761	90.93954
788	620 944	28.07134	88.76936	828	685 584	28.77499	90.99451
789	622 521	28.08914	88.82567	829	687 241	28.79236	91.04944
790	624 100	28.10694	88.88194	**830**	688 900	28.80972	91.10434
791	625 681	28.12472	88.93818	831	690 561	28.82707	91.15920
792	627 264	28.14249	88.99428	832	692 224	28.84441	91.21403
793	628 849	28.16026	89.05055	833	693 889	28.86174	91.26883
794	630 436	28.17801	89.10668	834	695 556	28.87906	91.32360
795	632 025	28.19574	89.16277	835	697 225	28.89637	91.37833
796	633 616	28.21347	89.21883	836	698 896	28.91366	91.43304
797	635 209	28.23119	89.27486	837	700 569	28.93095	91.48770
798	636 804	28.24889	89.33085	838	702 244	28.94823	91.54234
799	638 401	28.26659	89.38680	839	703 921	28.96550	91.59694
800	640 000	28.28472	89.44272	**840**	705 600	28.98275	91.65151
801	641 601	28.30194	89.49860	841	707 281	29.00000	91.70605
802	643 204	28.31960	89.55445	842	708 964	29.01724	91.76056
803	644 809	28.33725	89.61027	843	710 649	29.03446	91.81503
804	646 416	28.35489	89.66605	844	712 336	29.05168	91.86947
805	648 025	28.37252	89.72179	845	714 025	29.06888	91.92388
806	649 636	28.39014	89.77750	846	715 716	29.08608	91.97826
807	651 249	28.40775	89.83318	847	717 409	29.10326	92.03260
808	652 864	28.42534	89.88882	848	719 104	29.12044	92.08692
809	654 481	28.44293	89.94443	849	720 801	29.13760	92.14120
810	656 100	28.46050	90.00000	**850**	722 500	29.15476	92.19544
811	657 721	28.47806	90.05554	851	724 201	29.17190	92.24966
812	659 344	28.49561	90.11104	852	725 904	29.18904	92.30385
813	660 969	28.51315	90.16651	853	727 609	29.20616	92.35800
814	662 596	28.53069	90.22195	854	729 316	29.22328	92.41212
815	664 225	28.54820	90.27735	855	731 025	29.24038	92.46621
816	665 856	28.56571	90.33272	856	732 736	29.25748	92.52027
817	667 489	28.58321	90.38805	857	734 449	29.27456	92.57429
818	669 124	28.60070	90.44335	858	736 164	29.29164	92.62829
819	670 761	28.61818	90.49862	859	737 881	29.30870	92.68225

n	n^2	$\sqrt{n}$	$\sqrt{10n}$	n	n^2	$\sqrt{n}$	$\sqrt{10n}$
860	739 600	29.32576	92.73618	**900**	810 000	30.00000	94.86833
861	741 321	29.34280	92.79009	901	811 801	30.01666	94.92102
862	743 044	29.35984	92.84396	902	813 604	30.03331	94.97368
863	744 769	29.37686	92.89779	903	815 409	30.04996	95.02631
864	746 496	29.39388	92.95160	904	817 216	30.06659	95.07891
865	748 225	29.41088	93.00538	905	819 025	30.08322	95.13149
866	749 956	29.42788	93.05912	906	820 836	30.09983	95.18403
867	751 689	29.44486	93.11283	907	822 649	30.11644	95.23655
868	753 424	29.46184	93.16652	908	824 464	30.13304	95.28903
869	755 161	29.47881	93.22017	909	826 281	30.14963	95.34149
870	756 900	29.49576	93.27379	**910**	828 100	30.16621	95.39392
871	758 641	29.51271	93.32738	911	829 921	30.18278	95.44632
872	760 384	29.52965	93.38094	912	831 744	30.19934	95.49869
873	762 129	29.54657	93.43447	913	833 569	30.21589	95.55103
874	763 876	29.56349	93.48797	914	835 396	30.23243	95.60335
875	765 625	29.58040	93.54143	915	837 225	30.24897	95.65563
876	767 376	29.59730	93.59487	916	839 056	30.26549	95.70789
877	769 129	29.61419	93.64828	917	840 889	30.28201	95.76012
878	770 884	29.63106	93.70165	918	842 724	30.29851	95.81232
879	772 641	29.64793	93.75500	919	844 561	30.31501	95.86449
880	774 400	29.66479	93.80832	**920**	846 400	30.33150	95.91663
881	776 161	29.68164	93.86160	921	848 241	30.34798	95.96874
882	777 924	29.69848	93.91486	922	850 084	30.36445	96.02083
883	779 689	29.71532	93.96808	923	851 929	30.38092	96.07289
884	781 456	29.73214	94.02127	924	853 776	30.39737	96.12492
885	783 225	29.74895	94.07444	925	855 625	30.41381	96.17692
886	784 996	29.76575	94.12757	926	857 476	30.43025	96.22889
887	786 769	29.78255	94.18068	927	859 329	30.44667	96.28084
888	788 544	29.79933	94.23375	928	861 184	30.46309	96.33276
889	790 321	29.81610	94.28680	929	863 041	30.47950	96.38465
890	792 100	29.83287	94.33981	**930**	864 900	30.49590	96.43651
891	793 881	29.84962	94.39280	931	866 761	30.51229	96.48834
892	795 664	29.86637	94.44575	932	868 624	30.52868	96.54015
893	797 449	29.88311	94.49868	933	870 489	30.54505	96.59193
894	799 236	29.89983	94.55157	934	872 356	30.56141	96.64368
895	801 025	29.91655	94.60444	935	874 225	30.57777	96.69540
896	802 816	29.93326	94.65728	936	876 096	30.59412	96.74709
897	804 609	29.94996	94.71008	937	877 969	30.61046	96.79876
898	806 404	29.96665	94.76286	938	879 844	30.62679	96.85040
899	808 201	29.98333	94.81561	939	881 721	30.64311	96.90201

n	n^2	$\sqrt{n}$	$\sqrt{10n}$	n	n^2	$\sqrt{n}$	$\sqrt{10n}$
940	883 600	30.65942	96.95360	**970**	940 900	31.14482	98.48858
941	885 481	30.67572	97.00515	971	942 841	31.16087	98.53933
942	887 364	30.69202	97.05668	972	944 784	31.17691	98.59006
943	889 249	30.70831	97.10819	973	946 729	31.19295	98.64076
944	891 136	30.72458	97.15966	974	948 676	31.20897	98.69144
945	893 025	30.74085	97.21111	975	950 625	31.22499	98.74209
946	894 916	30.75711	97.26253	976	952 576	31.24100	98.79271
947	896 809	30.77337	97.31393	977	954 529	31.25700	98.84331
948	898 704	30.78961	97.36529	978	956 484	31.27299	98.89388
949	900 601	30.80584	97.41663	979	958 441	31.28898	98.94443
950	902 500	30.82207	97.46794	**980**	960 400	31.30495	98.99495
951	904 401	30.83829	97.51923	981	962 361	31.32092	99.04544
952	906 304	30.85450	97.57049	982	964 324	31.33688	99.09591
953	908 209	30.87070	97.62172	983	966 289	31.35283	99.14636
954	910 116	30.88689	97.67292	984	968 256	31.36877	99.19677
955	912 025	30.90307	97.72410	985	970 225	31.38471	99.24717
956	913 936	30.91925	97.77525	986	972 196	31.40064	99.29753
957	915 849	30.93542	97.82638	987	974 169	31.41656	99.34787
958	917 764	30.95158	97.87747	988	976 144	31.43247	99.39819
959	919 681	30.96773	97.92855	989	978 121	31.44837	99.44848
960	921 600	30.98387	97.97959	**990**	980 100	31.46427	99.49874
961	923 521	31.00000	98.03061	991	982 081	31.48015	99.54898
962	925 444	31.01612	98.08160	992	984 064	31.49603	99.59920
963	927 369	31.03224	98.13256	993	986 049	31.51190	99.64939
964	929 296	31.04835	98.18350	994	988 036	31.52777	99.69955
965	931 225	31.06445	98.23441	995	990 025	31.54362	99.74969
966	933 156	31.08054	98.28530	996	992 016	31.55947	99.79980
967	935 089	31.09662	98.33616	997	994 009	31.57531	99.84989
968	937 024	31.11270	98.38699	998	996 004	31.59114	99.89995
969	938 961	31.12876	98.43780	999	998 001	31.60696	99.94999
				1000	1000 000	31.62278	100.00000

Abridged from *Handbook of Tables for Probability and Statistics*, 2nd edition, edited by William H. Beyer (Cleveland: The Chemical Rubber Company, 1968). Reproduced by permission of the publishers, The Chemical Rubber Company.

Table 13 Random Numbers

Line/Col.	(1)	(2)	(3)	(4)	(5)	(6)	(7)	(8)	(9)	(10)	(11)	(12)	(13)	(14)
1	10480	15011	01536	02011	81647	91646	69179	14194	62590	36207	20969	99570	91291	90700
2	22368	46573	25595	85393	30995	89198	27982	53402	93965	34095	52666	19174	39615	99505
3	24130	48360	22527	97265	76393	64809	15179	24830	49340	32081	30680	19655	63348	58629
4	42167	93003	06243	61680	07856	16376	39440	53537	71341	57004	00849	74917	97758	16379
5	37570	39975	81837	16656	06121	91782	60468	81305	49684	60672	14110	06927	01263	54613
6	77921	06907	11008	42751	27756	53498	18602	70659	90655	15053	21916	81825	44394	42880
7	99562	72905	56420	69994	98872	31016	71194	18738	44013	48840	63213	21069	10634	12952
8	96301	91977	05463	07972	18876	20922	94595	56869	69014	60045	18425	84903	42508	32307
9	89579	14342	63661	10281	17453	18103	57740	84378	25331	12566	58678	44947	05585	56941
10	85475	36857	53342	53988	53060	59533	38867	62300	08158	17983	16439	11458	18593	64952
11	28918	69578	88231	33276	70997	79936	56865	05859	90106	31595	01547	85590	91610	78188
12	63553	40961	48235	03427	49626	69445	18663	72695	52180	20847	12234	90511	33703	90322
13	09429	93969	52636	92737	88974	33488	36320	17617	30015	08272	84115	27156	30613	74952
14	10365	61129	87529	85689	48237	52267	67689	93394	01511	26358	85104	20285	29975	89868
15	07119	97336	71048	08178	77233	13916	47564	81056	97735	85977	29372	74461	28551	90707
16	51085	12765	51821	51259	77452	16308	60756	92144	49442	53900	70960	63990	75601	40719
17	02368	21382	52404	60268	89368	19885	55322	44819	01188	65255	64835	44919	05944	55157
18	01011	54092	33362	94904	31273	04146	18594	29852	71585	85030	51132	01915	92747	64951
19	52162	53916	46369	58586	23216	14513	83149	98736	23495	64350	94738	17752	35156	35749
20	07056	97628	33787	09998	42698	06691	76988	13602	51851	46104	88916	19509	25625	58104
21	48663	91245	85828	14346	09172	30168	90229	04734	59193	22178	30421	61666	99904	32812
22	54164	58492	22421	74103	47070	25306	76468	26384	58151	06646	21524	15227	96909	44592
23	32639	32363	05597	24200	13363	38005	94342	28728	35806	06912	17012	64161	18296	22851
24	29334	27001	87637	87308	58731	00256	45834	15398	46557	41135	10367	07684	36188	18510
25	02488	33062	28834	07351	19731	92420	60952	61280	50001	67658	32586	86679	50720	94953

Abridged from *Handbook of Tables for Probability and Statistics*, 2nd edition, edited by William H. Beyer (Cleveland: The Chemical Rubber Company, 1968). Reproduced by permission of the publishers, The Chemical Rubber Company.

26	81525	72295	04839	96423	24878	82651	66566	14778	76797	14780	13300	87074	79666	95725
27	29676	20591	68086	26432	46901	20849	89768	81536	86645	12659	92259	57102	80428	25280
28	00742	57392	39064	66432	84673	40027	32832	61362	98947	96067	64760	64584	96096	98253
29	05366	04213	25669	26422	44407	44048	37937	63904	45766	66134	75470	66520	34693	90449
30	91921	26418	64117	94305	26766	25940	39972	22209	71500	64568	91402	42416	07844	69618
31	00582	04711	87917	77341	42206	35126	74087	99547	81817	42607	43808	76655	62028	76630
32	00725	69884	62797	56170	86324	88072	76222	36086	84637	93161	76038	65855	77919	88006
33	69011	65795	95876	55293	18988	27354	26575	08625	40801	59920	29841	80150	12777	48501
34	25976	57948	29888	88604	67917	48708	18912	82271	65424	69774	33611	54262	85963	03547
35	09763	83473	73577	12908	30883	18317	28290	35797	05998	41688	34952	37888	38917	88050
36	91567	42595	27958	30134	04024	86385	29880	99730	55536	84855	29080	09250	79656	73211
37	17955	56349	90999	49127	20044	59931	06115	20542	18059	02008	73708	83517	36103	42791
38	46503	18584	18845	49618	02304	51038	20655	58727	28168	15475	56942	53389	20562	87338
39	92157	89634	94824	78171	84610	82834	09922	25417	44137	48413	25555	21246	35509	20468
40	14577	62765	35605	81263	39667	47358	56873	56307	61607	49518	89656	20103	77490	18062
41	98427	07523	33362	64270	01638	92477	66969	98420	04880	45585	46565	04102	46880	45709
42	34914	63976	88720	82765	34476	17032	87589	40836	32427	70002	70663	88863	77775	69348
43	70060	28277	39475	46473	23219	53416	94970	25832	69975	94884	19661	72828	00102	66794
44	53976	54914	06990	67245	68350	82948	11398	42878	80287	88267	47363	46634	06541	97809
45	76072	29515	40980	07391	58745	25774	22987	80059	39911	96189	41151	14222	60697	59583
46	90725	52210	83974	29992	65831	38857	50490	83765	55657	14361	31720	57375	56228	41546
47	64364	67412	33339	31926	14883	24413	59744	92351	97473	89286	35931	04110	23726	51900
48	08962	00358	31662	25388	61642	34072	81249	35648	56891	69352	48373	45578	78547	81788
49	95012	68379	93526	70765	10592	04542	76463	54328	02349	17247	28865	14777	62730	92277
50	15664	10493	20492	38391	91132	21999	59516	81652	27195	48223	46751	22923	32261	85653
51	16408	81899	04153	53381	79401	21438	83035	92350	36693	31238	59649	91754	72772	02338
52	18629	81953	05520	91962	04739	13092	97662	24822	94730	06496	35090	04822	86774	98289
53	73115	35101	47498	87637	99016	71060	88824	71013	18735	20286	23153	72924	35165	43040
54	57491	16703	23167	49323	45021	33132	12544	41035	80780	45393	44812	12515	98931	91202
55	30405	83946	23792	14422	15059	45799	22716	19792	09983	74353	68668	30429	70735	25499
56	16631	35006	85900	98275	32388	52390	16815	69298	82732	38480	73817	32523	41961	44437
57	96773	20206	42559	78985	05300	22164	24369	54224	35083	19687	11052	91491	60383	19746
58	38935	64202	14349	82674	66523	44133	00697	35552	35970	19124	63318	29686	03387	59846
59	31624	76384	17403	53363	44167	64486	64758	75366	76554	31601	12614	33072	60332	92325
60	78919	19474	23632	27889	47914	02584	37680	20801	72152	39339	34806	08930	85001	87820
61	03931	33309	57047	74211	63445	17361	62825	39908	05607	91284	68833	25570	38818	46920
62	74426	33278	43972	10119	89917	15665	52872	73823	73144	88662	88970	74492	51805	99378
63	09066	00903	20795	95452	92648	45454	09552	88815	16553	51125	79375	97596	16296	66092
64	42238	12426	87025	14267	20979	04508	64535	31355	86064	29472	47689	05974	52468	16834
65	16153	08002	26504	41744	81959	65642	74240	56302	00033	67107	77510	70625	28725	34191

Line/Col.	(1)	(2)	(3)	(4)	(5)	(6)	(7)	(8)	(9)	(10)	(11)	(12)	(13)	(14)
66	21457	40742	29820	96783	29400	21840	15035	34537	33310	06116	95240	15957	16572	06004
67	21581	57802	02050	89728	17937	37621	47075	42080	97403	48626	68995	43805	33386	21597
68	55612	78095	83197	33732	05810	24813	86902	60397	16489	03264	88525	42786	05269	92532
69	44657	66999	99324	51281	84463	60563	79312	93454	68876	25471	93911	25650	12682	73572
70	91340	84979	46949	81973	37949	61023	43997	15263	80644	43942	89203	71795	99533	50501
71	91227	21199	31935	27022	84067	05462	35216	14486	29891	68607	41867	14951	91696	85065
72	50001	38140	66321	19924	72163	09538	12151	06878	91903	18749	34405	56087	82790	70925
73	65390	05224	72958	28609	81406	39147	25549	48542	42627	45233	57202	94617	23772	07896
74	27504	96131	83944	41575	10573	08619	64482	73923	36152	05184	94142	25299	84387	34925
75	37169	94851	39117	89632	00959	16487	65536	49071	39782	17095	02330	74301	00275	48280
76	11508	70225	51111	38351	19444	66499	71945	05422	13442	78675	84081	66938	93654	59894
77	37449	30362	06694	54690	04052	53115	62757	95348	78662	11163	81651	50245	34971	52924
78	46515	70331	85922	38329	57015	15765	97161	17869	45349	61796	66345	81073	49106	79860
79	30986	81223	42416	58353	21532	30502	32305	86482	05174	07901	54339	58861	74818	46942
80	63798	64995	46583	09785	44160	78128	83991	42865	92520	83531	80377	35909	81250	54238
81	82486	84846	99254	67632	43218	50076	21361	64816	51202	88124	41870	52689	51275	83556
82	21885	32906	92431	09060	64297	51674	64126	62570	26123	05155	59194	52799	28225	85762
83	60336	98782	07408	53458	13564	59089	26445	29789	85205	41001	12535	12133	14645	23541
84	43937	46891	24010	25560	86355	33941	25786	54990	71899	15475	95434	98227	21824	19585
85	97656	63175	89303	16275	07100	92063	21942	18611	47348	20203	18534	03862	78095	50136
86	03299	01221	05418	38982	55758	92237	26759	86367	21216	98442	08303	56613	91511	75928
87	79626	06486	03574	17668	07785	76020	79924	25651	83325	88428	85076	72811	22717	50585
88	85636	68335	47539	03129	65651	11977	02510	26113	99447	68645	34327	15152	55230	93448
89	18039	14367	61337	06177	12143	46609	32989	74014	64708	00533	35398	58408	13261	47908
90	08362	15656	60627	36478	65648	16764	53412	09013	07832	41574	17639	82163	60859	75567
91	79556	29068	04142	16268	15387	12856	66227	38358	22478	73373	88732	09443	82558	05250
92	92608	82674	27072	32534	17075	27698	98204	63863	11951	34648	88022	56148	34925	57031
93	23982	25835	40055	67006	12293	02753	14827	23235	35071	99704	37543	11601	35503	85171
94	09915	96306	05908	97901	28395	14186	00821	80703	70426	75647	76310	88717	37890	40129
95	59037	33300	26695	62247	69927	76123	50842	43834	86654	70959	79725	93872	28117	19233
96	42488	78077	69882	61657	34136	79180	97526	43092	04098	73571	80799	76536	71255	64239
97	46764	86273	63003	93017	31204	36692	40202	35275	57306	55543	53203	18098	47625	88684
98	03237	45430	55417	63282	90816	17349	88298	90183	36600	78406	06216	95787	42579	90730
99	86591	81482	52667	61582	14972	90053	89534	76036	49199	43716	97548	04379	46370	28672
100	38534	01715	94964	87288	65680	43772	39560	12918	86537	62738	19636	51132	25739	56947

Answers to Exercises

Chapter 1

1.4	$s' = 1.1$
1.5 (b)	$s' \approx 152$
1.6 (b)	$\bar{y} = 213.75$,
	$s' = 159.15$
1.7 (a)	$s \approx 2$
(c)	$\bar{y} = 6.8$,
	$s' = 2.04$
1.9	231, 323
1.10	.05

Chapter 2

2.1 (c)	$P(A) = 1/4$
2.2 (c)	$P(A) = 2/7$
2.3 (c)	$P(A) = 2/9$
2.4 (d)	$P(A) = 1/8$
2.5	42
2.6	294
2.7 (a)	36 (b) 1/6
2.9 (b)	16
2.10	120
2.11 (a)	40 (b) 320
2.12	$9(10)^6$
2.13	6
2.14	720

2.15 18
2.16 18!
2.17 16
2.18 (a) 21 (b) 6
2.19 $48/\binom{52}{2}$
2.20 $4\binom{13}{5}/\binom{52}{5}$
2.21 $P(A) = 1/323$
2.22 (a) .729
(b) .999
(c) .972
2.23 1/9
2.24 (a) 1/8 (b) 1/64
2.25 1/16
2.26 .5952
2.27 (a) 1/4 (b) 1/16
2.28 (a) $\binom{6}{3}(1/2)^6$
(b) $27(1/2)^{10}$
2.29 (a) 1/66 (b) 1/33
2.30 2/11
2.31 .149
2.32 (a) $(1/4)^7$
(b) $36(1/4)^4$
2.33 No
2.34 .8704
2.35 1/16
2.36 (a) 1/5
(b) 2/5
(c) 1/2
2.39 No
2.40 1/2
2.41 1/7

Chapter 3

3.1 .4752
3.2 .992
3.5 .32805, .99999

3.6 (a) .9606 (b) .9994

3.7 (a) 1.0

(b) .5905

(c) .1681

(d) .0312

(e) 0

3.9 (a) $n = 25,\ a = 5$

(b) $n = 25,\ a = 5$

3.10 (a) .151

(b) .302

3.11 (a) .081

(b) .81

3.12 .01536, .0256

3.13 $p(0) = .2,\ p(1) = .6,\ p(2) = .2$

3.15 $1 - (8/3)e^{-1}$

3.16 $1 - [(8/3)e^{-1}]^{10}$

3.17 $1 - (.99999)^{10{,}000}$

3.18 (a) $E(Y) = 2,\ V(Y) = 1.2$

3.19 $E(Y) = 1,\ V(Y) = .4$

3.20 (a) $p(0) = 1/8,\ p(1) = 3/8,$
$p(2) = 3/8,\ p(3) = 1/8$

(c) $E(Y) = 3/2,\ V(Y) = 3/4$

3.21 (a) $p(0) = .729,\ p(1) = .243$
$p(2) = .027,\ p(3) = .001$

(c) $E(Y) = .3,\ V(Y) = .27$

3.22 $E(Y) = 100,\ V(Y) = 90$

3.23

	Binomial	*Poisson*
$p(0)$	.358	.368
$p(1)$	.378	.368
$p(2)$	.189	.184
$p(3)$	.059	.061
$p(4)$	.013	.015

3.24 $1 - (.97)^{100}$

3.25 $E(Y) = 3,\ V(Y) = 2.91,$ no

3.26 $E(Y) = 1.625$
$V(Y) = .734$

3.27 $E(Y) = 1$

3.28 $V(Y) = \sigma^2 = .4$

3.29 $E(Y) = 3.5, \quad V(Y) = 2.917$

3.31 $E(Y) = \$8333, \quad \sigma = \$19{,}508$

3.32 $120

3.33 (a) $p(0) = (2/3)^4, \quad p(1) = 2(2/3)^4,$
$p(2) = (2/3)^3, \quad p(3) = 1/3(2/3)^3,$
$p(4) = (1/3)^4$

(b) 1/9

(c) 4/3

(d) 8/9

3.35 $[tp + (1 - p)]^n$

3.36 $3(.1)\,(.9)^{10}$

3.37 $\dfrac{pe^t}{1 - qe^t}, \quad V(Y) = \dfrac{q}{p^2}$

3.38 $V(Y) = \lambda$

3.43 .0837

3.45 (a) .119

(b) .117

3.46 .9665

3.47 $149.09

3.48 3

3.49 (a) $n[1 + k(1 - .95^k)]$

(b) 5

(c) Approximately $.57N$

Chapter 4

4.1 (a) $c = 1/2$

(b) $y^2/4, \quad 0 \leqslant y \leqslant 2$

(d) 3/4

(e) 3/4

4.2 (a) $c = 3/2$

(b) $(y^3 + y^2)/2, \quad 0 \leqslant y \leqslant 1$

(d) 0, 0, 1

(e) 3/16

4.3 (a) $c = 1.2$

(b) $.2(y + 1), \quad -1 \leqslant y \leqslant 0$
$.6y^2 + .2y + .2, \quad 0 < y \leqslant 1$

(d)	0, .2, 1
(e)	.25
4.4 (a)	.3849
(b)	.3159
4.5 (a)	.4279
(b)	.1628
4.6 (a)	.3227
(b)	.1586 (c) .3613
4.7	.7734
4.8	.9115
4.9	0
4.10	1.1
4.11	.2268
4.12	1.645
4.13	2.575
4.14	.1596
4.15	.2266
4.16	.2660
4.17	$(.2266)^3$
4.18	.0401
4.19	85.36
4.20	7.301
4.21	383.65
4.22	.073
4.23	$1 - (.927)^5$
4.24	$E(Y) = 4/3,\quad V(Y) = 2/9$
4.25	$E(Y) = .708,\quad V(Y) = .049$
4.26	$E(Y) = .4,\quad V(Y) = .2733$
4.27	$\dfrac{e^{t\theta_2} - e^{t\theta_1}}{t(\theta_2 - \theta_1)},\quad \dfrac{\theta_1 + \theta_2}{2},\quad \dfrac{\theta_1^2 + \theta_1\theta_2 + \theta_2^2}{3}$
4.28 (a)	1, 1/2
(b)	$(1 - t/2)^{-2}$
(c)	$c = 4$
4.29	$\mu_1 = 0,\quad \mu_2 = \sigma^2,\quad \mu_3 = 0,\quad \mu_4 = 3\sigma^4$
4.30	$b,\quad b^2,\quad (1 - bt)^{-1}$
4.31	$E(X) = \dfrac{\alpha}{\alpha + \beta},\quad V(X) = \dfrac{\alpha\beta}{(\alpha + \beta)^2\,(\alpha + \beta + 1)}$
4.32	.962
4.33	1.0

4.34 .9502

4.36 $f(t) = \lambda e^{-\lambda t}, \quad t > 0$

4.37 $m(t) = e^{\mu t + t^2\sigma^2/2}$

4.38 $\sqrt{2}$

4.39 $E(Y) = \Gamma(3/2)\,\alpha^{1/2}, \quad V(Y) = \alpha[1 - (\Gamma(3/2))^2]$

4.40 (b) $E(X) = 100(1 - e^{-2})$

4.41 $e^{-3} \approx .05$

4.42 .0235

4.43 $k = (.4)^{1/3}$

4.45 (a) $m(t) = e^{\mu t + t^2\sigma^2/2}, \quad \mu, \quad \sigma^2$

(b) $m(t) = e^{t^2/2}, \quad 0, \quad 1$

4.50 .05

4.51 $53.58

Chapter 5

5.1 (a) $k = 4$

(b) $f_1(y_1) = 2y_1, \quad 0 \leqslant y_1 \leqslant 1$

$f_2(y_2) = 2y_2, \quad 0 \leqslant y_2 \leqslant 1$

(c) $y_1^2 y_2^2, \quad 0 \leqslant y_1 \leqslant 1, \quad 0 \leqslant y_2 \leqslant 1$

(d) 9/64

(e) 1/4

5.2 (a) $f_1(y_1) = 3y_1^2, \quad 0 \leqslant y_1 \leqslant 1$

$f_2(y_2) = (3/2)(1 - y_2^2), \quad 0 \leqslant y_2 \leqslant 1$

(b) 23/64

(c) 0

5.3 (a), (b)

Y_1 \ Y_2	0	1	2	3	
0	0	1/28	1/14	1/84	10/84
1	1/21	2/7	1/7	0	10/21
2	1/7	3/14	0	0	5/14
3	1/21	0	0	0	1/21
	5/21	15/28	3/14	1/84	

(c) 9/16

5.4 $f(y_1|y_2) = 2y_1, \quad 0 \leqslant y_1 \leqslant 1$, yes

5.5 (a) $f(y_1|y_2) = \dfrac{2y_1}{1 - y_2^2}, \quad y_2 \leqslant y_1 \leqslant 1$

(b) $f(y_2|y_1) = \dfrac{1}{y_1}, \quad 0 \leqslant y_2 \leqslant y_1$

(d) 5/12

5.6 (a) $f(y_1, y_2) = \dfrac{1}{y_1}, \quad 0 \leqslant y_2 \leqslant y_1 \leqslant 1$

(b) 1/2

(c) ln(2)/ln(4)

5.7 (a) $f_1(y_1) = \dfrac{2}{\pi}\sqrt{1 - y_1^2}, \quad -1 \leqslant y_1 \leqslant 1$

(b) 1/2

5.8 (a) $f_1(y_1) = y_1 + 1/2, \quad 0 \leqslant y_1 \leqslant 1$

$f_2(y_2) = y_2 + 1/2, \quad 0 \leqslant y_2 \leqslant 1$

(b) No

(c) $f(y_1|y_2) = \dfrac{y_1 + y_2}{y_2 + 1/2}, \quad 0 \leqslant y_1 \leqslant 1$

5.9 (a) $k = 1$

(b) $f_1(y_1) = y_1/2, \quad 0 \leqslant y_1 \leqslant 2$

$f_2(y_2) = 2(1 - y_2), \quad 0 \leqslant y_2 \leqslant 1$

(c) $f(y_1|y_2) = \dfrac{1}{2(1 - y_2)}, \quad 2y_2 \leqslant y_1 \leqslant 2$

(d) $f(y_2|y_1) = 2/y_1, \quad 0 \leqslant y_2 \leqslant \dfrac{y_1}{2}$

(e) 1/2

(f) 8/9

5.10 (a) $f_2(y_2) = 2(1 - y_2), \quad 0 \leqslant y_2 \leqslant 1$

(b) $f_1(y_1) = 1 - |y_1|, \quad -1 \leqslant y_1 \leqslant 1$

(c) 1/4

5.11 (a) 2/3

(b) 1/18

(c) 0

5.12 (a) .019

5.13 (a) −.33

(b), (c) $E(Y_1 + Y_2) = 2.33, \quad V(Y_1 + Y_2) = .40$

5.14 (a) -.0069
(b) .5833
(c) 1.0764
5.15 (a) 2
(b) 2/3
5.16 1/4
5.17 $f_1(y) = \left(\frac{1}{2}\right)^{y+1}, \quad y = 0, 1, 2, \ldots$
5.18 $\left(\frac{1}{\theta}\right)^3 e^{-(y_1+y_2+y_3)/\theta}, \quad y_i > 0$
5.19 $E(G) = 42, \quad V(G) = 26,$ no
5.20 $E(Y) = np, \quad V(Y) = np(1-p)$
5.21 (a) $y_1/2$
(b), (c) 3/8
5.22 $E(Y) = 1$
5.23 3/8
5.24 300 hours
5.26 (a) $m(t_1, t_2, t_3) = (p_1e^{t_1} + p_2e^{t_2} + p_3e^{t_3})^n$
(c) $-np_1p_2$
5.30 $-\sqrt{\frac{p_1p_2}{(1-p_1)(1-p_2)}}$

Chapter 6

6.1 (a) $(1/2)(1-u), \quad -1 \leqslant u \leqslant 1$
(b) $(1/2)(1+u), \quad -1 \leqslant u \leqslant 1$
(c) $\frac{1-\sqrt{u}}{\sqrt{u}}, \quad 0 \leqslant u \leqslant 1$
6.2 (a) $u^2/18, \quad -3 \leqslant u \leqslant 3$
(b) $(3/2)(3-u)^2, \quad 2 \leqslant u \leqslant 4$
(c) $(3/2)\sqrt{u}, \quad 0 \leqslant u \leqslant 1$
6.3 Gamma $(\alpha = 1, \beta = 2)$
6.4 Normal, $E(U) = \mu \sum_{i=1}^{n} a_i, \quad V(U) = \sigma^2 \sum_{i=1}^{n} a_i^2$
6.5 (a) $\frac{1}{2\sqrt{u}}, \quad 0 \leqslant u \leqslant 1$

(b) $1/2, \quad 0 \leqslant u \leqslant 1$

$\dfrac{1}{2u^2}, \quad 1 < u$

(c) $ue^{-u}, \quad u > 0$

(d) $-\ln u, \quad 0 \leqslant u \leqslant 1$

6.6 Binomial $(n_1 + n_2, p)$

6.7 (a) Poisson (mean $\lambda_1 + \lambda_2$)

(b) Binomial $\left(m, \dfrac{\lambda_1}{\lambda_1 + \lambda_2} = p\right)$

6.8 (a), (b) $E(U_1) = -1/3, \quad E(U_2) = 1/3, \quad E(U_3) = 1/6$

6.9 $f(d) = 1, \quad 0 \leqslant d \leqslant 1$

6.10 (a) $f(u) = 2(1 - u), \quad 0 \leqslant u \leqslant 1$

(b) 1/3

(c) 1/18

6.11 Gamma ($\alpha = 4, \beta = 1$)

6.12 $e^{-(u-4)}, \quad u \geqslant 4$

6.13 .0228

6.14 Beta ($\alpha = 1, \beta = 1$) or $f(u) = 1, \quad 0 \leqslant u \leqslant 1$

6.15 (a) $f(u) = \dfrac{1}{\alpha} e^{-u/\alpha}, \quad u > 0$

(b) $E(Y^k) = \Gamma\left(\dfrac{k}{m} + 1\right) \alpha^{k/m}$

6.16 (a) $2(1 - u), \quad 0 \leqslant u \leqslant 1$

(b) $2u, \quad 0 \leqslant u \leqslant 1$

6.17 (a) $2u, \quad 0 \leqslant u \leqslant 1$

(b), (c) $E(U) = 2/3$

6.18 3/4

6.19 (a) $\dfrac{1}{4\pi} e^{-\frac{1}{4}[(u_1 - 2\mu)^2 + u_2^2]}$

(b) Yes

6.20 $f(u) = \dfrac{1}{\pi(1 + u^2)}, \quad -\infty < u < \infty$

6.21 $f(u) = 1, \quad 0 \leqslant u \leqslant 1$

6.22 $f(u) = \dfrac{1}{4\sqrt{u}}, \quad 0 \leqslant u \leqslant 1$

$= \dfrac{1}{8\sqrt{u}}, \quad 1 < u \leqslant 9$

6.23 $[1 - F(y)]^3[1 + F(y)]$

6.24 $n(n - 1)r^{n-2}(1 - r), \quad 0 < r < 1$

Chapter 7

7.1 3/4

7.2 ab

7.3 (a) $\theta/2$

7.4 No, $E(Y) = \infty$

7.5 $\dfrac{\bar{X}}{\bar{X} + \bar{Y}}$ converges in probability to $\dfrac{\lambda_1}{\lambda_1 + \lambda_2}$

7.6 $1 - 2e^{-1}$

7.7 .0668

7.8 .0548

7.9 .9544

7.10 $n = 157$

7.11 .0475

7.12 .7698

7.13 .0062

7.15 .0062

7.17 .1587

Chapter 8

8.2 $a = \dfrac{\sigma_2^2}{\sigma_1^2 + \sigma_2^2}$

8.3 1280 ± 28.4

8.4 $.92 \pm .0172$

8.5 (1256.64, 1303.36)

8.6 (.075, .125)

8.7 $5.4 \pm .277$

8.8 $1.7 \pm .494$

8.9 (1.294, 2.106)

8.10 $.11 \pm .09$

8.11 $n = 2400$

8.12 $n = 256$

8.13 $n = 40{,}000$

8.14 $-.04 \pm .104$

8.15 $n = 768$

8.16 .6170

8.17 $n = 44$

8.18 $n = 72$

8.19 795 ± 7.95

8.20 24.7 ± 1.05

8.21 11.3 ± 1.437

8.22 3.68 ± 1.56

8.23 4.9 ± 4.54

8.24 -5 ± 6.603

8.25 $-.75 \pm .771$

8.26 $.259 \pm .068$

8.27 (29.30, 391.54)

8.28 (1.41, 31.26), no

8.30 (a) $2\sigma^4 \dfrac{n-1}{n^2}$

8.31 $\text{MSE}(S^2) > \text{MSE}(S'^2) \quad (n > 1)$

8.32 (b) $\dfrac{2\sigma^4}{n_1 + n_2 - 2}$

8.33 $\dfrac{S_1^2}{S_2^2} \dfrac{1}{F_{\nu_1, \nu_2, \alpha/2}} < \dfrac{\sigma_1^2}{\sigma_2^2} < \dfrac{S_1^2}{S_2^2} F_{\nu_2, \nu_1, \alpha/2}$

where $\nu_i = n_i - 1, i = 1, 2$

8.34 $2^{3/2} t_{\alpha/2} \sigma \dfrac{\Gamma\left(\frac{n}{2}\right)}{\sqrt{n(n-1)}\, \Gamma\left(\frac{n-1}{2}\right)}$

8.36 $\overline{Y} \pm t_{\alpha/2} S \sqrt{\dfrac{n+1}{n}}$

Chapter 9

9.1 $\hat{p} = 1/Y$

9.2 $\hat{p} = 1/Y$

9.3 (a) $\hat{\theta} = \overline{Y}/\alpha$

(b) $E(\hat{\theta}) = \theta, \quad V(\hat{\theta}) = \theta^2/\alpha n$

(d) $\sum_{i=1}^{n} Y_i$

(e) $\frac{2\Sigma Y_i}{31.4} < \theta < \frac{2\Sigma Y_i}{10.85}$

9.4 (a) $\hat{\lambda} = \overline{Y}$

(b) $E(\hat{\lambda}) = \lambda, \quad V(\hat{\lambda}) = \frac{\lambda}{n}$

(d) 1.55

9.5 (a) $\sum_{i=1}^{n} Y_i^r$

(b) $\hat{\theta} = \frac{1}{n} \sum_{i=1}^{n} Y_i^r$

(c) Yes

9.6 (a) $\hat{\theta}_1 = \frac{1 - 2\overline{Y}}{\overline{Y} - 1}$

(b) $\hat{\theta}_2 = -\frac{n}{\Sigma \ln Y_i} - 1$

9.7 $\hat{p}_A = .30, \quad \hat{p}_B = .38, \quad \hat{p}_C = .32, \quad -.08 \pm .1641$

9.8 (a) $\hat{\theta}_1 = \overline{Y} - 1$

(b) $\hat{\theta}_2 = \min(Y_1, \ldots, Y_n)$

(c) $1/n$

9.9 $\hat{\theta} = 3\overline{Y}$, no

9.10 $3\left[\left(1 - \frac{1}{n}\right)\overline{Y} + \overline{Y}^2\right]$

9.11 (a) 63

(b) $E(\hat{\theta}) = \theta, \quad V(\hat{\theta}) = \frac{\theta^2}{2n}$

(c) 106.14

9.12 $\hat{\theta}^2 = \frac{\sum_{i=1}^{m} (X_i - \overline{X})^2 + \sum_{i=1}^{n} (Y_i - \overline{Y})^2}{m + n}$

9.14 $\hat{\theta} = (1/2)\,(Y_{(n)} - 1)$

9.15 $\frac{n+2}{3}$

9.16 No

9.17 $(1/2)(Y+1)$

9.18 .312

9.19 (a) $\hat{N}_1 = 2\overline{Y} - 1$

(b) $E(\hat{N}_1) = N, \quad V(\hat{N}_1) = \frac{1}{3n}(N^2 - 1)$

9.20 (a) $\hat{N}_2 = \max(Y_1, \ldots, Y_n)$

(b) $\hat{N}_3 = \frac{n+1}{n}\hat{N}_2$

(c) $V(\hat{N}_3) \approx \frac{N^2}{n(n+2)}$

9.21 252 ± 85.193

Chapter 10

10.3 $z = -2.53$

10.4 $z = 1.58$

10.5 $z = 4.71$

10.6 .1342

10.8 $z = -3.12$

10.9 $z = 1.52$, no

10.10 $n = 309$

10.11 (a) $H_0: p = .2, H_a: p > .2$

(b) $\alpha = .075$

10.12 (a) $H_0: \mu = 1100, H_a: \mu < 1100$

(b) $z < -1.645$

(c) Yes, $z = -1.90$

10.13 Yes, $z = 5.24$

10.14 .67

10.17 $t = -1.34$; do not reject.

10.18 $t = 2.635$; reject.

10.20 $t = 1.570$, no

10.21 $t = -1.71$, no

10.22 $\chi^2 = 12.6$; do not reject.

10.25 $F = 1.922$; do not reject ($\alpha = .05$).

10.26 $\chi^2 = 22.47$; reject.

10.27 $n = 16$

10.29 $-2\ln\lambda \approx 11$; reject H_0: $p_1 = p_2 = p_3 = p_4$.

10.30 (a) $\lambda = \dfrac{(\hat{\sigma}_1^2)^{n_1} (\hat{\sigma}_2^2)^{n_2} (\hat{\sigma}_3^2)^{n_3}}{(\hat{\sigma}^2)^{n_1+n_2+n_3}}$

where $\hat{\sigma}_1^2 = \dfrac{1}{n_1} \sum_{i=1}^{n} (X_i - \bar{X})^2$ and $\hat{\sigma}^2 = \dfrac{n_1\hat{\sigma}_1^2 + n_2\hat{\sigma}_2^2 + n_3\hat{\sigma}_3^2}{n_1 + n_2 + n_3}$

(b) Reject H_0 if $-2\ln\lambda > 5.99$.

10.31 (a) $\lambda = \dfrac{\bar{X}^m \bar{Y}^n}{\left(\dfrac{m\bar{X} + n\bar{Y}}{m+n}\right)^{m+n}}$

(b) $\dfrac{\bar{X}}{\bar{Y}}$ distributed as F with $2m$ and $2n$ degrees of freedom

10.32 $T = \dfrac{(\bar{X} + \bar{Y} - \bar{W}) - (\mu_1 + \mu_2 - \mu_3)}{\left[\left(\dfrac{1+a+b}{n(3n-3)}\right) [\Sigma(X_i - \bar{X})^2 + \frac{1}{a}\Sigma(Y_i - \bar{Y})^2 + \frac{1}{b}\Sigma(W_i - \bar{W})^2]\right]^{\frac{1}{2}}}$

with $(3n - 3)$ degrees of freedom

10.33 $t = 2.236$; reject H_0: $\mu_1 = \mu_2$.

10.34 $\chi^2 = \dfrac{(n-1)S_1^2 + (m-1)S_2^2}{\sigma_0^2}$ has $\chi^2_{(n+m-2)}$ distribution under H_0;

reject if $\chi^2 > \chi^2_\alpha$.

10.35 (a) $U = \dfrac{2}{\beta_0} \sum_{i=1}^{4} Y_i$ has $\chi^2_{(24)}$ distribution under H_0;

reject H_0 if $U > \chi^2_\alpha$

(b) Yes

10.36 (a) Reject H_0 if $\dfrac{2}{\theta_0} \sum_{i=1}^{n} Y_i^m > \chi^2_\alpha$, with $2n$ degrees of freedom.

(b) $n = 6$

10.37 (a) Reject H_0 if $Y_{(n)} \leqslant \theta_0 \sqrt[n]{\alpha}$.

(b) Yes

10.38 (a) Reject H_0 if $Y_{(n)} > \theta_0$, remainder of critical region unspecified.

(b) Yes

(c) No

10.39 (a) $R(d_1, \theta) = 2\theta - 5$
$R(d_2, \theta) = -(1/5)(3\theta^2 - 20\theta + 25)$
$R(d_3, \theta) = (1/5)(3\theta^2 - 15\theta + 25)$
$R(d_4, \theta) = 5 - \theta$
(b) d_2
(c) d_1
(d) d_4

10.40 d_2

10.41 d_1

Chapter 11

11.1 (a) $\hat{y} = 1.5 - .6x$
(b) SSE = .40, $s^2 = .133$, 3 d.f.
(c) $t = -5.20$; reject.
(d) $-.6 \pm .367$

11.2 $\hat{y} = 2.1 - .6x$

11.3 (a) $\hat{y} = -.71 - .14x + .14x^2$
(b) $t = 3.46$; reject.
(c) $.14 \pm .088$

11.4 (a) $\hat{y} = 1.4285 + .5000x_1 + .1190x_2 - .5000x_3$
(b) $\hat{y} = 2.0715$
(c) $t = -13.7$; reject H_0.
(d) $2.07 \pm .19$
(e) $2.07 \pm .34$

11.5 (a) $x_1 = \dfrac{T_1 - 60}{10}, \quad x_2 = \dfrac{P - 15}{5}, \quad x_3 = \dfrac{C - 1.5}{1.5}, \quad x_4 = \dfrac{T_2 - 150}{50}$
(b) $\hat{y} = 20.50 - 2.5125x_1 - 1.8375x_2 - .7875x_3 - .6500x_4$
(c) $t = -3.66$, reject; $t = -2.68$, reject; $t = -1.15$, do not reject; $t = -.95$; do not reject.
(d) 21.31 ± 2.75
(e) 21.31 ± 5.64

11.6 (a) $\hat{y} = 9.34 + 2.46x_1 + .60x_2 + .41x_1x_2$
(c) 9.34, 11.80
(d) $t = 2.63$; reject.
(e) $12.81 \pm .37$
(f) $12.81 \pm .78$

11.7 $x = \frac{T - 50}{15}$; -2, -1, 0, 1, 2

11.9 $n/2$ points at $x = -9$ and at $x = 9$

11.10 11/27

Chapter 12

12.15 $n_1 = 34, n_2 = 56$; signal amplification

12.16 $n_1 = n_2 = 48$

12.17 Three observations each at $x = 2, x = 5$; signal amplification

12.18 (a) 1.46 times as large

(b) 2.16 times as many observations

12.22 (a) μ_i

(b) $\mu_i, (1/n)(\sigma_p^2 + \sigma^2)$

(c) $\mu_1 - \mu_2, \ 2\sigma^2/n$, normal

12.25 $k_1 = 1/4, \ k_2 = 1/2, \ k_3 = 1/4$

Chapter 13

13.3 SSE = .020; $F = 2.0$.; do not reject.

13.4 $2.20 < \mu_A < 2.30$; $-.01 \leqslant \mu_A - \mu_B < .18$

13.5 SSE = 498.67; yes

13.6 $67.9 < \mu_A < 84.1$; $55.8 < \mu_B < 76.8$; $-3.6 < \mu_A - \mu_B < 23.0$; no, they are not independent.

13.7 SEE = .005; yes; yes; yes

13.8 $.02 < \mu_A - \mu_B < .14$; no

13.9 No, it is a completely randomized design; completely randomized design with nine treatments and two observations per treatment.

13.10 16; 135; 14.1

13.11 SSE = 4.99; no; no

13.12 $F = 5.14$; yes; blocking causes a loss of degrees of freedom for estimating σ^2. Blocking may produce a slight loss of information if the block-to block variation is small.

13.13 Randomized block design; 21; SSE = 58.91; F = 3.07; yes; no, the judge's responses may have been determined by a teaching characteristic other than the one under study.

13.14 At least 49

13.15 At least 22

13.16 (a) SSE = .571, SST = 1.212, F = 11.7; yes, evidence exists to indicate a difference in mean reaction times.

(b) t = −2.73 (based on 7 degrees of freedom); evidence exists to indicate a difference in mean reaction times for treatments A and D.

13.17

Source	*d.f.*	SS	MS	F
Blocks	3	.140	.047	6.62
Treatments	4	.787	.197	27.7
Error	12	.085	.0071	
Total	19	1.012		

F = 27.7, reject the hypothesis of no difference in treatment means.

13.18

Source	*d.f.*	SS	MS	F
Treatments	2	38.00	19.00	10.05
Blocks	3	61.67	20.56	10.88
Error	6	11.33	1.89	
Total	11	111.00		

(a) Yes, F = 10.05

(b) Yes, F = 10.88

(c) 3.50 ± 1.89

13.19

Source	*d.f.*	SS	MS	F
Rows	2	46.89	23.45	11.11
Columns	2	22.89	11.45	5.43
Treatments	2	91.56	45.78	21.70
Error	2	4.22	2.11	
Total	8	165.56		

7.33 ± 5.10

13.20

Source	d.f.	SS	MS	F
Rows	4	16.56	4.14	1.02
Columns	4	162.16	40.54	9.99
Treatments	4	187.76	46.94	17.32
Error	12	32.48	2.71	
Total	24	398.96		

(a) $F = 17.32$; evidence does exist to indicate a difference in mean response for the five cake mixtures.

(b) 4.20 ± 2.27

13.21

Source	d.f.	SS	MS	F
Blocks	7	18.96	2.71	1.38
Treatments	2	22.58	11.29	5.76
Error	14	27.42	1.96	
Total	23	68.96		

(a) Yes, $F = 5.76$

(b) 1.25 ± 1.50

13.22 Approximately 16 rats per chemical

13.23 $F = 1.535$; no evidence of a difference between treatment means

13.25 (a) $E(Y_{ij}) = \mu + \beta_i + \tau_j,\quad V(Y_{ij}) = \sigma^2$

(b) $E(\bar{T}_j) = \mu + \tau_j,\quad V(\bar{T}_j) = \sigma^2/b,$ yes

(c) $E(\bar{T}_j - \bar{T}_k) = \tau_j - \tau_k,\quad V(\bar{T}_j - \bar{T}_k) = 2\sigma^2/b,$ yes

13.26 (a) $E(\text{MST}) = \sigma^2 + \dfrac{b}{p-1} \displaystyle\sum_{j=1}^{p} \tau_j^2$ (b, blocks; p, treatments)

(b) $E(\text{MSB}) = \sigma^2 + \dfrac{p}{b-1} \displaystyle\sum_{i=1}^{b} \beta_i^2$

(c) $E(\text{MSE}) = \sigma^2$

Chapter 14

14.2 $X^2 = 24.48$; reject $H_0(\chi^2_{.05} = 7.81)$.

14.3 $X^2 = 10.40$; do not reject $H_0(\chi^2_{.05} = 11.1)$.

14.4 (a) $z = -2.30$; reject H_0.
(b) Formulate H_0 before collecting data.
14.5 $X^2 = 18.53$; reject H_0 ($\chi^2_{.05} = 3.84$).
14.6 $X^2 = 27.17$; reject H_0 ($\chi^2_{.10} = 6.25$).
14.7 $X^2 = 2.87$; do not reject H_0 ($\chi^2_{.05} = 5.99$).
14.8 $X^2 = 11.62$; reject H_0 ($\chi^2_{.05} = 9.49$).
14.9 $X^2 = 5.02$; do not reject H_0 ($\chi^2_{.05} = 5.99$).
14.10 $X^2 = 7.19$; reject H_0 ($\chi^2_{.05} = 5.99$).
14.11 $X^2 = 12.91$; do not reject H_0($\chi^2_{.05} = 15.5$).
14.12 $X^2 = 4.40$; do not reject H_0 ($\chi^2_{.05} = 9.49$).
14.13 $X^2 = 53.92$; reject H_0 ($\chi^2_{.05} = 15.5$).
14.14 Yes, $X^2 = 21.51$
14.15 Yes, $x^2 = 6.49$
14.16 $.09 \pm .10$
14.17 Yes, $X^2 = 6.18$
14.18 No, $X^2 = 1.884$
14.20 $X^2 = 8.56$; reject with $\alpha = .05$.
14.22 (a) $X^2 = 4$; reject the hypothesis that the model fits the data.
(b) $X^2 = 4$; do not reject.
14.23 $X^2 = .657$; the data are consistent with the model.
14.24 $X^2 = 3.26$; the data agree with the model.
14.26 $X^2 \approx 74.28$; reject the hypothesis.

Chapter 15

15.1 .014, $y \leqslant 6$ and $y \geqslant 19$; .044, $y \leqslant 7$ and $y \geqslant 18$; .108, $y \leqslant 8$ and $y \geqslant 17$
15.2 Rejection region: $y = 0$; $\alpha = 1/32$. Rejection region: $y \leqslant 1$; $\alpha = 7/32$.
Do not reject in either case.
15.3 (a) Rejection region: $y \leqslant 1$; $\alpha = .039$. Rejection region: $y \leqslant 2$; $\alpha = .180$.
Hence we may reject H_0 at $\alpha = .180$ but not at $\alpha = .039$.
(b) $|t| = .30 < 1.746$. Do not reject.
15.4 (a) Rejection region: $y \leqslant 1$; $\alpha = .0215$. Rejection region: $y \leqslant 2$; $\alpha = .1094$.
Do not reject.
(b) Rejection region: $y \leqslant 2$; $\alpha = .0547$. Do not reject.
15.5 Rejection region: $U \leqslant 21$; $\alpha = .094$. Sample $U = 32$. Do not reject.
15.6 Rejection region: $R \leqslant 7$; $\alpha = .109$. Sample $R = 13$. Do not reject.
15.7 (a) Rejection region: $T \leqslant 8$. Sample $T = 6$. Reject.
(b) Rejection region: $T \leqslant 11$. Sample $T = 6$. Reject.

15.8 (a) Rejection region: $y \leqslant 1; \alpha = .039$. Do not reject.

(b) Rejection region: $|t| \geqslant 2.306$ has $\alpha = .05$. $t = -1.65$. Do not reject.

15.9 Rejection region: $T \leqslant 6$. Sample $T = 10.5$. Do not reject.

15.10 Rejection region: $R \leqslant 2; \alpha = .029$.

Rejection region: $R \leqslant 3; \alpha = .114$.

Sample $R = 3$. Reject at $\alpha = .114$ but not at $\alpha = .029$.

15.11 (a) Rejection region: $z = \dfrac{U - 112.5}{24.1} \leqslant -1.645; \alpha = .05$.

Sample $z = -1.805$. Reject H_0.

(b) Reduction of the expected number of runs. Rejection region:

$$z = \frac{R - 16}{2.69} \leqslant -1.645; \alpha = .05.$$

Sample $z = -.37$. Do not reject.

15.12 Rejection region: $R \leqslant 3; \alpha = .040$.

Rejection region: $R \leqslant 4; \alpha = .167$.

Sample $R = 5$. Do not reject.

15.13 (a) .0256.

(b) An unusually small number of runs (judged at $\alpha = .05$) would imply a clustering of defective items in time.

15.14 (a) Rejection region: $R \leqslant 6; \alpha = .100$. Sample $R = 7$. Do not reject at $\alpha = .100$.

(b) $t = .57$. Do not reject.

15.15 (a) Rejection region: $U \leqslant 3; \alpha = .056$. Sample $U = 3$. Reject at $\alpha = .056$.

(b) Rejection region: $F = S_A^2/S_B^2 \geqslant 9.12; \alpha = .05$. Sample $F = 4.91$. Do not reject.

15.16 -.845

15.17 Rejection region: $r_s \leqslant -.564; \alpha = .05$. Reject H_0.

15.18 -.593

15.19 Rejection region: $r_s \leqslant -.497; \alpha = .05$. Reject H_0.

15.20 Use two-tailed test. Since $P[U \leqslant 13] = .0249$, rejection region is $U \leqslant 13$. Hence do not reject H_0, because $U = 17.5$.

15.21 Rejection region: $T < 25; \alpha = .05$. Reject H_0.

15.22 Use two-tailed test. Rejection region: $|r_s| \geqslant .441; \alpha = .10$. Reject H_0: $\rho = 0$.

15.23 .0159

15.24 .375

15.25 .10

15.26	1/63
15.27 (a)	.008
(b)	.040
(c)	.167

Index

*Derivatives

In the following formulas u, v, w represent functions of x, while a, c, n represent fixed real numbers. All arguments in the trigonometric functions are measured in radians, and all inverse trigonometric and hyperbolic functions represent principal values.

1. $\dfrac{d}{dx}(a) = 0$

2. $\dfrac{d}{dx}(x) = 1$

3. $\dfrac{d}{dx}(au) = a\dfrac{du}{dx}$

4. $\dfrac{d}{dx}(u + v - w) = \dfrac{du}{dx} + \dfrac{dv}{dx} - \dfrac{dw}{dx}$

5. $\dfrac{d}{dx}(uv) = u\dfrac{dv}{dx} + v\dfrac{du}{dx}$

6. $\dfrac{d}{dx}(uvw) = uv\dfrac{dw}{dx} + vw\dfrac{du}{dx} + uw\dfrac{dv}{dx}$

7. $\dfrac{d}{dx}\left(\dfrac{u}{v}\right) = \dfrac{v\dfrac{du}{dx} - u\dfrac{dv}{dx}}{v^2} = \dfrac{1}{v}\dfrac{du}{dx} - \dfrac{u}{v^2}\dfrac{dv}{dx}$

8. $\dfrac{d}{dx}(u^n) = nu^{n-1}\dfrac{du}{dx}$

9. $\dfrac{d}{dx}(\sqrt{u}) = \dfrac{1}{2\sqrt{u}}\dfrac{du}{dx}$

10. $\dfrac{d}{dx}\left(\dfrac{1}{u}\right) = -\dfrac{1}{u^2}\dfrac{du}{dx}$

11. $\dfrac{d}{dx}\left(\dfrac{1}{u^n}\right) = -\dfrac{n}{u^{n+1}}\dfrac{du}{dx}$

12. $\dfrac{d}{dx}\left(\dfrac{u^n}{v^m}\right) = \dfrac{u^{n-1}}{v^{m+1}}\left(nv\dfrac{du}{dx} - mu\dfrac{dv}{dx}\right)$

13. $\dfrac{d}{dx}(u^n v^m) = u^{n-1}v^{m-1}\left(nv\dfrac{du}{dx} + mu\dfrac{dv}{dx}\right)$

14. $\dfrac{d}{dx}[f(u)] = \dfrac{d}{du}[f(u)] \cdot \dfrac{du}{dx}$

* Let $y = f(x)$ and $\dfrac{dy}{dx} = \dfrac{d[f(x)]}{dx} = f'(x)$ define respectively a function and its derivative for any value x in their common domain. The differential for the function at such a value x is accordingly defined as

$$dy = d[f(x)] = \frac{dy}{dx}dx = \frac{d[f(x)]}{dx}dx = f'(x)\,dx$$

Each derivative formula has an associated differential formula. For example, formula 6 above has the differential formula

$$d(uvw) = uv\,dw + vw\,du + uw\,dv$$